THEORETICAL KNOWLEDGE

SYNTHESE LIBRARY

STUDIES IN EPISTEMOLOGY,

LOGIC, METHODOLOGY, AND PHILOSOPHY OF SCIENCE

VOLUME 326

Theoretical Knowledge

by

VYACHESLAV STEPIN
Russian Academy of Science, Moscow,
Russia

Springer

A C.I.P. Catalogue record for this book is available from the Library of Congress.

ISBN-10 1-4020-3045-2 (HB)
ISBN-10 1-4020-3046-0 (e-book)
ISBN-13 978-1-4020-3045-1 (HB)
ISBN-13 978-1-4020-3046-8 (e-book)

Published by Springer,
P.O. Box 17, 3300 AA Dordrecht, The Netherlands.

www.springeronline.com

Printed on acid-free paper

English translation by
A. G. Georgiev and E. D. Rumiantseva

Printed in the Netherlands.

CONTENTS

PREFACE
to the English Edition

The conception, which is posed in this book, has its own history. As I am seeing now the idea and some elements of the conception were outlined in the years of my postgraduate studies in the beginning of the 60th after the completion of my education at the philosophical department of Byelorussian State University. Choosing the critical analysis of Vienna Circle's positivism as the subject of my doctoral thesis I had not suspected at that time of the result of my future studies. With a knowledge of the papers of M. Schlick, Ph. Frank, V. Kraft, R. Carnap, C.G. Hempel and others I suddenly realized, that I have no objections against their position and that I agree in views with them. Being occupied next years by my duties as a teacher I still all the time pondered over issues of the methodology of science. Returned over the three years to my doctoral thesis I discovered that I had succeed in an elaboration of the new position overcoming an attitude of the standard conception of the analysis of science, which was shaped by the positivist tradition.

Positivism proposed a specific idealization of scientific knowledge. It was considered as an autonomous structure without any linking with the practical activity, philosophy, culture and, strictly saying, outside of the historical development. This development was approached narrowly, it was considered as a growing of knowledge, but not extended to include the methodology itself of its generating. In essence, positivism strove to discover the ultimate and strictly scientific methodology, which could provide an effective growth of knowledge at all times. In the framework of such an idealization the features of an empirical and theoretical language would be uncovered and differentiated; the two types of terms – observational and factual – in the system of an empirical language would be discovered.

However, the attempts to describe in the framework of this approach a knowledge dynamics and an elaboration of the new scientific theories led to the many known difficulties. In Western philosophy of science the passage to the post-positivist conceptions (K. Popper, T. Kuhn, I. Lacatos and others) was the result of their realization.

I proceed with my investigations after defending a thesis. My approach to the research of a structure and dynamics of science was connected with the comprehension of its historical development, in the course of which all basic components of the scientific activity – means and methods, systemic types of the studying objects, ways of the scientific communications, the functions of science in culture – would be changed. I begin to consider the scientific knowledge as a complicated historically developing system immersed in a changing sociocultural environment. The results of many years of these studies are posed in the book and the reader would evaluate their successfulness. I am happy that the circle of my readers will be broadening due to this English edition of my book. My thanks are due to the translators of the Russian text and to Progress-Tradition Publishers who published Russian version of this book and arranged its translation into English. I would like to express my gratitude to the scientific editor of English version of the book Prof. V. Vasyukov. My sincere thanks are due to prominent philosopher and logician Prof. J. Hintikka for the interest in my book, to the referees of English version, to all who help to bring about the publication of this book.

PREFACE

This monograph summarizes the results of more than twenty years of my investigations of structure and dynamics of scientific theoretical knowledge. I started this work at the end of the 1960s and the beginning of the 1970s. At that time in our science there was a transition from dominating ontological problems of natural history philosophy (discussing problems of development, causality, space and time under view of natural history's achievements in the 20th century) to intensive logical and methodological analysis of scientific knowledge construction and dynamics. These problems also became central in the Western philosophy of science. Critical rethinking of its results gradually led me to the image of scientific knowledge as a complex historically developing system, which represents a particular type of systemic organization. It differs from simple and mechanical and even from self-regulating systems with feedback. Historically developing systems include an aspect of self-regulation, but they feature transitions from one type of self-regulation to another. The grade hierarchy of elements is forming inside the systems and besides that, historical development appears with new grades of organization which influence the grades previously emerged. They transform these earlier grades, modifying the previous organization. Despite this the system finds new integrality each time, nevertheless increasing the variety of its relatively autonomous subsystems.

Such an approach sets a historical volatility problem of all components of scientific knowledge from empirical facts and theories to scientific methods, purposes and value substances, expressing a type of scientific rationality. However, at that time I had not had the idea of analyzing the types of scientific rationality. This came later but potentially it was planned to be accepted as a paradigm of science historical overview.

I linked an analysis of historical dynamics of knowledge with principles of active approach. G. Shchedrovitsky and E. Yudin played a significant role in the development of this approach in the 1960s and the 1970s. Their investigations influenced my understanding of science, as well as its development.

My major ideas on structure and genesis problems of scientific theory were established at the end of the 1960s and the beginning of the 1970s. An analysis of the history of electrodynamics at the period before Maxwell and also of classical mechanics history and some fragments of quantum theory, allowed me to work out a conceptual scheme the Minsk methodological school investigation program was later based. This school functioned successfully in the 1970-1980s together with other directions and schools in the Soviet philosophy of science. Those schools were in Moscow (the Philosophical Institute of the Academy of Sciences USSR, the Institute of Natural Science and Techniques History of the Academy of Sciences USSR, and also philosophers and logicians at Moscow State University, and works by the methodological club of Shchedrovitsky, etc.), in Leningrad (V. Bransky, A. Karmin, M. Kozlova and others), in Kiev (M. Popovich, S. Krimsky, P.Dishlevy and others), in Novosibirsk (I. Alexeev, M. Rozov), in Voronezh (B. Pahomov, A. Kravets), and in Rostov (M. Petrov and others).

By that time the conceptual scheme, developed by me, was implemented in research of a history of physics conducted together with L. Tomilchik (at that time the senior research worker of the theoretical physics lab (today its head) in the Byelorussian Academy of

Sciences, and now a corresponding member of the National Academy of Sciences in Byelorussia). We accomplished reconstruction of a history of Maxwell's electrodynamics and history of the earliest version of the relativistic theory of an electron (works of P. Dirac at the end of the 1920s). At the same time the analysis of conceptual structure of quantum mechanics from an active approach position was simultaneously attempted.

The results of these investigations were published at the beginning of the 1970s in a number of my articles and in our mutual book[i].

My further activity in the 1970s was connected with recess of initial conceptions of system dynamics of theoretical knowledge. Detection of the fact, that the fundamental theories are not the product of inductive generalization of experience, but they are formed in the beginning at the expense of translation of conceptual means borrowed from other areas of theoretical knowledge, and only then are substantiated by experience, has put forward a problem of selection of means and methods of theoretical synthesis. In the initial phase of our investigations we were not engaged in this problem. The concern was connected with understanding the hierarchy of theoretical models and their operational nature. But then the problem had arisen in a new form. It was seen as a problem of preconditions which determine the developing of scientific hypotheses, and as a problem of the ontological status of theoretical models.

Searching for the answer to these problems I came to think of the basis of science. First of all their components were selected and illustrated as a scientific picture of the world and the philosophical basis. Then ideals and standards of science were analyzed. Thus, the initial concept was developed. New points of view appeared concerning pattern of knowledge and operations of its generation. And from this viewpoint it was necessary to reexamine the results that were obtained in the first period of work. In particular the first variants, obtained in research together with L.Tomilchik, were certified and rewritten versions of reconstruction of Maxwell's electrodynamics history, as in them the operations, bound with interplay of idealized models and scientific picture of the world, were not taken into account[ii]. However, this is not surprising, as the idealized knowledge is rather a complex object and it is unlikely to reveal all basic features of its historical development. Therefore, expansion of the analytical area gives birth to a new vision of the old, apparently already solved, problems.

The field of methodological researches changed considerably, when in Russian literature on philosophy of science there was a shift of problems from the analysis of internal dynamics of science to the emphasis of its sociocultural dependence. This took place in the late 1970s-early 1980s.

Today I would explain the shifts of problems (which were characteristic of Western philosophy of science) by the questions, which had shown philosophy and methodology engendering post-non-classical rationality. But I began to reflect the types of rationality more recently, in the 1990s. Then it was important for me not only to emphasize and to describe separate plots and facts of the sociocultural context of scientific cognition, but also to try to discover mechanisms, due to which sociocultural influences are integrated in the internal for each scientific process of theoretical and empirical knowledge's growth. Purely, it was an old problem to overcome one-sided externalism and internalism in the description and explanation of the history of science.

I defended the point of view (I assert it even now) that, on one hand, the foundations of a science act as exactly a component of an inner pattern of science, and on the other hand, of its infrastructure, which indirectly influences scientific knowledge of the sociocultural factors and actuation of scientific knowledge in the culture of an applicable historical epoch.

All these discourses are presented in this new book with the results I achieved before as a base, of course. But this book is not only its compendium but also a new synthesis and re-understanding and supplement of previous ideas by new ones. Such ideas we can find in chapters dedicated to the foundations of science and in the other chapters. Specifically, in analyzing the scientific revolutions I paid attention not only to traditional investigation of how the revolutions take place in the frames of scientific discipline when new types of systemic objects insensibly are becoming involved. In this case if a picture of the world (disciplinarian ontology) and a "scheme of method" presented by ideals and norms of investigation are not corresponding to the new subjects, the facts and paradoxes that cannot be explained agglomerate in the system of knowledge. T. Kuhn named them as anomalies and crises. I tried to clear up the mechanisms of foundation and overcome such paradoxes and anomalies using the material of the relativity theory (analysis of which is still accompanied by numerous discussions).

But another variant of scientific revolutions exists where they take place in absence of internal crisis but are due to interactions between disciplines and "paradigmatic grafting" from one science to another. Thus the great revolution grew that led to discipline formation of organized science. Many sciences' foundation transformations passed in this way. They were connected with achievements of neighboring disciplines' influence (the examples of revolution changes such as these are shown in this book – in chemistry under impact of quantum physics and in modern biology under impact of ideas of cybernetics).

An ascertainment of the role of theory's interdisciplinary links and of the interactions between disciplines changed the approach to the methodological analysis of theoretical knowledge. In the traditional approach the starting point of analysis was the separate theory in its relation to experience[iii]. Nowadays it is necessary to view the scientific discipline as an initial unit. This scientific discipline would be considered as a system of complexly organized and developing theoretical knowledge in its links with experience, with foundations of given discipline, and via them with other sciences and with sociocultural context.

This approach had already been used in my investigations in the beginning of the 1970s although perhaps without enough meta-methodological reflection. The discovery of a heterogeneous block of theoretical knowledge (theories of various generality degree) in a separated science's branch (I analyzed, first of all, physics text) and ascertainment of the fact that theories are interconnected together and are developing as an integrated system, had emerged from frames of conceptions of a separate theory as an initial unit of methodological analysis. This was the first circumstance, which formed a new view that overcame standard conception's limitations. The second circumstance was the reflection of discussions of theoretical load of a fact. The analysis of the empirical level of knowledge's internal structure and of the fact forming procedures discovered that facts are not separate and independent atomic units but they are entered into the system of knowledge in scientific discipline. They are formed under influence of previously selected theories and they then become a basis for new theories. Finally, the third and determinant circumstance was connected with analyzing the structure of science's foundations (scientific picture of the world, ideals and forms of research, philosophic foundations of science). Their functions which generate the systems relating to theories and empirical knowledge determined a conception of systemic integrity of scientific discipline. Essentially, in the middle of the 1970s I had formed this conception for myself and had used in investigating the genesis of separated theories.

A few years later I discovered that something similar, though from my point of view with a smaller share of analytical detail of scientific discipline's structure, was implemented in a series of works of the Western philosophers and science methodologists in the same period. For example, the concept of scientific field was offered in the American philosopher D. Shapere's researches. This concept was considered as the ordered array of theoretical and empirical knowledge organized in specific blocks of scientific information. These blocks initially non-coherent then merge into a broader array (scientific field). Each theory that enters that field appears as its element and sets the problems, which stimulate the new theory's appearance. These new theories modify a configuration of scientific field and place its among the other ones[iv].

The Canadian philosopher C. Hooker developed a similar conception in the same period. He emphasized that scientific theories have an integral influence on the conditions of observation, choice of instrumental means and on events interpretation. From the other side, Hooker asserts that theories link to a "theoretical-worldview" vision of the world[v]. Theoretical-worldview in his opinion is a concept analogous to the scientific field idea by D. Shapere. Structure of "theoretical-worldview" as a whole knowledge's block is represented by a three-level hierarchy. There is a "coherent set of conceptual categories" on the top level which determines field of metaphysics, ontology used in research. Such spheres of knowledge as theory of methods, psychology of perception etc., are joined to it. Then the level of theories is situated and after that the level of experiments and observations. Thus "theoretical-worldview" as a scientific field, according to Hooker, appears in the role of integrated, conceptually organized manifold, oriented by determined cognition perspective[vi].

The methodological analysis unit presented by Hooker coincided in principle with a scientific discipline, though a scientific discipline's structure was given here only in a very first approximation (particularly it can be said of the block of science foundations to which a higher level of "theoretical-worldview" in Hooker's conception corresponds to).

A systemic organization of scientific discipline of knowledges and their structure's conception was set up in a book of mine *Becoming of scientific theory* (1976). In this book major attention was paid to investigation of method's operations and strategies determining the disciplinarian dynamics of theoretical knowledge as an integral, complexly organized and developing system. In further investigations I begin to consider the system of disciplinarian knowledge as a historical phenomenon, specified in its evolution by the sociocultural environment character in which science is immersed. Furthermore, the speech was not only on historical volatility of knowledge, forming a discipline, and of complicating their systemic organization by way of its development, but of historicity of disciplinarian organization of knowledge.

At this stage the problems of influence mechanisms on the science of different sociocultural factors and their integrity in a tissue of scientist investigation activity took a central place. Updated and systematized results are also expounded in this book.

Nowadays there is no need to demonstrate that science in its cognitive movement is constantly resonating with the development of other cultural fields (art, philosophy, religion, ordinary consciousness, etc.). A philosophy was closest to it. Strictly speaking, the "theoretical" concept which associates with science in this word in its own sense belongs to philosophy in many respects. Between those two types of "theoretical" not only a genetic link exists. Philosophic knowledge actively participates in the new scientific theories and scientific pictures of the world forming in a developed science also, mediating their entering in a cultural translation flow. In turn, philosophy experienced a huge influence of constituted science as an autonomous form of cognition. Patterns of scientific reasoning in a new European tradition served for a long time as an ideal to many philosophic schools.

But science interacts not only with philosophy in its historical development. All cultural spheres resonate with changes taking place in science. And those "cooperative effects" in a culture's development can be tracked particularly brightly on the turning stages when the type of scientific rationality is changing. In this book the reader will find presentation of my viewpoint on the problem of historical types of scientific rationality and their sociocultural contexts. But I would like to emphasize particularly that on the modern stage, when global crisis's exacerbation sets the values and choice of strategies of civilized development problems, the new science rationality contexts open unexpected opportunities for modern dialogue between cultures. In the final part of the book it is shown that if classical and neoclassical sciences were deeply oriented to new European cultural tradition's values (that synthesized the achievements of Antiquity and European Christian Middle Ages epoch), the post-non-classical science significantly widened a field of its world outlook's applications. It is starting to resonate not only with Western cultural tradition's values but also with many worldview's ideas of traditional Eastern cultures.

NOTES: PREFACE

[i] Stepin and Tomilchik (1970); Stepin (1970); Stepin (1971).

[ii] This refined construction was published in my book *Becoming of Scientific Theory* (1976) and is reproduced in this book with a little editing.

[iii] This approach was dominant in Western science's philosophy for a long time and was one of the crucial features of a so-called standard conception. For details of this conception see Sadovsky (1981).

[iv] Shapere (1974).

[v] Hooker (1975, p.155).

[vi] Ibid, pp.153-155.

CHAPTER ONE

SCIENTIFIC COGNITION IN A SOCIOCULTURAL CONTEXT

SCIENCE IN THE TECHNOGENIC CIVILIZATION CULTURE

Theoretical knowledge, along with its development, is an inherent feature of modern science, which constantly broadens our horizons in the cognitive and practical mastering of the world by man. Like science itself, theoretical knowledge is a cultural and historical phenomenon. It appeared within the context of historical development of civilization and culture, on certain stages of which were created theoretical science and the value of scientific rationality.

Modern civilization is inextricably connected with scientific achievements based on systematic deploying of theoretical investigations. Thanks to these achievements and their industrial implementation, astonishing technological progress became possible in the 20th century, which itself led to a new quality of life for the highly developed countries in the West and in the East. Science not only revolutionizes the sphere of production, but also influences other fields of human activity, regulating them and changing their means and methods.

It is not surprising that we cannot discuss the problems of modern civilization's future without an analysis of the contemporary tendencies in science and its perspectives. Even though in modern society there are some antiscientific movements, science is mainly considered as one of the highest values of civilization and culture.

But that is not the way it always was and science did not occupy such a high place in all the cultures within the scale of values priorities. That is why a question of the peculiarities of that civilization development type arises. This development stimulated a broad use of scientific knowledge in human life.

Traditional and technogenic civilizations

There were a lot of civilizations in the development of mankind after overcoming barbarity and savagery – specific kinds of society, each one of which had its own original history. A prominent philosopher and historian A. Toinby singled out and described 21 civilizations. They can be divided into two large parts: traditional and technogenic.

Technogenic civilization is more of a late human history discovery. For a long time it was a gathering of many traditional societies. Only in the 15th–17th centuries did a special development type form in the European region. It was connected with the appearance of technogenic societies, their world expansion and their influence on the traditional type. Some of the latter were simply absorbed by technogenic civilization; after going through stages of modernization they then transformed into typical technogenic societies. Others, having experienced Western technologies and culture, evolved into hybrid formations.

The differences between the traditional and technogenic civilizations are radical.

The traditional societies feature with impaired rate of social changing. Of course, innovations there also appear in production and in regulation of social relations, but progress is very slow compared with an individual or even a generation life circuit. In traditional societies several generations might pass living in the same public structures, replicating and conveying those to the following generation. Types of activity, its means and aims can exist as stable stereotypes. That is why such cultures' traditions are a priority, as are patterns and norms that accumulate the experience of ancestors. Canonized styles of thinking are preferred. Innovative activity is by no means recognized as the highest value; on the contrary, it is limited by centuries-old traditions. Ancient India and China, Ancient Egypt, the Muslim states of the Middle Ages period, etc. all had traditional societies. This kind of social arrangement still exists today in "third world" countries, though its conflict with modern Western (technogenic) civilization sooner or later will lead to a radical transformation of traditional culture and way of life.

Concerning technogenic civilization (which is often called "Western civilization", emphasizing the region of its appearance), it is a special type of social development, whose main principles are somewhat opposite to the ones of traditional societies. When the technogenic civilization was relatively complete, the rate of social changes started growing enormously. One can say that the extensive historical development is becoming intensive; spatial existence – temporal. Growth resources are not taken anymore at the expense of widening culture zones, but from rebuilding the old ways of living and creating completely new opportunities. The main and truly epochal world history change that had to do with the transfer from traditional society to technogenic civilization, is the appearance of a new human value system. It is considered as the innovation itself, originality, anything new (in a sense, the Guinness Book of Records can be a symbol of technogenic society. It differs from, say, the Seven Wonders of the World because it shows that every individual can become unique, can reach any extraordinary goal he sets up for himself, and the book also speaks up for it in its own way; the Seven Wonders of the World, on the contrary, were supposed to prove that the world is complete and that everything grand or truly unusual has already been done).

Technogenic civilization had begun long before computers and even the steam-engine. We could say that its predecessor and its first stage was Ancient History's development, first of all the policy culture, which gave the human race two great discoveries – democracy and theoretical science (the first example was Euclid geometry). These two discoveries in the sphere of social relations regulation and in the recipe of world cognition became important for the future, for a totally new type of civilization progress.

Second and very important epochs were the European Middle Ages with a special understanding of a human, made in God's own image; with a cult of the God son and a cult of love of a human for the God son, for Christ; with a cult of the human mind, capable of understanding and comprehending the heavenly creation mystery, also capable of decoding the writings that God had put in the world when he was creating it. The last circumstance should be especially noted: the goal of cognition was the decoding of God's deeds, of the heavenly creation plan completed in the world, – an awfully heretical thought from the point of view of traditional religions. But all of this is pre-history.

After that, in the age of Renaissance, many of the achievements of the Ancient tradition were rehabilitated, but at the same time the idea of godlikeness of the human mind was also assimilated. From this moment on the base is set for the cultural matrix of technogenic civilization, which starts its own development in the 17th century. It passes through three stages: first – pre-industrial, next – industrial and last – postindustrial. The most important basis for its life is, first of all, the technical and technological progress, not only by incidentally flowing innovations in the production sphere, but also by the means of generating all-new scientific knowledge and its implementation in the technico-technological processes. That is how a new type of development, based on the increasingly rapid changes in the environment and the objective world, where a human lives, appears. Changes made to this world lead to active transformations in social connections. In technogenic civilization the scientific technical progress constantly switches ways of communication, people's communication forms, types of personality and lifestyle. As a result, progress begins to be more oriented to the future. For technogenic culture it is common to think of the irreversible historical time, which flows from the past through the present to the future. For comparison we should say that in the majority of traditional societies other opinions dominated. Time was more often thought of as cyclic, when the world would periodically come back to the starting point. In traditional cultures it was considered that the "Golden Age" had already passed, it is behind, in the distant past. The heroes of the past had created the examples of deeds and actions, to which we should look up to. There is another orientation for technogenic culture. The idea of social progress stimulates the waiting for changes and the movement towards the future, which is a growing number of civilization conquests that make the world a more happier place.

Technogenic civilization has existed for a little more than 300 years, but it turned out to be pretty dynamic, mobile and very aggressive: it puts down, conquers, and actually swallows traditional societies and their cultures. We can see that everywhere today the process has taken over the whole world. This active interaction between technogenic civilizations and traditional societies usually ends up as a collision which leads to termination of the latter, to destruction of many cultural traditions and, in essence, to these cultures' death as original values. Traditional cultures are not only being moved to the periphery, but are also radically transformed by the time traditional societies start their way to modernization and technogenic development. Most often these cultures are saved as fragments, as historical rudiments. It has happened and still is happening with traditional Eastern cultures, that had gone through industrial development; the same can be stated concerning the people of South America, Africa, that stepped on the modernization path. Everywhere the cultural matrix of technogenic civilization is transforming traditional cultures, changing their vital attitudes for new worldview dominants.

These worldview dominants were settled in technogenic culture at the time of the pre-industrial stage of its development, during the Renaissance epoch and the European Age of Enlightenment.

They expressed cardinal worldview meanings of understanding of the human being, the world, the goals and the destination of a human life-circuit.

A human is considered as an active creature in active relations with the world. His activity should be directed outwards, for remaking and remodeling the outer world, primarily the environment which should be controlled by the human being. In its turn, the outer world is to be understood as an arena for human activity, like the world was created just to give mankind necessary resources, to satisfy all his needs. Of course, it does not mean that in the new-European cultural tradition other, even alternative worldview ideas do not appear.

Technogenic civilization in its own being is defined as a society constantly changing its basics. That is why permanent generation of new examples, ideas and concepts is highly supported and appreciated in its culture. Only several of them can come true in today's life, others are seen as possible programs for future life activity addressed for future generations. In technogenic societies, culture ideas and value orientations that happen to be alternative to the dominating values can be found. But in the real life-circuit of society they do not have to play the determining role, still being on the periphery of the social conscience and not moving people's masses.

An idea of remaking the world and controlling the environment has dominated in the culture of technogenic civilization over all stages of its history to the present days. It is well known that this idea was a prime constituent of "genetic code", determining existence itself and the evolution of technogenic societies. As concerns traditional societies, then here active relation to the life, which is admitted to be a generic human being property, has been understood and estimated from principally other viewpoints.

For a long time we thought that the activity worldview attitude was obvious. But it is hard to find it in traditional cultures common with traditional societies' conservatism in kinds of activity, slow speed of their evolution, and domination of regulative traditions constantly limited the display of active and reforming activity of man. That is why this activity itself was rather thought of not as directed outwards, for changing outside objects, but as oriented on the inside of a human, on self-contemplation and self-control which maintain the consecution of the tradition.[1]

The principle of the reforming deed, formulated in European culture during the epoch of Renaissance and Enlightenment, can be opposed as an alternative example of the principle of Ancient Chinese culture "wu-wei", which presumes non-interference in the flow of natural process and adaptation of an individual to the available social environment. This principle excluded aspiration for its purposive transformation, demanded self-control and self-discipline of an individual, including himself in this or that corporate structure. The "wu-wei" principle covered practically all most important aspects of human life activity. It expressed definite apprehension of specifics and values of agricultural labor, which a great deal depended from outside, natural conditions and which constantly demanded people to fit in with these conditions.

But the "wu-wei" principle also served as a specific way of incorporating an individual in the already established, traditional order of social connections. It oriented a person for such blending in the social environment, where freedom and self-realization of an individual are mainly reached in the self-change sphere, not by changing settled social structures.

The technogenic culture values start a whole new vector of human activity. Reforming activity is seen as the main human destination. The energetic ideal of how a human being looks at the world then is transferred into the social relations' sphere. They are also looked at as special social objects that can be purposely transformed by mankind. The cult of struggle and revolution as history locomotives is connected with it. We should mention that the Marxist concept of the fight of the classes, social revolutions and dictatorship as a way to resolve social problems was born in the context of technogenic culture values.

The understanding of activity and destination of man is closely connected with the second important aspect of values and worldview orientations, which is common for the technogenic world culture – understanding of the natural habitat as a regulative and normal field, where a smart creature, knowing the laws of nature, is capable of implementing his power over external processes and objects and of placing them under his control. It is only necessary to invent such a technology, which would artificially change the natural process and make it work for the benefit of human beings then the tamed nature would satisfy men's needs in constantly widening perspectives.

Concerning traditional cultures, we will not find in them such a conception of nature. It is understood as a live organism, which organically includes human beings, but not as an impersonal objective field controlled by objective laws. The law of nature concept itself, separated from the laws regulating social life, was alien for traditional cultures.

A well-known philosopher and scientist M. Petrov offered a distinctive thought experiment: how would a person look at new-European cultural ideals, if he was raised in the value system of traditional civilization. Referring to S. Powell's work *The role of theoretical knowledge in the European civilization*, he quoted missionaries who spoke of the reaction of the Chinese sages to descriptions of European science. "The sages found absurd the idea of science itself, because, even though the ruler of the Empyrean has the right to establish laws and demand their accomplishment under the threat of punishment, only those who "understand" laws can carry out and obey them. But "tree, water and rocks", which European deceivers are talking about, apparently, do not have the "comprehensive" feature: laws cannot be assigned to them and it cannot be demanded from them that they obey these laws".[2]

The inner force of conquering nature and reconstructing the world, which is common to the technogenic civilization, started a special attitude to the ideas of supremacy, strength and power. In traditional cultures they were firstly understood as direct power of one person over another. In patriarchic societies and Asian despotism, power and supremacy were distributed not only on the royal subjects, but were also implemented by the man, the head of the family, on his wife and children, whom he owned just like a czar or emperor owned the bodies and souls of his subjects.

In the technogenic world one can find many situations, when dominance is carried out as force of direct compulsion and power of one person over another. But relations of personal dependence are not dominant in this case and obey new social connections. Their nature is determined as a whole exchange of activity results that take on form of goods.

Power and dominance in this system of relations allow ownership and articles assumption (goods, people's flair, information as a selling value, which has a monetary equivalent).

As a result, in technogenic civilization's culture happens a peculiar shift of accents in the understanding of subject dominance, force and power – from a human being to the product made by him. In one's turn, these new meanings are easily connected with the ideal of an active-transforming nature of mankind.

The reforming activity itself is considered as a process, which gives man power over objects, and dominance over external conditions that a human is determined to conquer.

The human being must turn from being a slave of natural and social conditions into their master, and the process of this transformation itself was understood as overtaking of natural power and the power of social development. The characteristic of civilization achievements in terms of power ("labor force", "the strength of knowledge" and so on) expressed a setting for the human being to find all new capabilities which allow his horizons of reforming action to broaden.

Changing not only natural, but also social environment by way of using his gained force, a man is realizing his destiny as an author, and reformer of the world.

The ideal of a creative, sovereign, autonomous person occupies one of the priority places in technogenic civilization's system of values. We, born and living in the world of technogenic culture, take it for granted. But a human being, living in traditional society, would not accept these values. In traditional society a personality is realized only through belonging to some concrete corporation, being an element of a strictly operating system of corporate connections. If a human is not included in any corporation, he is not a person.

In technogenic civilization there appears a special type of personal autonomy: a human being can switch his corporate connections, he is not rigidly attached to them, he can and is capable of building flexible relations with other people, be involved in different social communities, and frequently in different cultural traditions.

As M. Petrov emphasized, since an individual who is forming in the womb of new-European culture and socialism is not rigidly attached to a family corporate system tradition of transmitting professional and social experience, that would be perceived by a person from traditional society as a sign of a European's evident disadvantage, whom from his childhood on "is used to an absurd idea that he is capable of doing anything, and when a European grows up, involves himself in a specialized practice, and until the end of his life he remains a disappointed human being, a carrier of pipe-dreams, which, of course, never came true, holding anger and grudge on fellow creatures, who, as he thinks, are engaged in something that he could do much better. Neither in his youth, nor in his adult years does a European know any orientations of his own life, he is not capable of comprehending its purpose, is unadvisedly rushing from one specialty to another, all his life he is getting familiar with something ...".[3]

This mental experiment suggested by M. Petrov, can be continued with precursory change of the frame of reference. We can look at the traditional cultures value system with the eyes of a person from technogenic culture. Then the attachment of a person from traditional society to strictly underlined, conservatively rendered types of activity and his rigid implementation from birth until death to some corporation, clan or caste, would be perceived by men raised in the new-European culture as a sign of non-freedom, absence of choice, disappearance of an individuality in corporate relations, and suppression of creative, individual beginnings in a human. Maybe this attitude in a somewhat keen way was expressed by A. Herzen when he wrote on traditional Eastern societies that a man there was never familiar with freedom and "didn't know his own dignity: that is why he was either a slave lying in dust, or an unrestrained despot".[4]

Life stability of traditional society from the position of this life meanings system is evaluated as stagnation and absence of progress, which are resisted by dynamism of the Western lifestyle. The whole technogenic societies' culture, which is oriented to innovations and traditions transformation, forms and supports an ideal of creative individuality.

Schooling, upbringing and socialization of an individual in the new-European cultural tradition contributes to forming a much more flexible and dynamic thought in him, than in a human from traditional societies. This also shows in a stronger reflex of the common conscience, in his orientation on the ideals of evidential and grounded judgment, and in a tradition of language games, which underlie European humor, and in the common conscience richness of guesses, prognoses, future anticipations as possible conditions of social life, and in its penetration of abstract logical structures, which organize a discourse.

Such logical structures often do not present person conscience in traditional societies. A study of the thought of the traditionalist groups in Middle Asia undertaken by A. Luria in the early 1930s, found that representatives from these groups cannot solve a problem that requires formal reasoning by the syllogism scheme. But those people of traditional societies, who had a school education which included mathematics and other sciences, solved these problems fairly easily.[5]

Similar results were obtained after researching thought of a traditional society person living in other regions (in particular, M. Cole's research of Liberian traditionalistic groups).[6]

All these peculiarities of conscience functioning in different cultural types are determined by inherent deep life meanings and values for these cultures.

In technogenic societies' culture this value system in based upon the ideals of a creative personality and original activity of a sovereign character. And only in this value system, scientific rationality and scientific activity gain priority status.

A special status of scientific rationality in the value system of technogenic civilization and extra importance of the science-technical view of the world are defined as scientific cognition of the world, is a condition for its reformation in extending measure. Scientific cognition creates a strong belief in that a human being is able to regulate the natural and social processes in accordance with his intentions after he discovers the laws of nature and social life.

That is why in the new-European culture and in the further development of technogenic societies, a category of scientific character is gaining a singular symbolic meaning. The culture is perceived as a necessary condition for prosperity and progress. A scientific rationality value and its active impact on other spheres of culture become a characteristic feature of technogenic societies' life.

Global crises and the problem of the scientific-technical progress value

The prestigious status of science stimulates development of a great variety of its advanced forms. Investigating them and analyzing how functions of science changed in the social life, we can reveal the main peculiarities of scientific cognition, its possibilities and limits.

The problem of these possibilities during the present time becomes especially acute. The whole matter is that technogenic civilization development itself came to its critical point, which showed the limits of this type of civilization growth. This became apparent in the second half of the 20th century in connection with the appearance of global crises and global problems.

Among many global problems generated by technogenic civilization and which threatened the existence of humans, we can distinguish three main ones.

The first one is the problem of survival under conditions of continuous improvement of weapons of mass annihilation. In the nuclear era, mankind ended up at the threshold of possible self-destruction, and this sad ending was a by-product of scientific and technical progress, which opens more new possibilities of military technology development.

The second and probably the most acute problem of modernity is the growth of ecological crisis on the global scale. Two aspects of human existence as part of nature and as an active creature who is reforming it, are coming to a conflict collision.

An old paradigm which stated that nature is an eternal reservoir of resources for human activity appeared to be wrong. The human was formed in the scope of biosphere, a special system emerged on the way of cosmic revolution. It presents not just the environment which can be looked at as a field for a human's reforming activity, but appears as a unified integral organism, which includes humankind as a specific subsystem. A human activity constantly changes biosphere dynamics, and on the modern stage of technogenic civilization the measure of human expansion in nature are such that they are beginning to destroy the biosphere as an integrated ecosystem. The threatening ecological catastrophe demands an elaboration of principally new scientific and technical strategies, and also social development of humankind, and activity strategies that provide co-evolution of a human and nature.

And finally, the third by counting order (last but not least!) problem, is the problem of preserving human personality, and the human as a biosocial structure in the conditions of growing and comprehensive processes of alienation. This global problem is sometimes defined as a modern anthropological crisis. A human who is complicating his world more frequently brings to life such forces which he cannot already control and which become alien to his nature. The more he transforms the world, the more he generates unexpected social factors, which begin to form structures, radically changing human life, obviously for the worse. In the 1960s a philosopher G. Markuse ascertained one of the consequences of modern technogenic development which is the appearance of a "one-dimensional man" as a product of mass culture. Modern industrial culture really creates large possibilities for consciousness manipulation, when a person loses the ability to rationally think of his entity. With all that, manipulators themselves and those who are manipulated become hostages of the mass culture, turning into persons in a gigantic puppet-theatre, whose performances play off the phantoms generated by the person himself.

The fast, forward development of technogenic civilization makes the problem of socialization and personality forming even more complicated. A constantly changing world breaks many roots and traditions, making a person live in different traditions and different cultures at the same time, and adapt to different, constantly renewing circumstances. A person's connections are becoming sporadic; from one point of view, all individuals are being pulled to a unified society of mankind, and from another point of view, humans are being isolated and atomized.

Modern technology allows people from different continents to intercommunicate with each other. It is possible to talk to colleagues from the USA over the telephone, then on TV find out what is going on in the far south of Africa, but at the same time not know your neighbors who you have lived next to for a long time.

The problem of saving your personality is acquiring in the modern world one more, totally new dimension. For the first time in human history appears a real danger of destroying that biogenetic basis which is a premise of individual being of a man and his forming as a personality. With this basis in the socialization process different programs of social behavior and value orientations come together. They are contained and elaborated in the culture.

We are talking about a threat of human physical existence which is a result of a million years of biological evolution and which is becoming deformed by the modern technogenic world. This world demands inclusion of humans in the growing diversification of social structures, which has to do with gigantic burden on the psyche, and stress that damages human health. Heavy loads of information, stress burdens, carcinogens, pollution of the environment, stockpiling of harmful mutations – all of these are problems of today's being, its day-to-day realities.

Civilization has notably prolonged the human's life circuit; developing medicine, which allows to heal many illnesses, but along with that it has taken away the act of natural selection, which in the early ages of mankind eliminated carriers of genetic errors from the chain of passing generations. With the growth of mutagenic factors in the modern conditions of biological human reproduction, a danger of incisive aggravation of the mankind gene fund arises.

Sometimes people can see an answer to all problems in the perspectives of genetic engineering. But in this case new dangers await us. If there is a possibility of interfering in the human genetic code, to change it, this way leads not only to positive results of treating a number of hereditary diseases, but also opens harmful perspectives of rebuilding the basis of human solidity. A temptation of "systematic" genetic perfection of nature created anthropological material emerges, adapting it to the new coming social charges. It is all written about not only in science-fiction literature. A similar perspective is seriously discussed by biologists, philosophers and futurologists. Undoubtedly, the achievements of scientific and technical progress will give mankind strong means with which to influence the depth of genetic structures, which are in charge of the human body reproduction. But when the human race acquires similar means in its disposition, it will gain something that is equivalent to atomic energy if we look at the consequences. With the modern level of moral development there will always be "experimenters" and volunteers for the experiments, who can make a slogan of human biological nature improvement realities of political struggle and ambitious aspirations. Perspectives of genetic rebuilding of human solidity are linked with no less dangerous perspectives of manipulating the human psyche by way of influencing his brain. Modern brain research discovers structures, the influence of which can lead to hallucinations, recollection of pictures from the past which are perceived as reality, and change emotional states of man, etc. Volunteers already appeared, the methods from this area are being practically applied. For example, they implant electrodes into the brain which give an opportunity to evoke unordinary psychic states, eliminate sleepiness, get a sense of cheerfulness and so on, with the help of soft electronic irritation.

The growing psychic charges, with which a human deals more often, contribute to saving negative emotions and often stimulate the use of artificial means to get rid of the tension. Under these circumstances a threat of distributing traditional (tranquilizers and narcotics) and new means of psychic manipulation arises. Mainly intrusion into the human solidity and especially an attempt to purposefully change the sphere of emotions and genetic bases of the human being even under strict control and weak variation, leads to unpredictable consequences. We cannot omit the fact that human culture is deeply connected with human solidity and the primary emotional tune which is dictated by it. Let us suppose that the character from Orwell's anti-Utopian "1984" accomplished a somber plan to genetically change the feeling of sexual love. For the people who would supposedly have lost this emotional sphere, they would not have any sense for Byron, Shakespeare or Pushkin. These people would have lost whole layers of humanitarian culture. Biological premises are not only a neutral background of social being, they are a ground for the growing humanitarian culture. If not for this culture humanitarian spirituality would never have been possible.

All these are problems of human existence which were brought about by technogenic culture. Modern global crises are questioning the type of progress, realized in the previous technogenic development.

It seems that on the frontier of the two millenniums, by the Christian calendar, humankind must radically turn to some new forms of civilization development.

Some philosophers and futurologists are comparing modern processes with the changes that the human race experienced under transition from the Stone Age to the Iron Age. This point of view has a deep foundation if we keep in mind that the solutions to global problems require radical transformation of the earlier accepted strategies of human life activity. Any new type of civilization development demands an output of new values, new worldview orientations. A review of the former attitude to nature, and the ideals of dominance which are oriented on forced transformation of the natural and the social world, is necessary. It is also necessary to develop a new ideal of human activity, a new understanding of human perspectives.

In this context a question of traditional values of science and scientific and technical progress for the technogenic civilization arises.

There are many antiscientific conceptions, which lay the burden of responsibility for the growing global problems on science and its technological application. Ultra antiscientism with its demands on restraining and even slowing down scientific and technical progress, essentially, is offering a return to traditional societies. But on these paths in modern conditions it is impossible to solve the problems of providing the constantly growing populace with elementary and vitally needed goods.

The answer is not to deny scientific and technical development, but to add to it a humanitarian dimension which, in its turn, sets a problem of a new type of scientific rationality, which includes humanitarian orientations and values in their true form.[7]

In this connection a series of questions emerge. How is it possible to include in the scientific cognition value orientations external for it? What are the mechanisms of this engagement? Won't a demand for measuring science with social values lead to deformation of the truth and keep ideological control over science itself? Are there any internal, growing in the science itself, premises for it to transform to a new condition? And how will this new condition reflect the fate of theoretical knowledge, its relative autonomy and its social value?

These are truly cardinal questions of today's science philosophy. The answer to them expects research of scientific cognition's peculiarities, its genesis, mechanisms of its development, finding out how it can historically change the types of scientific rationality and what are the modern tendencies of such change.

Obviously, the first step on this path shall be an analysis of science specificity, and elicitation of those invariant attributes which are solidly saved at historical shifts of scientific rationalization types.

During every concrete historical epoch these attributes can connect to special, incidental characteristics of scientific cognition. But if invariant features of science which distinguish it from other forms of cognition (art, common cognition, philosophy, religious apprehension of the world) will vanish, then this will mean the end of science.

SPECIFICITY OF SCIENTIFIC COGNITION

Main distinctive features of science

Intuitively, it seems clear how science differs from other forms of cognitive activity of a human being. But a legible explication of specific strokes of science in the form of attributes and definitions appears to be a hard enough task. This is attested by the diversification of scientific definitions, and unstoppable discussions over the problem of demarcation between science and other forms of cognition.

Scientific cognition, just like other forms of spiritual production, ultimately, is necessary to control human activity. Different types of cognition perform this role in various ways, and the analysis of this differentiation is the first and the crucial condition for elicitation of the peculiarities of scientific cognition.

Activity can be considered as a complicated web of different acts of transforming objects, when products of one activity transfer to another one and become its components. For example, iron ore as a product of mining extraction becomes an article when transferred to the steel-maker's activity. Machines that were made on a plant out of the steel dug by a steel-maker become the means of activity at another production. Even the subjects of activity – people, carrying through objects' reformation in accordance with the intents placed, can be to a certain extent introduced as results of schooling and upbringing, which provides a subject with adopted necessary examples of actions, knowledge and skills of using determinate means in their activity.

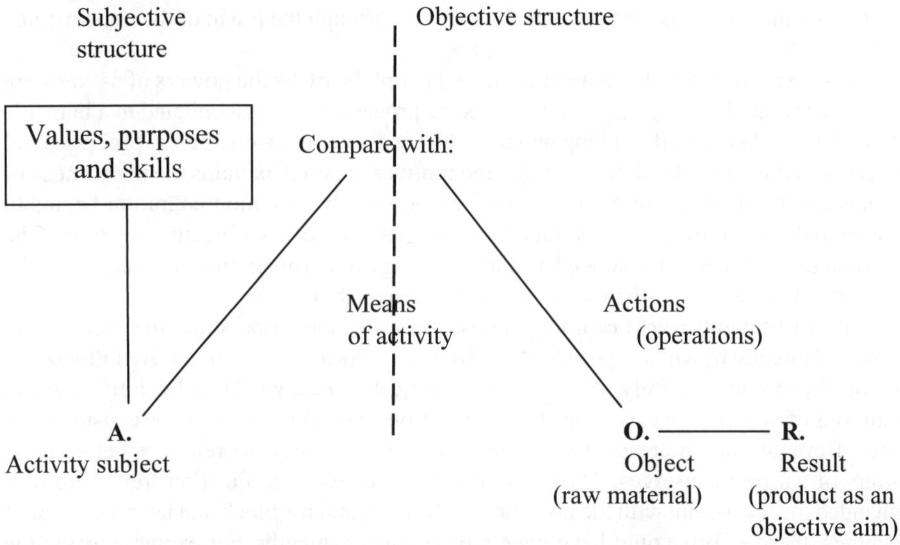

Structural characteristics of elementary acts of activity can be introduced with the help of the above scheme.

The right side of this layout shows objective structure of activity – relationship between means and the action object, and transformation of the latter into a product by virtue of performing definite operations. The left side introduces a subject structure which includes the action subject (with his aims, values, knowledge of operations and skills), who is acting rationally and using specific means of activity for his purposes. Both means and actions can be taken to the object and subject structures, because they can be looked at in two different ways. From one point of view, these means can be introduced as man-made organs of human activity. From another point of view, they can be looked at as natural objects, which are cooperating with other objects. Similarly, operations can appear in different considerations as man actions and as natural interactions between objects.

Activity is always regulated by certain values and purposes. Value answers a question: what is this or that activity needed for? Aim answers a question: what should we get in reality? Aim is an ideal product figure. It is embodied and objectified in a product, which displays a result of transforming activity product.

Since activity is universal, functions of its subjects do not have to be fragments of nature alone, which are transformed in practice, but also can be people, whose "qualities" are changed when they are included in different social subsystems, and also these subsystems themselves, cooperating in the frame of society which is considered as a whole organism. Then in the first case we are dealing with the "object side" of changes of the nature by man, while in the second case, with the "object side" of practice, directed at the change of social objects. A human, from this point of view, can appear as a subject and as an object of a practical activity.

In the early stages of society development subjective and objective sides of the practical activity are not divided in the cognition, but are taken as a unified whole. Cognition reflects the ways of practical change of objects, including in the characteristic of the latter intents, capabilities and deeds of the human being. Such vision of the reality objects is transferred onto nature as a whole, which is studied through the prism of the implemented practice.

It is known, for example, that in the ancient people's myths the powers of nature were always assimilated to a human's powers, and its processes were assimilated to a human's actions. Primordial way of thinking permanently used a comparison of these processes and powers with the humankind proceedings and motives, when it explains the phenomena of the outside world.[8] Only in the process of a long society evolution cognition begins to exclude anthropomorphic factors out of the characteristic of the objective relations. The historical development of practice has played an important role in this process, and in the first of hand perfection of the means and instruments labor.

As the instruments were becoming more complicated those operations that were earlier performed directly by a man, started to "materialize", appearing as consecutive influence of one tool to another and only after that on the object of change. Thereby, attributes and conditions of objects, appearing due to the operations indicated, ceased to be considered as direct efforts of man. Increasingly they appeared as a result of the relations between the powers of nature themselves. Thus, if in the early stages of civilization freight transfer demanded muscle strain, with the invention of the lever and the block and later the simplest machines, these efforts could have been replaced mechanically. For example, using the block system one was capable of balancing a large load with a small one, and by adding an insignificant weight to a small load to raise a heavy freight to the required height. Here the efforts of man are not needed to raise a heavy body; one load on its own can transfer another one.

A similar transfer of human functions to mechanisms leads to a new vision of the powers of nature. Earlier powers were understood only as an analogy with the physical effort of man while now they are considered as mechanical powers. Example adduced would be served as an analogy of that process of "objectification" of the objective relations in practice which apparently has already begun during the epoch of the first ancient urban civilizations. In this period cognition gradually starts to divide the object side of practice from the subjective factors and to look at this side as a special, independent reality. Such consideration of practice is one of the necessary conditions for the appearance of scientific research.

Science understands its final purpose as foreseeing the process of transformation of the practical activity's subjects (object in the beginning state) into corresponding products (object in the final state). This transformation is always determined by essential connections, laws of changing and developing objects, and the activity itself can be successful only when it is conforms to these laws. That is why the main problem of science is to bring out laws in accordance with which objects change and develop.

When we talk of the processes of nature reformation, this function is executed by physical and technical sciences. The processes of social objects changes are studied by social sciences. Since the most different objects can be transformed in activity, for example, nature articles, human beings (and the states of consciousness), society subsystems, sign objects functioning as cultural phenomena, and so on – in their own degrees they can all become the subjects of a scientific study.

Scientific orientation for exploring objects that may be included in the activity (either actually or potentially as possible objects of its future transformation), and their research as subordinate to objective laws of functioning and developing, constitute the first main peculiarity of scientific cognition.

This specialty tells it apart from other forms of cognitive activity of a human being. Thus, for example, in the process of artistic familiarization with the reality objects that are included in the human activity they are not separated from subjective factors, but are taken in a singular "conglutination" with them. Any reflection of the subjects of the modern world in art at the same time expresses the value attitude of a human to an object. Artistic image is such object reflection which contains a print of a human personality, its value orientations, which are fused into a characteristic of the reality being reflected. To eliminate this interpenetrating means destroying the artistic image. In science particularity of the life circuit of a person creating knowledge, his evaluations are not directly included in the composition of a generated knowledge (Newton's laws do njt allow us to judge what Newton loved and hated, whereas, for example, in the portraits by Rembrandt the personality of Rembrandt himself, his world attitude and his personal attitude towards the depicted social phenomena are imprinted; also any portrait made by the great artist always appears as a self-portrait).

Science is oriented on the subjective and objective research of reality. This does not mean, of course, that personal moments and value orientations of a scientist do not play any role in scientific creative work and do not influence its results.

The process of scientific cognition is specified not only by peculiarities of the studied object, but also by multiple factors of sociocultural character.

Looking at science in its historical development we can discover that by the way when the type of culture changes, the standards of reciting scientific knowledge change, the ways of seeing reality in science a change so, styles of thinking which are formed in the cultural context and are experiencing influence of its most variable phenomena. This influence can be introduced as inclusion of different sociocultural factors in the process of generating scientific knowledge, strictly speaking. But ascertaining that the object and the subject are connected in any cognitive process, and that a study of a complex science is necessary in its interconnection with other forms of spiritual human activity do not answer the question of the difference of science and these forms (common cognition, artistic thinking, etc.). The first and crucial characteristic of such difference is a sign of objectivity of scientific cognition.

In human activity science distinguishes only its subject structure, and everything is comprehended through the prism of this structure, such as King Midas, from a well-known ancient legend; everything he touched turned into gold. Similarly with science: anything it touches, is a subject for it, which lives, functions, and develops with objective laws.

Here a question arises: what to do with the action subject, his targets, values, and states of his conscience? All this belongs to the components of the subject structure of the activity, but science is capable of researching these components also, because it does not have any limits for researching any really existing phenomena. The answer to these questions is simple: yes, science can study any phenomena of a human life and his consciousness, it can study his activity as well as the human psyche, and culture, but only under one angle of vision, as special subjects that subordinate to objective laws. Subject structure of activity is also studied by science, but as a special object. And there, where science is not capable constructing a subject and introduce its "natural life", which is determined by its entity connections, that is where pretensions of science end. Thereby, science can study everything in the human world, but from a special angle of approach and with a special point of view. This special angle of approach of the subject specialty at the same time expresses that science is boundless and has boundaries, since a human as a self-confident and a conscious creature has a free will, he is not only an object, but also a subject of activity. And in this subject being of his not all states can be exhausted by scientific knowledge, even if we presume that through such all-embracing scientific knowledge of man, his life activity can be obtained.

There is no antiscientism in this assertion that science has its limits. It is just an ascertaining of an indisputable fact that science cannot substitute by itself all forms of world cognition, and all culture. And everything that slips away from its sight is compensated by other forms of spiritual apprehension: art, religion, morals, philosophy.

Studying objects, changed in reality, science is not limited by cognition of only those object connections that can be understood in the frameworks of existing, historically maintained activity types on the given phase of social development. The purpose of science lies in the anticipation of possible future change of objects, including those that would be in accordance with the future types and forms of practical world change.

As an expression of these purposes in science not only research that serves modern practice are maintained, but also layers of research, the results of which can be helpful only in future practice. The movement of cognition in these layers is already stipulated not so much by direct requests of modern practice, but by cognitive interests, through which society needs of prognosis of future ways and forms of practical familiarization with the world are appearing. For example, setting of internal scientific problems and their solution in the frameworks of fundamental theoretical studies in physics led to discovering the laws of electromagnetic field and to foreseeing electromagnetic waves, to discovering the laws of nuclear fission, and quantum laws of atomic radiation when one electron goes from one energy level to another, and so on. All these theoretical discoveries laid the basis for future ways of practical mass use of nature in production. After a few decades they became the basis for applied engineering and technical research and development, which inculcation in production at its turn revolutionized technique and technology, radio and electronic equipment, atomic power-plant, laser installations, and so on.

Major scientists, creators of new and original directions and discoveries always paid attention to this potential capacity of theories to include whole constellations of future new technologies and unexpected practical applications.

K. Timiryazev wrote on this subject: "In spite of the absence of the narrow utilitarian direction in modern science, specifically on its own as a free development, independent from the pointing of everyday wizards and moralists, it became, more than ever, a source of practical everyday usage. That astonishing technical development which blinded superficial observers who were ready to take it as the most prominent feature of the 19th century, is only a result for not obviously incredible in the history of development of science specifically, free of any utilitarian pressure. The development of chemistry is striking proof for that: it was both alchemy and iatrochemistry, it was a servant for mining and for the pharmacy, and only in the 19th century, "the century of science", being developed into chemistry, means clear science, it has become a source of uncountable applications in medicine, and technique, and in mining. It has enlightened the upstanding in the scientific hierarchy physics and even astronomy, and more young branches of knowledge, such as physiology, which was developed, can be said, in the flow on this century".[9]

Similar ideas were expressed by one of the creators of quantum mechanics French physicist Louis de Broglie. He wrote: "Great discoveries, even those made by researchers who didn't bear in their mind any practical usage and were engaged only with theoretical solution of problems, then quickly found their application in the technical field. Of course, Planck, when for the first time he wrote a formula, which presently bears his name, did not at all think of lighting techniques. But he did not doubt that the great amount of thought efforts he spent will allow us to understand and to foresee a large number of phenomena, which will be, promptly and in constantly bigger quantities, used by the lighting technology. Something similar also happened with me. I was greatly surprised when I saw that the principles developed by myself are very quickly finding concrete applications in the technology of diffraction of electrons and electronic microscopy".[10]

Science focusing on the study not only of objects, which are transformed in today's practice, but also those objects which can become a subject of mass practical involvement in the future, is the second distinctive feature of scientific cognition. This feature allow to distinguish the scientific and the common, elemental empirical cognition and to bring out a number of concrete definitions characterizing the nature of science. It allows us to understand why theoretical study is mainly characterizing developed science.

Scientific and common cognition

An aspiration to investigate real world objects and on this basis to foresee the results of its practical transformation is inherent not only for science, but also for common cognition, which is entwined into practice and is developed on its basis. As the practice development materialize in the working instruments human functions and creates conditions to eliminate subjective and anthropomorphic layers when the outward objects are being studied, in the common cognition appears some knowledge of the reality, in general similar to those that characterize science.

Germinal forms of scientific cognition evolved inside and on the basis of these types of common knowledge, and then separated from it (during the epoch of the first ancient city civilization). With scientific development and its transformation into one of the most important values of civilization, the way of thinking of this development begins to have a much more active impact on common knowledge. This impact develops, included in the common, elemental and empiric cognition elements of objective and subject reflection of the world.

Capability of elemental empirical cognition to bring forth subject and objective knowledge of the world sets a question on the differences between them and scientific research. Attributes which tell science apart from common cognition, are easily classified according to that category scheme in which structure of activity is characterized (seeing the differences between science and common cognition by the subject, means, product, methods and subject of activity).

The fact that science provides a more far practical prognosis and goes beyond the limits of existing stereotypes of production and common experience, means that it deals with a special assortment of reality objects, which cannot be reduced to objects of common experience. If common cognition reflects only those objects that principally can be transformed in the historically developed and settled ways and types of practical activity, then science is capable of learning those fragments of reality that can become subjects for familiarization in practice for the distant future. It permanently goes beyond the limits of objective structures of settled types and ways of practical familiarization of the world and opens new subject worlds for humankind's possible future activity.

These specialties of science objects are making insufficient for their familiarization, those means that are used in common cognition. Even though science uses native language, it cannot describe and study its objects only on the language basis. Firstly, everyday language is adapted for describing and foreseeing objects that are woven into existing human practice (science goes beyond its limits). Secondly, concepts of day-to-day spoken language are unclear and have many meanings, and most frequently their exact meaning is discovered only in the context of spoken interaction, which is controlled by everyday experience. Science cannot rely on such control, because predominantly it is dealing with objects that are not mastered by common practical activity. In order to describe studied objects science aims to fixate its concepts and expressions as precisely as possible.

Science output of a special language, suitable for describing objects that are not ordinary from the common sense point of view, is a crucial condition for a scientific research. Language of science is constantly developing by way of penetration into all new fields of the objective world. Furthermore, it pays a backward influence to a day-to-day native language. For example, the terms "electricity" and "refrigerator" were some time ago specific scientific concepts, and later were included in everyday language.

At the same time with artificial, specialized language, scientific research is in need of special system of practical activity means that by influencing a studied object allows emergence of its possible states under conditions that are controlled by a subject. Means used in production and at home, generally, are not suitable for this purpose, since objects studied by science and objects transformed in production and at home, most frequently are different by their character. Hereof a necessity for special scientific appliances (measuring instruments, device settings) arises which allow science to experimentally investigate new types of objects.

Scientific machinery and scientific language perform as expressions of already received knowledge. But as in practice, its products turn into the means of new types of practical activity, and in scientific research its products are also scientific knowledge that are expressed in language or materialized in devices becoming the means for further research.

Thus, from peculiarities of scientific research we had perceived differences in the means of scientific and day-to-day cognition as a special consequence.

By stating specifics of objects of scientific research we can further explain the main differences between scientific knowledge as a product of scientific research from knowledge received in the sphere of everyday, elemental and empirical cognition. The latter most often are not systemized; it is likely a conglomerate of information, prescripts, prescriptions of activity and behavior that were collected on the expansion of historical development common experience. Their authenticity is based due to immediate use of production and everyday practice in existing situations. What concerns scientific knowledge is, its authenticity cannot be explained only this way, because predominantly science studies such objects that are not yet mastered by production. That is why there are specific methods for justifying truth knowledge, and those are an experimental control of knowledge gained along with the knowledge deduced from one truth which is already proven. In its turn, the procedures of comprehension allow us to transfer truthfulness from one group of fragments of knowledge to others, due to which they become tied with each other and organized into a system.

Thus, we get characteristics of systematic character and legality of scientific knowledge, which tells its apart from the products of simple cognitive human activity.

From the main characteristic of scientific research can also be distinguished such a special feature of science when it is compared with common cognition, as a peculiarity of cognitive activity's method. Objects to which common cognition is directed, are forming in everyday practice. The ways which every object such as this uses are woven in common experience. The integrity of such ways, as a rule, is not made aware of by the subject as a method of cognition. Things are different when it comes to scientific research. Here the discovery of an object itself, whose specialties are subject to further study, is a very difficult task. For example, to discover short-living particles – resonances, modern physics conducts experiments on scattering bundles of particles and then uses complex calculations. Usual particles leave tracks in photoemulsions or in the Wilson's camera. Resonances, however, do not leave such tracks. They live for a very short time (10 seconds) and at this time period pass a distance smaller than size of an atom. Because of this, a resonance cannot cause an ionization of the photo emulsion molecules (or gas in the Wilson's camera) and leave a noticeable track. But when a resonance decays, particles which evolve with this are capable of leave tracks of the type shown. On a photograph they look like a set of rays and small lines which are coming out of one center. By the character of these rays a physicist determines that a resonance is present, using mathematical calculations. Thus, in order to work with one type of resonances, a researcher needs to know the conditions in which a corresponding object appears. It has to strictly define a method, using which a particle can be found in an experiment. On the outside of the method it will never tell the studied object apart from various connections and relations between nature objects. To fix an object a scientist must know the methods of such fixation. That is why in science the study of objects, and discovering their features and connections is always accompanied by understanding a method with which an object is researched. Objects are always given to a human in the system of determined recipes and methods of his activity. But these recipes are no longer evident in science, and they are infrequently repeated in day-to-day practice. Further science departs from customary things of everyday experience, deepening into researching "unordinary" objects, clearer and more distinctively displaying a necessity of creating and developing special methods, in the system of which enables science to study objects. At the same time with knowledge of objects science forms knowledge of methods. A need for unfolding and systematizing knowledge of the second type on the highest stages of scientific development, leads to forming methodology as a special branch of scientific study meant to actually direct scientific search.

Finally, an intention of science to investigate objects relatively independently of their comprehension in the available forms of production and common experience presumes peculiar characteristics of the scientific activity's subject. Scientific studies demand special training from the learning subject, in the process of which he masters historically formed means of scientific research and learns ways and methods of operating with these means. For common knowledge this preparation is not needed, it is implemented rather automatically, in the process of socialization of an individual, when the mind is formed and developed in him in the process of interacting with culture and the inclusion of an individual in multiple activity spheres. Scientific studies at the same time with the possession of means and methods also presume understanding a definite system of value orientations and value settings which are specific for scientific cognition. These orientations must stimulate scientific search, aimed at studying new objects independently of today's practical effect from obtained knowledge. Otherwise science will not perform its main function – to go out of the frames of subject structures during its epoch's practice, expanding the horizons of the possibilities of mastering the subject world by humans.

Two main science attitudes provide an aspiration for such a search: the self-value of truth and the value of novelty.

Any scientist accepts as one of the main attitudes of scientific activity the search for truth, and understanding truth as a highest value in science. This attitude is embodied in a whole range of ideals and norms of scientific cognition, which express its specifics (for example, a demand of logical consistency for a theory and its experimental verification), in the search for explanation of phenomena, proceeding from laws and principles that reflect essential connections of the studied objects, etc.

An equally important role in scientific research plays an attitude for constant growth of knowledge and special value of novelty in science. This attitude is expressed in the system of ideals and normative principles of scientific creative work (for example, banning plagiarism, admitting critical review of grounds for scientific search as a condition for comprehending more new types of objects, and so on).

Value orientations of science form the foundation of its ethos, which should be understood by a scientist in order for him to successfully engage in research. Great scientists left a significant trace in culture not only because of their accomplished discoveries, but also because of that their activity was a pattern of novelty and serving the truth for many generations. All backtracks from the truth for personal, mercenary purposes, or any display of unscrupulousness in science repulsed them.

In science as an ideal, a principle is proclaimed that before the face of the truth all researchers are equal and no past merits will be taken into consideration, if the question is of scientific proof.

One minor clerk of the Bern patent bureau A. Einstein, in the beginning of the 20th century, disputed with a famous scientist H. Lorentz, proving that his view of Lorentz's transformation was fair and correct. Ultimately it was Einstein who won this controversy. But Lorentz and his colleagues in this argument never fell back upon strokes that are widely used in arguments of common life. They did not assert that, for example, Lorentz's criticism was unacceptable on the basis that his status at that point in time was incommensurable with the status of the young physicist Einstein who was not yet known by the scientific community.

An equally important principle of scientific ethos is a demand for scientific fairness when results of research are announced. A scientist can be wrong, but he does not have the right to garble results, he can repeat previous discovery, but he does not have the right to engage in plagiarism. An institute of reference as a mandatory condition for formalizing scientific monographs and articles calls not only on fixating authorship of these ideas and scientific texts, it provides a well-defined selection of something that is already familiar to science and new results. Out of this selection there would not have been a stimulus for intense searches of the new, in science there would have appeared endless repetitions of past experience, and in the end its main quality to constantly generate growth of new knowledge, stepping beyond the boundaries of habitual and already known conceptions of the world, would have been broken.

Of course, demand of inadmissibility of falsifications and plagiarism appears as a peculiar scientific presumption which can be violated in real life. In various scientific communities different degrees of sanctions can be ascertained for violating ethical norms of science.

Let us look at one example from the life of modern science which can serve as an ideal of the community's nonconformity with these violations of principles.

In the middle of the 1970s in the sphere of biochemists and neurophysiologists, a so-called Gallis's case became very well known. Gallis was a young and promising biochemist, who worked on the problem of intercerebral morphines. He proposed an original hypothesis that morphines of vegetal origination and intercerebral morphines influence neural tissue in the same way. Gallis conducted a series of labor-intensive experiments, but could not convincingly justify this hypothesis, even though indirect material attested its prospects. Being afraid that other researchers would leave him behind and make this discovery, Gallis made a decision of falsification. He published fictional data of a test that ostensibly confirmed the hypothesis.

Gallis's "discovery" called for substantial interest in the neurophysiologists' and biochemists' community. But nobody could have proved his results by repeating the experiments following his published procedure. It was then proposed for the young and already well-known scientist to conduct these experiments in public, at special symposium in 1977 in Munich, with his colleagues watching. Finally Gallis had to admit falsification. The scientific community reacted to this confession with a strict boycott. Gallis's colleagues stopped scientific contacts with him and all his co-authors publicly stated that they had stopped all mutual articles with him. Ultimately Gallis published a letter, in which he apologized to colleagues and announced that he was discontinuing his work for science.[11]

Ideally, the scientific community should always deny some living privileges to researchers who were caught in deliberate plagiarism or intentional falsification of scientific results. The closest to this ideal stand the communities of mathematicians and naturalists, but humanitarians, for instance, because they feel much more pressure from ideological and political structures, impose sanctions for researchers who deviate from the ideals of scientific fairness.

It is revealing that for the common conscience confirmation with the main setting of the scientific ethos is not at all necessary, and sometimes undesirable. A man who tells a political anecdote in unfamiliar company, does not necessarily have to refer to the source of information, especially if he lives in a totalitarian society.

In their everyday life people exchange very different knowledge, share their life experience, but reference to the author of this experience in most of the situations is simply impossible, because this experience is anonymous and frequently is translated in culture for centuries.

Presence of specifics for science norms and purposes of cognitive activity and also specific means and methods which provide understanding of every new object, demands purposeful formation of scientific specialists. This necessity leads to the appearance of the "academic component of science" which is special organizations and institutions that provide schooling of scientific personnel.

In the process of such training future investigators must understand not only special knowledge, ways and methods of scientific work, but also general value orientations of science, and its ethical norms and principles.

* * *

Thus, when we find out the nature of scientific cognition, we can distinguish a system of differences of science. Among them the main features are: a) an attitude of investigating the laws of transforming objects, and subjectivity and objectivity of scientific knowledge that realize this attitude; and b) emergence of science beyond the barriers of object structures of production and common experience, and its study of objects relatively independently of today's possibilities of their production maintenance (scientific knowledge always belongs to a wide class of practical situations of present and future which is never set in advance). All other necessary features in which science differs from other forms of cognitive activity, can be introduced as dependent on and specified by the illustrated main characteristics.

GENESIS OF SCIENTIFIC COGNITION

Characteristics of advanced forms of scientific cognition to a large extent show the paths on which one should search for the answer to a problem with genesis of theoretical knowledge as a culture phenomenon.

Pre-science and developing science

In the history of formation and development of science two stages can be distinguished that follow two different methods of building knowledge and two forms of prognostication of the results of the activity. The first stage characterizes very young science (pre-science), the second one – science in its own true meaning. Science yet to be born studies predominantly those things and ways to change them, with which a human being had frequent contact in production and his everyday experience. He strove to construct models of such changes, in order to foresee the results of a practical activity. The first and necessary premise for it was researching the objects and their features and relations, distinguished by practice itself. These things, features and relations had been fixed in cognition in the form of ideal objects, which began to be operated by thinking as specific things that substitute objects of the real world.[12] This activity of the mind formed on the basis of practice and displayed by itself the idealized scheme of practical transformation of material objects. Combining ideal objects with matching operations of their changes, early science built this way a scheme of those changes of objects that could have been carried through in production of a given historical epoch. Thus, for example, analyzing ancient Egyptian tables of addition and subtraction of the integers, it is not hard to establish that the introduced knowledge in them constitutes in their contents a typical system of practical transformations realized over subject integrity.

In the tables of addition each one of the real objects (they could be animals gathered in a herd, stones put up for building, and so on) was substituted for an ideal object "unit", which was fixed with a sign I (vertical line). Collection of objects here was shown as a system of units (for "tens", "hundreds", "thousands", etc. in Egyptian arithmetic there were its own signs, fixing matching ideal objects). Operating with objects, joined into set (addition) and separating from set (subtraction) were shown in rules for activity on "units", "tens", "hundreds" and so on. Adding, say, three units to five units was done this way: a sign III was drawn (number "three"), then under it were written five more vertical lines IIIII (number "five"), and then all these lines were moved into one line placed under the first two. As a result eight lines appeared that denoted a matching number. These operations reproduced the procedures of appearing sets of objects in the real practice (real practical forming and separating object sets was based on a procedure of adding some single objects to the others).

Using this sort of knowledge it was possible to foresee the results of objects transformation, specific for different practical situations connected with joining objects into some set.

The same connection with practice can be found in the first knowledge concerning geometry. Geometry (Greek "geo" – land, "metria" – measurement) in the very first meaning of the term, discovers a connection with the practice of measuring territorial sites. Ancient Greeks took primary geometry knowledge from ancient Egyptians and Babylonians. The agricultural civilization of Ancient Egypt was based on the cultivation of fertile lands in the Nile Valley. Land sites, which were owned by various village communities, had their own borders. Where the Nile came out of its shores these borders became invisible under river sludge. Their reconstruction was an important task which was discussed by special state officials. Outlines of the sites and their size were drawn on plans on papyrus. Such plans were the models of land sites, and according to them the borders were reconstructed.

Besides reconstructing borders of the land sites there were practical necessities for measuring their area. This presented a new class of mathematical problems, whose solution demanded operations with plans. For this process the general geometric figures were used – triangle, rectangle, trapezium, circle, through combinations of which it was possible to show areas of the land sites of a complex configuration. In ancient Egyptian mathematics were found ways of calculating the areas of the main geometrical figures, and this knowledge begin to be used not only to measure land sites, but also to solve other practical problems, particularly for building different structures.

Operations with geometrical figures on the plans were connected with building and transforming these figures and were realized with the help of two essential instruments – a pair of compasses and a ruler. This is still the fundamental process in geometry. It is characteristic that this method plays a role of a scheme of real practical operations. Measuring land sites and also sides and planes of the created buildings in construction, was realized with the help of taut measuring rope with knots that showed a measuring unit (ruler) and a measuring rope, which had one end fastened with a peg, and the rod (peg) on its other side drew arcs (pair of compasses). Transferred to actions with plans, these operations appeared as the constructing of geometrical figures due to a pair of compasses and a ruler.

The way of constructing knowledge by means of abstraction and schematization of the object relations of the existing practice provided its results foreseeing within the limits of already established methods of the practical mastering of the world. However, as cognition and practice are developing together with the mentioned method a new method of knowledge structure in science is formed. It adumbrates a transfer to, strictly speaking, scientific research of the object connections of the world.

If on the stage of pre-science the first ideal objects and their relations (accordingly the meanings of the main language terms and the rules for operating with them) were directly taken out of practice and only after that inside the created system of knowledge (language) new ideal objects were forming, then cognition is now making the next move. It starts to build a quasi foundation "from the above" towards real practice and only after that, having done a great deal of mediations, checks construction, created from ideal objects, comparing them to object relations of practice.

When this method is used, initial ideal objects are not already taken from practice, but are borrowed from earlier developed systems of knowledge (language) and are put into practice as building material when new knowledge is formed. These objects submerge into a special "web of relations", which is a structure that is borrowed from another field of knowledge, where it previously maintained itself as an organized image of the object structures of reality. Combination of the beginning ideal objects with the new "web of relations" is capable of giving birth to a new knowledge system, in which frame essential features of earlier, not studied, sides of reality can be reflected. Straight or circumstantial grounding of the given system by practice turns it into authentic knowledge.

In developed science such method of research can be found to the letter on every step. Thus, for example, in mathematics' evolution, numbers began to be looked at not as a prototype of object sets, which are being operated in practice, but as relatively independent mathematical objects, whose attributes are subjected for systematic research. From this moment on, strictly speaking, begins mathematical investigation, in the course of which from earlier researched natural numbers, new ideal objects are constructed. Applying, for example, an operation of subtraction to any pair of positive numbers, it was possible to obtain negative numbers (where from a smaller number a bigger one is subtracted). With the opened from oneself class of negative numbers, mathematics is making its next move. It extends to them all these operations that were accepted for positive numbers and this way creates a new knowledge, which characterizes earlier uninvestigated structures of reality. Further on, new broadening of the class of numbers occurs: applying operations of extracting a root to negative numbers forms a new abstraction – an "imaginary number". And to this class of ideal objects are again extended all those operations that were used for natural numbers.

The described method of constructing knowledge is approved not only in mathematics, but is also extended to the sphere of natural sciences. In natural history it is known as a method for promotion of hypothetical models with their consequent justification in experience.

Due to the new method of constructing knowledge, science gets a chance to study not only those object connections that might meet in the maintained stereotypes of practice, but also to analyze the changes in objects which, in principle, would have been able to get familiar with the developing civilization. From this moment on ends the stage of pre-science and science begins in its true meaning. In science, at the same time with empirical rules and dependencies (which were known by pre-science also), forms a special type of knowledge – theory, which allows us to obtain empirical dependencies as a result of theoretical postulates. Also, there is a change of categorical status of knowledge, which can correlate not only with the carried through experience, but also with qualitatively different future practice, and that is why it is built in the limits of possible and necessary categories. Knowledge is no longer formulated only as a precept for existing practice, it performs as a knowledge of "all by itself" reality objects. And on its basis is worked out a compounding of the objects' future practical transformation.

Since scientific knowledge is beginning to orient on the search for object structures, which cannot be revealed in common practice and production activity, it can no longer develop standing on these sole forms of practice. A necessity in a special form of practice arises, which serves developing natural science. This form of practice becomes a scientific experiment.

Because a demarcation between pre-science and science is connected with a new method of knowledge generation, a problem of scientific genesis appears as a problem of preconditions, strictly speaking, a scientific method of investigation. These preconditions are maintained in culture in the kind of definite thinking attitudes that allow scientific method to appear. Their formation is a result of a continuous civilization development.

Cultures of traditional societies (Ancient China, India, Ancient Egypt and Babylon) did not create such preconditions. Even though there appeared a great variety of concrete types of scientific knowledge and recipes for problem solving, all this knowledge and recipes did not go beyond the limits of pre-science.

Transfer to science in the strictly speaking meaning of this word was connected with two critical conditions of developing culture and civilization. Firstly, with changes in the ancient world's culture that provided the application of scientific method in mathematics and rendered it to the level of theoretical investigation. Secondly, with changes in European culture that occurred in the period of Renaissance and the transfer to the New Age, when scientific method of thinking became a property of natural science (the main process here is accepted to consider experiment as a method for studying nature, combination of the mathematical method with the experiment, and observation and formation of theoretical natural science).

It is not hard to see that we are talking of those mutations in culture that provide ultimately the start of technogenic civilization. Developed science was ratified exactly at the point of civilization development, but the historical route to it was not simple and straightforward. Single premises and probes for explicating scientific method were repeatedly realized in different cultures. Some of them instantly struck in the flow of cultural translation, others somewhat drew back to the periphery, and then once again received a second wind, as it happened, for example, with many ancient ideas, recreated during the epoch of Renaissance.

To transfer to the scientific stage there was a necessity for a special way of thinking (world view) which accepted a view on existing situations of the being, including situations of social communication and activity, as on one of the possible displays of the essentiality (laws) of the world, capable of realizing itself in various forms, including very different ones from these already fulfilled.

Such a way of thinking could not have been confirmed, for example, in the culture of caste and despotic societies of the East during the epoch of the first city civilizations (where pre-science had begun). Domination in culture of these societies of canonized styles of thinking and traditions, oriented first of all on reproducing available forms and methods of activity and superimposed serious limitations on the prognostic potentially of cognition, disturbing its going beyond the limits of maintained stereotypes of social experience.

Knowledge was obtained of conforming to the laws of nature connections on earth, usually, spliced with notions of their past (tradition) or modern practical realization. Rudiments of scientific knowledge were elaborated and recited in Eastern cultures mainly as premises for practice and did not yet find a status of knowledge of natural processes, developing according to objective laws.[13]

Spiritual revolution of antiquity

In order to realize the transition to the scientific method of knowledge generation, with its intention of researching extraordinary from the point of view of the common experience object connections, it was necessary to have another type of civilization with a different type of culture. This kind of civilization, which created premises for the first step on the way to science, were the democracies of Ancient Greece. Precisely here happens a mutation of traditional cultures, and here social life is being filled with dynamism which was not familiar to agricultural civilizations of the East, with their stagnant and patriarchic life circulation. Domestic and political life of an ancient city-state was penetrated by the spirit of competition,[14] everybody competed with each other, showing activity and initiative. This necessarily stimulated innovations in multiple spheres of activity.

Norms of behavior and activity that defined the look of social reality, developed with the interest of various social groups and were confirmed largely through the struggle of the point of view of equal, free individuals on a people's assembly. Social climate of a city-state retrieved from the norms of activity the aureole of imperishable superhuman establishment, and formed an attitude towards them as to a people's invention that is subject for discussion and improvement if the need arises.[15] On this basis telescoped notions of forms of reality, of the possibilities of others, and more perfect forms in comparison with the already realized. This view can be denoted as an idea of "variable entity" which obtained its rational formalization and development in ancient philosophy. It stimulated an elaboration of a whole spectrum of philosophical systems, introducing different conceptions of world creation and different ideals of social arrangement.

Developing the models of "possible worlds", ancient philosophy, perhaps to the largest extent during this epoch, had realized the heuristic function of philosophical cognition, which served as a necessary premise for scientific making in the strict meaning of this word.

Precisely in philosophy for the first time were demonstrated the patterns of a theoretical discourse, capable of discovering connections and relations between objects that stand beyond the limits of common experience and connected to its stereotypes and archetypes of the common consciousness. Thus, when the problem of a part and a whole, single and plural was discussed, ancient philosophy takes a theoretical approach to it, looking at any possible variations of its solution. The world can be endlessly divided (Anaxagores), the world can be divided into parts insofar as a definite bound (atomistics of Democritus and Epicurus), and ultimately totally incredible from the point of view of the common sense solution – the world is not at all dividable (being is one and undivided – Eleatics).

Justification by Eleatics (Parmenides, Zeno) of this extraordinary idea set a series of problems that involved the attributes of space, time and motion. From the principle of being, indivisibility followed the impossibility of bodies motion, because body is a part (fragment) of the world, and its movement by itself presented a change in its location (place) in space at different moments of time. Body movement is impossible if the world is indivisible, and if space and time are also indivisible. But this contradicted the facts observed of bodies moving.

To these arguments a famous ancient Greek philosopher Zeno answered with a series of counterarguments that were named the aporiae of Zeno. They proved that from the positions of theoretical mind the representation of body motion leads to paradoxes. For instance, the aporia named "Arrow" demonstrates the following paradox: at every single moment of time the flying arrow can be considered as a resting one at some point of space. But the amount of rests does not give motion, and that means that the flying arrow is reposing. In other aporias Zeno brings out antinomies which are connected with conceptions of the endless ability of space to be divided. For example, in the aporia named "Achilles" is was stated that the fastest runner Achilles would not catch up with a turtle, because he would have to at first run half the distance between himself and the turtle, and during this time it would cover some distance. Then Achilles would have to overcome the half of a new distance and again the turtle will crawl away to a definite distance, and like so on *ad infinitum*.

Most interesting is that at first glance on these exotic discussions were set problems to which later, to the extent of more than two thousand years, philosophical and scientific thought had not once returned to. At the threshold of the appearance of mechanics the thinkers of the late Middle Ages discussed a question: Is it possible to talk of body motion in a point of space? If motion is characterized by speed, and speed is a passage divided by time, then in a point there cannot be speed, since a point is zero distance, and zero divided by t gives zero. That means that a moving body rests in a point.

After Galileo mechanics had emerged in the process of the search for generalizing theory of mechanical movements (that were finished by Newton's mechanics), people again had to solve this problem in connection with justifying a notion of instant speed. The problem that was brought up by philosophy transformed into concrete and scientific. Its solution was obtained due to the development in mathematics of a theory of limits and methods of differential and integral calculus that was applied in physics.

Its is also indicative that the paradoxes of infinite divisibility of space first formulated by Zeno were reestablished later as a problem of comparing infinite sets. In the aporia "Achilles" (and other aporiae), essentially, was discovered that any route (segment), if looked on as an infinite divisible, appears as an infinite set of points. Any part of this route is also an infinite set of points and from these positions it is possible to equate it to the whole. As was fairly underlined by a historian of science A. Coyré, this problem after almost two and a half thousand years, became one of the fundamental ones in mathematics. It was thought of by great mathematicians Bernard Bolzano and Georg Cantor, and it, to a substantial extent, stimulated a modern development of the theory of sets.

Of course, during the time of the Eleatics all these heuristic possibilities of philosophical cognition which reveal the problems of science in the future, were not known. But it is important that in philosophy of this time appeared patterns of a theoretical discussion, which were orienting not so much on the axioms of sensual experience, but more on the essential, given to the mind. Here priority was given exactly to theoretical thinking which is capable of going beyond the boundaries of the common sense of its time, and stereotypes worked out in a system of limited everyday practice.

In traditional societies of the East such kinds of theoretical functions of philosophy were realizing in a shorter version. Generation of nonstandard conceptions of the world in the philosophical systems of India and China was realized sporadically, coinciding with periods of massive social cataclysms (for instance, a period of "fighting kingdoms" in Ancient China). But generally philosophy was gravitating toward the ideological constructions which serve tradition. For example, Confucianism and Brahmanism were philosophical systems which at the same time played a role as religious and ideological doctrines, regulating behavior and activity of the people. What concerns Ancient Egypt and Babylon, where there was collected a great array of scientific knowledge and prescriptions of activity that are related to the stage of pre-science, the philosophical knowledge in them at best was in state of germination. It does not yet part from religious and mythological systems which were prevailing in the culture of these societies.

A principally new picture gives the social life of an ancient city-state. The peculiarities of this life created much more advantageous conditions for realizing theoretical functions of philosophy.

Ancient philosophy had demonstrated how it is possible to equally unfold a notion of different types of objects (frequently they are extraordinary from the point of view of available experience), and ways of their cognitive understanding. It gave patterns of knowledge construction of such objects. It is a search for a unified foundation (the origin and causes) and then inferring consequences (a necessary condition for theoretical organization of knowledge). These results had certainly effected the development of the theoretical layer of research in ancient mathematics.

An ideal of a well-founded and demonstrative knowledge evolved in ancient philosophy and science under the influence of the social practice of a city-state. Eastern despotisms, for instance, did not know of this ideal. Knowledge here had been worked out by the representatives of the ruling caste who were separated from the rest of the society members (priests and scriveners of Ancient Egypt, clerks in Ancient China etc.) and were assigned as a strict rule which was not a subject for doubt. A condition for knowledge acceptance formulated in a form of injunction, were the authority of their creators and the available practice constructed in accordance with the proposed norms. Proving knowledge by their inference from some reason was unnecessary (a demand for proof is justified only then, when a proposed injunction can be challenged and where a rival injunction can be interposed).

A series of knowledge in mathematics of Ancient Egypt and Babylon, as it seems, could not have been obtained without the procedures of inference and proof. M. Vygotsky thinks that, for example, such complex recipes e.g., as an algorithm of calculating volume of a truncated pyramid, were inferred on a basis of other knowledge.[16] But in the process of posing knowledge this inference was not demonstrated. Production and translation of knowledge in the culture of Ancient Egypt and Babylon were assigned to the caste of the priests and clerks, and had authoritarian character. Justifying knowledge by displaying proof did not turn in the Eastern cultures into an ideal of constructing and translating knowledge, which laid a burden of serious limitations on the process of transformation of "empirical mathematics" into a theoretical science.

Contrary to the Eastern societies, a Greek city-state took socially important decisions letting them go through a filter of competing propositions and opinions at a people's meeting. An advantage of one point of view over another was found out through demonstration, in the course of which arguments, such as authority, and special social rank of an individual proposing prescription for future activity, were not considered seriously. Dialogue was maintained between equal civilians and the only criteria was the well-foundedness of a suggested norm. This was also extended by the ancient philosophy to scientific knowledge. Precisely in Greek mathematics we come to see the posing of knowledge by way of theorems: "If it is given – it is to be proved – proof". But in Ancient Egyptian and Babylonian mathematics this form was not adapted, here we only find normative prescripts of problem solving, posed by a scheme: "Do so!" ... "Look, you are correct!".

It is typical that development in ancient philosophy of the methods for grasping and unfolding the truth (dialectics and logic) was flowing like a world reflection through a prism of social practice of a city-state. The first steps toward perceiving and cultivating dialectics as a method were connected to the analysis of encountering in an argument opposite points of view (a typical situation of developing and making decisions at a people's meeting). Concerning logic, its elaboration in ancient philosophy began with a search for criteria of a correct reasoning in elocution and the norms worked out here of a logical entailment were then applicable for scientific discourse.

Application of patterns of theoretical reasoning of pre-science mathematical knowledge gradually placed it on the level of theoretical cognition. Already in the source of ancient philosophy development, attempts were undertaken to systematize mathematical skills obtained in ancient civilizations and apply to them a procedure of proof. Thus, Thales, one of the early ancient Greek philosophers, is credited with proof of a theorem of the equality of angles at a base of an isosceles triangle (as a fact this knowledge was already obtained in Ancient Egyptian and Babylonian mathematics, but it was not proved to be a theorem). Thales's student Anaximander wrote a systematical essay of geometrical knowledge, which also contributed to a discovery of the collected recipes of problem solving that were subjected to substantiation and proving as theorems.

The most important stage on the way of creating mathematics as a theoretical science were the works by the Pythagorean school. The picture of the world was created by it. Even though this picture did include mythological elements, by its main components it was already a philosophical and rational figure of the universe. In the base of this picture lay a principle: the number is the beginning of everything. Pythagoreans thought enumerative relations were a key to understanding the way the world is made and this created special premises for the appearance of mathematics at theoretical level. The task was to investigate numbers and their relations not just as models of these or those practical situations, but as themselves alone, not relating to a practical application. Cognizing features and relations between numbers then was already thought of as cognizing the principles and the harmony of the cosmos. Numbers appeared as special objects that should be comprehended by the mind. Their features and connections should be studied. And after that, emanating from knowledge of these features and connections, observed phenomena can be explained. Precisely this attitude characterizes a transformation from purely empirical cognition of collective relations (cognition tied to the available experience) to theoretical investigation, which by operating abstractions and creating new ones on the basis of earlier obtained abstractions, is realizing a breakthrough to new forms of experience, opening previously unknown objects, their features and relations.

In Pythagorean mathematics at the same time with proof of a series of theorems, the most prominent one of which is the famous Pythagorean Theorem, were actualized important steps in combining a theoretical research of geometrical figures with the features of numbers. Connections between these two fields of evolving mathematics were two-sided. Pythagoreans wished not only to use numeric relations to characterize the features of geometrical figures, but also to exercise geometrical forms for investigation of numeric collections. Thus, number "ten" which was evaluated as the perfect number finishing tens of a natural row was correlated with a triangle, the general figure, to which, when the theorems demonstration was held, the philosophers aspired to reduce to other geometrical figures. Correlation of number "ten" and an equilateral triangle was displayed by the following scheme:

I
I I
I I I
I I I I

Here the first row corresponds with "one", the second one – with "two, the third one – with "three", the fourth one – with "four", and their sum gives us number "ten" (1 + 2 + 3 + 4 = 10).

We need to say that a connection between geometry and the theory of numbers specified a raising of perspective problems, which stimulated the elaboration of mathematics and led to a series of important discoveries. Thus, already in ancient mathematics when a problem of a numeric expression of relation of a hypotenuse to a leg of a right-angled triangle was solved, the irrational numbers were discovered. A research of "figurate numbers" which continued the Pythagorean tradition, also went on developing in subsequent history of mathematics.

The elaboration of theoretical knowledge of mathematics was held during the ancient epoch in close connection with philosophy and within the frameworks of philosophical systems. Practically all great philosophers of Antiquity – Democritus, Plato, Aristotle and others paid incredible attention to mathematical problems. They gave the ideas of Pythagoreans, which were heavily loaded with many mystical and mythological layers, a more strictly rational form. Plato and Aristotle in different versions fought for an idea that in the foundation of the universe lays a mathematical principle. These notions stimulated the development of mathematics itself, as well as its application in different fields of investigation of the world around us. During the ancient epoch already was formulated a thought that the language of mathematics must serve for understanding and describing the world. Like Plato stated, "Demiurge" (God) always geometrizes", that is to say geometrical figures perform as a basis for apprehending outer space. The development of theoretical knowledge in mathematics in ancient culture was honorably finished by creation of a first example of scientific theory – Euclidean geometry. In principle, its construction, which united into an integrated system, separated blocks of geometrical problems, which were solved in the form of a theorem proving, and adumbrated specification of mathematics into a special, independent science.

Along with that in Antiquity there were obtained various applications of mathematical knowledge to descriptions of natural objects and processes. First of all this concerns astronomy, where there were performed calculations of the planets locations, forecasts of solar and lunar eclipses, and brave attempts were undertaken to evaluate the size of the Earth, Moon, Sun and distances between them (Aristarchus of Samos, Eratosthenes, Ptolemy). In ancient astronomy were created two competing conceptions of the world structure: heliocentric notions of Aristarchus of Samos (coming ahead of the following discoveries by Copernicus) and geocentric system of Hipparchus and Ptolemy. And if the idea of Aristarchus of Samos, which assumed circular motions of a planet on orbits around the Sun, ran into troubles when it tried to explain the observed movements of planets in the sky, the Ptolemaic system with its notions of epicycles gave very correct mathematical forecasts of the observed positions of the planets, Moon and Sun. The main book of Ptolemy *Mathematical Construction* was translated into Arabic with the title *Al-Majisti* (the greatest) and then returned to Europe as "Almagest", had become the leading treatise of astronomy in the Middle Ages during fourteen centuries.

During the ancient epoch important steps were also made in application of mathematics for describing physical processes. Especially characteristic in this respect are the works by the great Hellenic scientists of the so-called Alexandrine period (about 300 BC-600 AD) – Archimedes, Euclid, Heron, Pappus, Ptolemy and others. In this period the first theoretical knowledge of mechanics appears. Firstly should be distinguished the discovery by Archimedes of the beginnings of statics and hydrostatics (the theory developed by him of the center of gravity, the theory of lever, discovery of the general law of hydrostatics and the work on the problem of stability and balance of floating bodies, and so on). In Alexandrine science a series of problems were formulated and solved connected with application of geometrical statics to balance and motion of weights on an inclined plane (Heron, Pappus); theorems were proved on volumes of rotating bodies (Pappus), and the main laws of geometrical optics were discovered– the law of rectilinear propagation of light, and the law of reflection (Euclid, Archimedes).

All this knowledge can be appraised as the first theoretical models and the laws of mechanics, obtained with the use of mathematical demonstration. In Alexandrine science one can already see expositions of knowledge, not strictly tied to nature's philosophical schemes and claiming self-importance.

But one step from the birth of theoretical natural science as a special and self-valuable field of human cognition and activity remained. It was needed to combine mathematical description and to systematically put forward these or those theoretical assumptions with experimental study of nature. But specifically this last step could not have been made by ancient science: it could not have developed theoretical natural science and its technological applications. A reason for that the majority of researchers see in slave ownership: use of slaves in the function of an instrument when these or those technical problems were solved. Cheap labor of the slaves did not create necessary stimuli for advancement of solid technique and technology, and consequently, to have working for it natural scientific and engineering knowledge.[17]

Indeed, an attitude to physical labor as the lowest sort of activity and strengthening by way of development of class layers of society, separating intellectual labor from physical labor, brought about in ancient societies a peculiar breakout between abstract theoretical investigations and practically utilitarian forms of applying scientific knowledge. It is known, for instance, that Archimedes who became famous not only because of his mathematical works, but also because he had applied their results in technology, thought that empirical and engineering knowledge is "a low and unhandsome deed" and only under the pressure of circumstances (the siege of Syracuse by the Romans) was forced to engage in military technologies and defensive structures' improvement. Archimedes did not mention in his works possible technical applications of his theoretical research, even though he himself engaged in such applications. On this matter Plutarch wrote that Archimedes was a human "of a dignified way of thinking and such deepness of the mind and richness of knowledge", that "looking on the building of machinery as on the low and uncivil process, had put all his eagerness into such exercises, where beauty and perfection are resting not mixed with a life necessity".[18]

But not only in these, generally outside relations to science, social circumstances lies a reason that ancient science could not manage to open for itself the experimental method and use it to apprehend nature. Ultimately the described social premises not directly and immediately determined the look of ancient science, but influenced it indirectly, through the worldview which expresses deep mentality of ancient culture.

Birth of the empirical sciences

It is important to fix that the idea of an experimental investigation itself not obviously presumed existence in culture of special notions of nature, activity and the cognizing subject – notions, which were not common for the ancient culture, but formed much later, in the culture of the New Age. An idea of experimental investigation considered the subject as an active beginning, opposing the natural substance, and changing its objects by their power enforcement. A natural object is cognized in experiment because it is placed in artificially made conditions and only due to this displays its invisible essential connections for a subject. Not without reason during the epoch of scientific development in the New Age in European culture, was spread a very diffused comparison of an experiment with torture of nature, through which a researcher must discover from nature its innermost secrets.

Nature in this system of notions is understood as a special composition of completely different things, which has a feature of homogeneity. It appears as a field of action of legally established connections, in which dissolve unrepeatable individuality of objects.

All these comprehensions of nature were expressed in the culture of the New Age with a category of "nature". But ancient Greeks did not have such understanding. They expressed a universal "nature" in categories of "physis" and "cosmos". "Physis" meant special, completely different specifics of every object and every essence, embodied in objects. This notion oriented a human for comprehending an object as a quality, as a figured substance, taking into account its purpose, aim and function. Cosmos was considered in this system of worldview orientations as a special self-integral essence with nature. In it each separate "physical entity" occupied a definite place and assignation, and whole cosmos appeared as a perfect completion.[19]

As A. Losev emphasized, unending movement of the cosmos was imagined by the ancient thinker as a peculiar eternal return, a movement in certain limits, inside which the harmony of the whole is constantly regenerating, and that is why the mobile and changeable cosmos at the same time was thought as some sculptural whole, where parts accompanying each other create a mature harmony. That is why the figure of eternal motion and change in the Greek conception had mixed with an idea of a conglobated form (the cosmos in almost all philosophical works looked like a globe).[20] A. Losev emphasized a deep connection of these special definitions of a universal "nature" with the basis of the city-state life itself, where diversity and dynamics of domestic activity and political interests of different social groups and separate civilians combined into a whole civic integrity of free inhabitants of a city-state.[21] Ideally a city-state was understood as unity in diversity, and as the reality of such unity cosmos was assumed. Nature for an ancient Greek was not an impersonal and inanimate substance, it was seen as a live organism, where separate parts – objects – have their own assignations and functions. To the ancient thinker the idea of apprehending the world through forcible preparation of its parts and their exploration in non-free, not common for their natural being circumstances, was strange. In his mind such way of investigation could only trouble the harmony of the cosmos, but the ancient researcher still could not find this harmony. That is why apprehension of cosmos, giving purposes for all "physically existing", could have been achieved only in conceptual contemplation, which was evaluated as the main way of searching for the truth.

Knowledge of nature (physis) in ancient Greek was the opposite of the knowledge of the artificial (tekhne). For antiquity, like the shifted European Middle Ages, was inherent a sharp separation of natural, unschooled and technical, and artificial. Mechanics during the ancient epoch was not thought of as a knowledge of the environment, but was related only to the artificial, made by the hands of man. And if we are evaluating experiments by Archimedes and his mechanics as a knowledge of the laws of nature, then in the ancient world it was related to "tekhne", the artificial but experimenting as a way of cognizing nature.

Theoretical natural science, which is based on the method of experiment, appeared only at the stage of development of technogenic civilization. Problems of culture transformation, which were realized during this epoch, are actively discussed in the modern philosophical and culturological literature.[22] Not aiming to have an analysis of these transformations in all aspects, we will only emphasize that their basis became a new understanding of man and human activity, which was called for by processes of great reformations in the culture of critical epochs – Renaissance and the transition to the New Age. At this historical period culture was maintaining an attitude to any activity, and not only to intellectual labor as a value and a source of public welfare.

This creates a new system of value orientations, which already begins to be revised in the culture of the Renaissance. On the one hand, contrary to the Middle Ages worldview, a new system of humanitarian ideas, connected with a conception of man who actively opposes nature in the quality of a thinking and active principle is asserted. On the other hand – an interest to cognizing nature is accented, where nature is looked at as a field of application of human powers. Already during the epoch of Renaissance begins a new understanding of connections between natural, and unschooled and artificial, generated by man's activity. Traditional Christian doctrine on the creation of the world by God obtains a specific interpretation. In relation to the divine mind, which created the world, environment is considered as artificial. Furthermore, activity of man is interpreted as a peculiar propinquity on a small scale of acts of creation. And the basis for this activity is assumed as an imitation of nature, figuring out its inner common sense (laws) and following reasonable harmony of nature in human art – science, artistic creative work, the technical inventions. Values of the artificial and the natural are balanced, and sensible change of nature in the process of human activity appears not as something that contradicts it, but as something that is consistent with its natural arrangement. Exactly this new attitude towards nature was fixed in the category "nature", which served as a premise for developing a principally new way of cognizing the world: there appeared an idea of the possibility to ask nature theoretical questions and receive answers to them by actively transforming environmental objects.

New meanings of the category "nature" were connected to formation of new meanings of categories "space" and "time", which was also necessary for becoming a method of experiment. Conceptions of the Middle Ages on space as a consequence of temporal moments completely different from each other, filled with hidden symbolic meaning, were a barrier on this passage.

As it is known, a physical experiment presumes its principal possibility of reproduction in different points of space and at different moments of time. It is comprehensible that physical experiments conducted in Moscow can be repeated in London, New York and in any other point of space and at any moment of time. If this reproduction did not exist, then physics as a science would not be possible. This also concerns reproduction of experiments in time. If an experiment, conducted in one moment of time, could not principally have been repeated in another moment of time, empirical science would not exist.

But what does this mean, what seems to be an obvious demand for the ability of an experiment to be reproduced? It means that all temporal and spatial points must be identical in the physical sense, that is to say in them the laws of nature must act in the same way. In other words, space and time are assumed homogeneous.

But in the culture of the Middle Ages a human never thought space and time as homogeneous, he assumed that different dimensional regions, like different moments of time have a different nature, have a different meaning and sense. Such understanding had penetrated all spheres of culture of the Middle Ages, like common thinking, artistic perception of the world, religious and theological and philosophical conceptions, physics and cosmology of the Middle Ages, and so on. It was a natural expression of the social relations system among humans in the given epoch, their way of life activity.[23]

Particularly, in science of that epoch this found its expression in the notions on qualitative difference between terrestrial and celestial space. In the worldview of the Middle Ages culture, celestial was always identified with "heavenly" and "spiritual", and terrestrial – with "corporal" and "peccant". It was reckoned that movements of celestial and terrestrial bodies have a principal difference, because these bodies belong to principally different spatial spheres.

Radical transformation of all these notions already began during the epoch of Renaissance. It was specified by many social factors, including influence of public consciousness by the great geographical discoveries, increasing migration of population during the epoch of primal accumulation, when bankrupt peasants were rushed from land, destroying traditional corporate connections and erosion of the lifestyle of the Middle Ages, based upon strict social hierarchy.

It is indicative that new concepts on space were evolving and developing during the epoch of Renaissance in very different fields of culture: in philosophy (concept of space infinity of the Universe by G. Bruno), in science (system by Copernicus which considered the Earth as a planet, turning around the Sun, and that alone already erased a strict border between the terrestrial and celestial spheres), in graphic arts, where there appears a concept of painting as a "window on the world" and where the dominating form of spatial organization of the showed becomes a linear perspective of homogeneous Euclid space.

All these notions, formed in the culture of Renaissance, maintain an idea of homogeneity of space and time and like that created premises for approving a method of experiment and combination of theoretical (mathematical) description of nature with its experimental study. By many means they prepared a revolution in science, realized during the epoch of Galileo and Newton and finished by creation of mechanics as the first natural scientific theory.

It is indicative that one of the fundamental ideas that led to its construction was the heuristic program formulated by Galileo– to study similarities of natural objects motion, also including celestial bodies, analyzing behavior of mechanical devices (in particular, weapons of the Venice arsenal).

In his own time Niels Bohr expressed an idea that a new theory that brings an overturn to the former system of notions on the world most frequently begins with a "crazy idea".

Speaking of Galileo's program, this would be correct. For many contemporaries this really was a crazy idea – to study the laws of motion, to which subordinate celestial bodies, by experiments with mechanical weapons of the Venice arsenal. But the source of this idea laid in the previous cultural revolution, when were overcome former conceptions of heterogeneous space of the universe, which sanctioned contradistinction of terrestrial and celestial spheres.

Incidentally, productivity of Galileo's program was demonstrated in the next period of mechanics development. Tradition, coming from Galileo and Huygens to Hooke and Newton, was connected with attempts of modeling in mental experiments with mechanical arrangement of interacting powers between celestial bodies. For example, Hooke examined the rotation of the planets by analogy with rotation of a body, fixed on a thread, and also a body tied to a rotating wheel. Newton used analogy between rotation of the Moon around the Earth and movement of a ball inside a hollow sphere.

It is characteristic that specifically on this passage was discovered a law of universal gravitation. To Newton's formulation of this law led a comparison of laws by Kepler obtained in a mental experiment over analogue mechanical model of mathematical expressions, which characterize movement of a ball under the effect of centrifugal forces.[24]

Theoretical natural science, evolved during this historical epoch, became the second (after the development of mathematics) most important milestone of science formation in the strict meaning of the word.

In the quality of consecutive historically significant stages, which determined its development and functions in culture, we can distinguish the development of the technical and social and humanitarian sciences. Their development as special subsystems of empirical science (at the same time with natural science) also had socio-cultural premises. It happened during the epoch of the entrance of technogenic civilization into a stage of industrialism and adumbrated an adding to science of new functions of being a productive and social power.

By the end of the 18th / beginning of the 19th century, science finally becomes an indisputable value of civilization. More and more actively it participates in the formation of the worldview, claiming for reaching objectively truthful knowledge of the world, and at the same time more and more distinctively detects pragmatic value, a possibility to constantly and systematically implement in production its results, which are realized by way of a new technique and technology. Examples of applying scientific knowledge in practice also can be found in previous historical periods, which gave impulses to comprehending practical importance of science (let us recall a famous Bacon saying: "Knowledge is power"). And still applications of results of science in production during the pre-industrial epoch had a more episodic than a systematical character.

In the late 18th / first half of the 19th century the situation changes radically. K. Marx fairly denoted that "scientific factor now for the first time is developing consciously and broadly, applied and is called for on such a scale, which previous epochs did not have any notion about".[25] Industrial development had set a fairly complex and multiple planned problem: not simply sporadic to use single results of scientific research in practice, but to provide a scientific basis of technological innovations, systematically including them in the system of production.

Specifically at this historical period begins a process of intensive cooperation of science and technology and appears a new type of social development which is more commonly known as scientific technical progress. Necessities of practice more and more distinctively showed tendencies to gradual transformation of science into direct production power. Scientific results implementation with broadening scales into production was becoming the main characteristic of social dynamics, and the idea of social progress was more distinctively connected with effective technological application of science.

An important role in scientific development, particularly in formation of new fields of knowledge, was the advent of a large machinery industry, which came as a change to manufacture production. It was not accidental in those countries where capitalism was obtaining more developed forms, that science received significant priorities. Its results implementation into production more frequently was evaluated as a condition for receiving profits by manufacturers, as a proof of power and prestige of the state. Value of science, its practical helpfulness connected with elicitation of dividends, was beginning to be distinctively perceived by those who invested money for conducting investigations.

Broadening application of scientific knowledge in production formed public necessity in appearance of a special layer of research, which could provide systematical use of fundamental natural scientific theories in the field of technique and technology. A scientific theoretical investigation of technical sciences is appearing as a special mediator for expressing this necessity between natural scientific disciplines and production.[26]

Their development in culture was specified by at least two groups of factors. On the one hand,, they were maintained on the basis of experimental science, when for the formation of technical theory it appears necessary to have its own "basic" natural scientific theory (in temporal relation it was the period of the 18th–19th centuries). On the other hand, a need for scientific-theoretical technical knowledge was initiated by practical necessity, when with solution of concrete problems engineers could not solely rely on the acquired knowledge, but were in need for scientific-theoretical justification of creating artificial objects, which is impossible to realize without a matching technical theory, worked out in the frameworks of technical sciences.[27]

Technical sciences are not just a simple continuation of natural science, they are applied researches, realizing conceptual elaboration of fundamental natural sciences. In the developed system of technical sciences its own layer is presented as fundamental, as applied knowledge, and this system demands specific object for investigation. The role of such object plays technique and technology as a special sphere of artificial, man-made and existing only due to his activity.

From the point of view of modern conceptions of the evolution of the Universe, appearance of man and society discovers a new line of evolution, in which are formulated objects and processes, highly unlikely for nature, and practically unable to evolve in it without purposive human activity. Nature creates neither a wheel, nor an engine of internal combustion, nor modern computers – all of these are products of human activity. At the same time all objects and processes created by man are possible only then, when the generating activity for them conforms with the laws of nature.

An idea of laws of nature becomes that foundation, which by saving a notion on the specifics of natural and artificial, connects them with each other. The idea itself historically formed via a basic worldview postulate and value during the epoch of developing technogenic civilization. It expressed a new understanding of nature and position of humans in the world, different from the notions that were characteristic for traditional cultures. An inextricable connection with this worldview idea, a notion of relativity of distinguishing artificial and natural, was one of the premises not only for development of natural science, but also for consequent formation of technical sciences.

First patterns of scientific technical knowledge, connected with application of the laws discovered by natural science under creation of new technologies and technical devices, appeared already on the early stages of natural sciences' development. As a classical example can serve a construction by Huygens of a mechanical clock. Based on the laws of downfall of bodies discovered by Galileo, he creates a theory of the pendulum's oscillation, and then objectifies this theory in the created technical device.[28] Furthermore, between theoretical knowledge of mechanics (law of bodies' free fall and the law of an ideal pendulum's oscillation), on the one hand, and a real construction of a pendulum clock, on the other, Huygens creates a special layer of theoretical knowledge, in which the laws of mechanics are transformed by taking into account technical demands of a created construction. The theory worked out by Huygens of an isochronal swinging of a pendulum as a free fall along an upturned cycloid, can be interpreted as one of the first examples of a local technical theory. Concerning systematical development of technical theories, this began later, during the epoch of developing industrial machinery production. Its needs, connected with replication and modification of different technical devices, and constructing of their new kinds and types, stimulated formation and transformation of an engineering activity into a special profession which serves production. Unlike technical creative work in the framework of an artisan's work, this activity is oriented on systematical application of scientific knowledge when technical problems are solved. Development of engineering activity in the 19th and 20th centuries led to differentiation of its functions, distinction into relatively independent specialization of engineering, and constructing and maintaining of technical devices and technological processes. With development of engineering activity scientific technical knowledge became more sophisticated. Within it formed empirical and theoretical levels; at the same time with applied technical theories appeared fundamental ones. Their formation was stimulated not only by progress of natural science, but first of all because of the needs of engineering practice. A characteristic example in this sense can serve a formation of a theory of machines and mechanisms. First steps to its creation were already taken during the epoch of a first production revolution and were connected to the problems of construction of relatively complex machines (ascensional, steaming, looms, spinners, etc.). Their development was based on the use as general components of the so-called simple machines (block, winch, screw, lever, and so on), investigation of which was an important starting point for discovering laws of mechanics (program of Galileo). But in the process of construction it became clear that the work of the most complex machines presumes transformation of motion with a change of its character, direction and speed. That is why the main problem was not so much in the distinction of "simple machines" as components of complex ones, but in development of theoretical schemes of their coupling and transformation of the types of motion common to them.[29] The need to resolve this problem firstly gradually led to creation of particular theoretical models, and then to fundamental theory of machines and mechanisms. The development of the latter was finished in the first half of the 20th century (V.Assur, V.Dobrovolsky, I.Artobolevsky).[30] Its characteristic specialty had become not only creation of methods of calculation of the available types of machines and mechanisms, but also the forecast of principally new types, not yet applied in practice (as the periodical system of elements created by D. Mendeleyev foresaw the existence of not yet discovered chemical elements, and fundamental theory of machines and mechanisms foresaw principally new families of mechanical devices, not familiar to the practical construction before its creation).

Appearing on the junction of natural science and production, technical sciences more clearly denoted their specific features, differentiating them from natural scientific knowledge. They obtained their own objective field, formed their own means and methods of investigation, and their own special picture of studied reality, that is to say all of which allows talk of the development of a certain scientific discipline.

After their formation technical sciences occupied a stable place in the system of developing scientific knowledge, and technical–technological innovations in production to a greater extent began to base upon application of the results of scientific–technical investigations. And if earlier science, as J. Bernal emphasized, gave little to production, then with approving technical sciences the situation has changed. They not only began to provide the needs of the developing technique, but even to outrun its development, forming the schemes of possible future technologies and technical systems.

Technical sciences, along with technical designing, beginning from the middle of the 19th century, started to perform as a connecting link between natural scientific disciplines, on one side, and production technologies, on the other.

The epoch of industrialism created premises not only for the appearance of technical disciplines as a special field of scientific knowledge. During the same historical period there started to develop a system of social and humanitarian sciences. Like other sciences, they had their sources already in antiquity, in the collected knowledge of man, different ways of social behavior and conditions for reproduction of these or those social integrities. But strictly speaking social and humanitarian sciences were constituted during the 19th century, when in the culture of the technogenic civilization there had distinctively formed an attitude to various human qualities and social phenomena as to objects of management and reformation. An attitude towards any investigated objects and processes as to objects is one of the necessary conditions of the scientific process of cognition, including social and humanitarian. That is why its premises were formation of practices and types of discourse, in which a human being, his qualities, his activity and social connections appear as special objects of purposely rational action. Precisely during the epoch of industrialism objective attitude towards man and human communities became dominant in technogenic culture. During this time was decisively formed a priority status of "relations of dependence on things" which subordinates itself and limits the sphere of "relations of personal dependence", acting as a basis of organization of social life in traditional societies. The main factor of such exchange of social and cultural priorities became the all-embracing development of marketable and monetary relations, when the capitalistic market turned multiple human qualities into goods, which have a monetary equivalent. K. Marx was one of the first scientists who analyzed the processes and social consequences of reifying human qualities in the system of relations in a developed capitalistic business. He interpreted these processes as alienation, which generates social powers not subjected to man and which turns people into objects of social manipulation. G. Simmel developed similar ideas later. Having in mind Marx's ideas, he worked out his own philosophical conception of money, which pays most attention to social and psychological aspects of monetary relations and their influence on the spiritual life of humans. Simmel considered money not only as a phenomenon of the economical social life, but as a universal way for exchange, determining the character of relations and communication in very different fields of human life activity. Simmel spoke of an idea of the meaning and symbolic role of money and their functioning as a special cultural phenomenon which mediates relations between human beings.[31]

Commenting on the book by Simmel *The philosophy of money* modern French psychologist Serge Moskovici wrote that Simmel did not discover money. But nevertheless he was the first one to grasp the philosophy of culture, begot by them and the first to formulate the whole theory of their power.[32] This power appeared in very different spheres of the human's being. It fixed a distance between an object and the human being consuming it. Precisely due to money as a mediator, not only material objects, but also spiritual essentials, ideas and values become a world which is in the same way autonomic and objective, as the physical world.[33] Money disintegrates and sterilizes (as something hampering in its way) that type of human connections, in the basis of which lies a mixture of feelings and interests, turn personal relations into impersonal, under which a human becomes a thing for another human.[34]

And another feature of money draws special attention from Simmel: its capability to turn individually unrepeatable things, conditions, human qualities into qualitative objects, which can be calculated.

After the works by Marx and Simmel this idea was developed by M. Weber in the frameworks of his conception of the capitalistic spirit. Weber especially emphasized the role of an ideal of purposefully rational action, in the developing and functioning new civilization, which was born during the epoch of Renaissance and Reformation. This ideal presumed a special type of rationality, based on principles of objectivity, legislative regulation, planning and calculation. New rationality was included in very different fields of human life activity organizing economy, law, science, arts, and everyday human life.

An attitude to man as an object of rationalistic regulation characterized vast variety of practices, maintained during the historical epoch of the advent and developing of technogenic civilization. In the famous researches by M. Foucault dedicated to forming clinical institution, history of prison and history of sexuality, it is fairly convincingly shown that in all these, at first sight, other spheres of human life not tightly connected, realized some common principle of "knowledge–power". The human appeared here as an object which should be investigated and regulated rationally. Foucault shows how this attitude came through in a historically appearing organization of surveillance and control in prisons, in a system of impersonal punishment in the name of the law, in rules of internal order in prisons, hospitals and educational institutions, in their architecture and planning of the inside space themselves. To the same class of phenomena appearing as peculiar culture symbols "knowledge–power", Foucault refers practice of a medical examination based on the inspection of the body, which appears as an object opened for observation; practice of testing and medical documentation; public discussion of problems regarding sexuality; periodical review examinations in educational institutions, when officials make a person–object in public demonstrate himself, and so on. Practices and discourses of such kind formed and fixed a new attitude to an individual as an object, being looked at, described and regulated by certain rules. Implied meanings were enrooted in the worldview universals of culture, in man and his social being's understanding, creating premises for appearance of social and humanitarian sciences. As Foucault emphasized, from that moment, "when "the norm" occupied a place of an "ancestor", and the measure of norm correspondence – place of a status, when the place of a *prominent* human individuality took an individuality of a *calculated* human, at this moment became a possible forming of sciences of man, because precisely then was launched a new technology of leadership and new political anatomy of the body".[35]

Appearance of social and humanitarian sciences completed science formation as a system of disciplines, covering all main spheres of world creation: environment, society and the human spirit. Science had obtained for us habitual features of universality, specialization and interdisciplinary links. Science's expansion into all new object fields, broadening technological and social regulative application of scientific knowledge, was accompanied by changes in the institutional status of science. In the late 18th / first half of the 19th century appears a disciplinary organization of science with common peculiarities of knowledge translation, their application and ways of reproducing a subject of scientific activity.

Development of natural scientific, technical and after them social and humanitarian knowledge resulted in a rapid growth of scientific information. Science in the end of the 18th / first half of the 19th century was featured by an increase of volume and variety of scientific knowledge, broadening differentiation of types of investigation and complication of their interconnections. All that led to changes in institutional forms of scientific cognition. A situation developed where became increasingly harder for a researcher to possess the collected scientific information, essential for successful investigations. If we use M. Petrov's terminology, we can say that for a concrete man were fairly distinctively determined new restraints of "informative capacity" connected to physiological, as well as mental limitations of man.[36]

An age of Encyclopaedists was gradually becoming a thing of the past. In order to professionally possess scientific information, it was necessary to limit spheres of investigation and organize knowledge in accordance with possibilities of "informative capacity" of an individual. All that fatally led to knowledge specialization. A researcher gradually became a specialist in s single, sometimes very narrow field of knowledge, growing as an "outside observer" in other fields of knowledge and not pretending to have all-covering knowledge. Growing specialization contributed to forming object fields of science, and led to differentiation of sciences, each one of which pretended not to be investigating the world as a whole and constructing some generalized picture of the world, but aiming at generalizing their own object of research, reflecting a special fragment or aspect of reality.

Fragmentation of the world was accompanied by peculiar splitting of earlier syncretic activity of a scientist-researcher in many different activities, each one of which was realized by a special researcher in accordance with a principle "informative capacity". Something that formerly was done by a single thinker, now presumes efforts of collective subject of cognition. Thence appeared a necessity to search for new forms of translating knowledge in culture, and also a necessity for a new type of reproduction of a subject of scientific activity.

In science of the 17th century the main form of retaining and translating knowledge was a book (manuscripts, folios), where fundamental principles and origins of "the nature of things" should be posed. It performed as a basis for education, supplementing the traditional system of direct communications "teacher–student", which provides transference of knowledge and skills of scientific work from the teacher to his students. At the same time, it appeared also as a main means of fixing new results of nature exploration.

The scientist of the 17th century faced a fairly complicated problem. It was not enough for him to get some particularistic result (solve a particularistic problem), his responsibilities included a construction of a whole picture of the universe, which should find its expression in a fairly bulky folio. A scientist was obligated not only with separate experiments, but also to engage himself with natural philosophy, correlating his knowledge with the existing picture of the world, entering into it corresponding changes. All prominent thinkers of this time worked in this way: Galileo, Newton, Leibniz, Descartes, and others.

During that time it was thought that without referring to fundamental bases it was not possible to give a full explanation even to particularistic physical phenomena. It is not accidental that Descartes wrote in a letter to Mersenne: "I would have with pleasure answered all your questions concerning candle flame and other similar things, but I can foresee that I can never fairly satisfactorily do it until the time when you get acquainted with all principles of my philosophy".[37]

But during the development of science and widening the field for research, activity more and more earnestly formed a need in such communication between scientists, which would maintain their collective discussion of not only final, but also intermediate results, not only "eternal" problems, but also final and concrete problems. As an answer to this social request in the 17th century appears a special form of fixing and transforming knowledge – correspondence between scientists. Letters that they exchanged, as a rule, contained results of an investigation, and description of the way by which these had been obtained. As such, letters turned into scientific communication, informing of the results of single researches, their discussion, argumentation and contra-argumentation. Systematic correspondence was prosecuted in Latin, which allowed reporting one's results, ideas and reflections to scientists who lived in very different countries of Europe. Thus appears a special type of community, which chose a letter as a means for scientific intercommunication and united the researchers in Europe into so-called "Republic of Scientists" (La Republique des Lettres).[38]

Correspondence between scientists appeared not only as a translation of knowledge, but also served as a basis for developing new means of investigation. In particular, it is assumed that a thought experiment obtained its fixation as a conscious investigative recipe precisely due to scientific correspondence, when during the process of describing a real object it turned into an idealized object, not coinciding with a real object.[39]

Means of intercommunication between researches and forms of knowledge translation, which emerged in the 17th century, provided a successful development of sciences of that historic epoch, but with an accumulation of the bulk of scientific information their change was required.

Already in the second half of the 17th century gradually began broadening of specialization of scientific activity. In various countries communities formed of scientists-specialists, frequently supported by public opinion and the state. As an example serves the community of German chemists – one of the first national disciplinary-oriented integrities of researchers, maintained in Germany by the end of the 18th century. As an historian of science Hufbauer writes: "In the end of the 18th century German chemists had developed their own community… They started to treat each other as necessary colleagues and basic arbiters in all that concerned scientific truth and personal achievements".[40] Communications between investigators were realized already in the national language (not in Latin). Exchanging results of researches mainly happens because of publication of single reports in the journal *Annals of Chemistry*.[41] This journal had played a special role in integrating German chemists, allowing intensive discussions of problems within its pages, encouraging German chemists "to look at each other as a general auditory", increasingly "feeling their own solidarity".[42]

A similar process featured formation of specialists' community in other fields of a swelling array of scientific knowledge.

Scientists did not already limit themselves only with correspondence between themselves and publications of books/folios as a main product of their scientific activity. Correspondence gradually loses its former status as one of the general integrations of researchers. And the "Republic of Scientists" is substituted for multi national disciplinary-oriented communities. Internal communication in these communities flows much more intensively than externally.

The place of private letters appearing as a scientific communication takes an article in a scientific journal. An article becomes especially important: unlike a book, it is not necessary to recite the whole system of points of view, that is why the duration of its appearance is reduced. But in an article this or that knowledge is not only fixed, it becomes a necessary form of fixing and translation of a new scientific result, determining a researcher's priority. In order for the new knowledge to enter culture, it is necessary to reify it and to fix it in a text, which was accessed for very different investigators, and an article successfully solves this problem. In this process an increasingly broader application national languages find. The former language of scientific communication – Latin – gradually assented its place to popular national language, which due to its special terminology, a special system of scientific notions transforms (modifies) into a language of scientific communication. It gives an opportunity for a wider circle of researchers to get acquainted with obtained scientific results and include them into the area of their own researches.

In contrast to the letter which is oriented for a concrete human being often personally known by the author, an article was addressed to an anonymous reader, which led to a necessity of more thorough choice of argumentation for justifying the proposed theses. Articles had not yet obtained all these necessary characteristics. Only by the middle of the 19th century (a period of intensive formation of disciplinary organization of science) did articles obtain those functions in which they perform in the modern scientific community: on the one hand, they appear as a form of knowledge translation, presuming successive connection with the former knowledge, because their form of writing presumes reference to the sources (institute of reference), and on the other hand is a request for new knowledge.[43]

Appearance of an article as a new form of fixing and translating knowledge was unbreakably connected to organizing and issuing periodical scientific journals. Firstly they served a special function of integrating researchers, in pursuit of showing *what* is done and *who* does it, but then at the same time with reviews, information on new knowledge was starting to be published, and this gradually became their main function.[44]

Scientific journals were becoming peculiar centers of crystallizing new types of scientific communities, appearing close to traditional integrations of scientists. During this historical period many previously evolved academic institutions were supplied with new communities with their own rules, where the purpose of science was defined. Unlike the "Republic of Scientists" where informal relations were maintained between scientists, such communities were formally organized, there were necessary meetings every week and regulations that determined life activity of such offices, etc.

It is indicative that in academies' regulations attention was drawn not only to a necessity of theoretical elaborations, but also to practical implementation of scientific research's results. This was a substantial argument through which scientists tried to get support from the government.[45]

In the late 18th / first half of the 19th century in connection with augmentation of the bulk of scientific, scientific-technical information, at the same time with academic institutions that already appeared in the 15th / beginning of the 16th century (London Royal Society – 1660, Parisian Academy of Sciences – 1666, Berlin Academy of Sciences – 1700, Petersburg's Academy of Sciences – 1724, and others), different kinds of new associations of scientists began to develop, such as the "French Conservatory (Storehouse) of Technical Arts and Crafts" (1795), "Assembly of German Natural Scientists" (1822), "British Association for the Advancement of Progress" (1831), and others.

Scientists who worked in various fields of knowledge began to integrate into scientific communities (physical, chemical, biological, and so on). New forms of organizing science also evolved new forms of scientific communication. More and more frequently as a main form of translating knowledge appear scientific journals, around which scientists integrated into interest groups.

A tendency for specialization served as an objective basis with which a scientist already did not set (or could not set) a problem of building an integral picture of world creation. More and more frequently his responsibilities included solving single problems, "puzzles"(T. Kuhn).

The situation connected to the growing volume of scientific information and the limits of "information capacity" of a subject, not only substantially transformed ways of translating knowledge, but also made more acute the problem of reproducing a scientific subject. A need for special training of scientists was appearing, when changing "amateurs or had been brought up as apprentices ... the university professor ... began to be the type of scientist".[46]

It is not accidental that during this period becomes increasingly diverse purposeful training of scientific specialists, when everywhere is developing a network of new scientific and educational institutions, including universities. The first universities had appeared already in the 12th-13th centuries (Parisian – 1160, Oxford – 1167, Cambridge – 1209, Padua – 1222, Naples – 1224, and so on) on a basis of spiritual schools and were created as centers for training for clergy. For a long time in teaching the main attention was drawn to the problem of humanitarian knowledge, but in the late 18th / beginning of the 19th century the situation changes: the necessity for broadening the network of learning subjects was beginning to be realized. During this historical period most of available and evolving universities natural scientific and technical disciplines are included into the number of subjects taught. New centers also opened for specialists' training, such as a well-known polytechnic school in Paris (1795), where Lagrange, Laplace, Carnot, Cariolis, and others taught.

A growing amount of scientific information led to a change in the whole system of education. Specializations of single fields of scientific knowledge emerge, education is beginning to be constructed as teaching groups of single scientific disciplines, obtaining pronounced features of disciplinary organized education. In its turn, this paid a back influence to scientific development and in particular to its differentiation and development of concrete scientific disciplines.

The process of teaching demanded not just acquainting listeners with an integrity of single data of the achievements in natural science, but systematical exposition and understanding obtained knowledge.

Systematization by an informal component and collection of methods, with the help of which such knowledge was obtained, began to be considered as a basis for certain scientific disciplines, distinguishing one collection of knowledge (scientific discipline) from another.[47] In other words, systematization of knowledge during the process of teaching appeared as one of the factors of forming concrete scientific disciplines.

Special training of scientific staff (reproducing a scientific subject) arranged a peculiar profession of a scientific worker. Science gradually maintained itself in the privilege of a strongly established profession, which demands specific education, and which has its own structure and organization.[48]

Disciplinary organized science with four basic blocks of scientific disciplines – mathematics, natural science, technical and social-humanitarian sciences – had accomplished much in science forming, in the strict meaning of the word. In science intra-disciplinary and inter-disciplinary mechanisms of knowledge generating had maintained, which provided its systematical breakthroughs into new object worlds. In its turn, these breakthrough opened new opportunities for technical and technological innovations in very different spheres of human life activity.

Growth of scientific knowledge appears as one of the most important factors of dynamism of modern civilization, featuring tendencies of permanent changes and renovations. But historically maintained mechanisms of this growth themselves are not given and unchangeable. They continue their intensive growth even during a stage when science obtains features of a mature organism. This development brings forth new types of scientific rationality, which actively influence fundamental worldview structures that determine the guise of modern culture. Historical dynamics of theoretical knowledge appears as a specific nucleus of these processes. Therefore an analysis of these mechanisms, and clarification of the ways of historical changes in these mechanisms is one of the most important conditions for understanding modern tendencies of scientific development.

In its turn, elaboration of a given problematic presumes preliminary study of the structure of theoretical knowledge.

NOTES: CHAPTER 1

---

[1] In culturological studies it was already emphasized that there are two types of culture: oriented on an object-activist way of life activity and on auto communication, introspection and contemplation (see, e.g., Lotman (1973)). Cultures of technogenic societies evidently gravitate towards the first type, and culture of traditional societies, to the second one.
[2] Petrov (1991, p.130).
[3] Ibid, pp.134-135.
[4] Hertzen (1946, p.84).
[5] See Luria (1974, pp.106-121).
[6] See Tulviste (1977, p.96).
[7] For more details see Frolov and Yudin (1986); Frolov (1989).
[8] As a confirmation for that serves a huge ethnographical material. Bushmen, for example, explain the origin of fire as a result of friction in such a way: "If you rub a tree for a long time, it perspires, smokes and becomes angry – and then flares up". For more details see Shachnovich (1961, pp.31-35).
[9] Timiryazev (1939, p.17).

[10] de Broglie (1960. p.178).
[11] The facts are stated in the paper "Mimicry in science", published in a journal "Techniques and Science", 1983, No 4, pp.31-32.
[12] An ideal object displays in conscience real objects, but not by all, just by some, strictly defined features. Since such fixation is realized through replacements of shown features with signs, then an ideal object appears as a meaning of the matching sign. An ideal object represents a simplified and systematized picture of a real object.
[13] See Neugebauer (1968).
[14] See Zaitsev (1985).
[15] See Kessidi (1972, pp.18-20).
[16] See Vygodsky (1967, p.237).
[17] See Doods (1951). See also *History of Ancient Dialectics* (1972, pp.61-63).
[18] See Plutarch (1961, pp.393).
[19] See Achutin (1988, pp.164).
[20] See Losev (1977, pp. 14-18).
[21] See Losev (1968, pp.21-22).
[22] From Russian works we will emphasize Achutin (1976), Bibler (1978), Gaidenko (1987), Kosareva (1989).
[23] For more details see Gurevich (1972, p.26). See also Stepin (1986).
[24] See Rosenfeld (1962-1965).
[25] Marx and Engels (1955-1981, vol.47, p. 556).
[26] On the development of technical sciences and their place in culture, refer to Gorochov (1984), Ivanov and Cheshev (1977), (1981) and others.
[27] Ivanov and Cheshev (1977, pp.97, 108, 126).
[28] For more details see *Philosophy of technics: history and nowadays* (1997, pp.128-129).
[29] For more details see Gorochov (1974, p.46), *Philosophy of technics: history and nowadays* (1997, pp.132-139), Stepin, Gorochov and Rozov (1996, pp.346-347).
[30] Gorochov (1974, p.51-53).
[31] Later, already during the second half of our century this thought was developed by T. Parsons, looking at money as on a special code of culture, "specialized language", and the transfer of money as "mailing messages" (see Parsons (1968)).
[32] Moscovici (1998, p.455).
[33] Ibid, p.398.
[34] Ibid, p.423.
[35] Foucault (1975, p.195).
[36] Petrov (1991, pp.73, 92).
[37] Quoted from *Philosophy of the epoch of early bourgeois revolution* (1983, p.303).
[38] Ibid, p.296.
[39] Ibid, pp.300-301.
[40] Hufbauer (1982, p.1).
[41] Ibid, p.62.
[42] Ibid, p.95.
[43] Price (1963).
[44] Ibid.
[45] Ibid.
[46] Bernal (1954, p.393).
[47] Mirsky (1980. p.60).
[48] Bernal (1954).

CHAPTER TWO

STRUCTURE OF THEORETICAL KNOWLEDGE

ABSTRACT OBJECTS OF THEORY AND THEIR SYSTEMIC ORGANIZATION

Analyzing the structure of theoretical knowledge, methodology is based mainly on empirical material: texts of historically formed scientific theories. Let us emphasize that methodology works first of all with highly developed (in the theoretical sense) disciplines, since in these fields it is easier to follow up all special features of theory structure. It is much more difficult in sciences which are only entering the stage of theoretical studies of the phenomena. This is caused by the fact that in a developing system (theory – in our case) the principles of functioning are more clearly visible at higher stages of development than in embryo. That is why methodologists take the structure of highly developed sciences as a kind of standard used for comparison with all other systems of theoretical knowledge.

Mathematics has been most often used as such a standard for logical and methodological investigations. Even today mathematics provide us with an important material for theoretical, educational and methodological analysis. But there is one aspect where the methodologist faces some difficulties. In "pure" mathematics it is impossible to discover any exact layer of empirical knowledge, so it becomes difficult to specify peculiarities of the structure and functioning of a scientific theory connected with its relation to the empirical basis. In order to investigate this aspect of theoretical knowledge epistemology and methodology have recourse to empirical sciences. Thereby physics is highlighted as a branch of natural science having all features of highly developed theoretical science along with broad empirical basis.

The historically formed knowledge of physics as initial material for methodological research gives us an opportunity to see clearly all features of the structure and functioning of theories in empirical sciences.

Notions and models of the dynamics of science deduced from this historical material may require some corrections when transferred to other disciplines. But it normally happens this way in the development of knowledge: notions deduced and approbated on one material are then transferred to another area and are transformed in case they do not completely correspond to the new material.

It is often believed that ideas of development of sciences should not be transferred to the sphere of social knowledge.

The reason for such limitation is division between knowledge of nature and knowledge of spirit, which was formulated in the 19th century. But we are to clearly understand that knowledge in humanities and in nature sciences have common features because of scientific nature of knowledge. The differences lie in the specificity of the objects domain. In social and humanitarian disciplines the subject includes a human being, his or her consciousness, and is often represented by a text which has some human sense. To fix such an object and to study it we need special methods and cognitive procedures. However, though the object of social and humanitarian disciplines is rather complicated, the attitude of its objective study and search for laws is a must for scientific approach. This item is not always taken into consideration by the supporters of the conception of "absolute specificity" of humanitarian, social and historical knowledge. Sometimes it is incorrectly opposed to natural sciences. Humanitarian knowledge is treated extremely widely. It is supposed to include philosophic essays, political writings, literary critique, fiction, etc. It would be correct to put the problem in a different way. We are to distinguish definitely "social and humanitarian knowledge" and "scientific social and humanitarian knowledge". The former includes the results of scientific study but is not exhaustive, as it also contemplates other, not scientific forms of creative work. The latter remains within frameworks of scientific research. Naturally, such a research is not isolated from other cultural spheres, it interacts with them, but there is no reason to identify science with other, though closely related with it, forms of human creative work.

If we confront studies of society and human beings, on the one hand, and studies of nature, on the other hand, we will have to admit that in their cognitive procedures there are both common and specific features. Methodological schemes developed in one field may express some common features of structure and development of knowledge in the other field; in this case methodology can develop its conceptions as is done in any other sphere of scientific knowledge, including social and humanitarian disciplines. It is free to transfer models solved in one sphere of knowledge to another sphere and then insert amendments, making them correspond to the specificity of the new object.

Meanwhile, we are to take into account at least two circumstances. First, the philosophical and methodological analysis of a discipline (equally natural or social and humanitarian) by itself belongs to the sphere of historical social knowledge. Even if a philosopher methodologist deals with specialized texts of science, its object is not physical fields, nor elementary particles, nor the processes of organisms' development, but scientific knowledge, its dynamics, and methods of research work taken in their historical development. It is clear that scientific knowledge and its dynamics are not natural but a social process, a phenomenon of human culture, so its study is a special kind of knowledge of spirit.

Second, we are to take into consideration that severe demarcation between knowledge of nature and that of spirit was reasonable for the 19th century but in many aspects is not valid for the science of the last third of the 20th century. Later we will have a chance to discuss this at greater length. As a preliminary, let us state that in modern science the role of research of complicated developing systems is constantly increasing. Such systems have "synergetical characteristics" and include people and their activity. Methodology of research of such objects draws sciences and humanitarian knowledge closer, erasing strict boundaries between them.

When we choose the theoretically developed sciences as our initial material, we make just the first step. One and the same material can be considered from different points of view, and different aspects of the structure of the theory may be revealed. Hence, it is necessary that we should determine the initial position for analysis of scientific texts, specify what aspects of the language of the science will be taken into account in the analysis and what aspects may be ignored.

Semiotics normally considers three aspects of language: syntactical, semantic and pragmatic.

The syntactical aspect supposes considering language only as some totality of signs which are transformed according to certain rules and form some linguistic system in their connections. When we study everyday language, we face this side when we consider word transformations in accordance with logical and grammatical rules of language.

In the language of science the syntactical aspect becomes most important when we make formal operations with symbols, for instance, with physical magnitudes (which enter mathematical expressions of physical laws), in accordance with rules of mathematics. Making such operations, a researcher disengages himself from the meaning of the linguistic terms and considers them only as signs which create formulae in their connections and then deduce other formulae according to the rules of the linguistic system given. For instance, integrating equations of motion in mechanics, a physicist operates with values m, F, x, t ("mass", "force", "space coordinate", "time") as with mathematical objects. Such operations clearly present the syntactical aspect of the language of physics.

The semantic aspect requires address to the contents of linguistic expressions. It assumes finding ideal objects and their connections which form the direct meaning of terms and propositions. Besides, semantic analysis requires that we should determine what sides of extra-linguistic reality are represented by those ideal objects. In physics, for example, this aspect clearly appears in interpretation of expressions which are results of a series of mathematical transformation of the initial formulae. In this case mathematical symbols of the expressions mentioned (functions, numbers, vectors, etc.) are regarded as physical magnitudes. Thus, connection of the magnitudes with real characteristics and relations of the material objects (which are distinguished from the universe by practical activity) appears.

Finally, the pragmatic aspect of the language assumes that we consider relation of linguistic expressions to practice and specificity of social communication which are characteristic for a certain historical period. It means that ideal objects and their correlations, which form the sphere of meaning of linguistic expressions, are taken in their relation to the sociocultural environment which has generated one or other "population" of scientific knowledge.

In the process of a scientist's cognitive activity all three aspects of the language of science interact. As to the texts which fix the results of cognition, they also represent all these aspects of language. Proceeding from our task (analysis of content structure of scientific knowledge), we shall consider the texts mainly in semantic and pragmatic aspects, i.e. single out ideal objects in expressions of the language of science and then analyze their intra-linguistic connections and their relation to practice.

Among ideal objects used in a scientific research at least two main types are traditionally singled out: empirical and theoretical objects.

Empirical objects are abstractions which fix features of real objects of experience. They are a kind of schematization of fragments of the real world. Any feature – the "carrier" of which is an empirical object – can be found in corresponding real objects (but not vice versa, since empirical objects represent only some, not all features of real objects, abstracted from reality in accordance with the aims of cognition and practice). Empirical objects make meaning of the terms of empirical language, such as "the Earth", "conductor with current", "distance between the Earth and the Moon", etc.

Theoretical objects, unlike empirical ones, are idealizations, "logical reconstruction of reality". They may be provided not only with features corresponding to the features and connections of real objects, but also with features not proper for any such object. Theoretical objects make sense of such terms as "point", "ideal gas", "black body" etc.

In logical methodological investigations theoretical objects are sometimes called theoretical constructs, or abstract objects.

Propositions in the theoretical languages are based on abstract objects, whose connections and relations form direct meaning of these expressions. That is why theoretical propositions become statements on natural processes only to such extent to which relations of abstract objects can be justified as substitutions of real features and connections revealed in practice. Thus, all theoretical statements in classical mechanics directly characterize connections, features and relations of idealized constructs, such as "material point", "force", "inertial spatiotemporal frame of reference", etc. All these are idealizations and cannot exist as real material objects. The latter truth is most evident for a "material point" which is defined as a body that has no size. But "force" or "spatiotemporal frame of reference" are also idealizations for which the real world has only prototypes but which cannot be identified with really existing objects.

"Force" in mechanics is defined as a special property of influence of one or several bodies on another body and change the state of its motion. This property is abstracted from the bodies themselves and turns into an individual object existing together with other bodies (material points) and having influence on them. Such transformation of a property into an individual object can be realized only as an abstraction.

It is easy to make sure that an inertial spatiotemporal frame of reference is also an idealized object – comparable to real objects of experience but not identical to them. An inertial frame of reference can be identified with, for instance, a real physics laboratory equipped with watches and rulers but only provided that such a laboratory is supplied with some features that do not exist in real life. It is supposed that it can be completely isolated from outer influences (inertiality). Then, it is assumed that we may ignore the influence of bodies being measured with watches and rulers. It means that the latter can be imagined as absolutely hard pivots with points and standard "hard" watches with constant length of period. Such idealization allows us to represent spatiotemporal measurements as transformation of points of Euclidean space and quasi-Euclidean time of an inertial frame of reference. Strictly speaking, in real life there are no such bodies which could be completely isolated from any influence. That is why an inertial frame of reference (characterized by Euclidean spacetime) is an idealized, theoretical construct.

Nevertheless, all these theoretical constructs of mechanics can be compared with some fragments of nature: "material points" – with bodies whose size may be ignored while resolving certain problems; "force" – with certain interaction of bodies which lead to changes in state of their motion; "inertial frame of reference" – with real objects and processes used in functions of rulers and watches, which motion can be regarded (in a certain approximation) as even and rectilinear. Due to the connection of theoretical constructs with reality of mechanics' statements formulated about the constructs, the statements appear to be descriptions of objective natural processes.

This kind of situation is characteristic for all spheres of theoretical knowledge. Fundamental definitions and postulates of Euclidean geometry have been formulated as characteristics of properties and relations of such objects as "point", "segment", "angle", "circle". The basic laws of Maxwell electrodynamics (Maxwell's equations) describe directly relations of such idealized constructs as vectors of magnetic and electrical intensity in a point, and vectors of density of current in a point at any moment of time. Only because the relations and connection of abstract objects may be justified as a representation of some real objective area, the statements of these theories receive objective value and importance.

Still, this does not mean that a theory gets objective foundation only in the case when its every abstract object can be compared with some fragments of reality. The connection between fragments of objective reality distinguished by people's practice, and the system of abstract objects of a theory are more complicated. It is well known that only some theoretical objects can be projected to reality by themselves. Most of them are related to reality studied only indirectly, by means of the first type of abstract objects.

This part of theoretical objects receives definition only within the theory, in the system of its meaning connections and relations of its statements. This fact is often reflected in the logic of science in the following way: not all, but only some terms of a theoretical system should have operational meaning, i.e. be connected by means of special rules (operational definitions) with the objects transformed in experience. The meaning of other terms is defined only within either system of theoretical language, within frameworks of linguistic contexts, where theoretical terms are related to each other and to the operationally meaningful terms. Sometimes the former connections are called intratheoretical, and the latter ones, going beyond the limits of theoretical language – epistemic.[1] Since the meanings of terms and statements are corresponding abstract objects and their correlations, then this specificity of theoretical knowledge witnesses that in theory there are abstract objects which have both intratheoretical and epistemic connections. Also there are abstract objects with only intratheoretical connections. These may be constructs that are extremely important for the theoretical system and – to a great extent – determine the peculiarities of its contents ("vector-potential" in classical electrodynamics, "charge" and "mass-energy" of a "bare electron" in quantum electrodynamics, etc.).

Existence of abstract objects which are justified only because of their intratheoretical connections presents us with evidence that abstract objects of a theory cannot be just a conglomeration of elements not attached to each other. They always form an integral system. Correlation of the elements in such a system is stipulated first of all by the fact that unfolding of a theory involves introduction of some objects on the basis of others. For instance, when in Newton mechanics one deduces equations of motion of a solid body or motion in central-symmetric field as consequences of the basic equations, it means that on the basis of fundamental abstract objects – "force", "material point", "spatiotemporal frame of reference" (whose correlations make the main sense of the basic laws of mechanics) – there are yielded new abstract objects, such as "absolutely solid body", "central-symmetric field", etc.

Constructing of abstract objects on the basis of others according to the rules of the language of this or that theory, should meet the principle of integrity of the system of theoretical objects yielded. Each newly constructed object enters into relations with theoretical constructs already built, and must correspond to them. It should not lead to emerging their new properties which are incompatible with those defined before. This is one of the basic stipulations which is fulfilled when the contents of a theory are being developed.[2] Clearly, in mechanics, while constructing abstract objects such as absolutely solid body or central-symmetric field, we should not get as a consequence, say, that the coordinate of a material point at a given moment is, in principle, uncertain. It would contradict to the initial features of a material point, since at any moment it – according to the definition – must be comparable with one and only one point of space.

In the final analysis, all abstract objects are justified within a theory by the fact that no object incompatible with the previously defined system cannot appear. As a result, we come to an idea of a special network of theoretical constructs, in which some elements are connected with empirical reality, while the others do not have such connections but are justified by their role of auxiliary elements, and the entire network exists because of them. The scheme offered by H. Margenau[3] (pic. 1) can illustrate such connections of theoretical objects with each other and with empirically investigated reality.

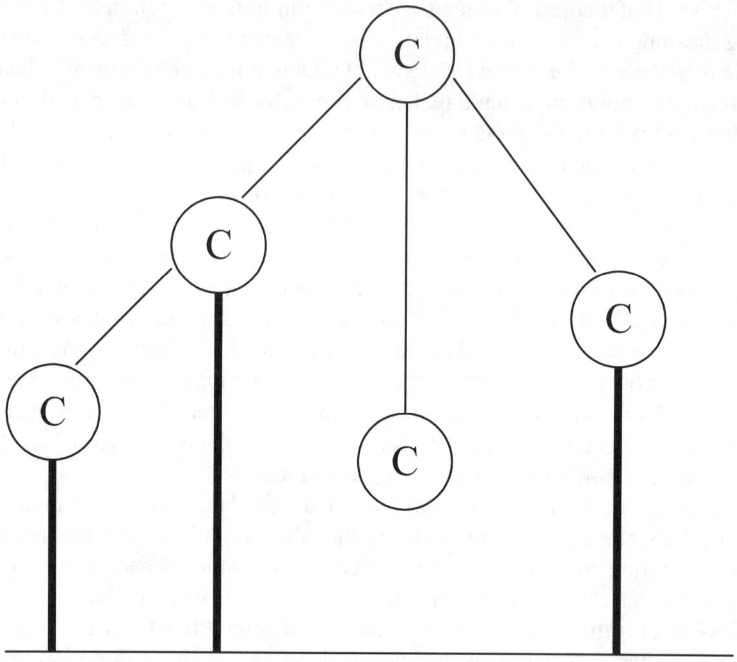

N

Pic. 1. C – theoretical constructs; N – directly given in observation and experiment studied reality; —, intratheoretical connections between constructs; ▬ *, connections of constructs with an empirical level (epistemic connections).*

This scheme reflects some quite general features of theoretical knowledge organization, but it is only a rough and, in a sense, quite limited approximation.

Further, more detailed analysis (not carried out by Margenau due to some reasons, including those connected with general epistemological attitudes) allows us to disclose the more complicated structure of theoretical knowledge and its interrelations with the empirical level.

First of all, we are to pay attention to internal organization of the network of theoretical constructs. Among them we may discover different subsystems, which are relatively independent and subordinate to each other. In the contents of the theory, first of all, it is necessary to distinguish correlations of fundamental abstract objects introduced through postulates and definitions of the theory. As an example: the above mentioned correlations of "force", "material point" and "spatiotemporal frame of reference", introduced in the framework of the initial definitions and axioms of motion of Newton mechanics.

It is significant that transformation or elimination of at least one of such objects immediately causes transformation of the whole theory. Suppose we exclude from mechanics such an object as "material point": mechanics would be destroyed. If we introduce a new fundamental object – "energy" – instead of "force", then instead of Newton mechanics we come to another theoretical construction – Hamilton mechanics. Expelling "energy" and "force" from the list of fundamental abstract objects, we may get the basic principles of H. Hertz mechanics, which is also another theoretical construction (different from Newton mechanics) that describes mechanical motion.

Thus, in the foundation of a developed theory we always can find a network of mutually corresponding objects which define specificity of this theory. We will call such network of objects a fundamental theoretical scheme. Initial features of its abstract objects and their main relations always characterize the most essential features in the object domain studied in the theory. A fundamental theoretical scheme can be regarded as a greatly abstract model of interactions studied in the theory. It reveals structural peculiarities of such interactions fixing in cognition their profound, essential characteristics features.

In our example, Newton mechanics, the fundamental theoretical scheme expresses the essence of mechanical motion as an abstract model, by means of which introduced the idea of shifting of a material point in the space of a frame of reference in the course of time and transformation under the force of states of motion of the material point. Presenting moving bodies as material points or systems of material points, by the use of this model we may describe and explain real mechanical processes.

The main features and relations of abstract objects forming the model given are fixed by the basic definitions of the theory and three Newton laws, which are a theoretical expression of objective laws of mechanical motion.

It would be fair to formulate a methodological thesis, universal enough: formulations of the theoretical laws directly refer to a system of theoretical constructs (abstract objects). Therefore, the corresponding laws can be implied to description of reality only to that extent, to which the theoretical schemes based on the theoretical constructs represent essential connections of the reality.

This peculiarity of theoretical knowledge can be traced not only in physics, though it appears here in the clearest way. It is traced in all spheres of science which have already achieved the theorization stage. Let us take, for instance, the well-known law of population genetics – the Hardy-Weinberg law; it characterizes the conditions of genetic stability of a population. This law belongs to a rather small group of laws of biology which have a mathematical formulation. It was formulated in accordance with Hardy and Weinberg's theoretical model (scheme) of distribution of mutant forms in a population. Population in this model was taken as a typical idealized object – it was a limitlessly large population with free interbreeding. It could be compared to real, large populations only in cases where migration and mutational processes were ignorable and we could abstract from the factors of natural selection and limitations for panmixia.[4] But it was due to these idealizing assumptions the theoretical model fixed essential connections which characterized the relative stability of populations, while the Hardy-Weinberg law formulated on the basis of that model has become one of the most important laws of population genetics.

Here it is easy to see a direct similarity to the developed forms of theoretical knowledge in physics. The idealized object which served for Hardy-Weinberg law played the same role as the model of ideal pendulum for discovery of the law of small oscillations, or the model of ideal gas for formulating laws of behavior of rarefied gases at relatively low pressure.

In the theories of social studies we can also find that the formulation of theoretical laws is connected with introduction of idealized objects which simplify and schematize empirically observed situations.

So, in modern neoclassical economical theories, one of the important laws is the famous law formulated by L. Walras – a Swiss economist of the late 19th century. This law is being concretized and modified with the unfolding and development of the theories mentioned. The law assumes that in scale of economy represented by various commodity markets, including money-market, the volume of surplus demand (the gap in value between demand for separate goods and their supply) always makes zero. It is easy to see that the Walras law describes an idealized model (scheme) of interrelations of various commodity markets, when their system is in balance (the demand for goods at each market equals to supply).[5] This is unreal, but also unreal are material points, absolutely solid body, and ideal gas.

Naturally, every theoretical scheme and every law formulated on its base have limits of their application. The law of ideal gas will not work for high pressures. In this case it is replaced by the van der Waals equation (law) which takes into consideration the forces of intermolecular interaction, while the ideal gas model abstracts from them. The same happens in the economical theory: the Walras law requires corrections for description of complicated processes of interaction of different markets, connected with breaches in the realization of goods and not approximated to equilibrium processes. These situations are expressed by more complicated theoretical models (for example, Keynes-Wicksell's model improved by J. Stein and G. Rose, which assume nonequilibrium of markets. Another example, the model offered in the 1960s/1970s by American economists D. Patinkin and H. Johnson; this model refers to the nonequilibrium of markets which take into account the effect of cash rest balance and the active part of the monetary market).[6]

The formulation of new theoretical laws allows us to widen our possibilities of theoretical description of the reality in research. But each time we have to introduce a new system of idealizations (theoretical constructs) which forms a corresponding theoretical scheme in their connections.

Even the "mildest" forms of theoretical knowledge (commonly among them we see study of literature, musicology, fine arts studies, as opposed to "hard" forms of mathematized theories in sciences) include a layer of abstract theoretical objects which form theoretical models of the researched reality. I would like to refer to works of V.M.Rozin who has applied my conception of theoretical knowledge to technical and humanitarian disciplines.[7] V.M. Rozin has analyzed the texts of works by M.M. Bakhtin and B.I. Bursov dedicated to Dostoevsky's prose, texts of theoretical musicology and a work of V.A.Plugin in which the author analyses the art of Andrei Rublev. In all the situations the author discloses the layer of theoretical knowledge and demonstrates that the development of the researcher's thought in this layer is based on constructing ideal theoretical objects and further operating them.[8] In particular, Bakhtin's main theoretical conclusions concerning peculiarities of Dostoevsky's "polyphonic novel" were made due to constructing a theoretical scheme which included such ideal objects as "heroes voices" and "author's voice".[9] These elements enter into a dialogue. Thus, we may deduce that ideal theoretical objects and integral theoretical models (schemes) based on them are an essential characteristic feature of structure of any theory, whether it belongs to the sphere of humanities, social studies or sciences.

THEORETICAL SCHEME AND MATHEMATICAL APPARATUS

In the theoretical language a theoretical scheme can be characterized by means of at least two types of expressions. First, it may be pithy descriptions like those regarded above: "a material point is moving along the continuum of points of a spatiotemporal frame of reference", "the force changes the state of motion of a material point", etc. Such expressions describe connections and relations of abstract objects forming a theoretical scheme. At the same time these connections can be expressed as mathematical dependencies. This can be reached through mapping abstract objects of a theoretical scheme onto the objects of mathematics. For instance, a frame of reference may be connected with coordinates (the inertial frame of reference in mechanics can be identified – within certain limits – with a system of rectangular, spherical or cylindrical coordinates in Euclidean space). Because of this it appears as a continuum of spatial and temporal points, and each of them has a corresponding certain number (or a system of numbers). Then, a material point in classical mechanics may be characterized by some constant magnitude indicating its mass. The location of a material point in a frame of reference can be described by means of spatial and temporal coordinates, and changes of the latter can be interpreted as characteristics of motion of the material point. Lastly, force may be presented as a vector.

Due to such reflection of theoretical schemes of physics to mathematical objects, we may express correlations among the elements of the theoretical schemes as a system of formulae. For instance, we may express the relations between a force, a spatiotemporal frame of reference, and a material point as mathematical formulation of Newton laws.

The features of abstract objects in passaging to such a description are fixed as physical magnitudes, and the connections of those features as connections of magnitudes in equations. As a theoretical scheme can be represented as an idealized image of the natural processes studied by the theory, the physical magnitudes and their connections in equations should express some characteristics of the processes which can be empirically ascertained. The equation in this case plays the part of expression of essential connections between physical phenomena and serves as formulations of physical laws.

Equations and abstract objects of a theoretical scheme can be regarded as relatively independent elements of theoretical knowledge. At least two factors justify such an approach. First, the same equations can be connected with different theoretical schemes and, if the latter reflect the corresponding fragments of physical reality, can present a description of various physical interactions (classical examples are: usage of equations of oscillation for theoretical description of both mechanical and electromagnetic oscillations, Maxwell's application of equations of hydrodynamics to description of electromagnetic interactions, etc.). Second, a theoretical scheme, fixed in the language of contents description can exist irrespective of the equations. So, describing the fundamental theoretical scheme of mechanics (motion of a material point in the space of a frame of reference under influence of a force), we may introduce an abstract model of real mechanical motions without using equations. Based on this model, we may also get qualitative characteristics of the laws of mechanics (for instance, Newton's *Mathematical principia of natural philosophy* presented the three basic laws of mechanics without formulae, in qualitative form).

However, emphasizing certain independence of equations and fundamental theoretical schemes, we are not to forget that this independence is relative and the specified elements of theoretical knowledge are tightly interconnected. On the one hand, out of the connection with the theoretical scheme the equations are no more than mathematical formulae, and not expressions of physical laws. In other words, the equations have no physical interpretation. Such interpretation is provided by the theoretical scheme, previously substantiated as an idealized model of some real area of interactions. On the other hand, the theoretical scheme without equations gives us only a poor and abstract idea of the studied reality. All the riches of the connections and relations of its abstract objects which characterize the natural processes in theoretical knowledge are revealed by means of equations. These unfold the contents of a theoretical scheme in the easiest way and in full measure. But the most important thing in interaction of equations and theoretical schemes is the fact that mathematical means take an active part in the very construction of abstract objects of a theoretical scheme, and determine their features. Even in the case when a researcher resorts to informal description of theoretical schemes, he or she latently uses mathematical ideas. He or she may speak, for instance, of motion of a material point in the space of an inertial frame of reference in the course of time, but it is assumed that the space has the qualities of Euclidean space, and the time has those of "quasi-Euclidean time" (uniform course of time in all frames of reference).[10] Characterizing the state of motion of a material point (a point mass) determined by its coordinates and velocity, the researcher assumes beforehand that the frame of reference is a coordinate system and, consequently, the relation of a material point to it can be expressed by coordinates and certain functions of coordinates and time.

Thus, the initial features of abstract objects of a fundamental theoretical scheme often carry the traces of influence of the mathematical structure used in the theory. They are introduced in such a way that it could be possible to use certain mathematical formalisms while theoretically describing natural processes. Here we can see the close liaison between mathematical means applied in the theory and relations of abstract objects forming a fundamental theoretical scheme. Such correlation lets us speak of a type of two-layer framework which is the foundation of a physical theory: the first layer consists of a mathematical formalism, and the second one of a fundamental theoretical scheme. Both layers are always correlated. Such correlation, in a narrow sense, can be seen in the fact that the main equations of the theory corresponding to the mathematical formulation of its basic laws serves as a sort of record of the basic relations among features of the abstract objects in the theoretical scheme. When we supply the objects with new features, we have to transform the equations, and vice versa. In a broad sense, correlation of the said layers is represented in the connection between the type of mathematical structure used for description of some area of physical processes, and the method of presentation of such processes in a theoretical scheme. The best way to illustrate this aspect of interplay of a theoretical scheme and mathematical means of description of physical processes, is considering historical examples.

When Newton began to create a theoretical scheme of mechanical processes, in which moving bodies were presented as material points changing their coordinates and impulses in a spatiotemporal point under influence of a force, this model of mechanical motion called for a special mathematical apparatus.

In the pre-Newtonian period mechanical processes used to be described by means of Euclidean geometry and ordinary algebra. Mechanics was satisfied with such apparatus because it represented real three-dimensional bodies as ideal geometrical bodies and considered their motions; it did not aim at describing the change in the point of impulse of the body and, consequently, the change in the point of its velocity.

When Newton tried to solve this problem, he found out that he had to describe the motion of a body and the change of its state in infinitesimal areas of space-time. In particular, to clear up the regularities of changes in velocity in a point under applied force, the researcher had to consider the relation between the contracting to point distance increment and contracting to point period of time increment. This led to transformation of the previously used apparatus of mechanics (Euclidean geometry) to a new apparatus, which became the earliest version of differential and integral calculus.

Thus, transition to a new theoretical scheme of mechanical motion required new mathematical structures for description of such motion (after Newton's development of differential calculus and, especially, after Leibniz's works, this apparatus became the main method of mathematical description of mechanical processes).

The example above illustrates changes of mathematical apparatus under influence of a new physical model of the processes researched. But there exists one more way, to some extent opposite to the one considered, when mathematical means involved in a shaped theory for solving some of its problems, led to reconstruction of the fundamental theoretical scheme. For example, Newton mechanics was reconstructed under influence of the apparatus of differential equations developed in 18th century mathematics and successfully used for solving theoretical problems of mechanics in its application to a wide scope of phenomena (including mathematical description of mechanical systems with large number of degrees of freedom). In order to ensure effective application of analytical methods while considering any mechanical phenomena, Lagrange, and then Hamilton and Jacobi, introduced new fundamental theoretical schemes of mechanics equivalent to Newton's (as to their ability to present the objective structure of mechanical motion in the form of an ideal model). So, Lagrange proposed to describe the state of motion of a material point not as changes of its coordinates and velocities in three-dimensional Euclidean space, but as transformation of generalized coordinates and generalized velocities in the configuration space.

Such reconstruction of an already shaped theoretical scheme under influence of new mathematical apparatus, is typical for the development of physics. In quantum mechanics, for instance, there first of all appeared two equivalent theories of quantum processes: Schrödinger wave mechanics and Heisenberg matrix mechanics. Each of them possessed its own mathematical apparatus and, correspondingly, its own theoretical scheme.

Further development of quantum mechanics led to synthesis of these two forms of theoretical description within a new description based on use of apparatus of infinite-dimensional Hilbert space. Transition to this apparatus required creation of a new fundamental theoretical scheme. In particular, the wave function in three-dimensional space which had been a part of the theoretical scheme of wave mechanics, was then considered as the state vector of the quantum system, but in Hilbert space. Its correlations to the state vector of the measuring instrument made it possible to represent profound characteristics of quantum processes in quantum mechanical description. Compared to the new theoretical scheme, previous views of Schrödinger and Heisenberg appeared to be "imperfect" theoretical models of quantum processes. The new theoretical scheme synthesized both models and gave scientists an opportunity to describe and explain the wide scope of physical phenomena in the atomic area.

Thus under influence of the new mathematical structures entering into the theory, a definite generalization of the theoretical scheme occurs. On the one hand, such generalization provides the most effective description and explanation of new facts. On the other hand, it prepared a base for transition to assimilation of the new types of theoretical objects in theoretical cognition. Developing the mathematical apparatus and filling it with a new physical content, it is as if cognition prepares the means for its future development. So the elaboration of mechanics by J. Lagrange and W. Hamilton functioned as a necessary base for further successful elaboration of electrodynamics and quantum mechanics, and R. Feynman's formulation of quantum mechanics was a preliminary and a necessary step to the newest deveopment of quantum electrodynamics (apparatus of path integrals developed by R. Feynman not only became an effective means for solving quantum mechanical problems in non-relativistic areas but assisted in building the relativistic invariant theory of interacting electromagnetic and quantized electron-positron fields, taking into account the higher approximations of the perturbation theory).

Thus, interaction of the accepted mathematical formalism in a theory and the fundamental theoretical scheme is not just a norm of the theory functioning, but a condition of theoretical knowledge development.

An active inverse impact of mathematical apparatus on the fundamental theoretical scheme leads to its elements (abstract objects) on the highest stages of theory's development appearing as specific equivalents to the abstract objects of mathematics. A number of features by which every abstract object of the theoretical scheme entered is captured, in form of any mathematical image, "filled with the physical sense". Some of these images may have visual analogues in an object world with which a human deals in his actual practical activity (for instance, a material point in classical mechanics can be easily compared with a real macroscopic body with which a human is operating everywhere in practice). But the largest part of them may have no such analogues. They are, for example, theoretical constructs, such as the state vector in Hilbert space (theoretical characteristic of a microscopic object in quantum mechanics), and vectors of electrical and magnetic fields in a space-time point, which interact with a vector of charge-current density in a point (a theoretical characteristic of electromagnetic interactions in classical electrodynamics). In this case the attributes of abstract objects already have no analogues as a separately taken object, selected from nature by means of practical activity. The main form of objectiveness which unifies and consolidates these attributes is a mathematical image.

Mathematical form of the abstract objects expression allows to enter by the means of their correlations the generalized model of the reality under consideration, even when scientific cognition is beginning to study unusual, from the point of view of the ordinary common sense. In this case it is often impossible to imagine every abstract object of a theoretical scheme as an analogue of objects with which we operate in practice. Abstract objects function as complex substitutes for such objects' interactions in practice. But a mathematical form allows us to express these interactions as a particular ideal object which becomes an element of a more complex structure. It is the theoretical scheme representing in cognition the investigated reality.

So, the analysis of theory's structure requires us to single out as its basis a particular organization of abstract objects which is a fundamental theoretical scheme with which it is related according to it mathematical formalism.

Being an ideal model of the investigated processes, the theoretical scheme provides a mathematical apparatus of theory's interpretation and functions as a specific intermediate between this apparatus and experimentally fixed properties and relations of physical objects.

In distinction from the formalized theories of mathematics, where theory (calculus) is separated from models interpreting a calculus (theory has an interpretation field), in physics the models which determine a physical sense of equations are included in the theories content.

We called such models theoretical schemes to distinguish them from the other types of models which are applied in a theoretical investigation. Some of them function as a mean of theory building but are not entered in its composition. Theoretical schemes are always included in theory as the most fundamental component of its content.

Together with equations a fundamental theoretical scheme forms a physical theory's foundation, based on which an investigator can obtain the new characteristics of investigated reality, not appealing to its experimental studying. Such characteristics can be obtained resulting in a deductive development of a theory, revealing the new attributes of theoretical scheme's abstract objects based on the primary attributes.

The deductive development of a theory is fulfilled as a deduction of the consequences from the basic postulates and definitions. Methods of this deduction may be rather different. They are the formal and logical devices of deductive inference of one statement from the other, the methods of equations' solving, and, finally, the mental experiments with the objects of theoretical scheme. For instance, using the mathematical apparatus of mechanics and based on mental consideration of the links between the objects of its fundamental theoretical scheme, we can obtain on the basis of the main attributes of pointed objects the new attributes, such as the property of forces to perform a work, and the property of a material point to have the potential and kinetic energy, etc. These properties of forces and material points function as specific characteristics of mechanical motion. In the process of developing a theory such attributes are fixed in the form of conceptions, and their links are expressed as a corresponding theoretical statement. In the mathematical apparatus they act as the new physical magnitudes interrelated with the other magnitudes.

At first glance it seems that it is enough to have a composition of abstract objects, forming a fundamental theoretical scheme, to construct relatively to them the new statements, and to develop a theory not entering the new abstract objects. But in the actual theory development the new attributes of a fundamental theoretical scheme are often transformed into independent abstract objects. For instance, when in motion in the mathematical apparatus, the pointed attributes are operated as with the independent formations, seeing them as the appropriate physical magnitudes. And only when interpreting the results, the physical magnitudes are considered as the characteristic of objects of a fundamental theoretical scheme. But such interpretation is not the only possible mode of explication of the physical magnitudes' theoretical sense. Often to develop the theory successfully it is important to imagine a physical magnitude as a term fixing a specific abstract object that exists side by side with the fundamental abstract objects of theory, and which can be operated in the same way as the investigator operated the fundamental objects of the theoretical scheme. In such case theoretical concept is turned into an appropriate abstract object. For instance, in mechanics, when analyzing the fundamental theoretical scheme the attribute of a material point is obtained as an ability to own an energy, and if this attribute is fixed in a concept, "an energy", one can then form a specific theoretical construct which represents a result of abstracting an appropriate attribute of material points. Mental experiments can be implemented with this construct considering the processes of energy exchange between the mechanical systems, processes of transforming energy of one type into another, etc. A content analysis of such situations and applying to them the means of mathematical description, allows us to obtain the new characteristics of the objects' motion.

The development of a theory is always a creation, on the base of fundamental attributes and relations, of theoretical scheme abstract objects whose attributes and correlations are fixed in a system of the appropriate statements. Then one can imagine that in a network of interrelated theoretical constructs, forming a theory content, the major subsystem is singled out (a fundamental theoretical scheme). Other constructs are formed around it as the theory is developing. However, more detailed analysis shows that such conception of a content structure of theory need further sharpening and concretizing.

"Daughter" (in relation to fundamental) theoretical constructs are also organized in a specific subsystem, such as the constructs that form a fundamental theoretical scheme. Such subsystems can be independent of one another and submitted only to a fundamental theoretical scheme. Each such subsystem is characterized by its relatively marked in theory collection of statements and concepts, which form a specific theory part. Thus, in mechanics some relatively independent parts appear distinctly: mechanics of small oscillations of a point, mechanics of motion in central forces' field, mechanics of solid rotation, etc. Each part like this is formed by a system of statements entering a collection of their own, specific abstract objects (for example, "the oscillation period" and "an amplitude" in the mechanics of small oscillations or "the relative torque", "the instantaneous axis of rotation", "the resultant moment of inertia" in the mechanics of solid). Among these collections, in turn, the systems of fundamental abstract objects and its derivatives can be singled out. Thus in a theory of small oscillations a "material point", a "quasi-elastic force" and a "frame of reference" (for example, a fixed straight line that allows the deflection of a point to register from an equilibrium position) appear as a system of objects that have an independent status (in a framework of a given mechanics part). They are entered relatively independently from the other abstract objects of theory of oscillations at the time when, for example, an "oscillation period" has already appeared as a theoretical construct, justified only by virtue of correlation of objects listed above.

On that ground they can be singled out as a fundament of a mechanics theory of small oscillations. It is indicative that when expounding this part of Newton's mechanics a specific status of correlation of a "material point", a "quasi-elastic form" and a "frame of reference" is necessarily fixed. They form a theoretical model of small oscillations which is named as a linear harmonic oscillator and is linked with a fundamental equation of oscillation.

The mechanical oscillations model (an oscillator) is entered in mechanics relatively independently from other systems of abstract objects to which it is similar, but it depends on a fundamental theoretical scheme of mechanics. In relation to it an oscillator functions as a particular case.[11]

It is not difficult to ensure that based on a mechanics fundamental theoretical scheme one can build not only an oscillator but some other similar system of abstract objects (for example, to form a model of an absolute solid linking the material points by reaction forces, to build a model of elastic collision of bodies, etc.).

As a result one can draw the conclusion that in a developed theory's content, except for its fundamental scheme, one more layer of the abstract objects' organization can be singled out – a level of particular theoretical schemes. The latter concretize a fundamental theoretical scheme as applied to the different theoretical tasks situations, and provide a transition from the analysis of common characteristics of the investigated reality and its fundamental laws to consideration of individual concrete types of interaction in which the pointed laws appear in a specific form.

Thus, when considering a scientific theory in respect of its internal semantic ties of its terms and statements, the complex organization of theoretical knowledge's content is revealed. In a theory there is no linear row of abstract objects consistently being constructed one from another (as H. Margenau represented it). More likely it is better to talk of some key systems of such objects around which the directly related to them "daughter" constructs are forming. An original carcass, linking all these elements in one organization, consists of a fundamental theoretical scheme and the local theoretical schemes, which are formed on the basis of a fundamental theory and together with it are entered into a scientific theory. The content structure of a developed theory is characterized by the fact that the constructs entered in a theory are organized not as a simple but as a complex system that includes relatively independent subsystems, interrelated themselves by the principle of the level hierarchy (the subsystems of a lower level are coordinated with each other and at the same time are submitted to the subsystems of a higher level).

A THEORETICAL SCHEME'S ROLE IN DEDUCTIVE UNFOLDING OF A THEORY

In a logic-philosophical analysis of a science language, the marked particularities of the theoretical constructs system organization are missed fairly often and theoretical schemes are not fixed as a specific theory component. In our opinion, this is caused by a widespread approach in logic and in methodology to any scientific theory only as to a knowledge that is built at the rates of axiomatic and deductive organization.[12] When considering a scientific theory from these positions they see in it only a derivation of some other statements in accordance with logic rules, that in respect of the theoretical content can be interpreted as a forming of the new, and new abstractions that are requested to give a characteristic to an object domain under consideration. These abstractions appear as an integral system within which any levels of organization can hardly be singled out.

However, natural science theories (as like many theoretical systems of mathematics), generally speaking, only conditionally may be accepted as axiomatic and deductive systems. When analyzing theoretical texts it is discovered even in highly developed theories that widely use methods of the formalized axiomatic, some principal informal reminder exists besides the formal and axiomatic part furthermore and this reminder is not organized according to the norms of axiomatic and deductive construction.

It is becoming clear that in the process of deductive development of a theory, together with the axiomatic methods of discussion, the genetically constructive method of knowledge construction plays a great role, and further more it appears in a form of its content variant.[13] Unlike the axiomatic method, when "some system of statements describing a field of objects, and the system of logical actions on the statements are taken as initial",[14] the genetic method presupposes operating directly with theory's abstract objects that are fixed in the appropriate symbols.[15] The process of reasoning in this case appears "in a form of thought experiment of objects that were taken as concretely available".[16]

One example of such theory's development can be Euclidean geometry.[17] Euclid's postulates introduced the new abstract objects – a "point", a "straight line", a "circle", a "segment" – as being determined by means of an ideal pair of compasses and a ruler. All subsequent discussions were conducted based on building different geometrical figures from the fundamental objects. Thought experiments with figures (their partition and transformation and also their superposition on each other) functioned as a basis for obtaining the knowledge that as fixed in a system of appropriate statements of Euclidean geometry.[18]

The genetic-constructive approach makes evident the fact of the theoretical scheme's existence right away. Such schemes (that are introduced in a theoretical language in the form of drawings supplied with appropriate explanations or through the system of statements which characterize the construction methods and the main correlation of some set of abstract objects) appear as a foundation that provides the development of theoretical knowledge.

If we consider the process of conclusion of consequences from the fundamental definitions and the physical theory axioms, it will become clear that together with the methods of knowledge's development at the expense of motion in mathematical formalism and of formal logical operations with theory forms and statements, the thought experiments with the theoretical schemes' abstract objects play a big role. It is not difficult to make certain of this in a concrete example. So, returning to the case of description of the small oscillations' process in a framework of previously examined Newton's mechanics, it is possible to determine that the statement for the law of small oscillations cannot be obtained if using only the formal logical conclusion and the means of mathematical formalism. To deduce the law of small oscillations a row of substantial assumptions is necessarily needed: to concretize a type of force, to set a concrete type of the frame of references, and to examine a character of a material point's movement under the influence of the quasi-elastic force in a given frame of reference. Such concrete definitions take into account the specifics of the actual oscillations that are fixed when experienced. Only then the equation of an oscillation can be derived from the fundamental equations of the mechanic's motion. All these operations that are usual for a physicist mean the constructing of a small oscillations model (an oscillator) on the basis of a fundamental theoretical scheme of mechanics, and derivation of the equation of an oscillation by way of mental observation at the fundamental ties of the given model's abstract objects. In the process of such deduction the handling of elements of an "oscillator model" is started at the moment when a type of force's concretizing is performed when applied to a task of small oscillations in the mechanics equations. The definition of quasi-elastic force as a "force that tends to get return a material point to the equilibrium position" as itself, explicates an oscillator as a small oscillations model. Only in a framework of relations between this model's elements a fundamental attribute of quasi-elastic force may be introduced – "to be a magnitude that is in proportion to the magnitude of the point's deviation from an equilibrium position". Denoting a force as F, a deviation from an equilibrium position as x, the statement $F = kx$ for a force is obtained, where k – is a constant of proportionality. Substituting this statement into the equation $F = \frac{d^2x}{dt^2}$, one obtains the equation of small oscillations $m\frac{d^2x}{dt^2} + kx = 0$. This details a procedure of the oscillations' equation's deduction from the fundamental mechanics laws.

In physicist discussions an oscillator plays approximately the same role that a geometric figure plays in mathematician discussions. It allows us to establish a link between force and magnitude of a material point's deviation from an equilibrium position that, in turn, leads to concrete definition of Newton's second law and to its transformation into an oscillations equation.

Even if the most contemporary mode of physical theory's exposition with the developed mathematical means application is used, one cannot avoid an appeal to the substantial operations with abstract objects included in theoretical schemes. Exactly owing to such operations the limitations are imposed on a theory's fundamental equations and local laws applicable to this or that concrete theoretical task solving are formulated. Thus, in a classical field theory (that has achieved an extremely high mathematization level in a modern exposition) to obtain, for instance, from fundamental electromagnetic field equations (Maxwell's equations) a statement for the Coulomb's and Biot-Savart's laws, it is necessary to carry out previously a row of thought experiments with a fundamental theoretical scheme, which characterizes the electromagnetic interactions structure by means of a connection between electrical and magnetic field vectors and density of charge-current vector.

To deduce Coulomb's law, initially the classical electrodynamics fundamental theoretical scheme is concretized and on its base the theoretical model is created that characterizes the electrostatic field of a point source. It is presumed that field is created by a point charge, e.g., it is directed along a radius–vector that had been traced from a point where a charge e is located; then a flux of electric field through the spherical surface with a radius *R* around a charge *e*. In accordance with this model Maxwell's equations are transformed. In the beginning they are re-written in a form that is appropriate to laws' expression for a constant electric field and then they are applied to the concrete situation when the magnitude of a field flux through the ball surface is calculated. Only thanks to all these operations on the basis of Maxwell's equations, is Coulomb's law[19] obtained.

In the same way Biot-Savart's law is deduced. It cannot be obtained only by way of Maxwell's equation mathematical transformations. In the beginning it is necessary to transform the classical electrodynamics fundamental theoretical scheme into its "daughter" model which characterizes the constant magnetic field generated by a stationary current. So deduction of the Biot-Savart's equation is begun with the assumption that the charges which create an electromagnetic field make only a "finite motion, when particles always remain in a finite space region. The impulses also always remain finite".[20] Then, based on the pointed concretizing preconditions, they modify the Maxwell equation $rot\ \mathbf{H} = \frac{1}{c}\frac{\partial \mathbf{E}}{\partial \mathbf{t}} + \frac{4\pi}{c}\mathbf{J}$ into an equation $rot\ \overline{\mathbf{H}} = \frac{4\pi}{c}\overline{\mathbf{J}}$. And, following this using a number of mathematical transformations,[21] Biot-Savart's law is deduced. This law in a mathematical form expresses a correlation between abstract objects of a theoretical scheme that characterizes a magnetic effect of stationary current.

Thus, if any natural sciences theory is not previously adapted to the ideal of axiomatic-deductive knowledge's construction, not only theoretical scheme existence can be fixed within it, but also its important role in the process of theoretical content's development can be discovered.

In physics such development may be at least realized by the two interrelated ways. First, is the way of formal operations with theoretical language symbols (for example, operations with physical magnitudes in accordance with mathematical rules). The second is provided at the expense of an investigation by a method of thought experiment to the correlation of objects that are unified into theoretical schemes. In the first case no attention is paid to the symbols' sense[22] and they are operated with a usage of some given rules that form an accepted theoretical language syntax. In the second case the content of the appropriate symbolic statements is explicated and a conception of abstract objects that are in the strictly defined ties and relations between each other is introduced. In this case, knowledge's development is implemented by way of thought experimenting with abstract objects. Their links investigation allows us to discover the new attributes of abstract objects and to introduce the new abstractions, thus moving forward on a theoretical content's plane without addressing the modes of formalized thinking. It is indicative that in a developed scientific theory these two modes of knowledge's conclusion supplement each other. In any event, the analysis of physical theory's development procedures shows that the race in a mathematical sphere, which sets the modes of a "formal work" with the physical magnitudes, is always combined with progress in the theoretical schemes. These schemes are explicated from time to time in a form of particular model conceptions.

This, of course, does not mean that when developing, the theories operate with theoretical schemes' abstract objects only when the mode of informal-genetic conclusion is applied. A motion in a mathematical formalism plane also acts as a particular mode of investigation of the properties and relations of the theoretical scheme abstract objects. Since such properties and relations are represented in equations as physical magnitudes and their relations, then solving these equations can be considered only as an original procedure of the theoretical schemes' appropriate abstract objects operating. In this sense the description of possible modes to solve an equation is correctly considered as characteristic of operations that can be carried out with abstract objects.

The theoretical scheme abstract objects may be explicated by way of informal definitions and of substantial description of their correlation. But together with this, they may be entered during theory unfolding by way of replacing a part of informal definitions with mathematical statements, and of the further operations with the given statements in accordance with the mathematical rules when the theory is developing. It is indicative that the connection between the informal-genetic and the formal methods of theory's development appear in a constant shift from one form of the "symbolic being" of the theoretical scheme to the other.[23] In a conclusion process an investigator operates both with mathematical language and informal descriptions. From time to time he corrects the motion in a mathematical formalism by the operations with the theoretical schemes' abstract objects. Then he again turns to a formal way of operating with the given objects, investigating their links at the expense of transformation of the mathematical language symbols in accordance with its norms.

Since physical theory's development always presupposes a reduction of a fundamental theoretical scheme to local ones, then a question arises about this reduction's modes and ways. If the genetic and constructive mode of theory building is used, it is necessary not only to determine the initial abstract objects but to set a way of building on their base the new abstract objects. The building procedures provide of such a transition from a fundamental theoretical scheme to a local ones.

The specifics of theoretical knowledge's complex forms, such as the physical theory, is that the operations of building the local theoretical schemes based on fundamental theoretical scheme objects are not described in an explicit form in the theory postulates and definitions. These operations are demonstrated in the concrete examples of a fundamental theoretical scheme's reduction to a local one. Such examples are included in theory's composition as a kind of sample situation that shows how the consequences' inference from a theory's fundamental equations is implemented. In mechanics these sample situations are the derivation of the law of small oscillations from Newton laws, derivation of the law of body motion in a central forces field, derivation of the laws of a solid rotation, etc. In classical electrodynamics they can be a derivation of Biot-Savart, Coulomb, Ampere, Faraday, and the other laws from Maxwell equations. If all these forms of derivation are analyzed it will be discovered that the building of a local theoretical scheme on the basis of a fundamental one presumes the addressing to an object under consideration in theory and the discovering of its new links every time. When doing that from the very beginning the specifics of those actual processes are taken into consideration for explanation of which an according local theoretical scheme must be introduced. An investigator views these processes through the prism of a fundamental theoretical scheme (for example, he sees the oscillating body motion as the movement of a material point in a frame of reference) and then implements a series of thought experiments in the course of which he imposes constraints on a fundamental theoretical scheme. These limitations meet the investigated processes' particularities (for example, he marks, that the process of oscillations is linked with the influence of forces that each time return a material point to an equilibrium position). At the expense of such limitations the concretizing of a fundamental theoretical scheme occurs and it transforms into a local theoretical scheme.

The informal character of all these procedures and the necessity every time to address the investigated object and to take into consideration its particularities when constructing the local theoretical schemes, transforms an inference of every next consequence from the theory fundamental equations into a specific theoretical task. A deductive development of a theory is implemented in a form of such tasks' solving. A solution of some of them is described in theory from the very beginning and is offered as the pattern. In accordance with this pattern all other tasks should be solved. The mode of construction of local theoretical schemes' abstract objects on the base of the fundamental theoretical scheme objects is necessary for each new theoretical task solution and is demonstrated in the samples of the tasks that have already been solved.

The mentioned particularity of the physical theory deductive development was fixed by T. Kuhn when he considered the so-called ordinary situations of scientific investigation that are linked with the appliance of theory that have already been built to events' explanation and prediction.

T. Kuhn emphasized that theoretical description and explanation of each new physical situation is implemented in accordance with a scheme of vision (a paradigm) that allows us to study a situation in the image and likeness of another one. He picked out the patterns of tasks'solutions as a paradigm fundamental component. Owing to them the transition from fundamental theory's laws to their consequences that are used to characterize these or those concrete situations is implemented.[24]

However, in Kuhn's works a concept of patterns is not clearly determined. Only from the context of Kuhn's work can it be established that under the patterns he sees the modes of operating with the model conceptions that provide a derivation of some mathematical apparatus' formulas from the others. In the framework of our terminology this activity can be described as a reduction of a fundamental theoretical scheme to a local one. But the latter is not clearly expressed in Kuhn's work (there are neither any characteristics of theoretical models' structure, nor their typology in his works). In this sense the foregoing analysis can be considered as a clarification of a subject and logical base of the activity that T. Kuhn called the "pattern" usage.

Theoretical schemes and thought experiments with them are a foundation of deductive development of a theory and its appliance to description and explanation of different displays of the reality under theoretical consideration.

In light of what have been discussed a series of the more precise definitions can be given into a conception of physical theory as a mathematical apparatus that has received a physical interpretation.

First of all, the apparatus cannot be understood as a formal calculation that develops only in accordance with the rules of mathematical handling. Only the individual fragments of this apparatus are built in such a way. Their "coupling" is implemented at the expense of turning to the theoretical schemes that are explicated in the form of specific model conceptions. This allows us to correct the transformations of the accepted formalism's equations, carrying out the thought experiments of the schemes' abstract objects.

Secondly, it is needed to specify the concept of an interpretation itself. It is known that the equation interpretation is provided by its connection with a theoretical model among the objects of which the equation is fulfilled and by the equation's connection with an experience. The latter aspect is called an empirical interpretation. There is no definitive term for the first aspect's designation; sometimes it is called a semantic interpretation. The equation's semantics is determined by both interpretation's aspects and these aspects are interrelated between themselves (in the following text it will be shown that the empirical interpretation construction presupposes a projection of a theoretical model on the actual experience objects).

The fundamental equations of a theory acquire a physical sense and status of the physical laws because of their projection on a theoretical scheme. But it would be an oversimplification to think that a physical sense of theoretical consequences that are deduced from the fundamental equations are also provided this way. To provide this sense it is needed to know how to construct the local theoretical schemes on the basis of a fundamental theoretical scheme. It is not difficult, for example, to ascertain that mathematical statements for the laws of Ampere, Biot–Savart etc., that were deduced from Maxwell equations cannot already be interpreted by means of the electrodynamics' fundamental theoretical scheme. They contain specific magnitudes which are identical to the attributes of the appropriate theoretical schemes' abstract objects, in those vectors of electrical and magnetic strength and of the current density at a point are substituted by the other constructs: the current's density within a certain volume, the field strengths that are taken at some finite space domain, etc.

Since the local theoretical scheme's construction on the basis of a fundamental one presupposes the usage of the sample situations of theoretical tasks' solutions, then the interpretation of the mathematical apparatus of a developed theory presupposes including the initial set of such situations in a theory. The origins of their forming and entering in a theory can be discovered only if the regularities of a fundamental theoretical scheme's genesis are investigated.

Thus, a fundamental theoretical scheme and its derivative formations represent a kind of an inner skeleton of a theoretical knowledge that determines both an informal specifics of a theory and the procedures of its development. Taking this into consideration it is no exaggeration to say that the problem of theory genesis first of all acts as a problem of its theoretical scheme's formation. In favor of a decisive role of such schemes in a theoretical knowledge's genesis, this circumstance speaks just for that they provide a particular status of the necessity which is peculiar to the theoretical laws and which distinguishes the latter from the empirical generalizations that are represented by the empirical dependencies.

It is always possible to find in science such sorts of empirical generalizations that are expressed in the form of quantitative dependencies, and in their mathematical form completely coincide with an appropriate statement for a theoretical law. Though outwardly such statements are identical, there is a great distinction between them: the first of them possess only the probabilistic truth, the second one represents feasible knowledge.

A totally new meaning of an empirical formula that emerges when it is transitioned in a class of a theoretical law provides its connection with a theoretical scheme. That means its substantiation as a mathematical expression of the correlation between the abstract objects that compose the given scheme. It is easy to be sure of the latter if any concrete example of an empirical relation of a formula to the theoretical law is investigated.

Let us assume that we repeated Boyle's experiments and determined a dependence between gas volume and pressure. From tabular data obtained on the basis of the real experiment the formula $pV = \text{const}$ would be inferred, where p means pressure and V means gas volume. But it is no matter how many experiments with gases are conducted, there is no guarantee that in the next series of experiences the found dependence will not be broken. Moreover, it will surely be broken when we come to the experiences with large pressures, because in this case those forces will play a considerable role which is not taken into account in Boyle's law but only in van der Waals equation. This means that the increase in a number of observations does not necessarily give status to the investigated empirical dependence. This requires particular evidence that is implemented in the following way.

The abstract objects system is introduced. The next objects figure as the abstract ones: a) an ideal gas that is represented as a set of ideally elastic and extremely small particles that collide with each other; b) an ideal vessel that contains these particles; and c) an ideal piston that compresses an ideal gas when moving inside a vessel. In a process of thought experiment the following relations of given objects are established: the ideally elastic particles that are moving in accordance with the mechanics laws, strike the vessel's walls in such a way that the total impact value of all their blows per square unit characterizes the gas pressure. The mathematical expression of these statements which is based on application of the fundamental laws of mechanics allows us to deduce a dependence $pV = \text{const}$, previously fixed when experienced.

As a result of these procedures the formula obtained inductively becomes a law that describes the rarified gas behavior. Thus to obtain a law that characterizes a relation between the gas volume and gas pressure it was needed to construct a theoretical scheme, which is known in science as the ideal gas model. This model was fixed in a particular sign form (for example, in the form of a draft that contained the appropriate explanations and depicted an ideal vessel of a variable volume and a set of the gas particles encased in a vessel). Then when links and relations between the given scheme's objects were expressed by means of a mathematical language, the formula was obtained. This formula now appears as a theoretical statement. Though its form remained the same as an expression by means of mathematical language of the dependence that was received from experience, the magnitudes p and V acquired another physical meaning. They started to express not the correlation of the real, empirically fixing vessels and gases but the relations of theoretical language abstract objects by means of which the ideal gas model is built. As a result of this the formula pV = const was raised into a class of a theoretical law and acquired the attributes of universality and necessity.

The considered example of a principal distinction between empirical dependence and theoretical law has its roots in the actual science's history. It reproduces in a condensed form a logic of one of the fundamental law of the gas theory discovery. The history of this discovery is interesting and instructive in itself. As an empirical dependence, the formula pV= const was obtained accidentally in many respects. It was a by-product of a discussion between two famous physicians of the 17th century, R. Boyle and F. Linnus.[25] This discussion concerned interpretation of R. Boyle's experiments that discovered the appearance of barometric pressure. R. Boyle conducted the following experiment: he immersed a tube soldered from above and filled with mercury into a cup with mercury. According to the principle of communicating vessels the mercury level equalization in a tube and in a cup is expected, but the experiment showed that only some part of the mercury flows into a cup and the remaining part stands as a column above the mercury surface in a cup. R. Boyle interpreted this experiment in the following way: the air pressure at the mercury surface in a cup holds a column of mercury above this surface. The column height is an indicator of an atmospheric pressure magnitude. Thus the principle of a barometer as a device that measures atmospheric pressure was offered.

But F. Linnus raised the following objections: air consists of light particles, it is similar to thin and compliant liquid that cannot stand under the heavy mercury particles' pressure. Therefore, air cannot hold a column of mercury. It is held by the mercury's gravity to the top part of a barometer tube. F. Linnus wrote that when he stopped up a barometer tube above with his finger, he felt a filament of tension when inserting it into a cup. This historical fact is itself rather demonstrative. It is evidence that the same result of an experience can be interpreted in different ways and be used for the different conceptions' confirmations.

R. Boyle conducted a new experiment to prove to F. Linnus that air is able to hold a column of mercury. He took a bent siphon glass tube with a soldered short neck and little by little started to fill it with mercury. As the column of mercury increased, the air in the neck was compressed but was not forced out totally. R. Boyle charted a table of relations between air volume and the column of mercury's magnitude and sent it to F. Linnus as justification of his own interpretation's correctness.

It seemed that history regarding barometric pressure's explanation was settled but unexpectedly it continuated. R. Boyle had a follower, a young man whose name was Townley. R. Boyle taught him the basics of physics and mathematics. Townley, studying Boyle's table of experiences, noticed that values of the air under compression were proportional to the height of the column of mercury that pressurizes the air. Following this, Boyle saw his experiments in a new perspective. A column of mercury is an original piston that compresses the air and the column weight is in accordance with pressure. So the proportion in a table's data means the dependence between the pressure and the gas volume. Thus the proportion was obtained which Boyle confirmed by many experiments with pressures that were larger or smaller than an atmospheric one.

About the same time Mariotte repeated Boyle's experiments using small pressures when experimenting with different gases and received the same result.

The apparatus, that both Boyle and Mariotte used, did not allow the implementation of the experiments pressures that were bigger than the atmospheric one. But if they had the possibility to conduct such experiments they could find a disturbance of a discovered dependence which nobody would interpret as a law.[26]

Once again let us underline that that dependence, discovered by Boyle, was a probabilistically true knowledge, a generalization of the same type as a statement that "all swans are white", which was true until black swans were discovered. Theoretical law $pV=$ const was obtained later, when the ideal gas model had been built.

A physician D. Bernoulli (an academic of the Saint-Petersburg Imperial Academy) deduced this law in 1730. He issued from atomistic conceptions about a gas and presented gas particles as the material points that collide, like elastic balls.

To the ideal gas that is in an ideal vessel under pressure, Bernoulli applied the laws of Newton's mechanics and by means of calculations obtained a formula $pV=$ const. This was the same formula that R. Boyle had offered earlier, but its meaning had already changed. Boyle's $pV=$ const corresponded with a scheme of actual experiments and a table of their results. Bernoulli linked it with a theoretical model of an ideal gas. In a framework of this model the essential characteristics of any gas's behavior were expressed when relatively small pressures presented. And the law, directly describing these essential links, functioned as reliable, true knowledge.

The character of a general statement can be given to the things said.

Prognostic strength of inductive generalizations always has a stochastic nature because a simple broadening of observations class, which are in accordance with an empirical dependence, does not take it from the rank of hypothetical assumption about a law and does not give it the necessity attribute. This transition is possible only when the link between the quantities represented in empirical dependence will be obtained within a system of operations on the theoretical scheme's abstract objects. This scheme is an idealized model of the reality under consideration.

Thus, the problem of theoretical scheme origin is raised as a fundamental problem of epistemology and methodology of science. From the first view it seems obvious that the source of their origin should be sought in the experience generalization, because they are created to describe the already known data of experience and for the new results' prediction. The task is only to reveal how this generalization is implemented.

But here the main difficulties appear. That characteristic particularity of theoretical knowledge emergence is related firstly to theoretical schemes, which lies in impossibility of their deduction from the experience by a purely inductive way.

In the simplest case with the law it is already seen that the model which was used in a process of theoretical proof could not be taken directly from Boyle's and Mariotte's experiments, although it was necessary for these experiments' description. In this model the interaction of gas molecules was presented as a collision of absolutely elastic and unlimitedly small bodies. More clearly this particularity of theoretical schemes' construction is tracked in contemporary physics. Even cursory acquaintance with its history allows us to discover a specifics of building of fundamental abstract objects that form its theoretical schemes. It is not difficult to verify that the objects such as, for example, an electron-positron field, the energy of vacuum in a quantum electrodynamics, or the four-dimensional space-time continuum in electrodynamics of Einstein and Lorentz etc., were originally introduced from theoretical considerations and only later were given an empirical substantiation. But in this case a task appears before the epistemology and methodology of science to explain why the abstract objects system (theoretical scheme) may serve as a basis for prediction of the experimental data. Just in this way the key to understanding the methods of theory's building should be sought.

In our opinion, the first steps in this direction must be connected with an analysis of a role of theoretical schemes in a framework of knowledge already existing when the latter are used for explanation and prediction of real events. As in the process of explanation and prediction the theoretical schemes correlate with the reality under consideration, as the mentioned analysis will allow us to reveal the attributes which guarantee an objective theoretical scheme's value, that in turn can become a starting base for an elucidation of their genesis.

THEORETICAL SCHEMES AND EXPERIENCE. OPERATIONAL STATUS OF THEORETICAL SCHEMES

Theoretical knowledge is created solely to explain and predict the results of experiment and so they should be compared with empirical material. However, this comparison itself is not a simple procedure.

Let us assume, that using the Biot-Savart formula expressing the law of magnetic activity of current, there is a need to calculate an angle of the magnetic needle's deviation which is close to a rectilinear wire when the current of a defined strength is going through it (Biot-Savart experiment).

As the sense of formula expressing the Biot-Savart law is related with correlation of abstract objects which form a theoretical scheme ("differential-small current" and "magnetic field generated by a current"), this formula may not be applied for calculations in the empirical field. In such cases it is sought previously to interpret the appropriate magnitudes of mathematical law's formulation as correlating with a concrete experimental situation. With this purpose an intermediate consequence – an empirical formula is deduced from the Biot-Savart law. Against the magnitudes that characterize the differential-small current and the magnetic field intension, the new magnitudes are introduced into it, which characterize a magnetic needle deviation at the given angle and the wire configuration determining an integral allocation of current. Only with this empirical formula, but not the Biot-Savart law can the empirical dependencies obtained in the actual experience be compared.

Let us consider, where the meaning of mentioned empirical consequence that was deduced from theoretical law lies. It turns out that specific constructs appeared within an empirical formula, which in contrast to the theory's abstract objects are not idealizations and can be directly compared with real objects which interact when experienced. These constructs are empirical objects. In their links they introduce a particular conception of experimental situations which we will call an empirical scheme.

Empirical objects, though, are compared with real subjects of experience, but are not equal to the latter. They are abstractions that exist only in the ideal context, as a meaning of symbols of science empirical language. Thus the real magnetic needle and a wire with a current possess a great number of properties and attributes but in the framework of an empirical scheme they are represented only on the basis "to be oriented by a magnetic field" and, accordingly, "to carry a current of a defined power" and "to have a definite configuration". All remaining properties of the given objects are eliminated from the consideration. In this connection every element of an empirical scheme is compared not just to one object which is operated by an investigator during an experiment, but to the whole class of such objects. This means that the scheme corresponds not to every actual experimental situation in a given time interval but to a type of situation like this (for example, an empirical scheme of experience with a wire and a magnetic needle is related to any experiment with any current of given power within a rectilinear wire and with any diminutive magnetic needle). In the empirical scheme the main characteristics of objects interacting when really experienced are represented. This side of an empirical scheme is particularly clearly tracked in that case, if it is taken into account that it can be obtained not only "from above", when empirical dependence is deduced from a theoretical law, but "from below" also – as a content of an empirical dependence that emerged as a result of a statistical treatment and interpretation of the observation data. This problem needs to be dwelt on especially because here we face the complex organization of an empirical level of investigations and appropriate forms of empirical knowledge.

For a long period in the philosophy of science the observations were laid as its foundation on which the scientific theories grow and correspond to. Observational data were called the experimental data or the experimental facts. However, in the 1930s the discussion in a positive philosophy about the problem of protocol statements discovered the inadequacy of these, seemingly evident conceptions. It was elucidated that the empirical knowledge represented by the protocol statements – expressions fixing data of direct observation in a language form – are not the empirical theory's basis and are not equal to empirical facts as they are to a specific kind of empirical knowledge.

In the observation protocol it is stated who observed, time of observation, the devices are described if they were used at the time of observation, and protocol statements are formulated as the statements such as: NN observed that after the current was switched on, the device's needle had showed a figure 5; NN watched with a telescope on the sky (with the coordinates x, y) a bright small spotlight, etc.

If, for example, a sociological poll was conducted, the questionnaire with the respondent answer acts as a protocol of observation. If the metrologies were implemented during an observation's process, every measuring result fixation is equal to a protocol statement.

The analysis of protocol statements' meaning showed that they consist not only of information about the events under consideration but, as a rule, include the observer's mistakes, extraneous features of external perturbation actions, and accidental devices errors, etc. However, it became evident that data of observation cannot serve as a basis for the theoretical constructions as a result of the fact that they are burdened with subjective extraneous features.

As a result of this the problem of revealing such empirical knowledge's forms was set, which could have an intersubjective status, and could contain objective and reliable information about the studied events.

During the discussions it was stated that the empirical facts are the knowledge such as these. Exactly, they form an empirical basis on which the scientific theories rely.

Facts are fixed in a scientific language by means of such statements as: "the current strength in a circuit depends on the conductor's resistance"; "a supernova lit up in Virgo's constellation"; "more than half of the city's respondents were dissatisfied with the ecology of the city", etc.

The character of the fact expressing statements itself underlines their particular objective status as compared with protocol statements. But then a new problem emerges: how the transition from data of observations to empirical facts is implemented and what guarantees a scientific fact's objective status?

This problem's setting was a significant step on the way to revealing the structure of empirical cognition. This problem was actively elaborated in the scientific methodology of the 20th century. In a competition between different approaches and conceptions it revealed many important characteristics of scientific empiricism, though today this problem is far from its final solution.

A determined contribution to its development was introduced by a positivism, though it is not out of place to underline that its tendency to confine itself only by the studying of scientific knowledge internal links and to abstract from interrelations between science and practice, sharply narrowed the abilities of a description adequate to the research procedures and modes of formation of the empirical science basis.

It seems to us that the active approach gives more abilities for analysis. From this approach we will consider structure and functions of every mentioned layer of the empirical level of cognition. Let us begin with the more detailed analysis of observations subtotal, which provides a direct contact of subject with the investigated processes. It is important to immediately clarify that scientific investigation has an active character and presupposes not just the passive contemplation of investigated processes, but their particular preliminary organization providing a control of their passing.

The empirical investigation's active nature at the observations' level becomes most clearly apparent in situations when observation is carried out in a process of real experiment. It is expedient in the beginning to view in more detail what the particularity of experimental investigation as a practical activity lies in, and which structure really reveals these or those connections and reality conditions that are interesting for an investigator.

A subject structure of experimental practice can be considered in two aspects. Firstly, as the objects' interaction proceeding in accordance with the natural laws. Secondly, as an artificial, human-organized action. In the first aspect we can consider the objects' interaction as an aggregate of the reality links and relations, where none of these links is pointed as the investigated one. In principle, every of them can be a cognition's object. Only the second aspect's taking into account allows us to select this or that link in relation to the cognition's aims and thus to fix it as an object of investigation. But then, evidently or implicitly, an aggregate of objects interacting when experienced, as if organized in a system of a definite chain of relations, a whole series of their actual connections turns out as non-essential, and functionally only some groups of relations characterizing the investigated "cut" of reality are picked out.

Let us illustrate this in a simple example.[27] Assume that in a framework of classical mechanics the motion of a massive body of small dimensions that is suspended on a long non-stretched thread is studied relative to the ground. If we consider this motion only as the natural objects' interaction, it appears as a total result of the very different laws' demonstration. Here, such nature's links "are laid" onto one another as the laws of oscillation, free fall, friction, aerodynamics (streamline of a moving body by a gas), and laws of the motion within non-inertial frame of reference (presence of Coriolis' forces due to the Earth's rotation), etc. But as soon as the described interaction of natural objects begins to be considered as an experiment of studying, for example, the laws of oscillatory motion, thus a definite group of properties and relations of these objects is singled out.

First of all the interacting objects – the Earth, the moving massive body and the suspension – are considered as the carriers of solely defined properties which are functionally, by the way of their "inclusion" into the "experimental interaction", singled out from the other properties. The suspension and the body suspended on it appear as one object – pendulum. The Earth is fixed in a given experimental situation as a reference body (for this gravity's direction is picked out which sets an equilibrium line for a pendulum) and as a source of strength that makes a pendulum move. The latter, in its turn, presupposes that the Earth's gravity must be considered only in a determined aspect. Concretely: as, according to the aim of the experiment, the pendulum's motion is presented as a particular case of a harmonic oscillation, thereby only one constituent of gravity which returns a pendulum to an equilibrium position is taken into account. The other constituent is not taken into account because it is compensated with strength of the thread's tension.

The described properties of interacting objects, coming in the act of experimental activity in the foreground, thus enter a strictly defined group of interrelations which is functionally singled out from all other relations and links of natural interaction. Essentially, the described motion of the massive body suspended on the thread in the Earth's gravity, appears as a process of a periodical motion of this body's center of mass under the influence of quasi-elastic force. One of the Earth's constituents of gravity figures as this force. This "network of relations" selected within the framework of investigated nature interaction is that objective practice structure in a framework of which the oscillatory motion's laws are studied.

Assuming that, however, the same motion of a body suspended on the thread in the Earth's field of gravity functions as an experiment with the Foucault's pendulum. In this case another nature link becomes an object of study. This object is the laws of motion in a non-inertial system. But it is then required to pick out the other properties of interacting nature fragments.

In fact the body fixed on the thread functions nowadays only as the moving mass with a fixed relativity to the Earth's direction of movement. Strictly speaking, the system "body plus thread in a field of gravity" already is not considered as a pendulum (because the main pendulum characteristic – its oscillation period – becomes non-essential here from the point of view of the investigated link). Furthermore, the Earth, to which the body movement is considered relatively, is nowadays fixed according to other attributes. From the variety of its properties in the framework of this experiment, the direction of the Earth's axis of revolution and the magnitude of the angular velocity of rotation become essential. Their assignment allows us to determine Coriolis forces. Gravity, in principle, does not play a significant role for the purposes of Coriolis forces experimental investigation. As a result the new "network of relations" is picked out. It characterizes the cut of reality that is investigated within a framework of a given experiment. At the first stage the body motion is situated with a given speed along the radius of a uniformly rotating disk, which role the plane plays that is perpendicular to the axis of the Earth revolution and passes the point where an investigated body is at the moment of observation. This is the structure of the experiment with the Foucault's pendulum which allows us to study the motion laws in a non-inertial (uniformly rotating) frame of reference.

By the analogous mode in a framework of the analyzed nature's interaction, the object structures of the other type can be singled out if the given interaction is imagined as the variety of experimental practice of studying, for example, the laws of free fall or, let us assume, the aerodynamics laws (of course, in the actual experimental activity experiments like these are not used for the given purpose). Such abstract situations' analysis illustrates this circumstance well, that the real nature interaction can be imagined as kind "superposition" of the different type of "practical structures", which number may be, in principle, unlimited.

In a system of scientific experiments each of these structures is distinguished because of the interacting object's fixation at the rigorously determined properties. This fixation, of course, does not mean that all other properties of nature objects disappear, except those in which the investigator is interested. In real practice the necessary properties of objects are picked out by a character of operating with them itself. For this purpose the objects, brought to the interaction in process of experiment, must be previously adjusted by the practical usage with the object to existence of their properties which are substantially represented in conditions of future experimental situation. Thus, it is not difficult to see that the experiment with the pendulum oscillation could be implemented only because it was strongly revealed by the previous practice's development, that, for example, the Earth's gravity in a given place is constant, that any body that has a point of suspension, will make the oscillations relative to equilibrium position, etc. It is important to underline that separation of these properties became possible only due to the appropriate practical functioning of investigated objects. In particular, the Earth's property to be a source of constant gravity was used many times in human practice, for example, when moving different objects, when making piles of fallen weight, etc. Such operations allowed the singling out of a characteristic property of the Earth "to be a source of constant gravity".

In this sense, in the experiments of studying the law of pendulum oscillations, the Earth acts not only as a natural body, but as an original "artificially made" human practice object, because for the natural object the "Earth" in this property has no "particular privileges" in comparison with the other properties. It exists really but as a particular separated property and functions only within a system of determined human practice. The experimental activity represents a specific form of nature interaction. And the most important feature, determining the specifics, is just that nature's fragments, interacting at the experiment, always appear as the objects with the properties functionally picked out.

In developed forms of experiment objects like these are made artificially. Among these are, in their turn, the instrument plants which help the experimental investigation to be conducted. For example, in contemporary nuclear physics these can be plants that prepare the beams of particles stabilized by the defined parameters (energy, pulse, polarization); targets which are bombed by these beams; devices that register the results of the beam's interaction with the target. For our purposes it is important to clarify that the production, adjustment and usage of such plants are analogous to the operations of functional properties' separation at the nature's objects with which the investigator operates when undertaking the above-described experiments with a pendulum. In both cases from all selection of properties that the material objects possess, some properties are singled out and given objects function in the experiment only as their carriers.

From these positions it is quite right to consider the natural objects, introduced in an experimental situation, as the "quasi-instrumental" devices obtained independently by an artificial way or appearing naturally in nature independently of human activity. Thus in the experimental situation of the laws of oscillations studying the Earth "functions" as a particular instrumental subsystem, as if which "prepares" constant gravity (it is analogous to the accelerator, created by a human; it will generate the impulses of charged particles with the parameters given if the operating mode is strictly fixed). The pendulum itself here plays the role of a working device whose functioning gives an ability to fix the oscillation's characteristics. As a whole the system "the Earth plus pendulum" can be considered as an original quasi-experimental plant, which "work" allows us to investigate the laws of a simple oscillatory motion.

In light of what has been discussed the experiment's specifics that distinguish it from the interaction in nature "by itself", can be characterized in the way that at the experiment the interacting nature's fragments always function as the instrumental subsystems. The activity of "providing" the natural objects with the functions of the instruments we will further call a creation of an instrumental situation. We understand an instrumental situation itself as the quasi-instrumental devices' functioning in the system of which some nature fragment is experienced. And as the character of the experienced fragment interrelations with the quasi-instrumental devices functionally singles out in it some aggregate of characteristic properties, the presence of which, in their turn, defines a specifics of interactions in a working part of a quasi-instrumental plant, the experienced fragment is entered as an element into an instrumental situation.

In the experiments with the pendulum's oscillations considered above we had to deal with essentially different instrumental situations' independence if the purpose of investigation was to study the laws of oscillations or the laws of motion within an uniformly rotating system. In the first case the pendulum is included in an instrumental situation as the experienced fragment. In the second one it carries out the absolutely other functions. Here it appears in three relations. The first of them is that the motion itself of a massive body (an experienced fragment) is included into the working system's functioning as its essential element (along with the Earth revolution). The second is that a periodicity of the pendulum's motion, which played the role of the researched property in a previous experiment, is now used only to maintain the stable conditions of observation. In this sense the fluctuating pendulum already functions as the preparing instrumental subsystem. The third of them is that the pendulum's ability to keep the plane of vibration allows it to be used as part of a regulating device. The plane of vibration itself appears here as the original needle, which turns relatively to the plane of the Earth revolution and fixes the Coriolis forces presence. Such functioning of natural fragments, interacting at the experience in a role of instrumental subsystems or their elements, actually marks out and as if "pushes out" some properties of these fragments to the foreground. All this leads to the functional separation from multitude of potentially possible practice's object structures only that structure which represents the researched nature's link.

A link of this kind is the object of investigation, which is studied both at empirical and theoretical levels of a cognitive activity. A separation of the investigation's object from an aggregate of all possible nature's links is determined by the cognition's purposes and finds its expression in formulation of different cognitive tasks at different levels of cognition. At the level of experimental investigation such tasks act as a requirement to fix (to measure) any characteristic property's availability at the experienced nature's fragment. However it is immediately important to clarify that the investigation's object is not always represented by an individual element (a subject) inside of the instrumental situation but by all its structure.

In the samples, analyzed above, it was essentially shown that appropriate object of investigation – the process of harmonic oscillation or a motion within the non-inertial frame of reference – may be discovered only through the structure of relations participating in the experiment of natural fragments.

In a similar way the matters remain the same with more complex cases relating, for example, to the experiments in atomic physics. Thus in the famous experiments of Compton effect discovery the subject of investigation – "the corpuscular X-radiation's properties that are scattered on the free electrons" – was determined through the interaction of the X-radiation stream and the graphite target scattering it on the conditions of the emanation registration with a particular device. And only the structure of all these objects relations (including the device for registration) represents the investigated reality's cut. Such fragments of the real experimental situations, which usage the subject of inquiry sets, we will further call the objects of handling. The given distinction will avoid ambiguity when using the term an "object" in a process of description of the science's cognitive operations. In this distinction the essential fact is fixed that the object of inquiry does not coincide with any separately taken objects of handling in any experimental situation. It is underlined also that the objects of handling, by definition, are not equal to the "natural" fragments of nature because they act as the original carriers of some functionally selected properties in a system of experiment. As it was shown above, the objects of handling are usually provided with the instrumental functions and in this sense, being the actual nature's fragments, at the same time act as the products of "artificial" (practical) human activity.

The observations in this case are not just a fixation of some properties of the experienced object. They implicitly carry the information about those links, which gave birth to the observed phenomena.

The final goal of the natural-science investigation is to find the laws (the essential links of objects) which manage the natural processes and to predict on this basis the future possibilities of these processes' state. So if to issue from the global cognition's aims, the object of investigation is needed to consider the essential links and natural objects' relations.

But on the different cognition levels such links are studied in different ways. At the theoretical level they are reflected "in a pure form" through the system of appropriate abstractions. At the empirical one they are studied by their appearance in directly observed objects. So the global purpose of cognition is concretized as applied to each of its levels. In experimental research it appears in a form of specific tasks, which add up to establish how the primary state of the experienced nature's fragment gives birth to its finite state on fixed conditions. In relation to such local cognitive task the particular subject of study is introduced. It is the object whose change of state is tracked when experienced. In contrast to the object of cognition it can be called the empirical knowledge's object in a global sense. A profound internal connection exists between it and the object of cognition that is one for both the theoretical and empirical levels.

When in the process of experiment and observation an investigator registers the finite state O_2 of an experienced object, then in a presence of a fixed instrumental situation and initial O_1 state of object it is equivalent the last missing link's discovering. This link allows the characterization of the structure of experimental activity. If this structure is defined, an investigator thus singles out implicitly among the numerous links and relations of natural objects the links (the regularities) which manage the states' changing of the empirical knowledge's object. The transition of an object from the state O_1 to the state O_2 is not arbitrary but is determined by nature laws. So, if an investigator registered the changing of object's states in an experiment and in an observation many times, he implicitly fixes an appropriate nature law by the activity structure itself.

The empirical knowledge objects here are the specific indicators of the subject of investigation, which is common both for empirical and theoretical levels.

Of course, it becomes possible only when the unregulated perturbation actions, distorting the result of experiment, are absent.

But in the real investigation, even if the experimental purity conditions are met, there are no guarantees that the occasional disturbance, distorting the passing of investigated process, will not appear. Then a separately taken observation can appear as a result of this distorting mistake's influence. Moreover, accidental and systematical mistakes of devices, used in experiment and in observation, are possible. And, finally, human errors of an observer are possible.

By virtue of all these contingencies and subjective layers the data of observation cannot be a direct empirical basis for a theory. Such basis is composed of the empirical knowledge of other types. They are empirical dependencies and facts which form a particular layer of the empirical science's level that dominates above the layer of observation's data.

The transition from data of observation to empirical dependencies and scientific fact presupposes the elimination of subjective moments from the observation where they were present (these moments are connected with possible mistakes of observer, occasional disturbances distorting the passing of investigated events, and devices' mistakes) and the obtaining of reliable objective knowledge about events.

Such transition presupposes enough complex cognitive procedures. To obtain an empirical fact it is necessary to implement at least two types of operations. First of all, this is the rational processing of data of observation and searching for stable invariant content in them. To form a fact it is necessary to compare a great number of observations between themselves, to mark the repeated attributes in them and to eliminate the accidental perturbations and errors connected with observer mistakes. If measuring is carried out in a process of observation, the data of observation are recorded as the numbers. Then the definite statistical treatment of the measurement's results and search of average statistic quantities in a multitude of this data are required to obtain an empirical fact.

If, during an observation process, the instrumental plants were applied, then together with the observation protocols the protocol of the device's check test is made up, in which all possible systematic mistakes are fixed. When the observation's data are under statistical treatment these mistakes are also taken into consideration. They are eliminated from observations in a process of searching of their invariant content.

The search of invariant as a condition of the empirical fact's forming is peculiar not only to natural-scientific but also to social-historical cognition. Let us say, a historian, prescribing a chronology of past events, always aspires to discover and compare a multitude of independent historical evidences, appearing for him as a function of data of observation.

Secondly, for ascertainment of a fact, it is necessary to interpret an invariant content, which is discovered in the processes of observations. In the process of such interpretation theoretical knowledge previously obtained are widely used.

Let us consider two concrete situations that illustrate this role of theoretical knowledge when the transition is from observations to fact.

It is known that one of the fundamental physical discoveries at the end of the 19th century was a detection of cathode rays, which (as was clarified in a process of further investigations) represent an electron stream. Experimenting with cathode rays, W. Crookes registered their deviation under the magnet influence. Data of observation, obtained from this experiment, were interpreted as proof that cathode rays are a stream of charged particles. Theoretical knowledge about the interaction between charged particles and a field, received from classical electrodynamics, served as a foundation of such interpretation. Preciously their application led to a transition from the observations' invariant to an appropriate empirical fact.

The procedure of interpreting the observation data should not be connected with the process of formulation of theory, which must give an explanation to the obtained fact. The ascertainment of the fact that cathode rays are electrically charged particles is not a theory, although it was obtained using theoretical concepts.

But then a very complex problem appears which is discussed in methodological literature today. It turns out that to ascertain a fact theories are needed, but they must be checked by facts. This problem can be solved only if the interaction between theory and fact is considered historically. Unconditionally, the reliable theoretical knowledge obtained before and substantiated by the other facts is used to ascertain an empirical fact. But this can only be theoretical knowledge that was previously checked independently. As regards the new facts, they can serve as a basis for development of new theoretical ideas and conceptions. In turn, the new theories, transformed into reliable knowledge, may be used in interpretation procedures when other fields of reality are empirically investigated and new facts are formed.

Thus, when investigating the empirical cognition's structure, it is discovered that no scientific empirism exist which does not contain a touch of the theoretical. But this is not an obstacle to the formation of objectively true empirical knowledge but a condition of such forming.

Empirical dependencies and facts, in contradistinction to the observation's data, are not already correlated directly with the concrete instrumental situations of concrete, single experiments. Their relation to the actual experimental situations is mediated by empirical schemes which represent a particular kind of model conception expressing the typical features of some actual experimental situations class and their subject structure. Only with these schemes are empirical dependencies and empirical facts directly correlated.

Usually, previous hypothetical variants of empirical schemes are forming at the stage of experimental project. But after its implementation and in process of transition from observation protocols to empirical dependencies and facts, the basing of hypothetical variants of empirical schemes as the expression of essential features of some series of actual experiments occurs.

In the process of statistical treatment of the observation's data, the observation protocol and protocol, and fixing the average statistical data of an instrumental plant's behavior, are compared between themselves. So all objects, as a result of such comparisons, interacting at the experiment – experienced fragment and quasi-instrumental subsystems, – are found to be defined only by the statistically invariant attributes. On this basis an empirical scheme is built which generalizes the determined experimental interaction's class. In this sense it is literally a scheme of such interaction that depicts its typical features. These features are realized in every concrete experimental-measuring situation. Together with this an empirical scheme may be considered not only as a model conception of the experiment's activity and measurement but also objectively, as an inartificial natural interaction process's depiction in which the experienced object passes from the state O_1 into the state

O_2 under given conditions. Such angle of approach emerges in an interpretation process of the data of observation's invariant when fact is forming.

So, empirical schemes act as an important mediating element between theoretical schemes and instrumental situations of the actual experiments. They can be obtained as "from above", when empirical consequence is deduced from theoretical laws, and as "from below" as a result of transition from the observation's data to empirical dependencies and facts. The relation of theoretical schemes to empirical ones and the possibility of the latter's consideration in two angles of approach (as the model of experimental situations and as the image of an inartificial nature process) also allows the consideration of theoretical schemes' nature in a new light. Each of them may be compared with some empirical schemes' class (in an example with the Biot-Savart law not only the scheme of experiment with the rectilinear wire and the magnetic needle belongs to this class, but the schemes of the experiments with any types of conductors, through which the current goes, and also with any types of magnets).

From these positions the theoretical scheme can be considered as the invariant content of the empirical schemes.[28] Taking into account the latter function as a depiction of the typical features of experimental-measuring situations, the relations of the theoretical scheme's abstract objects can be rightfully considered in this aspect. They then will appear in a form of a particular idealized experiment that expresses the most general and essential features of the real experimental practice.

When analyzing the theoretical schemes from this point of view, their "operational" side is discovered immediately. The oscillator's scheme, for example, appears as a model which expresses the essential features of experiments with the oscillations of the real pendulums, of a tight string, with the periodical compression and stretching of a spring, etc.

The subject side of all these real experiments in the theoretical scheme is represented in a form of mental experiment with a material point which deviates from an equilibrium position and returns to the initial position under the influence of quasi-elastic force. Fundamental schemes, laying in the basis of a developed theory, may also be interpreted as ultimately idealized depictions of the typical features of experimental situations, which are generalized and predicted in a framework of this theory. So, Maxwell's theoretical scheme can be considered as a mental experiment, accumulating in itself the essential characteristics of experimental procedures, that are generalized in the schemes of Ampere's electrodynamics, Coulomb's electrostatics and magnetic statics, and Faraday's induction, etc.

The fundamental theoretical scheme of Newton's mechanics, describing mechanical motion as the material point's displacement at the continuum of space and time points of the frame of reference under the forces' influence, represented itself as an original thought experiment. This experiment contained the most general and essential experiences' features of mechanical motion's studying of different sides. The practical operations of the bodies displacement at the inclined plane, of the pendulum oscillations, of the bodies collision and the operations of the potential energy's transition into the kinetic one when the engines are working, etc. were generalized in this experiment.

This side of theoretical schemes is often not paid attention because in a majority of cases the theoretical model form itself disguises its operational nature. However, if the appropriate analysis is conducted this nature will appear in a clear form. We used to, for example, consider Thompson's and Rutherford's models of an atom only as a depiction of some sides of an atom's structure. But the attentive analysis shows that each of these models together with the depiction of an atom's structure introduces an ultimately abstract scheme of experimental situation, in the framework of which an atom was singled out and studied as a particular nature's fragment.

In Thompson's model an atom is depicted as an oscillator (positively charged sphere with electrons immersed in it, which are able to deviate from an equilibrium position), which interacts with the radiation falling on it and is able to generate radiation. All fundamental attributes of abstract object of Thompson's model are defined through their relation to ideal test radiation. This radiation represents the real beams of light at the theoretical model's level, which are fixed in experiments of studying the regulations of interaction between a light and a substance. Hence, Thompson's model may be represented as an abstract and schematized depiction of such experiments' essential features.[29]

Rutherford's planetary atom's model can be considered from the same positions. It represents a theoretical scheme that is formed from the following interrelated abstract objects: a "center of potential repulsive forces" (an atomic nucleus) and "elementary negative charges" (electrons). In this model the abstract object, an "atomic nucleus" was determined by two attributes: "to carry a positive charge" and "to be the center of potential repulsive forces".[30] It is principally important that the latter attribute has its sense only because the presence of a test body is presupposed. This test body is an ideal alpha particle scattering on the "centre of potential repulsive forces".

In this way, the main distinctive characteristic of Rutherford's atom model is the conception of an atomic nucleus that was introduced through depiction of thought experiment of an ideal alpha particle scattering on the kernel-electrons system. This experiment expressed the essential particularities of real experiments of heavy particles scattering on an atom. These were the experiments by means of which the real particularities of an atom's structure were discovered.

Rutherford's model implicitly contained an idealized scheme of mentioned experiments and this model's particularity appeared directly in those physical laws that could be obtained on its basis. The main equations, that Rutherford obtained based on the planetary model of an atom and that allowed explanation and prediction of the results of real experiments, were the laws of scattering of hard charged particles on an atom.

In this way, Thompson's and Rutherford's models can be imagined in a form of thought experiments with an atom as with an oscillator, and with an atom as with a system, dispersing the heavy particles. Each of these experiments accumulates in itself the essential features of real experimental-measuring practice in a framework of which the appropriate properties of a real atom were discovered. They were an object of study in Thompson's and Rutherford's investigations and were represented in the appropriate models of an atom.

As a result, we came to the important conclusion according to which theoretical schemes possess two indissoluble sides connected between themselves. The first side is that they appear as a particular model of experimental-measuring practice. The second one is that they serve as a systemic depiction of investigation's object and as a depiction of the essential connections of investigated reality.

This conclusion is illustrated only by the physical material. However, its formulation in a general form is justified completely because it can be shown that the given assertion is valid in respect to all empirical sciences. But, at first glance, it is imagined that theoretical statements' content of such sciences as astronomy may not be interpreted as a practice scheme, because there is no subject active interference in the natural processes passing as a necessary condition of practical activity. However, when analyzed closer, it is revealed that astronomical observations, carried out for the purpose of checking these or those theoretical schemes, have a character of specific quasi-experimental procedures. In the process of such observations the nature objects are applied in a function of quasi-instrumental devices. An instrumental situation results from this which is typical for the experimental-measuring activity to be created.

To understand this circumstance better, let us remember, that any experimental activity is characterized by such natural fragments' interaction in that they appear as the objects-carriers of the functionally selected properties. Such subjects are made artificially in a developed physical experiment. They can be the plants, preparing the beams of particles with the given parameters (the preparing subsystem of the experimental device), targets which are bombed by these particles (the working part), or devices that register the results of interaction between the particles and the target (a registering part of the experimental device).

However, the inartificial nature's objects, considered only from the side of their experimental properties, can also be applied in a function of the experimental activity's means. In an example with the studying of oscillations' processes in the experiments with a pendulum considered above, the Earth was used as a source of gravity in a quasi-instrumental subsystem function, providing the appearance of the quasi-elastic (returning) force.

A similar situation appeared in Faraday's experiments with electromagnetic induction, when the property of the Earth's magnetic field to give birth to the *emf* in the conductors which crossed its magnetic power lines occurred. Here the Earth was also used as a specific quasi-instrumental plant. It was considered only as a source of magnetism, combining functions of preparing and working parts of the "instrumental plant". This property of the Earth was discovered in previous experiments with the orientation of magnetic needles. It was functionally singled out among all other numerous properties of the Earth in the considered experiments. Due to this our planet usage in a function of a particular object of the instrumental situation became possible.

The analogous usage of nature objects in a function of specific instrumental devices can also be found in many contemporary physical experiments. Thus in experiments of research of neutrino, and radiation by the Sun, the latter was considered as the neutrino generator (the preparing subsystem). The research of neutrino's properties presumed that they need to be marked among other constituents of space radiation. For this purpose the registering devices were immersed into a shaft and then the Earth crust was used as a specific screen which detained all the space radiation particles, except for neutrino.

Systemic observations in astronomy are based on the same principle of application of inartificial nature's fragments in the instrumental subsystem function.

With the aim of illustrating of what have been discussed let us consider a concrete example. This is the X-radiation of the Crab nebula observation which was accomplished in 1964.[31] Its purpose was to reveal what is the source of this radiation. On the basis of the hypothesis regarding the neutron star's existence, the assumption was made that the neutron star, located within a Crab nebula, can be a source of radiation (practically a point source for the Earth observer). However, the source of radiation could be another; the distant radiation source related with nebula. To reveal the character of radiation source the coverage of the Crab nebula by the moon's disc was used; at this moment the change of the signal strength was fixed. This signal was from the X-ray source (the X-ray counters, lifted by the rockets, registered a number of γ-quanta per a time unit). The empirical dependency, revealed by the statistical treatment of the observations data, showed that the radiation intensity decreased not abruptly but gradually.

It is not difficult to see that in a framework of considered investigation the researcher could receive information on the radiation character of the Crab nebula only because he constructed an instrumental situation from the natural processes of environment. The X-radiation source, this radiation itself, and the Moon used as a specific screen appeared in the function of preparing and working subsystems of an "instrumental device". The registering part was played by a device which was artificially created in practice. The whole system, which contained the "X-radiation source in the Crab-like nebula", the "Moon" and "registering devices on the Earth", represented a specific giant experimental plant whose functioning allowed us to reveal the investigated dependency.

The instrumental situation's creation in a process of empirical investigations in astronomy can also be illustrated on the other facts. It is indicative in this relation that, for example, the observation of a star's light polarization is conducted with the purpose of studying the galactic magnetic field. The instrumental situation, which characterized this experiment, was built by the mode of marking the three components in a system of nature's interactions. The first is the galactic magnetic field and the particles oriented by it in the clouds of interstellar dust. The second is the light emitted by the star and passing through the interstellar dust. The third are the devices registering the polarization effects. The relations between all these object aggregates can be considered as the giant quasi-experimental device, whose "work" allowed us to reveal the empirical dependencies characterizing the galactic magnetic field (an object of investigation). In a framework of a given situation this "work" lay in the interaction of the light, and the oriented particles of interstellar dust gave birth to the light polarization, to the extent of which it became possible to conclude about the intensity of the galactic magnetic field.

It is rather difficult to settle how the instrumental situation was constructed in empirical investigations of astronomy at the early stages of its development. However, here everything also occurred in the same way. Thus, even a simple visual observation of a planet's motion in the vault of heaven supposed that the observer should previously mark the skyline and the markings at the vault of heaven (for example, the stars) in which background a planet motion is observed. These operations in themselves, essentially, presented the vault of heaven as a specific graduated scale on which a planet motion as a lighting point was fixed. Moreover as mathematical methods penetrated into astronomic science the value of heaven's graduation became more exact and convenient for the conduction of measurements. Zodiac, which consisted of twelve parts of 30 degrees each, had already emerged at the 4th century as a standard scale for description of motion of the Sun and planets.[32]

Any systematical scientific observation supposes the instrumental situation's construction independently of its accomplishment in a process of experiment or outside of experiment. The systematical observations can be considered as a quasi-experimental activity in this situation. Concerning the occasional observations, they are insufficient for the scientific investigation. They can become a primary impulse to the new investigations, but if such investigations are established they must overgrow into the systematical observations. In occasional observations, as a rule, some extraordinary effect is registered but it is unknown which objects participate in the interaction that gives birth to the given effect. The instrumental situation's structure is not determined here and object of empirical investigation is unknown. The transition from occasional to systematical observations presupposes the instrumental situation's building and a clear fixation of object, which states' changing is studied in the experiment. An example of this when K. Yansky in his experiments of thundery impediments in the intercontinental radiotelephonic transmissions occasionally came upon stable cosmic noise which was not connected with any earthbound sources. This occasional observation gave an impulse to the series of systematical observations, the final result of which was the development of radioactivity of the Milky Way. The characteristic moment of these observations' establishment was the construction of instrumental situation.

The major task here was in determination of a stable cosmic noise's source. After the ascertainment of its extraterrestrial origin, the demonstration that the Sun, the Moon and the planets are not such a source had become crucial. The observations which allowed us to make this conclusion were based on application of two types of instrumental situation. First of all, the Earth revolution was used. The Earth' thick layer was applied in observation as a function of a screen overlapping the Sun, the Moon and the planets at a definite time of day (observations showed that cosmic noise does not disappear at the moments of this overlaying). Secondly, a behavior of the source of cosmic noise was investigated in the observation when movement of the Sun, the Moon and the planets at the vault of heaven was relative to the skyline and the motionless stars. The latter were used in this situation as the fiducial points (the means of observation), in relation to which the possible movement of source of cosmic noise was fixed. These series of experiments allowed, at the final result, the identification of the source's position with the positions in the Milky Way firmament which are observed at every moment of a day and a year.

It is characteristic that on the last step of K. Yansky's investigations the observation's subject structure had already been marked clearly. In its framework the investigated effect (the cosmic noise) was presented as the Milky Way radio emanation. The primary state, the final state and the instrumental situation of the empirical knowledge's object were marked. The primary state was the position of the source at the moment T_1. The final state was the position of the source at the moment T_2. In the instrumental situation the following objects were fixed as means of investigation: the vault of heaven with the stars' position marked on it, the skyline, the Earth whose revolution maintained the changing of radio source position relative to the observer, and finally, the devices which were the radio waves recorder. The observations with the toughly fixed structure of the mentioned type allowed the discovery of the nature of occasionally detected effect of the Milky Way's radiation.

Thus, the process of an accidental registration of a new event to revealing the main conditions of its origin and its nature goes through the series of observations which distinctively appear as the quasi-experimental activity.

The analysis of systematical observation's situations, established outside of an experiment, allows us to unify the approach to the theory empirical foundations and to the theoretical scheme's operational treatment. Then theoretical models of astronomy may also be rightfully considered not only as a reflection of investigated object, but as a generalized scheme of the observation's subject side, appearing in a function of experimental-measuring situations in a framework of which the given object was revealed.

As in any cognitive activity, a fundamental principle is revealed here in accordance with which the cognition's object is defined only relatively to some system of activity. An object of investigation is always given in a form of practice to the subject that fulfils a cognition activity and so it has no way of vision of reality except as through the prism of this practice. Thus, the schematized and idealized depiction of the practice essential features is contained in all layers of scientific knowledge. At the same time this depiction (or rather by virtue of it) serves as a depiction of the reality under consideration. This depiction appears in a particular form at every level of investigation. Thus, the actual experiment a subject of investigation is represented through the correlation of objects interacting in the experiment. For example, a current magnetic action, which is studied in Biot's and Savart's experiments, is adjusted through the relation of the real wire to the real magnetic needle, which acquires an angular momentum at a period of a current's passing in a wire.

At the next stage of investigations, at the layer of empirical schemes, an investigated object is represented through the correlation of empirical constructs, which form an empirical scheme. Thus, a current's magnetic action in Biot's and Savart's empirical schemes was depicted by means of such constructs as a rectilinear wire with a current and a test magnetic needle, with indication of their relations as a meaning of appropriate empirical formula. Then an investigated object is introduced to the local theoretical scheme's layer through the correlation of abstract objects. In our example they will be the following abstract objects: a "magnetic field averaged over some volume" and "density of the charge-current that generates it" (the relations of these objects make up a sense of the Biot-Savart law). Finally, at the level of a fundamental theoretical scheme, which lies in a foundation of a developed theory, a subject of investigation is represented through the correlation of abstract objects of a given scheme. For example, a current's magnetic action at the level of Maxwell's theory is represented through the relation between the "vector of a current density in a point" and the "vector of a magnetic field in a point". These vector connections make meaning of the second pair of Maxwell's equations.

Each of the selected levels of the investigated object representation constitutes a particular layer of the science language where the ideal schemes of experimental-measuring practice's subject side appear as a substantial plane expressed in appropriate sign form (see pic. 2). Any one of these layers of language has its norms of construction and lives by its relatively independent life where the new content can appear at the expense of internal laws of operating with the symbols. An example of this can be the introduction of new abstract objects at the expense of operations in a framework of the theory's mathematical formalism.

The connection of the mentioned levels of science language allows the introduction of underlying levels accordingly to the new content of each top level. Thanks to this, it became possible to forecast the practice predicting the future experiments' results. In existing theory the connection between the different levels of language is achieved at the expense of particular language expressions which also belong to a theory's content. By means of these expressions the mode of theoretical objects' reduction to the objects of underlying levels is described. The expressions of this kind are the essence of the correspondence rule (the operational definitions).

If to take into consideration that the objects of each top level's schemes appear as invariant content of correlation of objects of underlying layer, the description of the appropriate objects' attributes in terms of correlation like this makes up an essence of operational definitions.

This side is not paid enough attention because when the links between theory and empirism are analyzed, at best only two levels are singled out – empirical and theoretical – but these levels themselves have already been considered as undifferentiated.

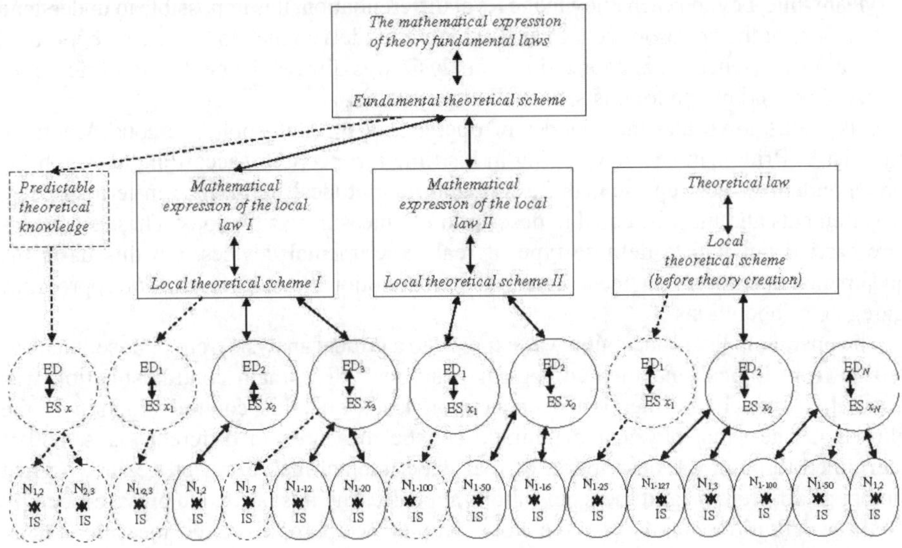

Pic. 2. ED - *empirical dependence* (1,2 - *assimilated by theory;* N - *is not assimilated by theory*)*;* ES - *empirical scheme;* $N_{1\text{-}n}$ - *observations;* 1-*n* - *a conditional number of observations;* IS - *instrumental situation;* [ ] - *predictable theoretical law and appropriate theoretical scheme;* ( ) - *predictable empirical dependence, empirical scheme, observations and instrumental situation.*

Meanwhile, beyond each knowledge level differentiation, it is impossible to understand the structure of the correspondence rules, which provide a connection between theoretical terms and experience. This connection's analysis was always the center of attraction of philosophers and methodologists, as well as physicists.

It is well known that the founder of operationalism philosophy, famous American physicist P. Bridgman was developing in his time a conception, according to which the correspondence rules represented the physical magnitudes' definitions in terms of real measurements and must be equal to description of measuring situations. These situations were carried out with a definite type of real experimental devices. On this basis the fundamental operationalism thesis emerged that "A notion is synonymous of an appropriate aggregate of operations".[33]

The operationalism conception was exposed to a critical analysis from philosophical, as well as from logical and methodological positions.[34] The main contradistinction was revealed, to which the mentioned conception leads and this contradistinction is the following. The same physical magnitude can be measured in different ways, and if determined through the description of real experimental-measuring procedure, a great number of different magnitude's definitions appears. And it is needed to prove especially that these definitions are of one value. For example, receipts of distances' measurement by way of using a tough ruler and by radiolocating are different. But the physical magnitude that marks a distance is the same in both cases.

In the direction itself of the physical magnitude determination through the real procedures of measurement, there are no rules of such measurements' equalization. Therefore, to accept an operationalism conception, it is needed to consider that the same magnitude, measured by different modes, is in essence two different and unequal magnitudes.

In addition to what have been discussed let us note that the determination of magnitudes by a recipe, originally offered by P. Bridgman, can determine only the meaning of these magnitudes inside an empirical layer of investigations. But this recipe does not allow us to solve the major task: the transition from empirical level to theoretical and vice versa.

In developed science, which formed the theoretical level of investigations, the magnitudes, which figure in theory and are related with experience, have two meanings – empirical and theoretical. Their theoretical meaning corresponds to attributes and correlation of theory's abstract objects. Their empirical meaning corresponds to correlation of empirical objects represented to the real subjects of the experimental-measuring situations. For example, in Maxwell's theory the magnitude H, designating the magnetic field strength at a point, receives definitions through the relations to vectors E (the electric field strength), j (density of charge-current), B (magnetic induction), and D (electric induction). At the empirical level the magnetic field magnitude is determined otherwise. It can be adjusted, for example, through the turn of a magnetic needle in the Biot-Savart experiment or through the turn of Helmholtz's coil, when magnetic field magnitude is measured in other analogous experiments. To relate these two meanings of a magnitude, it is needed to be able to move from the theory's abstract objects to the objects which are operated by an experimenter. Such transitions are not supplied only with the operations of real measurement's description. It is characteristic that Bridgman was forced to acknowledge that theoretical term meaning may not be reduced to description of the measurements conducted at the real experiment.[35] He then widened a comprehension of operational definitions and included the so-called "paper-pencil" operations into their composition (calculations, made in a framework of thought experiment and necessary for transition from theoretical level magnitudes to the results of experiment). But at the same time the comprehension itself of the operational definitions became indistinct and their structure remained uncertain. The shortcomings of Bridgman's conception gave birth to a suspicious attitude of some philosophers and logicians toward the notion of the "operational definitions". Even the opinion was expressed that operational definitions do not generally exist in scientific theory. A conception like this was developed, for example, by M. Bunge. Correctly marking a limited nature of Bridgman's approach to analysis of the correspondence rules and criticizing the operationalism philosophy, Bunge concluded that "a human and its operations... have no place at the reflection of a physical reality in notions" and that operational definitions "have not existed in theory at all".[36] So "radical criticism" of Bridgman's concept of operational definitions supposes that there are no rational moments in it at all. In our opinion, such assessment is one-sided. It does not follow from the fact of irreducibility of theoretical notions to the operations of measurement that those notions do not contain any operational multiplier at all and that theory can manage without the operational definitions.

Rejecting the term "operational definitions", M. Bunge speaks about the rules of theory's mapping on the objects of experience, and about the agreement of a theory as a whole with the experience data. But then a question arises, where do such rules lie and how is the connection provided between theory and experience?

The matter is not in changing the term "operational definitions" with another term, which characterizes a compounding of connection of these or those theoretical notions with experience, but the matter is to analyze the operational definition's structure, to discover a nature of the correspondence rules connecting a theory with an experience.

Bunge, in essence, left these questions unsolved. And in many aspects that is because that "philosophy of realism" itself, which he upholds and develops, suffers from the series of limitations in spite of its positive content (the recognition of objective reality and its reflection in notions). One of these limitations is an ignorance of practical-active foundation of objects theoretical setting or that an object is always given to the investigator in a form of practice. By this Bunge's conclusion is dictated, that the reflection of physical reality in notions has no relation to human activity operations. In its turn, such setting stopped the way to analysis of the correspondence rules structure.

Insufficiently detailed analysis of the structure of correspondence rules of theoretical quantities and observations often leads to inaccuracies of methodological character, even in extremely known and competent works. Thus, L. Mandelstam pointed out fairly in his interesting and rich profound philosophical reflections lectures on quantum mechanics, that every physical theory includes not only a theoretical apparatus but the recipes of connection of physical magnitudes with experience. However, he made an inaccuracy characteristic of such recipes. The physicist's intuition prompted to him that the connection between theory and experience cannot be fulfilled outside of taking into account specificity of real experimental-measuring activity. Therefore Mandelstam determined recipes of connection with experience of magnitudes which are represented in theory equations as "concrete operations with concrete things". It is when "concrete things are selected as the standards" and "concrete measuring processes" are applied, i.e. "determination of a coordinate, of time etc, using tough scales, clocks, etc.".[37] Determination such as this is acceptable only as a direction to take into consideration in theory the actual experiment particularities. But without further specifications it becomes identical to the definition given in Bridgman's early works. Real measuring procedures are actually supposed in the recipes of connection of magnitudes with experience but such recipes are not brought to the mentioned procedures. Really, if acquainted with the concrete examples of mentioned recipes, which Mandelstam quotes for explanation of thesis advanced by him, an extremely interesting and important moment is revealed. It turns out that the recipes of connection of theory's physical magnitudes with experience are the description of not real, but idealized measuring situations, which correspond to the real situations of experiment and measurement.

In this one of the important particularities of the correspondence rules lies. The connection between the real measurements and theoretical objects is established only at the expense of such thought experiments and idealizations.

The key to decoding the correspondence rules and the meaning of operational definitions lies in taking into account the major levels of the experiment's schematization in the science language, and in comprehension of the fact that the object of every top level appears as a characteristic of the object's correlation of "underlying" language layer appropriate to it. Moreover, a transition from the empirical level to the theoretical one always presupposes an idealization and replacement of the real scheme of experiment with the idealized one. Idealization allows the separation of essential characteristics of interactions studied in the experiment from occasional and replacing factors. Due to this, the operationally-defined theory terms appear as an expression of attributes and interrelations of mentioned interactions. The tracks of all these sufficiently complex operations can be discovered when analyzing the sense of concrete operational definitions of physical magnitudes. Thus, the operational definition of a magnetic field strength as in Maxwell's equations is given not through description of measurements with application of a real device, for example, an electrometer (as it is often considered) but is given through a description of relations of an electrical field at a point to a test charge. In its turn, an "electrical field at a point" and a "test charge" are the constructs which are typical for the local theoretical schemes of Maxwell's theory. In these schemes the relations between charges and electrical field are characterized. As regards to the given constructs' definitions (such as a test charge), these definitions appear as a characteristic of the particular correlation of empirical objects. For example, a test charge "is such an influence of one massive charged body on another one, at which this influence may be neglected owing to smallness of a counter influence of the second body on the first one" (an idealization of a real experiment). Only the empirical object's definition may be given through the description of arrangement of real devices and real procedures of measurement. The operational definitions of physical magnitudes include all this hierarchy of definitions in a compressed form. Due to this, they characterize the way of reflection of theory abstract objects on the real relations between the objects of experiment and the measurement.

Thus, if to sum up what have been discussed, we have a right to consider the science theoretical schemes as the original models of practical situations, on which explanation and forecasting they do appear to be. But the theoretical schemes have not only an operational status, they are always perceived by an investigator, accepting this or that theory, as a conception of the investigated object area as an image of its essential links. And then a particular problem of the theoretical scheme's objectification emerges.

When analyzing this problem (of relations between the theoretical knowledge and investigated reality), it is important to take into consideration the fact of existence of two levels of the theoretical knowledge's organization. One of them is formed by a developed theory. The second is a theoretical knowledge represented by the local theoretical schemes and the laws related to them. As the history of science shows, they can genetically precede to a developed theory and appear as independent formations that are fixed in appropriate theoretical language. Thus, the scheme of current's power interactions was introduced by Ampere long before the Maxwell electrodynamics' creation. And the simple oscillation model was built by Huygens long before the Newton mechanics creation. By the analogous way the different aspects of quantum mechanical processes were depicted before quantum mechanics creation. They were described using such theoretical schemes as Bohr's atomic model, photoeffect model, Compton effect, the blackbody radiation, etc. When the developed theory is built the local theoretical schemes, preceding it, are transformed and included in a theory's composition as the components of its content.

Taking into consideration the two-level organization of theoretical knowledge and the presence of genetic links between the levels, let us consider how theoretical schemes are related to the investigated reality.

The most important theoretical schemes' particularity consists in that they are an idealized model of interactions studied in a theory. Due to the theoretical scheme, a particular vision of investigated activity forms in a science. This activity is represented in a theory in a form of an idealized object possessing a strictly determined structure. This side of theoretical assimilation of reality was described in enough detail in methodological and philosophical investigations.

It is ascertained that the model of investigated reality, supplied with "a small number of properties and a simple structure", lies in a theory's base. The main function of this model is to serve as an idealized presentation of the investigation's object and to be a means of obtaining theoretical knowledge.[38]

The separation of theoretical schemes as a systemic depiction of the reality under theoretical investigation, continues the already existing philosophical and methodological tradition. The new moment of analysis is not a discovery itself of the marked scheme, but an attempt to view its internal structure more concretely.

Just this structure, i.e. relation of a theoretical scheme's abstract objects represents in a framework of a theory an objective reality studied in it. Therefore, when an investigator characterizes a subject of this or that theory, he describes its in a terms of the abstract objects of appropriate theoretical scheme. If, for example, we ask a physicist what he means by electromagnetic processes as a subject of investigation of classical electrodynamics, his answer will be that they are the interactions of electrical and magnetic fields between themselves and electric charges (differentially small currents). In the further specification of this definition the interaction of electrical and magnetic fields and charges will be characterized as a change in time of vectors of electrical and magnetic strength and current density at a point. Thus the differentiated description of a theoretical scheme lying in a foundation of classical electrodynamics will be given. This description outlines a subject that is investigated in theory, characterizing its essential parts and relations.

We will distinguish the given subject from those abstract objects which are the elements of a theoretical scheme. No abstract objects of a fundamental theoretical scheme of electrodynamics that are isolated from the others represents in cognition the electromagnetic processes' structure. Only the network of links and relations of mentioned abstract objects represents it, i.e. the theoretical scheme as a whole.

The same particularity of a theoretical scheme can be revealed if addressed to the examples of theoretical knowledge that have already been considered. It is not difficult to ascertain, for example, that only the whole set of relations between the oscillator elements (a material point, quasi-elastic force and a frame of reference) can serve as a depiction of a "process of a simple oscillation" as an investigated subject of the mechanical oscillations theory. But none of these objects taken separately can serve in this way. To make distinct the abstract objects, which an investigator operates with in a framework of theoretical scheme, from the investigated subject, which systemic and structural depiction is given by a scheme as a whole, we will call the first of them the objects of handling and the second ones the subjects of investigation.

Theoretical schemes are perceived as a depiction of a subject of investigation owing to a particular procedure of their objectification. To reveal the particularities of this procedure and its role in theory construction, let us turn in the beginning to concrete historical example.

Thus, at the first acquaintance with H. Hertz's mechanics, the impression is created that theoretical scheme, applied here, is an extremely artificial depiction of mechanical processes. Hertz built all the mechanics on the basis of such system of abstract objects where only material point's (masses) correlation to a time and space are given. A state of the material system motion is characterized in Hertz's theory as displacement of material points at a constant speed on the course of geodesic lines ("the straightest ways").[39] Hertz's idea was to describe any mechanical system motion as a free motion on one of the possible "straightest ways". In this description a force is substituted with a connection between interacting systems and is expressed through the characteristic of a track curvature along which the system moves, limited by the constraints.

Force and energy in Hertz's mechanics are not the fundamental notions by means of which the system state and its change are described, they are the secondary notions and may be, in principle, eliminated at the expense of reduction to the main notions ("mass", "space" and "time").[40]

Hertz shows that from his theoretical scheme, the famous mechanics laws can be obtained, and the principle of Hamilton and the principle of least action of Euler-Lagrange can be proved as the theorems both in a classical form and in a form of Jacobi's principle. It seems, these arguments are sufficient to prove a theory as expression of essence of mechanical processes. Nevertheless, Hertz included one basis into his theory statement. He mentioned that after successes of electromagnetic field's theory, the conception strengthened in physics that the natural processes were the interaction of "weighty bodies" (atoms, molecules, macro-bodies) with a universal environment-ether, which was an intermediate in the interactions' transition from one body to another.[41] All that was called in physics the force transmission, was the motion in a worldwide environment (ether).[42] Therefore the power influences of one observed body on another can be imagined as motion of particles-masses of the worldwide environment). According to Hertz, if every observed mechanical system is supplemented with a hidden material system, whose carrier is an ether, it is possible in any case to consider the system's motion as a free (natural) motion on one of the possible "straightest ways".[43]

Owing to these explanations, Hertz's theoretical scheme of mechanical motion is coming to be perceived as an adequate and natural view of the nature of mechanical processes.

It is not difficult to see that the objectification of theoretical scheme was achieved at the expense of its connection with some system of general conceptions of nature's "organization". By means of these conceptions all natural processes were depicted as interaction between bodies and ether. The system of such conceptions is a particular component of scientific knowledge and forms a physical picture of the world. Here we came to a specific problem of science's foundations and of scientific picture of the world as a component of these foundations.

NOTES: CHAPTER 2

[1] See, e.g., Margenau (1950. p.86).
[2] If one considers that correlation of theory abstract objects are identical to the meaning of its statements, the impossibility of appearance of such objects' attributes, which are incompatible with the initially introduced ones, will correspond to a well-known requirement of consistency of a theory.
[3] Margenau (1950,p.85).
[4] Rokitsky, Savchenko and Dobina (1977, p.12).
[5] See Harris (1990, pp.139-156).
[6] Ibid, pp. 578-579, 580-595.
[7] Rozin (1989).
[8] Ibid, pp.40-46, 48-65.
[9] Ibid, pp.48-58.
[10] Newton had noted even in the initial definition of space and time in his "Principia" that in mechanics physical space is identical to "true" or "mathematical space", and physical time is identical to "true" or "mathematical time".
[11] It will be said below about the procedures of constructing the small oscillations model on basis of mechanics' fundamental theoretical scheme.
[12] An investigator, applying a mode of knowledge analysis like this, can realize its conditional character and even stipulate specially for its limited nature. But if at investigator disposal there are no other standards of theoretical organization, then voluntarily or not he will look at the scientific theory as at the axiomatic-deductive system and try to understand what corresponds in it to an accepted standard. The remainder is out of the sphere of concrete analysis.
[13] The distinction between the axiological and genetically constructive developments of a theory in Russian logic and methodological literature was for the first time drawn in works by V. Smirnov (see Smirnov (1962)).
[14] Ibid, p.269.
[15] Ibid.
[16] Hilbert and Bernays (1934, S.20).
[17] As it was shown by V. Smirnov, the traditional interpretation of the Euclidean "Elements" as a pattern of theory's axiomatic construction did not consider that the deductive development of theories could be established not only in a form of axiomatic construction, but also in a form of genetically-constructive one too. "The conception realized in "Elements" – he mentioned, – was not an imperfect attempt of implementation of the axiomatic method's ideal in its contemporary comprehension, but an attempt of constructive (genetic) building of theory" (Smirnov (1962, p.278)).
[18] See Yanovskaya (1972), Rozin and Moskayeva (1967).
[19] Landau and Lifshitz (1960, pp.111-112).
[20] Ibid, p.127.
[21] Ibid, p.127-128.
[22] In the case of usage of a mathematical formalism in physics, the appropriate symbols and their relations are perceived only from the direction of their mathematical sense (as the mathematical

objects and the operations at them). Physical sense of given terms is not explicated at the motion like this.

[23] Consequently, in a process of philosophical analysis of the science language, theoretical scheme can be singled out and fixed first of all at the turn of different language contexts, but not by the way of pointing at some form of expression.

[24] Kuhn (1970).

[25] See Rosenberger (1937, p.136).

[26] In this case the intermolecular interactions become essential, which may be neglected when investigating the gases under small pressure. And the dependence discovered by Boyle and Mariotte, is generalized in van der Waal equation.

[27] This example, showing how the structure of experimental activity reveals an object of investigation, was analyzed in detail in a book by Stepin and Tomilchik (1970, pp.19-31).

[28] The term "invariant content" should not be perceived as indication of the inductive mode of theoretical content obtaining. To obtain the theoretical invariant it is needed to know in advance that this or that group of empirical schemes forms a class. This class can be easily determined "from above" at the reduction of theoretical schemes to empirical ones. But at the movement "from below", looking from a position of one theoretical scheme to another, it is not at all obvious that they possess a common content. It is indicative, for example, that, according to pre-Faraday's theory of electricity, the different electrical events, fixed at experience, were considered as the events of principally different natures (at that time frictional electricity, galvanic electricity, etc., were considered). So, the empirical schemes, which were unified in a common class in Faraday's theory, had been perceived before as uncoordinated conglomerate.

[29] See Stepin and Tomilchik (1970, pp.35-36).

[30] To avoid misunderstandings it is necessary to underline that in this case an "atomic nucleus" is in essence an ideal object appearing as a carrier of a limited number of toughly fixed attributes (real object is inexhaustible in its attributes).

[31] Ginzburg and Syrovatsky (1967, p.36).

[32] Neigebauer (1968, p.111).

[33] Bridgman (1954, pp.32-33).

[34] In Russian literature the criticism of an operationalism was most thoroughly and sufficiently correctly given in the works of D. Pivovarov (see Pivovarov (1971)).

[35] Bridgman (1955, p.153).

[36] Bunge (1966, p. 74).

[37] Mandelstam (1972, p. 327).

[38] Kuznetsov (1975, pp. 30-31), Schtoff (1966, pp. 155-157). Compare Wartofsky (1966, pp. 29-31).

[39] Hertz (1900).

[40] Ibid.

[41] Ibid.

[42] Grigorian (1971, p.284).

[43] Hertz (1900).

CHAPTER THREE

THE FOUNDATIONS OF SCIENCE

In philosophical and methodological literature of the last decades the fundamental ideas, notions and concepts more often become the subjects of investigation, which form the relatively stable foundations on which the concrete empirical knowledge and theories, which explain them, develop.

Discovery and analysis of these foundations presupposes a consideration of scientific knowledge as an integrated developing system. In Western philosophy such vision of science formed relatively late, for the most part in a post-positivistic period of its history. Concerning the stage, on which the concepts of science, developed in the framework of positivistic philosophy, were dominant, then so-called standard conception of structure and growth of knowledge could be considered as their most outstanding manifestation.[1] Separately taken, theory and its interrelation with experience functioned as a unit of analysis in it. Scientific knowledge appeared in it as a set of theories and empirical knowledge considered as a basis on which theories develop. However it became clear that empirical basis of theory is not a pure theoretically neutral empirism, and that the facts, but not data of observation represent the empirical basis on which theories rely. And facts are theoretically laden because other theories take part in their formation, and then the problem of interaction between a separate theory and its empirical basis appears as a problem of this theory correspondence with other theories which have been formed before generating composition of theoretical knowledge of definite scientific discipline.

Somewhat from the other side this problem of theories interrelation was revealed when their dynamics had been investigated. It was revealed that the growth of theoretical knowledge was carried out not just as a generalization of experimental facts, but as usage in this process of theoretical notions and structures developed in previous theories and applied to generalize the experience. Thus theories of an appropriate science appeared as some dynamical network, an integrated system interacting with empirical facts. Systemic influence of the scientific discipline knowledge set a problem of main factors determining the integrity of an appropriate knowledge system. In this way the problem of science foundations became visible due to which the different knowledge of scientific discipline was organized into systemic integrity at each stage of its historical development.

Finally, consideration of knowledge growth in its historical dynamics revealed particular states connected with crucial epochs of the science development, when radical transformation of its most fundamental notions and concepts was going on. These states were called scientific revolutions and they can be considered as the reconstruction of science foundations.

Thus the broadening of the field of methodological, problematic in post-positivistic philosophy of science put forward as the actual methodological problem the analysis of the foundations of science.

These foundations and their individual components were fixed and described in the following terms: "paradigm" (T. Kuhn), "core of research program" (I. Lacatos), "ideals of natural order" (S. Toulmin), "major themates of science" (G. Holton), and "research tradition" (L. Laudan).

In the process of discussions between the followers of different conceptions the problem arose abruptly of a differentiated analysis of science foundations. The discussions around the key notion "paradigm" in Kuhn's conception are indicative in this respect. Its extreme polysemy and indistinctness was marked by a great number of Kuhn's opponents.

Under the influence of criticism Kuhn tried to analyze the paradigm structure. He singled out the following components: "symbolical generalizations" (mathematical formulations of the laws), examples of concrete tasks solutions, "metaphysical parts of paradigm" and values (value directions of science).[2] This was a step forward in comparison with the first variant of conception. However, at this stage the structure of science foundations remained unclear. First of all, it was not shown in what relations the marked components of paradigm were located. This meant, strictly speaking, that its structure was not revealed. Second, according to Kuhn, as the components related with deep foundations of scientific search as the forms of knowledge which had been built on these foundations were included into paradigm. For example, mathematical formulations of local scientific laws (such as the formulas expressing Joule-Lenz's law, the law of mechanical oscillations etc.) are included in the composition of "symbolical generalizations". But then it turns out that discovery of each new local law must mean a change of paradigm, i.e. scientific revolution. Thus distinction between the "normal science" (the evolutionary stage of knowledge growth) and scientific revolution disappears. Third, marking such scientific components as "metaphysical parts of paradigm" and values, Kuhn fixed them ostensibly through the description of appropriate examples. It is seen from Kuhn's examples that he understood "metaphysical parts of paradigm" either as philosophic ideas or as the principles of concrete scientific character (like a principle of short-range action in physics or principle of evolution in biology). Concerning the values, Kuhn's characteristic looks to be only the first and extremely approximate draft. In essence, here the ideals of science are intended at that taken in extremely limited range as ideals of explanation, prediction and application of knowledge.

In principle, it may be said that even in the most advanced investigations of the foundations of science, to which T. Kuhn's works can be attributed, Western philosophy of science is insufficiently analytical. It has not yet established what are the fundamental components of science foundations and their links. The links between science foundations and theories relying on them and empirical knowledge have not yet been clarified. This means that the problem of the foundations' structure, their place in a system of knowledge and their functions in its development requires further deep discussion.

In the current and developed system of disciplinary scientific knowledge the foundations of science are discovered, first, when analyzing the systemic relations between the theories of varying degree of generalization and their respect to different forms of empirical knowledge in a framework of some discipline (physics, chemistry, biology, etc.). And second, when investigating interdisciplinary relations and interactions of different sciences.

As the most important components forming the foundations of science one may single out the following: 1) scientific picture of the world; 2) ideals and norms of scientific cognition; and 3) philosophical foundations of science.

The mentioned components express general views on specificity of the scientific investigation's subject, on particularities of cognition activity setting this or that type of objects, and on character of relations between science and culture of appropriate historical epoch.

SCIENTIFIC PICTURE OF THE WORLD

Worldview, philosophy, scientific picture of the world

Analysis of the picture of the world as a particular component of scientific knowledge presupposes a preliminary clarification of meaning of the initial terms the "world" and the "picture of the world". There should be a distinction made of the category the "world" in its philosophical sense, when it is used regarding the world as a whole, from those notions of the world that are formed and used in concrete sciences, when it is used regarding the "world of physics", the "world of biology", the "world of astronomy", etc., i.e. about the reality which makes up a subject of investigation of appropriate concrete scientific discipline.

The picture of the world like any cognitive image simplifies and schematizes the reality. The world as endlessly complicated and developing reality is always larger then the views on it formed at the definite stage of social and historical practice. Together with this, at the expense of simplifications and schematizations, the picture of the world singles out from the endless variety of the real world just those essential links, whose perception makes the major purpose of science on this or that stage of its historical development.

When describing the picture of the world these links are fixed as a system of scientific principles, which the investigation relies on and which allow it to actively construct the concrete theoretical models, to explain and predict empirical facts. In its turn, the practical application field of these moments contains the potentially possible spectrums of technical and technological phenomena which human activity is able to generate relying on theoretical knowledge. This aspect of the picture of the world attitude towards the world itself requires a particular understanding. It is necessary to consider that owing to human activity the lines of development are realized which are possible and are not contradictory to nature laws, but at the same time are improbable for it. The overwhelming majority of objects and processes generated by human activity belongs to the area of being artificial and not emerging in nature itself without humans (nature created neither the steamship, the car, the computer, nor cities architecture). But since science creates preconditions for appearance in technical and technological applications a wide spectrum of such "artificial" objects and processes, then one may suppose the scientific picture of the world as an extremely abstract "matrix" of their generation. In this sense one can say that the scientific picture of the world, being a simplification, includes together with this a far reaching content in comparison with the actually existing world of natural processes, because it opens the abilities to actualize directions of evolution improbable for nature itself (though not contradicting its laws).

The further substantial explication the notion of the "scientific picture of the world" presupposes clarification of the major meanings in which term "the picture of the world" is used, taking into account its extreme polysemy.

In contemporary philosophical and special scientific literature it is applied, for example, for designation of worldview structures lying in a fundamental of culture of a definite historical epoch. In this meaning such terms as "image of the world", "model of the world", and "vision of the world" characterizing the integrity of the worldview are used. The picture of the world's structure in such approach is given through the system of so-called categories of culture[3] (universalities of culture).

Broad interpretation of the term the "picture of the world" gave reason to a series of investigators to equate the notions of the worldview and the picture of the world. So, for example, A. Chanyshev had mentioned that "under the worldview we mean the general picture of the world, i.e. more or less complicated and systematized aggregate of images, concepts and notions in which and through which the world is realized in its integrity and unity. And (what is most important) the place of such most important (to us) part of the worldview as humanity, is also realized in this way".[4]

However in this case it is very important to bear in mind that the worldview image of the world is not only a comprehension of the world and the knowledge about it, but at the same time it is a system of values determining the character of the attitude, of human experience of the world and a defined evaluation of these or those events and occurrences and, accordingly, the active approach of a human to these events.

In A. Chanyshev's definition the accent is made on the cognitive aspects of the worldview, and value and active aspects of the picture of the world as a worldview's image are not fixed in a clear form. If we take them into consideration, the notion of the "picture of the world", used in the meaning of the worldview as the image of the human's world, then acquires a more adequate determination.

Application of the term the "picture of the world" in this meaning can be found not only in Russian, but in foreign investigations, including those dedicated to philosophical science problems.

One can mark that in Western philosophy of science in the 1980s a rehabilitation of its kind was going on with the notions of the "worldview" and the "picture of the world". G. Holton paid attention to this aspect of the problem. He noted that philosophy of science was compelled to appeal to these phenomena when the necessity of complication of science's methodological analysis arose and accordingly the need for a more delicate methodological toolkit appeared.[5] Together with this, it was practically identified with the worldview when there was a discourse about the picture of the world. Notion of the picture of the world as synonymous with notion of the worldview is used exactly in G. Holton's conception. It appears in his works as a model of the world which "generalizes the experience and innermost human views, and plays a role of an original mental map with which a human checks his actions and orients among things and events of real life".[6] Its main function is to be a connecting force directed to human society consolidation.

Together with the picture of the world's understanding as the worldview, G. Holton also uses the notion of the "scientific picture of the world". It seems he is close to distinguishing the picture of the world as the worldview and the scientific picture of the world. However, judging by the context, the term the "scientific picture of the world" is also used by him in the meaning of the worldview. And the adjective "scientific" is used to underline that the human worldview must be based on a collection of received scientific results but not on possible cults, astrological prophecies, etc.

G. Holton does not only fix the picture of the world presence but aims to discover its thematic core. He mentions that at the center of each picture of the world, forming the most important in the epistemological sense cognitive structure, the collection of thematic categories and assumptions is situated. They bear a character of unconsciously accepted, non-checkable, quasi-axiomatic basis statements which became firmly established in practice of thinking as its guiding and supporting means.[7] Giving the examples of thematic preconditions, Holton had already spoken about the scientific picture of the world and had called its thematic categories "hierarchy/reductionism – integrity/holism", "vitality – materialism", "evolution – statism – regress".

One can evaluate as positive the striving of Western philosophy of science latterly to include into the arsenal of methodological analysis the new categorical means. But together with this let us note that no clear differentiation between the notions of the "picture of the world" and the "scientific picture of the world" yet has been made.

In Russian philosophical-methodological literature the term the "picture of the world" is applied not only for designation of the worldview, but in a more restricted sense, when talking about scientific ontology, i.e. those conceptions of the world which are a particular type of scientific theoretical knowledge.

The scientific picture of the world appears in this meaning as a specific form of scientific knowledge's systematization, giving a vision of science's subject world according to the definite stage of its functioning and development.

This sense of notion of the "picture of the world" was not noted right away. Only as philosophical-methodological reflexion on scientific activity was developing, did the ability appear to fix some integrative system of conceptions of the world as a particular science's component. This system is worked out as a result of knowledge's synthesis, which had been obtained in different fields of scientific investigations, and afterwards this system acquired the name of the scientific picture of the world.

With the appearance of science and the gradual increase of its influence on social life the worldview meanings began their formation in many respects under the scientific picture of the world's impact. This picture is beginning to appear as a component of the scientific worldview which in many aspects makes the investigator's activity purposeful. This component fixes only the one block in the worldview, which represents knowledge about the world's structure which was received at some stage of historical development of science. And as the scientific picture of the world appears only as the worldview component, there is no reason in this sense to speak about coincidence of the worldview and the scientific picture of the world. But at the same time it is impossible to draw a tough demarcation line between them. It is better to talk about interrelation between the worldview and the scientific picture of the world. One may note that prominent naturalists, comprehending science history, faced this problem. For example, V. Vernadsky paid sufficiently great attention to analysis of interrelation between the scientific picture of the world and the scientific worldview. He underlined that scientific worldview, which certainly includes as a component the general-scientific picture of the world and also its philosophical foundations, was developing in a close interaction with all aspects of society's spiritual life. The fruitful attempt was undertaken in Vernadsky's works to track the mutual influence of scientific world view and different forms of spiritual life which is a necessary nutrient medium for developing science.

Enough stable dependence of scientific conceptions of the world (the scientific picture of the world) from wider fields of culture, in which science functions, and backward influence of science in other spheres of contemporary culture, have been noted by the other naturalists. E. Schrödinger carried out the analysis of interrelation between the picture of the world which was introduced in quantum-relativistic physics, and the culture of contemporary technical civilization. The latter appeared as a tendency to purposefulness of subject forms, simplicity, "predilection to deliverance from traditions" as the expression of social life dynamism, "mass governance methodic oriented to search for the invariant within a set of possible decisions", etc.[8]

This aspect of mutual influence of the scientific picture of the world and the worldview's structures, forming a fundament of technogenic culture, is extremely actual, because it allows to concretize the problem of internal and external scientific development's factor's correlation. And, what is especially important, not only in separately taken sciences but in science as a whole, the periods of intensive reorganization of scientific worldview take place together with quiet states.[9]

At least the following interrelated aspects may be distinguished in the worldview itself: axiological, epistemological and ontological.

Scientific picture of the world can have a significant influence on formation of the worldview's ontological components. It is clear, this relates only to particular types of cultures and civilization development. In traditional civilizations science did not make significant influence on dominated worldview structures. An influence like this is peculiar only to non-traditional societies that have begun the ways of technogenic development.

The scientific picture of the world interacts with the worldview structures, which are forming culture's fundament, both directly and indirectly through the system of philosophical ideas, which appear as a rational explication of corresponding worldview meanings.

Thus the problem of correlation between the scientific picture of the world and the worldview transforms into the problem of interrelations between scientific, philosophical picture of the world and basic worldview images of culture.

To discuss this problem it is necessary to previously specify the appropriate notions. In the beginning it is purposeful to concretize the notion of the worldview as an integral image of the human world, and to clear up its correlation with the system of world's conceptions creation in philosophy. As this theme has been discussed intensively enough in our philosophical literature in recent years,[10] I will only briefly reproduce the most important results, which are related to the problem raised.

The categories the "world" and the "human" are the fundamental categories of the worldview. They can be concretized through the system of categorical meanings of other culture universalities expressing human attitudes to nature, society, other people and to himself (meanings of categories "nature", "space", "thing", "attitude", "Myself", "others", "freedom", "conscience", and others). All these worldview categories always have the sociocultural dimension and in many aspects determine the character of human vital activity and consciousness at this or that historical stage of social development.

Categorical structures of the worldview determine the mode of the world comprehension and understanding by a human. They specify an integral image of a human vivid world, a picture of this world. And if at the early stages this picture had the anthropomorphic, mythological character, the worldview along with emergency of philosophy acquired the status of theoretical nature.

Philosophy exactly constitutes the theoretical core of the worldview. Carrying out the reflection under the culture universalities, it reveals and expresses them in logical-conceptual form as philosophic categories. Operating them as with particular ideal objects, philosophy is able to construct new meanings and thus new categorical structures.

As the result of analysis of correlation between philosophy and the world-view the new meanings of the notion the "picture of the world" are revealed. Philosophical cognition also strives to build such a picture, explicating and developing senses of culture universalities in a form of philosophical categories. But the actual worldview structures represented with a network of culture's categories, and their philosophical explication are not identical. Philosophy as theoretical core of the worldview not only schematizes the images of the world represented by meanings of culture categories, but constantly invents the new non-standard conceptions going out of the framework of these images.[11]

As a result the analytical differentiation of problem of interrelations between the worldview, philosophical world's images and the scientific picture of the world takes place.

Science experiences the influence of philosophical principles and regulations from the beginning of its formation and in its development. In our time, philosophers of different orientations recognize their value and heuristic capacity for the development of scientific knowledge.

Although a series of investigators in Western philosophy noted the philosophical idea's productivity in the development of scientific knowledge, nevertheless, the mechanism of this influence had not received sufficient basis in their investigations. In this respect the results obtained in Russian philosophical literature look more preferable. It is related in many aspects with the revealing, on the one hand, of a particular layer connecting the worldview and philosophy, and concrete-scientific knowledge on the other. The scientific picture of the world appears just as this layer in relation to the system of ontological conceptions. Yet in the 1960s, investigating the mechanisms of philosophy's influence on physical knowledge formation on the material of physics, M. Mostepanenko had emphasized that a particular intermediate existed between physical theory and philosophy. Through this intermediate, on the one hand, philosophy influenced physics, and on the other hand, physics influenced philosophy. This intermediate is the "system of physical conceptions and notions called the physical picture of the world".[12] V. Chernovolenko developed the analogous point of view. In his opinion, the "scientific picture of the world is such horizon of knowledge systematization, where theoretical synthesis of results of concrete sciences' investigation with knowledge of the worldview character occurs. The latter represent integral generalization of aggregate practical and cognitive experience of humanity. The scientific picture of the world is joint both with theoretical systems of lesser generality extent (concrete sciences, generalizing theories of natural science, etc.) and with maximally wide form knowledge and experience systematization which is called the worldview".[13]

The scientific picture of the world always bases on definite philosophical principles but they do not exhaust and surrogate it yet. The mode of generalization and synthesis of most significant scientific achievements form this picture inside of science. Philosophical principles orient this synthesis process and substantiate the results obtained therein.

The scientific picture of the world can be viewed as a form of theoretical knowledge, representing the science subject of investigation according to a definite historical stage of its development. The concrete knowledge obtained in different fields of scientific search are integrated and systematized by means of this form.

As the different levels of knowledge systematization exist in a framework of the scientific picture of the world, three major types are distinguished. Accordingly, one can point at three major meanings in which the notion of the "scientific picture of the world" is applied when characterizing the processes of science's structure and dynamics. First of all, it designates the particular horizon of systematization of knowledge obtained in different sciences. In this meaning we talk of a general scientific picture of the world which appears as an integral image of the world including the conceptions of nature as of society. Secondly, the term the "scientific picture of the world" is applied to designate the system of conceptions of nature, which are formed as a result of the synthesis of achievements of natural science's disciplines. In an analogous way this notion may denote an aggregate of knowledge obtained from humanitarian and social sciences. Thirdly, the horizon of knowledge systematization in a separate science is denoted by it, fixing an integral vision of the given science's subject. This vision forms at a definite stage of this science history and changes when in transition from one stage to another. In accordance with the noted meanings the notion of the "scientific picture of the world" is splintered in the row of interrelated notions. Each of these notions means a particular type of the scientific picture of the world as a specific level of scientific knowledge's systematization. These are the notions of general scientific, natural scientific, social, and, finally, local (special) picture of the world.

In the latter case the term "world" is applied in a particular, narrow sense as the world of a separate science ("world of physics", "biological world", etc.). In this connection the term "picture of the reality under consideration" is also applied in Russian literature to designate the disciplinary ontology. Under the "reality under consideration" a universal set fragment or aspect is understood which is studied by methods of corresponding science and which forms a subject of its investigation.

Each of these types of scientific picture of the world experienced the impact of the worldview structures at the different stages of science functioning, and together with this made its contribution in their formation and development.

The worldview can influence the "scientific picture of the world" development both directly and indirectly through the philosophy which subjects the worldview categories to reflection.

Interrelation between the worldview, philosophy and the "scientific picture of the world" fixes the infrastructure of the developing knowledge system, which determines strategy of investigations and includes their results in culture. At the same time the scientific picture of the world belongs to the inner structure of science represented by the interrelations between theoretical and empirical knowledge.

Historical evolution of the notion of the "scientific picture of the world"

The notion of the scientific picture of the world was introduced into the composition of conceptual apparatus of philosophy and methodology of science in many respects due to investigation of mechanisms of formation of scientific theories and empirical facts, taking into account the processes of scientific knowledge differentiation and integration. Scientists-naturalists and philosophers made their contribution to the elaboration of this notion.

The important incentive to analyzing the scientific picture of the world's place and functions were the revolutionary improvements of natural science on the boundary of the 19-20th centuries, when the problem of choice and justification of physics' ontological postulates was set keenly enough. As one of this problem's aspects, the question of ontological status of fundamental abstractions arose. Investigators perceived the latter as adequate reflection of the objective reality's fragments. A number of such abstractions (indivisible atom, world's ether, absolute space and time) turned out to be idealizations that had a limited field of application. Therefore it was necessary to clarify to what extent do physical notions express the essence of investigated objects and processes.

Different approaches have existed for considering the problem of notion's ontological status and scientific conceptions. In a classical epoch the majority of naturalists shared the point of view according to which the complete correspondence of fundamental notions confirmed by experience to the external world's elements had existed. Supposing that experimental confirmation of fundamental abstractions allowed the discovery of all attributes of these abstractions in reality itself, that guaranteed exact and exhaustive reflection of essence of the investigated processes in science. Yet, in the second half of the 19th century this position had been undermined by a series of facts. It became clear, for example, that abstractions of phlogiston and thermogen, allowed for the time being to describe and to explain the experience, did not have correlates in nature, although earlier they had been identified with particular substances. Revolution in science in the $19–20^{th}$ centuries revealed the limitation of mode of thought at which fundamental scientific abstractions appeared as definitive and invariable and demonstrated flexibility and variability of scientific notions.

Discussion of the problem of correlation between science's fundamental notions with the investigated reality led to discovery of important characteristics of the scientific picture of the world. Thus, M. Planck insisted that the objective picture of the world construction was an ideal of natural science, and set a question: what is that we call the physical picture of the world? Is the picture of the world only more or less a spontaneous creation of our mind, or, vice versa, are we obliged to admit that it reflects the real, absolutely not dependent on us natural phenomena?[14] From this point of view, the striving to find a constant scheme, not dependent on the change of times, is characteristic of natural scientific investigation. And in this sense yet the contemporary picture of the world, although glittering with different colors depending on investigator personality, nevertheless contains in itself some features which whatever revolution in nature or in the world of human thought can erase anymore. This constant element, not depending on any human or even on any thinking individuality, constitutes what we call the reality.[15]

Planck emphasized that change and development of the scientific picture of the world did not abolish these constant elements but preserves them, adding new elements to the existing ones. In this way the succession in development of the scientific picture of the world and the increasingly detailed reflection of the world in scientific cognition are carried out.

The presence of elements in every picture of the world corresponding to objective reality allows identifying this picture at once with the world itself. Ontologization of the picture of the world, according to Planck, has vital importance in the process of scientific creative work. He mentioned that outstanding investigators (Copernicus, Kepler, Newton, Huygens, Faraday) had made their discoveries, only due to the fact that the basis of all their activity was stable confidence in actuality of their picture of the world.[16]

Together with this, the change of physical pictures of the world shows that not all their elements may be compared with objective reality. In this relation new questions arose: what were the arguments for ontologizing our conceptions of the physical world, how were the ascription of the picture of the world elements to objective reality carried out? Planck did not formulate these questions in an explicit view, but the determined preconditions for their setting were laid out in his works. Further discussion of the given problematic required consideration of physical knowledge in a particular aspect. This aspect was from the point of historical development of conceptual scientific means and their role in empirical and theoretical investigation of physical objects. A great work in this direction was made by A. Einstein in connection with the notion of the "physical reality". The term "physical reality", introduced by Einstein into physics methodology for designation of the physical cognition basis, had several meanings. As a minimum, one can point to two main Einstein interpretations of this term. In the first meaning he used the term "reality" to characterize an objective world existing outside and independently from human consciousness. Einstein mentioned that trust in the external world existence that is independent from perceptive subject, lies in the foundation of all natural science.[17] However, the way we perceive the world under investigation and how the structure of this world is seen by us depends on the level of cognition and practice development, and from the system of conceptual means applied when describing the world.

With their assistance we single out some aspects and structural characteristics of the objective world and build a theoretical conception in which the world is reflected, simplified and schematized. In such an approach at different stages of science development, investigators may uncritically identify the conceptions of the world with the world itself. So when analyzing physical theory it is necessary to take into consideration the distinction between objective reality, which does not depend on any theory, and those physical notions with which theory operates. These notions are introduced as the elements which must correspond to objective reality and with the assistance of these notions we imagine this reality.[18]

Here Einstein came to the second aspect of the physical reality consideration. The term "physical reality" is used in this meaning for "consideration of a theorized world as an aggregate of theoretical objects representing the properties of the real world in a framework of given physical theory".[19] In this plan the "physical reality" is given by means of scientific language with the assistance of which a physician comprehends the essence of investigated objects. But the same reality can be described with the assistance of different language means.

Einstein considered this circumstance and fixed the difference in reality description on empirical and theoretical levels of scientific cognition. Corresponding to that he noted the distinction in vision of the physical world at the different levels of its cognition. Einstein talked about the different pictures of the physical world: the picture of physicist-experimentalist's world and the picture of the physicist-theorist's world.

Making a comparison of these pictures of the world, he gave preference to the picture of the physicist-theorist's world. He did so on this foundation that "due to the usage of mathematics this picture satisfies the highest requirements with respect to strictness and accuracy of expression of interrelations"[20] and that this picture exposes the regularities of the physical world. But, talking about the picture of the physicist-theorist's world, Einstein did not make a detailed analysis of theoretical language itself. In a framework of this language's system he did not mark those statements, which could represent the picture of the world, in distinction from separate theories which were connected with it. And he did not set a question about distinction between theory and the picture of the world in this system. Einstein applied the notion of the "physical picture of the world" in different senses. Together with the senses yet to be noted, he talked about the picture of the world as "a minimum of primary notions and correlation of physics which provide its unity". Evidently, this sense is closer to the characteristic of the physical picture of the world as a particular component of scientific knowledge, which differs from concrete physical theories and at the same time unifies them providing their synthesis. However, we do not find in Einstein's works the stricter definition of the physical picture of the world taken in this meaning. He distinguished the picture of the world from theory, most likely, at the level of methodological intuition.

On the heels of Planck, Einstein emphasized that every picture of the world simplifies and schematizes the reality, but at the same time it reveals some essential features of reality. It allows identifying the picture of the world with the world itself to the definite moment (until the investigator discovers the new, earlier unknown aspects of reality). "A human strives to create by some adequate mode the simple and clear picture of the world within himself, to try fairly to replace this world with the picture created by this mode".[21]

The ideas of the schematizing role of the physical picture of the world were noted by many creators of contemporary physics (N. Bohr, M. Born, W. Heisenberg). They considered the physical picture of the world development as a result of discovery in a cognitive process of the new properties and aspects of nature, which had not been considered in the previous physical picture of the world. In this case insufficiency and sketchiness of previous conceptions of nature was clearly discovered, and they were reconstructed into the new physical picture of the world. N. Bohr had written that Planck's discovery, which was saying that all physical processes were characterized by features unusual for the mechanic picture of nature discontinuity, revealed the fact that the laws of theoretical physics were idealizations. And these idealizations were applicable to the event descriptions only when the magnitudes of action dimension participating in them were sufficiently large to neglect the quantum's value. At the time when this condition is performed with a large reserve in events of usual scale, we face the regularities of quite another type in atomic processes.[22] Exactly this circumstance required rejection from the mechanical picture of the world. M. Born, generalizing the experience of physics historical development, noted that every physical picture of the world has its limits but until the consciousness has faced the outside world obstacles, these borders can not be seen. They are discovered by physic development itself, by revealing the new facts displaying the action of the new natural laws.[23] Discovery of such borders of the previous picture of the world leads to widening and deepening of knowledge and opens the new ways to studying nature.[24]

Classics of contemporary natural science showed that to create each new picture of the world, as a rule, the elaboration of definite categorical apparatus is required. This categorical apparatus acts as a base of its kind on which the scientific picture of the world is created. Thus, N. Bohr and A. Einstein emphasized that the mechanical picture of nature was based on notions of indivisible corpuscle, absolute space and time, and Laplacian causality. And the physical reality was imagined, after Maxwell, as continuous fields which cannot be explained from the mechanical point of view.[25]

Further physics development, as N. Bohr mentioned, led to the classical picture's modifications. Specifically, general theory of relativity elaborated the new notions, widened our scope with their assistance, and gave to our picture of the world the unity which could not have even been imagined before.[26] It led to an absolutely new picture of the world, modifying its Newton's construction,[27]

Classics of natural science fixed the circumstance that the great revolutions in physics had always been related with reconstruction of the picture of the world. Noting that mechanics creation was a revolution in science, many of them evaluated Newton's conception of nature as the first scientific picture of the world.[28]

Revolution, during which the transition from classical physics to the contemporary was carried out, was also related with radical reconstruction of the picture of the world. Creators of quantum relativistic physics paid much attention to analysis of preconditions which provided reconstruction such as this. In this analysis they picked out the extremely important circumstance that transition to the new vision of the physical world required the changes of deep orientations of physical investigation.

The understanding of dependence of our conceptions about the physical world from the position of cognizing subject in the Universe and from the specifics of its cognitive means, due to which it marks in nature these or those of its objects and links, is expressed clearly in works of A. Einstein, M. Born, W. Heisenberg, and particularly of N. Bohr.

This new mode of thinking appeared as a condition for building the new, adequate to nature, picture of physical reality.

In works of contemporary physics creators the point of view is expressed clearly that changes, which occurred in our understanding of the world owing to the theory of relativity and quantum mechanics, did not mean the introduction of some subjective element into science and refusal from building of an adequate picture of nature. They meant only collapse of the picture of the world and creation of another representing more thorough understanding of "reality" nature.[29]

Evaluating the statement of contemporary physics from these positions, prominent naturalists pointed out that it represented only one of the steps of evolution of our nature picture and it is necessary to wait, that this evolution would not stop.[30]

Selection and investigation by natural science classics of different aspects of complex and many-sided problems of the scientific picture of the world, were related in general with analysis of the physical picture of the world. By virtue of a prolonged leading position of physics in natural science and owing to fundamental nature of knowledge received in this science, attempts were made repeatedly to explain from positions of the existing physical picture of the world such appearances, which did not relate to the subject of physical science. But the physical picture of the world did not contain in itself all knowledge about the world, therefore it could not give the adequate interpretation of all natural phenomena. A situation such as this required the introduction of another world vision, a particular picture of it (irreducible to the physical one), which contained the conception of those objects also, which did not include in physics the subject of investigation.

This aspect of the problem was analyzed sufficiently in detail by V. Vernadsky, N. Wiener and M. Born.

Thus, Vernadsky considered the physical picture of Cosmos only as one of the ways of the world's description. The investigator deals with it only with conceptions of ether, quanta, electrons, lines of force, curls and corpuscles.[31] But knowledge about the world must not be confined only by the knowledge about fragments obtained with assistance of these physical notions. The world around us has a huge variety of appearances and an important place in it belongs to a particular element. This element is the element of the natural, which the physical picture of the world does not describe. Therefore, according to V. Vernadsky's opinion, together with the physical, the "naturalistic" conception of the world exists ("the naturalist's picture of the world"), which is "more complicated and closer and actual to us, that is still associated not with the whole Cosmos, but with its part that is our planet, the conception of environment that every naturalist studying the describing sciences possesses. The new element, which is absent in constructions of cosmogonies, theoretical physics or mechanics, is always included in this conception. This is the element of living substance".[32] Actually, Vernadsky fixed clearly enough one of the types of scientific picture of the world – the natural science picture of the world – as a particular form of systematization and synthesis of knowledge obtained in sciences of natural-science cycle.

One can find in his statements the important idea that there are also the foundations about the general scientific picture of the world, which organically joins the conceptions of inorganic matter's development and the conceptions of biological and social evolutions.[33] This arterial way of science's development must provide in future the construction of a united picture of the world in which "separate local events are combined together as the parts of a whole. And in the end the one picture of the Universe, Cosmos, in which the motions of celestial bodies and the structure of smallest organisms, and the human societies' transformations are included".[34]

The same ideas were expressed by other prominent naturalists of the 20th century. Thus, N. Wiener wrote about the necessity of such picture of the world building, which would join together the achievements of physics, biology and other sciences.[35]

This integrative picture of the Universe (general scientific picture of the world) was considered by naturalists as the scheme of the world.

"In the 20th century a human tried again, based on data which natural science had accumulated by our epoch, to create the general picture of the world, but of the extremely schematized and simplified world".[36] Thus, the idea that our picture of reality was only an approximation to the objective world and that this picture contains relatively true conceptions about it, was developed by the classics of natural science not only in relation to the physical picture of the world but also to the general scientific one.

Considering the general picture of the world as the reality's schematization, eminent naturalists noticed that together with facts of science, some other extraneous features, which certainly could not be referred to scientific facts, might be included in it. These extraneous features "sometimes represent the real "fictions" and simple "prejudices" which disappear from the scientific picture of the world as time goes by. But at the definite stage they can assist science's development because they stimulate setting of such tasks and questions, which serve as a kind of scaffolding of scientific building: they are necessary and inevitable when this building is being created, but then they disappear without a trace".[37]

Thus, the methodological analysis of science history during the period of transition from the classical natural science to the contemporary, carried out by eminent naturalists of the 20th century, revealed a series of important characteristics of the picture of the world as a particular form of knowledge, which brings together a variety of the most important facts and most significant theoretical results of science.

First of all, it was fixed that fundamental notions and fundamental scientific principles generate the picture of the world. Their system introduces an integral image of the world in its main aspects (objects and processes, nature of interrelation, and spatial-temporal structures).

Secondly, the important characteristic of the picture of the world is its ontological status. The idealizations (notions) of which it is composed are identified with reality. The true knowledge moment contained in these notions is a basis for this. Together with this, such identification has its limits which are revealed when science discovers objects and processes not beyond the frameworks of idealized admissions implicitly contained in the picture of the world. In this case science creates the new picture of the world considering the particularities of new types of objects and interrelations.

Thirdly, in classics' methodological generalizations the important question was set about correlation between disciplinary ontology, such as the physical picture of the world with the general scientific picture of the world elaborating as a result of interdisciplinary synthesis of knowledge.

Unfortunately, all these important methodological results were not assimilated by Western philosophy of science for a long time. The cause of this was dominance of positivistic attitudes of methodological analysis. These attitudes entered extremely narrow idealization of scientific knowledge, considering it outside of links with practical activity and culture. Knowledge also was analyzed outside consideration of historical development of means and modes of scientific investigation. The separately taken scientific theory and its correlation with experience, but not the system of scientific theories and disciplines interrelating in a process of science historical development, were selected as a primary unit of methodological analysis. In such an approach it was extremely difficult to fix the scientific picture of the world as a specific form of knowledge because it was discovering, just when analyzing the processes of intra-disciplinary and interdisciplinary synthesis of knowledge, knowledge relation to the reality under consideration (the ontological capacity problem), links of empirical and theoretical knowledge with philosophy, the worldview and culture.

Only after the collapse of positivism and critical overcoming of its principles were definite preconditions created in Western philosophy of science for investigation of the scientific picture of the world. Those preconditions were the sufficiently substantiated refusal from positivistic requirement to eliminate the "metaphysical principles" from science language and the recognition of philosophy's heuristic role in the development of scientific knowledge; the knowledge analysis, taking into account its history, refusal to consider knowledge only from the side of its formal structure and of studying its substantial aspects including general cultural and philosophical determinants; choice of series of scientific theories in their relation to metaphysical statements as a primary unit of methodological analysis. As a result of this the means of methodological analysis were significantly broadened and certain steps were made towards the studying of superior forms of systematization of knowledge to which the scientific picture of the world also belonged.

The most significant improvements in investigation of highest norms of knowledge's systematization forming deep scientific structures were established in conceptions of T. Kuhn, I. Lacatos, G. Holton, and L. Laudan. The truth is that the scientific picture of the world as a particular form of knowledge was not fixed evidently in any of these conceptions, but some elements of science's foundations, functionally coinciding with this form of knowledge were described in postpositivistic researches. So, in Kuhn's conceptions the key notion of paradigm was determined in the beginning as "...some accepted examples of actual scientific practice – examples which ... provide models from which spring particular coherent tradition of scientific research".[38] Subsequent attempts to concretize this notion led to fixation of a particular block marked by Kuhn as "metaphysical parts of paradigm".[39] He understood them at least in two senses: as philosophical ideas participating in forming of scientific knowledge and as principles having a concrete-scientific character and lying in foundations of scientific theories. In the latter case the matter is, in essence, about the system of ontological postulates constituting the scientific picture of the world.

If one takes into consideration that "metaphysical parts of the paradigm" really belong to the deep structures of science and its foundations, even their preliminary fixation could stimulate new task setting that is a more detailed analysis of science's foundations. If one differentiates the knowledge block which Kuhn marked as "metaphysical parts of the paradigm" and singles out the scientific picture of the world, distinguishing it from philosophical foundations of science, the paradigm's functions fixed by Kuhn should be referred also to the scientific picture of the world. The scientific picture of the world appears as such vision of investigated reality, which determines a set of admissible tasks and orients when choosing the modes to solve it.

The idea of anomalies and crisis as the preconditions to changing the paradigm is very important to Kuhn. If one considers the development of the scientific picture of the world in concordance with this idea, the problem arises of mechanisms of correlation between empirical facts and concrete theories and distinguishing between two types of situations: when facts and new theoretical corollaries are coordinated with the picture of the world and when mismatch arises between them expressed in accumulation of inexplicable facts and appearance of paradoxes.

Thus, in spite of the insufficient preciseness and insufficient differentiation capacity of Kuhn's analysis of knowledge dynamics, there was in it the hidden positive content which was necessary to assimilate when investigating the structure and dynamics of science's foundations and the scientific picture of the world as their most important component.

Analogously we should have regard to the conception of "research programs" by Lacatos. The main notion of his conception was polysemantic as the notion of paradigm. Under "research program" Lacatos, for example, appreciated a concrete theory such as Sommerfeld's theory of the atom. He also talked about the Cartesian and Newtonian metaphysics as two alternative programs of mechanics construction. Finally, he wrote about the science as a whole like the global research program.[40] However at the same time the problem of revealing the hierarchy of research programs of science was hidden in polysemy and uncertainty of initial term. But to do this the more greatly differentiated analysis of scientific knowledge structure was necessary than the one that had been represented in Western philosophy of science.

If we were to apply the characteristics of research programs marked by Lacatos to analysis of the scientific picture of the world, they would allow the revealing of its new functions in dynamics of science. Firstly, the consideration itself of the picture of the world as a research program includes a specific content (which was noted also in Kuhn's conception) which means that the picture of the world must determine the frame of admissible theoretical and empirical tasks and the choice of means for their solution.

Secondly, the specific feature of a rigid program's core to preserve oneself at the expense of protective hypothesis's belt, even in conditions of its mismatch with facts, was noted in Lacatos' conception. This circumstance throws light on well-known situations where even the appearance of paradoxes to explain the new facts does not bring about refusal from the previous picture of the world, but stimulates the attempts to explain facts at the expense of engaging in an additional hypothesis.

Thirdly, the specific feature, marked by Lacatos, of majority of research programs' development presupposing its competition, allows us to clarify the important aspects concerning reconstruction of pictures of reality under consideration (special scientific pictures of the world). It requires us to pay attention to existence of pictures of reality that are often alternative to each other. Their competition characterizes development of science at the stage of scientific revolutions.

When investigating the transformation processes of the scientific picture of the world, the problem of succession in development becomes important. This problem was not considered by I. Lacatos and, in essence, was eliminated by T. Kuhn, who interpreted the changing paradigms as Gestalt-switching.

G. Holton made the significant contribution in solution of this problem. He considered the history of science as translation and the meeting of different thematic ideas (themes), which were realized through the categorical structures, principles and concrete knowledge about the appropriate subject area and the methods of its investigation.[41]

In the content of these themes G. Holton separately noted the fundamental ideas of structure of investigated reality as like ideas of atomism, conceptions of space and time, principles of Laplacian and quantum mechanical determinism, principles of evolution of organisms and species[42], etc. Taking into consideration that ideas, principles and conceptions such as these constitute the scientific picture of the world, in Holton's conception, in essence, the succession was revealed accompanying the change of scientific pictures of the world. In this point Holton's conception had something in common with ideas expressed by classics of natural science, which noted the assimilation of elements belonged to objective content in the process of historical evolution of the scientific picture of the world.

A series of interesting ideas regarding dynamics of deep research traditions of science can be found in L. Laudan's conception.

Analyzing a science as a historically evolving process, he consecutively developed the idea of theoretical weight of scientific problems. Their field is determined by theoretical vision of the world which, according to Laudan, represents the most important aspect of research tradition.

From his point of view, the history appears as the history of becoming, functioning and changing of research traditions.

The notion of research tradition in its semantic content is close to Kuhn's "paradigm", Lacatos's "research program", and Holton's "theme". Laudan noted the ontological admissions as the essential component of scientific tradition. This is the particular layer of knowledge which in many aspects coincides in its functions with characteristics of the scientific picture of the world.

According to Laudan, science in a greater extent deals not with facts, but with problems whose solution depends on accepted methodological and ontological norms. They are formed based on the theoretical vision of the world and are the assumptions about an essence of reality under consideration as about the methods of theories' construction and checking. These assumptions form a definite research tradition which represents a "series of ontological and methodological "permissions" and "prohibitions".[43]

If we differentiate between the methodological and ontological norms, which conceptions Laudan developed, in their system the collection of ontological principles may be singled out, which sets a conception of reality under consideration (the picture of reality under consideration).

From these positions many characteristics of research traditions considered by Laudan can be applied to analyzing the scientific picture of the world.

So, to Laudan's opinion, a stable invariant presents in research traditions which does not permit the varying principles to modify the previous tradition. Together with this Laudan noted that "there was no such research tradition in the history of scientific reflection which had been characterized by a permanent series of principles during all its evolution".[44]

These ideas become important when understanding the particularities of the scientific picture of the world's evolution. Their changing is a condition of the scientific progress, but in their content some objective knowledge can always be found which cannot be eliminated on the next stages of its historical evolution.

Further, Laudan noted a particular role of anomalies in rational assessment of a theory and besides, from his point of view, anomalies are not only reduced to contradictions between theoretical knowledge and its empirical foundation.

Extending the class of anomalies he introduced the notions of a conceptual anomaly and of a conceptual problem, which was formed, on the one hand, between knowledge and methodological attitudes, and on the other hand, between knowledge and the worldview. And in the latter case this contradiction existed not as much within "a framework of science as between science and extra-scientific views".[45]

These ideas of Laudan allow us to consider the functioning of the scientific picture of the world in a broad context of its sociocultural determination, when its evolution may be represented as the one implementing not only at the expense of interaction between theoretical knowledge and newly discovered facts, but also at the expense of links with worldview structures dominating in the culture of this or that historical epoch.

All these results obtained within a framework of Western philosophy of the latest decades, pertinent to science's structure and historical dynamics, were assimilated and developed in Russian methodological investigations. And moreover, many ideas here were formulated independently and received a more detailed elaboration.

The studying of scientific knowledge structure and dynamics in Soviet methodological literature of the 1970-80s led to revealing the series of components and structures which had not been analyzed in Western philosophy. Precisely in those investigations' framework was the question about position and functions of the scientific picture of the world in a system of theoretical and empirical knowledge and about its role in forming the new knowledge.[46]

After the scientific picture of the world was fixed as the form of knowledge's systematization, which mediated the influence of philosophical categories and principles on concrete scientific theories, the question had arisen about its relation to theory and experience and the mechanisms with which the scientific picture of the world influenced their formation.

Originally these mechanisms were considered based on the material of the history of physics. The following scheme of interactions between picture of the world and theories and experience was offered (the works by M.V. Mostepanenko). Based on the productive philosophical ideas and the new facts considering the picture of the world was created in science (in the considered case, the physical one), which represented the "ideal model of nature including the most general notions, principles and hypotheses of physics and those ones characterizing a definite historical stage of its evolution".[47] This picture sets the targets to the construction of theories. Each new theory is based on the picture of the world corresponding to it. For example, mechanics building was preceded by the appearance of a series of the mechanical picture of the world's fundamental notions, such as force, gravitation, inertia, mass, etc. Under pressure of new facts and theoretical results the created picture of the world may be further built and broadened. However, such situation is possible when the limits of broadening will be exhausted and then the old picture of the world begins to hinder the evolution of science. In this case the necessity arises to reconstruct the existing picture of the world itself. And here the philosophical ideas and principles are starting to play the active heuristic role.

Some real, specific features of physical knowledge's dynamics found their reflection in the described methodological scheme but there were many weak spots in it. Their discover in a process of critical analysis brought about the shift of problems and to setting the new research tasks. In this way the limitation of conceptions that the scientific picture of the world always forewent the theories and was a condition of their forming had been fixed. This situation was endorsed only by classical physics material, but in an evolution process in contemporary physics the situation occurs when theory starts its creation before the construction of its adequate picture of the world. And only then the building of the new picture of the world starts as the final stage of formation of a theory. For the first time P.Dishlevy paid attention to this specific feature; in his works the other important problem was also set. This was the problem of distinction between the picture of the world and theory. He proposed to distinguish the physical picture of the world and theory by the following attributes. First, by the notions with which the physical picture of the world and the physical theory operate. In his opinion, the notions of the picture of the world are modified substantive order philosophical categories (movement, interaction, causality, etc.), which are transformed into the fundamental physical notions characterizing the physical objects independently from conditions of cognition (substance, particle, field, vacuum). As regards the physical theories then they are based on another conceptual structure. They consist along with the means for explanation of behavior of definite systems of physical objects, the using such means the description of experimental investigations' procedures and results is provided.[48]

Second, the physical picture of the world, when representing the physical world, is distracted from the process of obtaining knowledge; the physical theory includes logical means providing obtaining knowledge as verification of their objective character. Third, and finally, one of the distinctions between the physical picture of the world and a theory is in their different historical destinies. If the appearance of each new theory brings about only specifying the scope of the "old" theories' application, the appearance of the new physical picture of the world is linked either with negation of relevancy of the previous picture of the world, or with attempts to somehow unite these pictures into a whole.[49]

The mentioned attributes contained the series of constructive moments, making clear the correlation between a theory and the scientific picture of the world, but together with this they needed some correction.

First of all this concerns the problem of historical destinies of the picture of the world and theory. The formed fundamental theories are actually preserved as the new fundamental theories appear, but they do not only specify their application field, but, as a rule, change their primary form, and are reformulated many times in the process of science evolution.

In the process of the picture of the world changing, the definite succession always exists between old and new systems of conceptions of reality under consideration. Thus, breaking of the mechanical picture of the world did not cancel the idea itself of atomistic structure of a substance, although it changed the old conceptions of atoms as indivisible corpuscles. When in transition from the mechanical picture of the physical world to the electrodynamic, the conceptions about the interactions (the idea of short-range action became firmly established) were radically modified. But at the same time the conceptions of absolute space and time remained. In the contemporary physical picture of the world the conceptions of the physical object/s typology became significantly broadened, but the conceptions that the specific aggregative states of a matter existed have also been preserved at the contemporary stage. Later the idea of succession in evolution of the scientific picture of the world was tracked not only on the physical material but also on the material of the other sciences, and thus it was substantiated in general.[50]

The distinction between the picture of the world and theory on the assumption of their conceptual structure's particularities also required serious specifications. The greatest degree of generality of conceptual structure of the picture of the world in comparison with concrete theories is expressed in its direct closeness with philosophical categories, though in a definite sense (if we take into consideration that philosophical categories express the universal forms of thinking) any scientific notions function as specific concretizing of philosophical categories. But the major difficulty is that on the notion's level it is impossible to distinguish clearly where the notions of the picture of the world begin and where the notions of theory end, because the conceptual structure of theory always includes some determined notions characterizing the picture of the world. In other words, theory may not be considered outside in relation to the picture of the world because it cannot be formed without using the language which describes the picture of the world.

The picture of the world in a system of scientific knowledge

The new opportunities for resolving the question of correlation between the scientific picture of the world and a theory, were being discovered in a process of analyzing the structure of science at the angle of the ideal object's organization which formed meaning of different types of statements of its language.[51] In this approach the scientific language was often considered as a heterogeneous hierarchically-organized system where the statements were directly formed in respect to the ideal objects representing inside of cognition the real objects, their properties, links and relations. Then the different types of ideal objects which appear as the abstractions characterizing the reality under consideration must correspond to different layers of empirical and theoretical language. All these ideal objects are organized into systems: they form a complicated hierarchical system tracing its roots back to practice.

At the empirical level the subject area under consideration is primarily represented with the structure of real experiments and observation situations which implicitly single out the separate links, being the subject of investigation from the mixed-up multiplicity of connections and relations of reality. Then the same links are fixed by the empirical scheme by means of relations between the empirical objects and the fact-fixing statements formulated relatively to these objects.

The same links are represented in theoretical language by relations of the constructs of local and fundamental theoretical schemes and the formulations of appropriate symbols.

It turns out that the same reality appears in qualitatively specific images and forms of description at the different levels of investigation.

Further cognition moves from the real experiments and observations to their theoretical descriptions, and the language of these descriptions becomes more complicated and specific.

And here the important epistemological and methodological problem arises: what allows us to correlate these different descriptions and models with the same reality under consideration? What connects all these languages of description into an integral system of science language?

The answer to these questions leads to discovering in a system of scientific knowledge the particular subsystem of the ideal objects forming in their relations the disciplinary ontology (the special scientific picture of the world).

It introduces conceptions about the major systemic and structural characteristics of a subject belonging to an appropriate science. Representation of the empirical schemes as of the theoretical ones provides the connection of different patterns of reality represented in these schemes and their attribution to the unified object region.

The most studied pattern of the picture of the reality under consideration is the physical picture of the world. But pictures such as these present in any science as soon as it is constituted as an independent field of scientific knowledge.

The generalized characteristic of the investigated subject is introduced in a picture of reality by means of the following conceptions: 1) about the fundamental objects on which all the rest of the objects investigated in a framework of another science are supposed to be built upon; 2) about typology of the objects under consideration; 3) about the general regularities of their interactions 4) about the space-time structure of reality. All these conceptions can be described in a framework of a system of ontological principles by means of which the picture of reality under investigation is explicated, and which appear as a basis of scientific theories that belong to an appropriate discipline. For example, the principles: the world consists of indivisible corpuscles; their interaction is performed as the momentary translation of forces along straight line; corpuscles and bodies formed from them are moving in the absolute space during the absolute time – describe the picture of the physical world which was formed at the second half of the 17th century and subsequently acquired the name of the mechanical picture of the world.

Analogously, when after the success of Maxwell's theory the electrodynamic picture of the world had strengthened itself in physics and had replaced the mechanical one dominating in science for more than two and a half of centuries, all natural processes in it were described by means of introduction of a particular system of abstractions (the ideal objects). The following elements appeared as these abstractions: indivisible atoms and electrons (atoms of electricity); the world ether, whose states were considered as electrical, magnetic and gravitational forces extending from one point to another in accordance with the principle of short-range activity; absolute space and time.

This picture can be considered as the extremely generalized model of those natural objects and processes, which were the subjects of physical investigation in the last third of the 19th century.

At the expense of attribution to this picture of empirical and theoretical schemes of classical electrodynamics, they acquired an objective status and were perceived as the reflection of nature's characteristics.

The transition from the mechanical to the electrodynamic (the last quarter of the 19th century) and then to the quantum-relativistic picture of the physical reality (the first half of the 20th century) was accompanied by changing of physics, and ontological principles. It was being more radical than ever during the developing period of quantum-relativistic physics (revision of principles of indivisibility of atoms, of the absolute space-time existence, and of Laplacian determination of physical processes).

By analogy with the physical picture of the world one can also single out the pictures of reality in other sciences (chemistry, biology, etc). Among these the types of pictures of the world exist historically changing one another. This is revealed when analyzing the history of science. For example, the image of the world of chemical processes accepted by chemists in Lavoisier's time had little in common with the contemporary one. Only few known chemical elements of today were supposed as the fundamental objects. The series of complicated combinations (for example, lime) which had been attributed to the "simple chemical substances" was added to them. After the appearance of Lavoisier's works the phlogiston had been eliminated from the number of substances such as these but the thermogen had yet to be reckoned in this row. It was considered that interaction of all these "simple substances" and elements, expanding in the absolute space and time, gave birth to all known types of complicated chemical compounds.

Such picture of the reality under consideration at the definite stage of history of science seemed true to most the of chemists. It set purposes as to the search for the new facts as to the construction of theoretical models explaining these facts.

Each of the concrete-historical forms of the picture of reality under investigation may be realized in a series of modifications expressing the main evolution stages of scientific knowledge. The lines of succession in evolution of this or that type of picture of reality (for example, evolution of Newtonian conceptions of the physical world by Euler, evolution of electrodynamic picture of the world by Faraday, Maxwell, Hertz, Lorentz, when each of them introduced new elements into this picture) can be among modifications like these. But other situations are possible. Those are when the same type of picture of the world is realized in a form of competing and conceptions alternative to each other of the physical world, and when one of them wins as a "true" physical picture of the world (the samples of this may be the struggle between Newtonian and Cartesian conceptions of nature as two alternative variants of the mechanical picture of the world, and the competition of two main directions in development of the electrodynamic picture of the world – Ampere-Weber's program on the one hand, and Faraday-Maxwell's program on the other).

Revealing complicated and historically developing organization of the ideal objects of science language allows us to formulate the problem of correlation between theory and the scientific picture of the world in a new fashion. Nowadays it is concretized as the questions about distinction between the picture of the world and theoretical schemes as the core of theory, and the specific features of their interaction.

One may point to two major attributes by which this distinction is fulfilled. Firstly, by the character of ideal objects forming the picture of the world and theoretical schemes, and consequently, by the specifics of language means used when describing the same reality. Secondly, by the scope of envelopment and the character of generalization of the events under consideration.

Theoretical schemes' abstract objects and constructs of the picture of the world are the different types of ideal objects. If the laws are formulated relatively to the first of these, the principles are formulated relatively to the second ones. Theoretical schemes' abstract objects represent idealizations and their inequality is obvious, whereas constructs of the picture of the world, also being idealizations, are ontologized and identified with reality. Every physicist understands that material point is an idealization because nature has no bodies without dimensions. But physicists of the 17-19th centuries, who accepted the mechanical picture of the world, supposed that indivisible atoms actually existed in nature and was its first fundamental building block.

Analogously, the abstractions of a point charge and of vectors of electrical and magnetic strength at a point quite clearly appear as idealizations. But an electron (atom of electricity) represented in the electrodynamic picture of the world as a very small charged spherical body, and the electromagnetic field as a state of ether - all these objects were perceived by the majority of physicists at the end of the 19th century as the real substances, the fragments of nature itself existing independently from human cognition.

Meanwhile, these abstractions, functioning as the elements of the physical picture of the world in the last third of the 19th century, also represented the idealizations not identical to reality, but schematizing it. Their borders were discovered in the development of quantum and relativistic physics. It was revealed that the world ether, as physicists at the end of the 19th century conceived it, was the same invented essence as thermogen or phlogiston. The conception about the pure continuity of electromagnetic field and about the pure discontinuity of electrons also underwent changes – the ideas of corpuscular-wave dualism as of particles as fields were included into the physical picture of the world.

Theoretical schemes, although distinguished from the picture of the world, at the same time are always linked with it. These links are provided by the particular mapping procedures. In the process of these procedures the correspondence is established between the attributes belonging to the ideal objects of theoretical schemes and those belonging to the picture of the world. One can illustrate such correspondence by example of correlation between the core of classical theory of the electromagnetic field with the electrodynamic picture of the world.

ABSTRACT OBJECTS OF A THEORETICAL SCHEME OF MAXWELL-LORENTZ'S ELECTRODYNAMICS	THE CONSTRUCTS OF THE ELECTRODYNAMIC PICTURE OF THE WORLD
Vector of the electric strength in a point	Electrical field as a state of the world ether
Vector of magnetic strength in a point	Magnetic field as a state of the world ether
Vector of current density in a point	Movement of electrons
Spatial-temporal frame of reference	Absolute space and time

Owing to links between the constructs of the picture of the world and the theoretical schemes' abstract objects, they can often be named by one term which acquired different senses in different contexts.

For example, the term "electron" in laws of Maxwell-Lorentz's electrodynamics denoted an elementary point electrical charge. But as a description of an appropriate element of the physical picture of the world it was entered on the basis of "to be an extremely little electrically charged particle which presents in all bodies",[52] "to be a spherical body on the volume of which the electrical charge is assigned uniformly",[53] "to interact with ether in the way that ether remains immovable when electrons are moving".[54] Images of the electron as of the point charge and of the little spherical charged particle ("atom of electricity") corresponded to different ideal objects and different senses of the term "electron".

Description of links between the attributes of theoretical schemes' abstract objects and of the ideal objects, forming the picture of the world, is included in the scientific notions' content as one type of definition. The example of this can be definition of the mass as the quantity of matter in Newtonian physics, because it was supposed that in the indivisible corpuscles (atoms), of which the bodies were built, the matter quantity was preserved in accordance with the attribute of atoms' indivisibility and indestructibility. Scientific notions include the multiplicity of notions and their development is fulfilled as interaction of all types of definitions, including those emerging when theoretical schemes are brought into correlation with the scientific picture of the world.[55]

That is why on the notions level one can not make a clear distinction between the picture of the world and theory, but it can be made taking into consideration the specifics of ideal objects of theoretical schemes and the picture of the world which interrelations between themselves and with experience influence the development of scientific cognitive apparatus in a crucial way.

Procedures of mapping of theoretical schemes on the picture of the world are the obligatory conditions of theory building and provide its further functioning, its application to explanation and prediction of new facts. In the case when theory laws are formulated on mathematical language, the mapping of theoretical schemes on the picture of the world provides their semantic (conceptual) interpretation, and the mapping on situations of actual experience provides empirical interpretation of equations.

Empirical interpretation sets the compounding of links with the experience of magnitudes appearing in equations. But this interpretation alone is not sufficient for theory recognition. It is considered to be incomplete without conceptual interpretation of its mathematical apparatus.

In classical physics these two types of interpretation emerged together because theory was created based on a previously introduced and substantiated picture of the world. In contemporary physics they can be divided in time. This happened, for example, when quantum mechanics was being constructed. The fundamental construct of its theoretical scheme, a "state vector" (ψ-function) had no empirical interpretation for some time, but which later was found by M. Born. But exactly after this the discussions became keener, in which the problems of corpuscular-wave dualism, of electron's nature and the questions of what did ψ-function reflect in physical reality, were discussed. All of them were concerned with problematic of conceptual interpretation and stimulated the evolution of the quantum-relativistic picture of the physical world.

The picture of the world is always characterized with a wider scope of envelopment of investigated events than any separately taken theory. Thus several theoretical schemes constituting the cores of different theories, including the fundamental ones, may be mapped on the same picture of the world.

Thus, the fundamental theoretical schemes lying in the foundations of Newtonian mechanics, electrodynamics, thermodynamics, and Ampere's electrodynamics were related with the mechanical picture of the world. Theoretical schemes of Maxwell-Lorentz's electrodynamics and Hertz's mechanics were correlated with electrodynamic pictures of the physical world. The contemporary quantum-relativistic picture of the world unifies all accumulated varieties of fundamental physical theories, classical and quantum mechanics, special and general theories of relativity, thermodynamics, and classical and quantum electrodynamics.

The special scientific picture of the world (the disciplinary ontology) is indirectly related with experience through the theoretical schemes, but it also has direct links with the empirical level of knowledge.

The situations of experiment in which these or those events are discovered and studied represent types of human activity. To interpret this activity in terms of natural process it is necessary to be viewed as the interaction of natural objects existing independently of a human. The picture of reality under consideration adjusts exactly such a vision. Through relation with it, the situations of real experiment and their empirical schemes acquire an objective status. And when, for example, Biot and Savart discovered in experiments with a magnetic needle and rectilinear conductors with current that a magnetic needle reacts to current, they interpreted this phenomenon as generating the magnetic forces by current. Thereby, when interpreting the results of experiment they applied the conception of the physical picture of the world on existence of electrical and magnetic forces and their propagation in space.

Links with relations that the picture of reality under consideration possessed consist not only in interpreting and explaining the results of experience, but also in that this picture is directly substantiated with the experimental facts.

The main attributes of its ideal objects must necessarily receive an experimental confirmation and this is one of the conditions of their ontological capacity. Even if there is a discourse about the indivisible attributes, for example, about the indivisibility of the atom, or about the absolute space and time in the mechanical picture of the world, one can in principle discover some conditions of experience in which these assumptions have a sense. Over the range of the mechanical influence energies, with which the physics of the 17th–19th centuries operated, it was really impossible to discover the divisibility of the atom.

With regard to the conceptions of absolute space and time, they had foundations in numerous observed facts of studying mechanical motion being evidence of preservation of space and time intervals when passaging from one inertial frame of reference to another. Later it was determined that measuring procedures with the help of clocks and rulers, in the framework of which the characteristics of space and time intervals were fixed, were based on the idealizing assumption of momentary translation of a signal applied by observers when synchronizing the clocks. Such assumption was an idealization which had its foundation in that the speed of passing mechanical processes was significantly less than the speed of light, which was applied implicitly as a signal carrying to observers the information about the clock swing in different frames of references. Owing to this one could neglect the finite velocity of spreading interaction.[56]

The clarification of the place of the special scientific picture of the world (the disciplinary ontology) in the structure of scientific knowledge (its relation with theories and experience) introduces the conception of integral knowledge system in scientific discipline. The special picture of the world appears as a particular link forming the system in multiplicity of theoretical and empirical knowledge which compose this or that discipline (field of science). Only the links of the picture of the world with all types of this knowledge allow the consideration of it as a particular form of their systematization.

Pictures of reality, which are being developed in different scientific disciplines, are not isolated from each other; they interact between themselves. In this connection the question arises: do the wider horizons of knowledge systematization and this knowledge systematization's form's integration regarding the special pictures of reality (the disciplinary ontology) exist? In methodological investigations such forms have already been fixed and described. The general picture of the world, which appears as a specific form of theoretical knowledge, belongs to them. It integrates the most significant achievements in natural, humanitarian and technical sciences. These achievements are such conceptions as the non-stationary Universe and the Big Bang, about quarks and synergetic processes, genes, ecosystems and biosphere, about society as an integral system, about formations and civilizations, etc. In the beginning they develop as fundamental ideas and conceptions of corresponding disciplinary ontology and then are included into the general scientific picture of the world.

And if the disciplinary ontology (special scientific pictures of the world) represent the objects of each separate science (physics, biology, social sciences, etc.), the most important systemic-structural characteristics of subject area of scientific cognition as a whole, taken at the definite stage of its historical evolution, are represented in the general scientific picture of the world.

Revolutions in different sciences (physics, chemistry, biology, etc.), changing the vision of the corresponding science's subject area, permanently give birth to mutations of natural scientific and general scientific pictures of the world, and result in revision of conceptions of reality which were formed earlier in science. However, the link between changes in pictures of the world and cardinal reconstruction of natural scientific and general scientific pictures of the world is ambiguous. It is necessary to take into consideration that the new pictures of reality in the beginning are laid down as the hypotheses. The hypothetical picture passes the substantiation stage and may for a very long time co-exist side by side with the previous picture of reality. More often it is strengthened not only as a result of the prolonged verification of its principles by the experience, but also owing to these principles serving as a base for the new fundamental theories.

Entering of new conceptions of the world elaborated in this or that field of knowledge into the general picture of the world does not eliminate, but presupposes, a competition between the different conceptions of reality under consideration.

The conceptions of the world which are introduced into the pictures of reality under investigation always experience the definite impact of analogies and associations gathered from the different spheres of cultural creative work, including the ordinary consciousness and the factory floor experience of a definite historical epoch.

It is not difficult, for example, to discover that the conceptions of electrical fluid and thermogen, included in the mechanical picture of the world in the 18th century, were being formed in many aspects under the influence of subject images gathered from the sphere of everyday experience and production of an appropriate epoch. It was easier for the common sense of the 18th century to agree with existence of non-mechanical forces, conceiving them in the image and likeness of the mechanical ones (for example, conceiving the flow of heat as the flow of weightless liquid – thermogen falling like a water flow from one level to another and making, at the expense of this, the work such as water does in the hydraulic units). But together with this the introduction into the mechanical picture of the world of conceptions about different substances – carriers of forces – also contained the moment of objective knowledge. The idea of qualitative different types of forces was the first step on the way to acknowledgement of irreducibility of all types of interaction to the mechanical one. It contributed to the formation of the particular, different to the mechanical, conceptions of structure of each type of interaction such as this.

Forming the pictures of reality under consideration in every field of science always proceeds not only as a process having the scientific character, but also as the interaction between science and the other fields of culture.

Science always draws these or those fragments, which are entering into the substance of its pictures of investigated reality, from the field of significant obvious images elaborated in different cultural spheres. The images of the Universe as of a simple machine dominated in development of the mechanical picture of the world in the 17th–18th centuries (the world as clocks, the world as mechanism) having something in common with usual conceptions of object structures of techniques in the epoch of the first industrial revolution.

In contemporary scientific pictures of the world the images emerge increasingly of the self-organizing automatic machine, which appear as a specific appellation to obviousness of technical devices that are the complicated self-organizing systems which are applied in different fields of techniques in the second half of the 20th century.

The combination of heterogeneous, but also inter-consistent substantiation (empirical, theoretical, philosophical, worldview) determines the admission of the special scientific pictures of the world by culture of an appropriate historical epoch and their functioning as the scientific ontology.

The obviousness of conceptions of scientific pictures of the world provides their comprehension not only by specialists in a given field of knowledge, but also by scientists specializing in other sciences, and even by the widely-educated people who are not concerned directly with the scientific activity. When we talk of the scientific achievements' influences on the culture of an epoch, first of all the talk is not about the special results of theoretical and empirical investigations, but about their accumulation in conceptions of the scientific picture of the world. Only in such form can they acquire the general cultural, worldview meaning.

Even if taking the ideas which the historical retrospection allows us to fix as significant to the worldview, many of them in their primary formulation emerged as specialized theses, understood only in the narrow circle of scientists.

Let us take, for example, an assertion: in the formula $ds^2 = \Sigma g_{mn} dx_m dx_n$, the magnitudes g_{mn}, representing the continuous functions of coordinates and determining the metrics of four-dimensional manifold (space-time), at the same time also describe the field of gravitation[57]. This assertion expresses the major physical idea of the general theory of relativity (GTR). But if formulated in thisway, it will not arouse the broad human interest of those who are not concerned with theoretical physics. Only the translation of this statement in the language of the physical picture of the world and its further philosophical interpretation discover the deep worldview senses contained in Einsteinian discovery. Joining the scientific picture of the world and getting the philosophical interpretation, the conceptions of GTR about the mutual correlations between the geometry of the physical space-time and the character of gravitation field begin to confront the common sense comprehension of the spatial-temporal structure of the world. They require the reconstruction of the deep-rooted ideas from the conceptions belonging to Galileo's and Newton's times of the homogeneous, infinite Euclidean space and the homogeneous quasi-Euclidean time of the Universe. These are conceptions which turned into an original worldview postulate of ordinary consciousness through the system of education and training.

Special scientific notions and conceptions can acquire worldview status and then resonate in other cultural spheres only through the procedures of their correlation with the scientific picture of the world, at the same time resulting in its reconstruction.

Thus, it was not only with the theory of relativity but also with all other scientific discoveries which modified the scientific picture of the world and influenced through it the system of worldview attitudes orienting human vital functions.

General cultural meaning of special scientific pictures of the world and possibilities of their understanding by investigators working in the different sciences, appear as the condition of their synthesis into the whole general scientific picture of the world.

IDEALS, NORMS AND PHILOSOPHICAL FOUNDATIONS OF SCIENCE

The ideals and norms of investigation

Let us now turn to analyzing the second component of the foundations of science that represents the ideals and norms of scientific cognition.

As any one activity, scientific investigation is regulated by the definite rules, patterns, and principles which express the ideals and norms accepted in science at the definite stage of its historical development. The value orientations and the aims of scientific activity, and also the general conceptions regarding the ways to achieve these aims are expressed in their system.

One can single out two interrelated "blocks" among the ideals and norms of science: a) properly, the cognitive attitudes which regulate the process of reproduction of an object in the different forms of scientific knowledge; and b) social standards which fix the role of science and its value to public life at the certain stage of historical development, manage the process of communication between investigators, and the relations of the scientific associations between themselves and with the community as a whole, etc.[58]

These two aspects of ideals and norms of science correspond to the two aspects of its functioning: as the cognitive activity and as the social institute.

In Western philosophy of science the analysis of the normative structures regulating scientific activity was originally conducted in the course of discussing the specifics of scientific method, and searching the stable foundations which separate science from the extra-scientific knowledge. The ideal of strict scientific method that must lead to the truth was advanced by Bacon and Descartes. This ideal expressed the claims of scientific reason on autonomy and priority in searching for the truth and on the position of superior judgment regarding different spheres of human activity.

In the classical period of development of philosophy and science this ideal was dominant on the whole, although in philosophy the critical attitude existed relatively to it, represented first of all by the trends of agnosticism and skepticism. In the end of the 19th century and in the beginning of the 20th century the empirical criticism and then the logical positivism interpreted the ideal of scientific character in the manner of requirements of the strict demarcation between science and metaphysics. Correspondingly, the accent was made on seeking such system of norms which could allow us to draw this demarcation and to refine science from the metaphysical statements. As the pattern of building a science, logical positivism offered the formalized systems of mathematics and logic. It was supposed that all other sciences could be reduced to these patterns, only entering small corrections for the empirical sciences related with the experimental verification of their theories.

But as the inefficiency of the proclaimed ideal was revealed, the problems of pluralism of science ideals and norms have arisen. It was revealed that different disciplines had their specifics regarding norms and they were irreducible to the one, previously selected pattern. From another side this problem arose when considering the development of scientific knowledge in the historical context. Such approach was accomplished, as known, in the post-positivistic philosophy of science. T. Kuhn, P. Feyerabend, L. Laudan, and a number of other investigators fixed the historical variability of the ideals and norms of science, existence of competing normative structures in the same historical epoch, which different scientists may hold when creating theories and estimating the empirical facts.

Resulting from this the problem emerged: did the refusal from the positivistic methodological fundamentalism and reductionism mean the transition on the positions of absolute pluralism and relativism? P. Feyerabend's conception is the closest one to this extreme point of view. He, establishing the relativity of any methodological prescriptions and their historical variability, supposed that there were no firm regulations of scientific investigation and the only "rule" the assertion that could be "everything is permitted". But if accepting this point of view it is necessary to recognize that one cannot make any distinction between science and extra-scientific forms of knowledge. Feyerabend was consistent in this respect and upheld thesis about equivalence of science and myth, and about the principal inability to draw the line between them.

Feyerabend fairly emphasized that one could discover in scientific creative work the influence of images, ideas, and the worldview attitudes exceeding the bounds of science. These images and ideas are borrowed from the other cultural fields and often become the impulse to forming of new conceptions, notions and methods in science. Nowadays it is unlikely that any philosopher would call into question that science has no absolute autonomy regarding other spheres of cultural creative work and that it is developing in interaction with them. There is no doubt that in our time science, together with the other cultural spheres (and maybe even more than some of them) has an active influence on people's worldview. And the worldview projection of science presupposes persuasion and propaganda of scientific ideas, which is not necessarily based on reproduction of the whole complicated system of evidence and substantiation owing to which these ideas strengthened themselves in science and entered the scientific picture of the world. Most people orient to the scientific images of the Universe (the conceptions about the Big Bang and emergence of metagalaxy, about quarks and genes, about the evolution of life on Earth, etc.), not because they know all discussions and argumentation regarding that these images acquired a status of substantiated and reliable knowledge, but because they trust science and are convinced of its ability to obtain the truth. In other words, the belief in science plays the crucial role in acceptance of the fundamental conceptions of the scientific picture of the world as the worldview images of the ordinary thinking people.

P. Feyerabend particularly accentuated this circumstance, emphasizing the role of persuasion, propaganda and belief in spreading of the scientific conceptions of the world and their striking roots in culture. But it is still not a foundation for identification of science and myth.

As the forms of worldview knowledge they can have common features but the latter does not eliminate their distinction. Besides, the comparison itself of science and myth already presupposes their preliminary distinction. Feyerabend, of course, drew this distinction intuitively. Otherwise, there is no sense in talking about the similarity of these two phenomena if they are completely identical. In this case they will only flow together in one indistinguishable whole. Feyerabend's position was that he consistently criticized the explication of different attributes of science and myth, showing their insufficiency. And it ought to be said that in this aspect he discovered the actual weaknesses of contemporary methodological investigations, which after refusal from the positivistic ideal of "strict demarcation" could not achieve agreement in determination of the attributes distinguishing science from the other forms of cognition. But the problem of revealing in these attributes the ideals and norms of science and content connected with them does not disappear but only becomes keener. One may blame Feyerabend not for stipulating for himself the position of a critic in respect to suggested solutions of a problem (a position like this can be useful to a certain extent and even necessary, compelling deeper consideration of the problem), but that he generally tried to eliminate it.

Nevertheless, many representatives of the postpositivist philosophy of science, agreeing with the thesis of pluralism and historical mutability of scientific normative structures, do not agree with Feyerabend's position in his radical overthrow of the scientific method.

Following the classification offered by W. Newton-Smith, one can single out two approaches to the problem in the Western philosophy of science. The first of these is targeted at building the rational models of changes in science, including changing of its rules and norms regulating an investigation (C. Popper, L. Laudan, I. Lacatos, J. Agassi, W. Newton-Smith, and others). The second, defending the irrational models of growth in knowledge and changes in science (the most significant representatives of this approach are T.Kuhn and P. Feyerabend).[59]

The first approach evidently recognizes the problem of searching the stable attitudes of scientific rationality in the changeable context of regulative rules and values accepted by the scientific community. But this problem is not rejected by everyone, even in the framework of the second approach. Kuhn's position is characteristic in this respect. He, marking the values as the most important part of a paradigm, implicitly set a problem of how the scientific values were being changed in the epoch of changing paradigms. Discussion of this problem required the differentiated consideration of values. In the works appearing after the well-known book *The structure of scientific revolutions*, Kuhn made an attempt to distinguish the values as maxims adjusting some general strategy of investigation, and the methodological regulations which concretize the values.

Thus, considering the ideal of theoretical knowledge, he marks the following features as the set of values: 1) exactness of a theory (theory consequences must discover an accordance with experiments and observations); 2) the consistency; 3) the broadening field of application (theory consequences must spread by far more than the limits of those facts and subtheories, to which explanation it was originally oriented); and 4) the fruitfulness of a theory (it must discover the new events and correlation, which previously had not been marked).[60] Historical analysis shows that if these criteria are considered as the strict regulative rules, they are not always kept. Copernicus' system before Kepler gave a less exact coincidence with observations than Ptolemy's, though on the other criterion (for example, a simplicity) it excelled. The scientists, as Kuhn emphasized, may differently interpret these values and give greater preference to some of them in comparison with other. The principle of simplicity was interpreted in different ways. The comprehension of the exactness value was being changed. It increasingly accented the quantitative or numeral agreement, sometimes to the detriment of the qualitative one.[61] Before the advent of natural science in the 17th century, as Kuhn marked, exactness in this comprehension had been applied only in astronomy. During the 17th century, however, the numeral agreement criterion had spread into mechanics, and during the 18th century and beginning of the 19th century it had spread into chemistry and the other fields, such as electricity and heat.[62] Kuhn noted that in this connection the interpretation of such value as the broadening of the field of theory's application changed historically. Before Lavoisier, chemists accented their attention on explanation of such qualities as color, density, roughness, and concentrated on explanation of qualitative changes. T. Kuhn wrote that together with acceptance of Lavoisier's theory, explanations such as these lost their value for chemists for some time: the possibility of explanation of qualitative changes had not by now been a criterion relevant to estimation of the chemical theory.[63]

Thus, Kuhn fixed that there is a historically variable content, which is presented in each of the values selected by him, and this set a problem of an invariant, stable content, which corresponds to ideals of scientific character notwithstanding the variability of those ideals themselves. Kuhn acknowledged the possibility of such an approach when he said that changing the criteria of theories' choice did not repeal the definite canons which made a science itself, though the existence itself of canons such as these was still insufficient to be the criterion of choice in every concrete historical situation.[64]

The values which distinguish scientific investigation, according to Kuhn, function not as regulations or criteria which determine a choice, but as the general strategies influencing the choice. And Kuhn saw in this one of the most important characteristics of science, because the joining of the general value attitudes with the concrete norms and regulations, which could change in its historical development, occurred therein.

In an approach such as this the problem of selective analysis of the content of the ideals and norms of investigation, and separation of different organization levels in this content arises. These levels are from the general invariant attributes expressing the essence of scientific cognition and its distinction from other forms of cognitive activity, to the concrete characteristics of norms being accepted by the community at the definite stage of historical evolution of this or that scientific discipline.

In the 1970s–80s in Western philosophy of science, definite steps were taken towards the elaboration of this problem. In discussions about the characteristics of scientific rationality between the adherents of rational models and their opponents, the different variants of such characteristics were proposed. First of all, it is necessary to note the development of tradition ascending to C. Popper's ideas. He advanced the attribute of growth of knowledge, based on permanent criticism and correction of discovered mistakes, as the main characteristic of scientific rationality.[65] The attempts to concretize this ideal were related with the striving for escaping the evident introduction of the notion of truth, taking into consideration the factor of relative truthfulness of knowledge and historical mutability of ideal truth. Replacing the notion of truth of a theory with the notion of its credibility, most representatives of the rational approach to the problem of general characteristics of science and scientific method, content themselves with conceptions of knowledge's growth as the setting and solving of scientific problems mainly at the expense of internal factors. As W. Newton-Smith mentioned, most accepted the rational model, considering the history of science from these positions as, for example, I. Lacatos, who strove to demonstrate that "those changes in science which explanation was originally related to external factors, do not really require these factors for their explanation".[66] This sufficiently strict position was softened under the influence of criticism from the adherents of the irrational models.

Accenting of the historical situations in which cognitive norms were changing required consideration of the influence of external, sociocultural factors. L. Laudan attempted to make this step when he allowed the inclusion into the rational model of knowledge growth the investigation of ways of consensus and dissensus of community relative to the ideals and norms. Ascertaining that a high degree of accordance in respect to basic theoretical principles and methods existed in the developed sciences, Laudan noted that change in basic explaining ideas and rules of scientific search led to mismatching; dissensus which, however, was replaced by consensus. And this circumstance, as Laudan emphasized, connected with formulations and reformulation of a consensus, as a matter of fact was surprising if one takes into consideration that as distinct from religion, science was not based on the dogmatic frame of doctrines.[67] Solution of the consensus problem in early variants of the rational approach was connected with the hierarchic model of substantiation which, as Laudan thought, was advanced as basic in the so-called theory of instrumental rationality (C. Popper, K. Hempel, G. Reichenbach were the most influential adherents of this model).[68] This model was built hierarchically: the factual (the lowest level) – the theoretical (the middle level) – the methodological (rules and norms as the highest level regulating the relationship between theory and facts). Disclosure of historical mutability of methodological rules and norms set the problem of a consensus regarding acceptance by the scientific community of these or those methodological principles. Laudan, from this point of view, modified the hierarchical model. He represented it as the model of consensus of a community at three levels: factual, methodological and axiologic. Here Laudan noted as the factual not only the assertions about the directly observed events, but also the announcements about what went on in the world, including announcements about theoretical and unobserved essences.[69] In other words, he unified the empirical and the theoretical and their interrelations into one level. And discussions regarding what empirical data and facts and also what theories were accepted by a community, Laudan marked as "factual disagreements" and "factual consensus".[70]

The methodological level represents the regulative rules, prescriptions which determine some strategy and tactics of acceptance of theories and facts by a community. As these rules may change historically, methodological discussions exist regarding to those rules themselves.

The axiologic level fixes the fundamental cognitive purposes and values of scientific cognition. Laudan pointed out that in the framework of this modified hierarchical model, corresponding, on the whole, to the classical rational approach, it was supposed that factual disagreements were regulated by the methodological level and methodological disagreements were regulated by the axiological level.[71]

But the historical analysis of science witnesses that in the scientific community disputes can emerge relatively to the comprehension of purposes and values of science. And this circumstance, as Laudan fairly noted, is not taken into consideration in the hierarchical model.

Laudan made some other claims regarding this model. He emphasized that feedback was not taken into consideration and the direct links were interpreted as too rigid dependencies in this model: it supposed that one might not settle the disagreements at the lowest level having no consensus at the highest. Laudan gave the historical examples, indicating that on the different comprehension of methodological principles and rules and on the different interpretation of scientific purposes it was possible to achieve an agreement relative to the factual situations.

On this foundation Laudan rejected the hierarchical model and proposed the "screen model" of scientific rationality instead of it. He emphasized that the screen model differs greatly from the hierarchical because it shows that the complicated process of substantiation runs through all three levels of scientific states. Substantiation flows upwards and downward through a hierarchy, relating purposes, methods and factual assertions.

There is no sense in interpreting any one of these levels as more privileged or more fundamental than others. Axiological, methodological and factual statements inevitably interlace in relations of mutual dependence.[72]

Laudan's reflections about historical mutability and the mutual influence of values represent sufficiently important steps in investigation of the ideals and norms of science. One can conclude from the text of the quoted Laudan's book that he interpreted values and purposes as the ideals of science, and the rules, concretizing it, as the network of norms determining the historically mutable scientific method.[73] The progressive changes in science are expressed not only in creation of new theories, but also in changing the methods and shifts in cognitive values.

And nevertheless, Laudan kept aside the question of the internal structure of the ideals and norms of science and of the possibility, though all their mutability, to select in them that layer of the invariant content which distinguished science from other forms of cognition. In some of his works he addressed the problem of the specifics of science, but the attributes which he selected as the fundamental, were obviously insufficient to characterize the specifics. Continuing the line marked by C. Popper's works, L. Laudan gave the main attention to such characteristic of science as the persistent growth of knowledge presupposing the setting and solution of problems. Defining science as a problem solving activity, he interpreted its historical development as the increase of ability of investigation programs to solve empirical and theoretical problems.[74]

As Newton-Smith noted regarding this, Laudan's assumption was that the development of science might be described in terms of solution of problems, not using the attribute of truth.[75] Newton-Smith concluded that Laudan did not deny the existence of truth but aspired not to use it when analyzing scientific activity, supposing that one could escape the knotty problems, replacing the notion of theories' truth with the opinions about their abilities to solve problems.[76]

However, if to content oneself with only this attribute of science, serious methodological difficulties arise. Newton-Smith, to my mind, earnestly revealed them in the course of critical analysis of Laudan's conception. If one accepts this conception, it would be difficult to answer the question: why are not all problems accepted by science? If someone would like to work, for example, on the problems: why does sugar not dissolve in hot water? why are swans green? why does matter repel? why is the freely moving body accelerated when force is absent? – then, naturally, the question arises whether these problems are correct. Newton-Smith wrote, "the desire appears for noting that these are not genuine problems, because the opinion settled in every case as a question is false, and it is known that it is false".[77] Truth, as he emphasized, plays a regulative role in science, and if refusing that, the prohibitions on the voluntary formulation of problems disappear. But in the practice of scientific activity "theories oriented to solve problems, about which it is known that the statements appropriate to them are false, are rejected extremely for this reason".[78]

Newton-Smith was right when he defended the rational conception of science in which the statements must figure about the truth as the relation of science to the reality under investigation.

Of course, this approach needs to be specified, and, as it seems to me, they can be obtained when analyzing a science as a particular kind of cognition, considered in relation to the requirements of practice. In the analysis conducted above (see chapter I) I have selected two major characteristic attributes of science: the attitude on obtaining the subject and objective knowledge about the world, and the attitude on the growth of this knowledge. These attitudes allow us to leave the frameworks of the subjective structures of the present activity, and open the possible worlds of the future in mastering the practical. With these major attributes those attributes are correlated which express the specifics of means, methods, procedures of scientific activity, and also of the subject of science and scientific ethos. I think that historical evolution of means, methods, research procedures and forms of scientific communication (determining the type and specifics of subject of scientific activity) does not change these two major attributes, which can be considered as the invariant core of the ideal of scientific character. And in principle the different fundamental and anti-fundamental, reductionist and anti-reductionist versions in the contemporary methodology of science somehow or other are forced to take into consideration these invariant features of the ideal of scientific character. In the frameworks of the rational models the criterion of science's ability to solve the problems (Popper, Lacatos, Laudan, and others) appears as a "variation on the theme" of the second attribute (feature), when the attribute of objectiveness and truthfulness, taken as the regulative criterion, is taken into consideration in different conceptions of theories' plausibility (Popper, Newton-Smith, and others).

I would like to reemphasize that the major characteristics' attributes, taken as the invariant of the ideals of scientific character, are expressed in the most clear form in a developed science. They are based in many aspects on values of technogenic civilization's culture, and to some degree support these values. This, of course, does not eliminate the revealing of their preconditions in ancient and medieval cultures which are the genetic cradle of the technogenic civilization culture. And also it does not eliminate the setting of the problem of possibility to reconcile them with some traditionalistic values still featuring the cultures of modernizing societies.

Speaking about the major attributes of science as about the values, I turned my attention on their organic connection with the ethical maxims regulating the relations of investigators in the scientific community (the prohibition on intentional distortion of truth and on plagiarism). In this sense one may say that in the invariant features of the ideal of scientific character the cognitive values connect with the institutional values determining the functioning of science as a social institute.

Of course, it is hopeless and senseless to consider that the invariant exists as itself in the variable mutability of the ideals of scientific character, separately from specific disciplinary and historical demonstrations. So its fixation neither deny the variety of its historical appearances nor the multiplicity of the local ideals of scientific character forming in different scientific disciplines, nor the dependency of the ideals of scientific character on sociocultural values.[79]

In this context the problem emerges of the internal structure of the ideals of scientific character and norms of scientific cognition realizing it. In my opinion, this problem was not set in a clear form in Western philosophy of science and Russian investigators have made a greater contribution in its elaboration.

In Russian philosophical and methodological literature this problem of cognition's ideals and norms started to be discussed sufficiently strongly in the 1970s–1980s, i.e. in the period when keen interest emerged in this problem in Western philosophy of science.

An analysis of the ideals and norms firstly was carried out in respect of investigation of the regulative role of methodological attitudes and principles in the process of theoretical search and forming of new scientific theories (P. Dishlevy, E. Chudinov, N. Ovchinnikov,

V. Kuptsov, and others). At the same time the problem began to be discussed of the selection of theory and functions of the methodological principles in situations of choice (E. Mamchur).

In the second half of the 1970s and the beginning of the 1980s Russian investigations appeared dedicated to the analysis of interaction between cognitive and institutional ideals and norms (N. Motroshilova, A. Ogurtsov, B. Yudin). The consideration of social and cultural preconditions and determinations of the ideals and norms of science was a particular theme which acquired broad recognition and attracted growing circle of investigators. This problem was elaborated the methodological schools in Moscow, Kiev, Minsk, Leningrad, Novosibirsk, and Rostov which had already been formed at that time.[80]

In my works of those years the active and cultural-historical paradigm of philosophy of science was accented. In the course of foreshortened investigations of structure dynamics of scientific knowledge, the tasks emerged before me: to clear up how the ideals and norms of science are built into its structure, what was their internal systemic organization, how did they correlate with empirical knowledge, theories, scientific picture of the world, and what did their historical and cultural dimension consist of.

Searching for answers to these questions led to development and significant concretizing of the conceptions about the structure of the ideals and norms of science and their functions in the system of developing knowledge.[81]

In what follows I will use the results obtained in those years and also their elaboration in the subsequent development of my conception of structure and dynamics of scientific knowledge.

First let us dwell on the problem of structure of the ideals and norms of investigation.

The cognitive ideals and norms of science have a sufficiently complicated organizational structure. The following major forms can be singled out in their system: 1) the ideals and norms of explanation and description; 2) the ideals and norms of proof and justification of knowledge; and 3) the ideals and norms of construction and organization of knowledge. In aggregate they form a specific scheme of the method of research activity ensuring the settling of objects of a definite type.

The ideals of theory and fact, and also the normative principles and rules regulating their formation may be represented as a complex of characteristics assigned to the named major forms. For example, the principles, describing the "solid theory" (the ideal of a theory according to Kuhn), such as broadening of the field of theory application, and exactness, appear as a variant of the ideal of explanation and description, and the simplicity as the expression of the ideal of theoretical knowledge organization.

Science creates different types of schemes of a method represented by the system of the ideals and norms of investigation at the different stages of its historical development. Comparing them, one can single out both general, invariant and particular features in the content of the cognitive ideals and norms.

If the general features characterize the specifics of scientific rationality, the particular ones express its historical types and their concrete disciplinary varieties. In the content of any one of the kinds of ideals and norms of science (explanation and description, proof, substantiation and organization of knowledge) selected by ourselves one can fix at least three interrelated levels.

The first level is represented by the attributes which distinguish a science from other forms of knowledge (ordinary, spontaneously-empirical cognition, art, religiously-mythological assimilation of the world, etc.). For example, in different historical epochs the nature of scientific knowledge, procedures of its substantiation and the standards of proof were understood differently. But that scientific knowledge differs from the opinion, that it must be substantiated and proved, that science cannot be confined by direct verifications of events, but must reveal their essence – all these normative requirements were fulfilled in ancient and in medieval sciences and science of our times.

The ideal of knowledge growth (the accumulation of new knowledge) was also accepted at the different stages of development of science. The talk is, of course, not about the prescience, but about the science in the true sense of the word, forming the level of theoretical knowledge. Yet in ancient mathematics the intention is clearly traced on investigating the properties of numbers and geometrical figures and obtaining more knowledge about these objects. In the new European science this ideal has been formulated in a clear form and appears as a fundamental value determining the strategy of scientific creative work.

The second level of content of the ideals and norms of investigation is represented with the historically changeable attitudes which characterize the style of thinking dominant in science at the definite historical stage of its development.

Thus, comparing ancient Greek mathematics with mathematics of ancient Babylon and ancient Egypt, one can discover the distinctions in the ideals of organization of knowledge. The ideal of account of knowledge as a set of recipes of tasks' solutions, accepted in mathematics of the ancient East, is replaced in Greek mathematics with the ideal of knowledge organization as the deductively expanded system in which the consequences are deduced from the initial premises-axioms. The brightest realization of this ideal was the first in the science history theoretical system – Euclidean geometry.

When comparing the modes of justification of knowledge dominant in medieval science with the normative of investigation accepted in the science of the New Time, the changing of the ideals and norms of proof and justification of knowledge is discovered. In accordance with general worldview principles, with value orientations and cognitive attitudes formed in the culture of its time, scientist of the Middle Ages distinguished the correct knowledge, checked by the observations and bringing a practical effect, and true knowledge, revealing the symbolic sense of things, allowing the seeing of a macrocosm through the sensible things of a microcosm, and touching the world of the celestial essences through earthly objects. Thus, when knowledge was substantiated in medieval science, the references to experience as to the proof of correspondence of knowledge to the properties of things, at best meant the revealing of only one sense of a thing from many others, and besides, far from the major sense.

The advent of natural science at the end of the 16th century and in the beginning of the 17th century approved the new ideals and forms of knowledge justification. In accordance with the new value orientations and worldview attitudes, the major aim of cognition was determined as studying and discovering the natural characteristics and links of subjects, and revealing the natural causes and natural laws. Hence, as the major requirement of justification of knowledge on nature, the requirement was formulated to check it experimentally. Experiment came to be considered as the most important criterion of truthfulness of knowledge.

It can further be shown that after the arrival of theoretical natural science in the 17th century, its ideals and norms endured the essential reconstruction. It is unlikely, for example, that a physician of the 17th–19th centuries could be satisfied by the ideals of the quantum-mechanical description in which the theoretical characteristics of object were given through the references on the character of devices. And instead two supplementary pictures were proposed of the whole picture of the physical world. One of them gave the spatial-temporal, and another, the cause-effect description of events. The classical physics and the quantum relativistic physics are different types of scientific rationality which find their concrete expression in different understanding of the ideals and norms of investigation.

Finally, one may single out the third level in the content of the ideals and norms of scientific investigation, in which the attitudes of the second level are concretized with reference to the specifics of subject area of each science (mathematics, physics, biology, social sciences, etc.).

For example, the ideal of the experimental verification of theories is absent in mathematics, but it is obligatory for the empirical sciences.

In physics the particular standards exist of justification of its developed mathematized theories. They are expressed in principles of observation, correspondence and invariance. These principles regulate the physical investigation, but they are redundant for the sciences which are only entering in a stage of theorization and mathematization.

Contemporary biology cannot manage without the idea of evolution and so the methods of historicism are organically included in the system of its cognitive attitudes. Physics does not resort to these methods in a clear form. Unlike biology where the idea of development is spread into the laws of animate nature (these laws emerge together with the becoming of life), physics till the last time have not generally considered the problem of origin of the physical laws governing the Universe. Only in the last third of the 20th century owing to development of theory of elementary particles in a close relation with cosmology, and also with the achievements of thermodynamics of nonequilibrium systems (the conception of I. Prigogine) and of synergetics, the evolutionary ideas began to penetrate into physics, causing change of the disciplinary ideals and norms formed earlier.

The specificity of investigated objects certainly affects the character of the ideals and norms of scientific cognition, and every new type of objects' systemic organization, involved in an orbit of research activity, as a rule, requires transformation of the ideals and norms of scientific discipline.

But their functioning and development are stipulated not only by a specificity of an object. The definite image of a cognitive activity, the conception of the obligatory procedures, which provide the comprehension of truth, are expressed in their system. This image always has a sociocultural dimension. It is formed in science under the influence of social needs, experiencing the impact of worldview structures lying in the fundament of culture in this or that historical epoch. These impacts determine the specifics of the above-mentioned second level of content of the ideals and norms of investigation. This level functions as a basis for the forming of normative structures expressing the particularities of different subject areas of science. On this level the dependency of the ideals and norms of science from the culture of epoch, from the worldview attitudes and values dominating within, is tracked more clearly.

Let us elucidate the above with examples.

If we turn to the works of the famous chemist and medical man of the 16th century Paracelsus and his followers, one can see in them a great number of echoes of science ideals of scientific explanation dominanting in medieval. In Paracelsus' epoch the recipe was known well, according to which the extract of walnut on wine vinegar was effective when a headache. In our time science gives an explanation of this fact: the extract, described in ancient recipe, includes substances which reduce arterial pressure, and so in some cases (for example, during hypertensive illness) it actually could have a healing influence. But in the Middle Ages such influence was explained as the attraction of the walnut substance to the head substance, the "sympathy" between these "things". To prove it they referred to the "signs" allowing us to ascertain such sympathy: like a nut grows on the top of a tree, so the head crowns a body; a nut has a structure similar to the structure of a skull and is covered by skin; finally, the core of a nut is very similar to cerebral hemispheres. The following conclusion was drawn from these things: since there was attraction of its kind between two things, then one thing could be useful to another.[82]

From positions of the ideals which became consolidated in natural science of the New Time, the explanation that was given during the epoch of Paracelsus now looks especially unscientific. But medieval science is full of explanations such as this and they were found, as we see, even in science of the Renaissance.

The same things can also be said about the ideals and norms of description, which were transformed in the advent of natural science in the 17th century. When the famous naturalist of the 18th century G. Buffon saw the treatises of the Renaissance naturalist Aldrovandi, he expressed extreme perplexity about the unscientific mode of description and classification of events in his treatises.

For example, Aldrovandi's treatise on snakes, along with the knowledge that the naturalists of the next epochs would like to refer to a scientific description (the kinds of snakes, their reproduction, the effect of snake poison etc.), included the descriptions of miracles and prophecies related secret signs of the snake, legends about dragons, data of emblems and heraldic symbols, data of the constellations of Serpent, Ophiucus, Dragon, and astrologic predictions related with them, etc.[83]

Such modes of description were the relicts of cognitive ideals characteristic to the culture of the medieval community. They were generated by the worldview attitudes dominating in this culture, which determined perception, comprehension and cognition of the world by a human. In the system of attitudes the cognition of the world was interpreted as decoding of the meaning enclosed in things and events by the act of divine creation. Things and events were considered as dually splintered. Their natural characteristics were perceived at the same time as the signs of the Providence embodied in the world. In accordance with these worldview attitudes the ideals of description and explanation, accepted in medieval science, were formed. To describe a thing or an event meant not only the fixation of attributes, which were qualified in later epochs (in science of the New time) as the natural characteristics and qualities of things, but also the discovery of the "sign-symbolic" attributes of things, their analogies, "consonance" and "interchange" with the other things and events of the Universe.

As things and events were perceived as signs, the world was treated as a specific book written in "God's letters", because the verbal or written symbol and the thing which it marked could be assimilated to each other. So in the descriptions and classifications of medieval science the real thing's attributes are often unified into a common class with the symbolic indications and language symbols. It is quite acceptable from these positions, for example, to group in one description the biological characteristics of a snake, heraldic symbols and legends about snakes, and to interpret all this as different kinds of signs designating some idea (an idea of a snake) which is introduced into the world by the Providence.

Explanation of events, in medieval science was imagined as groping for the law of creation which consisted in analogy between microcosm and macrocosm. For a medieval scientist this "law" was the deep essence of things and events, and the search for its demonstrations and its effect was the ideal of explanation accepted in medieval science. This ideal accumulated a whole system of norms: it was considered that the analogies between things, their "sympathies" and "antipathies" to each other, their "inclinations" and "repulsion" are required to be exposed for its explanation, because in these inclinations, and sympathies and antipathies the law of creation was expressed.

Reconstruction of the ideals and norms of medieval science, which started in the Renaissance, was being established during a prolonged historical period. In the beginning the new content was enveloped with the old form, and the new ideas and methods adjoined with the old ones. Therefore, in the science of the Renaissance, along with the principally new cognitive attitudes (the requirement of experimental confirmation of theoretical constructions, the attitude on the mathematical description of nature) we also meet the widespread modes of description and explanation borrowed from the past epoch.

It is indicative, that in the beginning the ideal of the mathematical description of nature was strengthened in the epoch of the Renaissance going from traditional to the medieval culture conceptions of nature as a book written in "God's letters". Then this traditional worldview construction was filled with the new content and received the new interpretation: "God wrote the book of nature in the language of mathematics".

Thus, ideals and norms of investigation form an integral system with complicated organization. This system, if A. Eddington's analogy is used, can be considered as the "network of a method", which science "throws into the world" to "fish out of it the determined types of objects". The "network of a method" is determined, on the one hand, by sociocultural factors determined by the worldview presumptions dominant in the culture of this or that historical epoch, and on the other hand, by the character of investigated objects. This means that the "network of a method" changes along with transformation of the ideals and norms, and, therefore, the possibility is opened for the cognition of new types of objects. All that kept within the frameworks of a given scheme of method is a subject of investigation in the respective sciences.

Since a special picture of the world expresses the general systemic-structural characteristics of the subject of investigation, it must be introduced correlative to the scheme of method expressed in the ideals and norms of cognition. The latter obtain their realization and concrete embodiment in the picture of the world.

The statements, describing the picture of the world and fixing it as a component of knowledge, represent the principles on which basis an investigator constructs an explanation of events.

Thus, physicists of the 18th century, accepting the mechanical picture of the world, strove to explain all physical events as the interaction of atoms and bodies (the principle of atomistic structure of a substance), taking place in consequence of momentary delivery of forces along straight line (the principle of long-range action), in a way that the state of movement of atoms and bodies at a moment of time t_1 uniquely determined their state at the later moments of time (the principle of Laplacian determinism). These principles of the event's explanation were used not only in mechanics, but also in classical thermodynamics and in Ampere-Weber's electrodynamics.

The ideals and norms of scientific cognition regulate the advent and development of special pictures of the world in different sciences. They also orient their synthesis into the general picture of the world. Moreover, the ideals of explanation and description, corresponding to which the special pictures of the world in the leading fields of science were created, acquire a universal character and appear as the basis of construction of the general scientific picture o the world.

Cybernetics took a place among the leaders of science in the middle of the 20th century. The discussions of those years about the possibilities of applying its principles of explanation not only to events in the technical world, the biological and social worlds, but also to the processes of inorganic nature, to the Universe as a whole, transferring on it the images of the self-organizing automatic machine, can serve as a characteristic evidence to that.

A series of principles expressing the specificity of the contemporary physical picture of the world (the conservation laws, the principle of complementarity, etc.) enters the general scientific picture of the world exercising its rights on the universal principles of explanation and description. Characteristically, for example, after N. Bohr's works, in which the possible extrapolation of the principle of complementarity in the field of biological and social processes was justified, the research programs appeared in biology, oriented at the description of biological objects from the positions of conception of subsidiarity. Finally, biology going out in a number of leading fields of natural science was accompanied with extrapolation of such fundamental principles as the principle of integrity, the principle of evolution, etc., on the other fields of natural science.

The ideals and norms of science regulate the advent and development not only of the picture of the world, but also of concrete theoretical models and laws connected with it, and also the carrying out of observations and formation of historical facts. They are as though engrained in the appropriate patterns of knowledge and thus are assimilated by the investigator. In this case the investigator can not realize all the normative structures accepted in a search, many of which he sees as obvious. More often he assimilates them orienting on the patterns of already conducted investigations and on their results. In this sense the processes of constructing and functioning of scientific knowledge demonstrate the ideals and norms corresponding to which scientific knowledge was created.

In systems of knowledge such as these and the modes of their yielding, the original sample forms emerge on which the investigator orients. So, for example, for Newton the ideals and norms of theoretical knowledge organization were expressed by Euclidean geometry, and he created the mechanics orienting on this pattern. In its turn, Newtonian mechanics was a specific standard to Ampere, when he set the task of creation of the generalized theory of electricity and magnetism.

The fundamental nature of theory is determined in many respects on dependence on how it is perceived as a pattern demonstrating the ideals of explanation, proof and structure of knowledge. And the fundamental theories of leaders of science may perform the function of patterns to the adjoining scientific disciplines. Thus the ideals and norms, realized in these theories, are extrapolated in the other fields of scientific knowledge. Those programs of theorizing biology may serve as a typical example, in which the mathematized deductive system, analogous to the physical theory, is proposed as the ideal of theory organization. Functioning of knowledge as the patterns demonstrating the ideals and norms of science, defines the unconscious usage of these norms in research practice.

The problem of correlation of conscious and unconscious in the regulations of research activity was discussed in Russian as in the foreign literature of the philosophy of science. Particularly, M. Polanyi drew a distinction between the "know-how" and the "know what", emphasizing the existence in science of unconscious forms of using the modes and methods of investigation (the "know how"). Lacatos and J. Agassi also pointed to the frequent application of normative knowledge in scientific practical activity without its explication in the form of principles and regulations. Agassi, reproducing Lacatos' metaphor that "a fish swims well though it does not know the hydrodynamics", noted that for many Newton followers his teaching was perceived as something very natural, such as a fish's swimming, but not as the system of methodological rules,[84] though Newton had formulated such rules (the famous Newton's "I don't contrive the hypotheses"). But for a great number of naturalists of his time the pattern of a theory itself meant more than the methodological rule formulated by its creator. Comparing Bacon's and Descartes' statements, who thought that a scientist must realize his method, with Duhem's and Popper's statements, who supposed that a scientist seldom realized the things he did, Agassi defended the palliative point of view. In his opinion, the evolution of science includes unconscious as well as conscious application of methods, and the acts of reflexion upon the method are built into the substance of the concrete scientific knowledge's development as their composite element.[85]

In these reflections the problem was implicitly set of determining those situations in which the transition is necessary from the unconscious application of some ideals and norms to their understanding and methodological explication.

This problem was set in rather another perspective in Russian methodological literature. It emerged when the questions about the role of philosophy in science dynamics were discussed. A task was set of showing that philosophical ideas and principles appear as a necessary condition of breakthrough to the new theoretical ideas in natural science and social sciences (this task itself, being a methodological one in its nature, was also stimulated by a social order, if one takes into consideration that the opposition of dialectical materialism to positivism was first of all expressed in criticizing the positivistic idea of the necessity of separation of one science from another). It is necessary to say that in Russian literature the facts of heuristic function of the philosophical-methodological principles in scientific search were demonstrated earnestly enough. But as those facts assimilated, it became increasingly clear that realized application of such principles, as a rule, is related to situations of revolutionary transformations in science. The distinction by T. Kuhn of stages of normal science and scientific revolution set the problem: how did the methodological principles function at the stage of normal science? On this angle, in the end of the 1970s and the beginning of the 1980s I analyzed the different situations in history of science and at the end I came to the following solution of the problem.[86] Before science faces the objects, requiring radical changes in the picture of the world and the accepted standards of investigation for the settling of them, the system of these standards may not be explicated.

The different layers of content in the ideals and norms of cognition are as if pasted together in the consciousness of investigators, are not separated and are not analyzed critically. The ideals and norms work and thus they may be perceived as something obvious. In situations such as these the usual patterns of knowledge and activity serve as a basis of scientific search, and the methodological rules, presuming the reflexion on patterns, may be used only as an additional means confirming confidence in the right of a chosen way.

Another situation occurs at the stage of scientific revolution, related to discovery of the discrepancy of the formed picture of the world and the ideals and norms accepted in science to the character of the new object which the investigation has faced.

In this situation one must often modify the previous standards regulating the search.

Then the critical analysis of traditional ideals and norms acquires a particular importance and it becomes necessary to find the new scheme of method providing the discovery of new objects. Many things in this process can be inadequately understood by scientists. But the general impulse of search has its essence in separation of different levels of content of the ideals and norms, in elaboration of the new specific concretizing of the scientific method, and then in their relating with the stable accepted content expressing the most general characteristics of scientific cognition. Comprehension and criticism of previous samples can be accompanied with formulation of new methodological regulations already in the early stages of scientific revolution. But they may be formed as principles also on its final stage, when new theoretical patterns appear and the problem of their inclusion into a culture emerges. The conscious application of the new methodological regulations, their explication in a form of principles and their justification, support the new patterns demonstrating the new standards of investigations.

I will further analyze this process in more detail in the example of development of a special relativistic theory. In the same part one may confine oneself with a thesis that the historical mutability of ideals and norms, and the necessity to produce new regulations of investigation gives birth to the need for their comprehension and rational explication. The methodological principles, in which system the ideals and norms of investigation are described, appear as the result of such reflexion on the normative structures and ideals of science.

Philosophical foundations of science

In a system of the foundations of science, along with the scientific picture of the world, and the ideals and norms of investigation, one may single out one more extremely important component that is known as the philosophical foundations of science.

In Western methodological investigations the prolonged domination of the positivistic tradition has almost eliminated the problem of philosophical foundations of science from the sphere of methodological analysis.

Only in the investigations alternative to positivism, and then also in the post-positivistic philosophy of science, the problem of functions of metaphysics in the processes of growing of scientific knowledge was rehabilitated.

The revaluation of the problem of "metaphysical preconditions of cognition" first of all expressed in the most significant schools and conceptions, refused the ideas of a strict demarcation between philosophy and science, emphasizing the inclusion of philosophical ideas and principles in the context of scientific search. Thus, M. Wartofsky, opposing the neo-positivistic conception of science logic, noted more than once that metaphysical terms possess the same value as scientific-theoretical terms, and any attempt to separate them from each other did not lead to success. He wrote, "We cannot have any doubt in that in the history of science the "metaphysical models" played an important role when constructing the scientific theories and in the scientific discussions regarding the alternative theories. It is enough to refer to the notions of matter, motion, force, field, elementary particle and to the conceptual structures of atomism, mechanism, intermittence and continuity, evolution and leap, a whole and a part, invariability in change, space, time, and causality which originally had "metaphysical" nature and made a great influence on the most important constructions of science and on its theoretical notions".[87]

The analogous approaches are characteristic for K. Popper, T. Kuhn, I. Lacatos, G.

Holton, and others.

Popper, who in the 1930s–1950s tried to draw strict line of demarcation between science and "metaphysics" on a basis of principles of falsification, in the 1960s-70s softened his position, openly recognizing, that distinction proposed by him between science and metaphysics was unrealistic and formal.[88] Marking the important role of philosophy in formation of new knowledge about the world, he emphasized that philosophical ideas were the source from which fundamental scientific theories subsequently grew. And these ideas often stimulated the scientific search and showed the way to new scientific investigations. "...It would be *inadequate* to draw line of demarcation between science and metaphysics so as to exclude metaphysics as nonsensical from a meaningful language".[89]

In Kuhn's conception the philosophical statements are also considered as the important preconditions of forming the "disciplinary matrix", accepted by the scientific community and orienting the solution of scientific tasks. He wrote, "It is no accident that the emergence of Newtonian physics in the seventeenth century, and of the relativity and quantum mechanics in the twentieth should have been both preceded and accompanied by fundamental philosophical analysis of the contemporary research tradition".[90]

Lacatos noted in his investigations that philosophical principles were included in the composition of a core of scientific investigation programs and might be considered as heuristics placed in every kernel like that. In general, science as a whole appears as a huge research program based on "metaphysical principles".[91]

Considering the history of science as translation of relatively stable structures - "themes" and the reconstruction of thematic field at the expense of formation of new themes, G. Holton pointed out that the appearance of any theme in science presupposes the inclusion of philosophical analysis into the process of scientific search.[92]

In the opinion of one of the famous historians of science A. Coyré, the history of scientific reflection teaches us that, first, it was never completely separated from philosophical reflection; second, the great scientific revolutions were always determined by changing of philosophical conceptions; third, scientific reflection always took place not in a vacuum: this development always happened in the frameworks of determined ideas, fundamental principles provided with axiomatic obviousness, which, as a rule, were considered as properly philosophical.[93]

In Russian philosophy of science the problem of philosophy's role in scientific cognition traditionally occupied one of the central places. In the 1960-80s several directions of investigation of this problem had formed. In the beginning the main attention was paid to analyzing the two-sided link between philosophy and science. On the one hand, the changes were analyzed which entered the fundamental scientific theories of the 20th century into the content of philosophical categories (causality, development, space, time, etc.). On the other hand, the heuristic functions of philosophy in the formation of these theories were analyzed.[94]

In the 1970s–1980s the themes of analysis were broadened resulting in the synthesis of methodological and historical-scientific investigations. The types of interaction between philosophy and science in the epoch of ancient mathematics and natural science's emergence in the New European culture (the works by P. Gaidenko, L. Kosareva, and others) were analyzed. Comparison of different historical stages of interaction between philosophy and science, and consideration of the social and cultural context of this interaction revealed new aspects of the problem. The questions emerged: how and why were the heuristic functions of philosophy possible in scientific cognition, and how did philosophy influence the process of acceptance of the new scientific knowledge by culture? The search for answers to these questions was related with the principally important distinction between a philosophy as a whole and its particular part which formed the philosophical foundations of science. This search led to setting the question about the cultural and historical dimension of philosophical foundations.

In my works of the second half of the 1970s and the beginning of the 1980s I analyzed the problem just in those aspects.[95]

Inclusion of scientific knowledge into a culture always supposes its philosophical substantiation. It is carried out by means of philosophical ideas and principles which substantiate the ontological postulates of science and also its ideals and norms.

The justification by Faraday of the material status of electrical and magnetic fields by reference to the principle of unity of matter and force may serve as a characteristic example in this respect. Faraday's experimental investigations confirmed the idea that electrical and magnetic forces are transmitted in a space not momentary along a straight line, but along lines having different configuration from point to point. These lines, filling the space around charges and sources of magnetism, influenced charged bodies, magnets and conductors. But forces cannot exist separately from matter. So, as Faraday emphasized, the lines of forces ought to be related with matter and to be considered as a particular substance.[96]

Bohr's substantiation of standards of the quantum mechanical description is no less indicative. Bohr's argumentation here played the crucial role, especially his considerations about the principal "macroscopic capacity" of the cognizing subject and about the measuring devices applied by him. Basing on analyzing the cognition process as an activity, whose character was stipulated by nature and the specificity of cognitive means, Bohr substantiated the principle of quantum mechanical description, which later acquired the name of principle of relativity of description of object in respect to the means of observation.

As a rule, in the fundamental fields of investigation developed science operates with the objects that have not been assimilated either in production or in the ordinary experience (sometimes the practical assimilation of objects such as these is fulfilled not even in the historical epoch in which they were discovered). To ordinary common sense these objects can be unusual and incomprehensible. Knowledge about them and the methods of obtaining such knowledge may not coincide essentially with the standards and conceptions about the world of the ordinary cognition in an appropriate historical epoch. Therefore, the scientific picture of the world (a scheme of an object) and also the ideals and normative structures of science (a scheme of a method) need a specific joining with the dominant worldview of this or that historical epoch, and with the categories of its culture not only during the period of their forming, but also during the next periods of reconstruction. Philosophical foundations of science provide such "joining". The ideas and principles providing the heuristics of searching are included in their composition together with the substantiating postulates. These principles usually orient the reconstruction of normative structures of science and pictures of reality, and are then applied for justification of obtained results - the new ontology and the new conceptions about a method.

We will settle on this particularly, because the development of heuristic and prognostic components of philosophical understanding of the world is the necessary condition for the evolution of science. It is the precondition of science motion in the field of theoretical operating with the ideal objects, providing the comprehension of subject structures have not yet been assimilated in a practice of this or that historical epoch.

The permanent emergence of nature outside the frameworks of the subject structures, assimilating in the historically organized forms of production and common experience, sets a problem of categorical foundation of scientific search.

Any cognition of the world, including a scientific one, is performed in every historical epoch in accordance with the definite "network" of categories which fix the determinate mode of articulation of the world and a synthesis of its objects.

In the process of its historical development science had studied the different types of systemic objects: from the component subjects to the complex self-developing systems being assimilated at the contemporary stage of civilized development.

Each type of objects' systemic organization required a categorical network in accordance with which the development of concrete scientific notions, characterizing the details of structure and behavior of the given objects, then took place. For example, when assimilating small systems one may consider that parts are additively assembling in a whole; understand the causality in the Laplacian sense and identify it with a necessity; consider thing and process as the mutually incompatible characteristics of reality, imagining a thing as relatively immutable body, and a process as motion of bodies.

Just this content was inserted into the categories of a part and a whole, causality and necessity, thing and process by natural science in the 17th–18th centuries, which was oriented mainly to description and explanation of the mechanical objects representing small systems.

But as science turns to assimilation of large systems, the new categorical outline must enter the substance of scientific reflection. The conceptions of correlation between the categories of a part and of a whole must include the idea of irreducibility of a whole to a sum of parts. The category of contingency, interpreted not as something external regarding necessity, but as a form of its demonstration and development, starts to play an important role.

The prediction of big systems' behavior also requires the usage of categories of the potentially possible and the actual. The categories "quality" and "thing" are filled with new content. If, for example, in the periods of dominant conceptions about the natural objects as the simple mechanical systems a thing was considered as a permanent body, then now the insufficiency of interpretation such as this has been revealed. It is required to consider a thing as a specific process reproducing certain stable states, and at the same time variable in a series of its characteristics (the big system can be understood only as a dynamical process, when in a mass of occasional interaction of its elements some properties characterizing a system's integrity are reproduced).

Originally, when natural science had only begun to study large systems, it tried to consider them as already studied objects, i.e. small systems. For example, in physics for a long time they were trying to imagine solids, liquids and gases as the merely mechanical system of molecules. But already with the development of thermodynamics it has been revealed that this conception is not sufficient. Gradually the persuasion began to form that the stochastic processes in thermodynamic systems were not something external regarding a system, but were the internal essential characteristics determining its state and behavior. But especially encouraging, the inadequacy of approach to the objects of physical reality only as to small systems was revealed along with the evolution of quantum physics. It turned out, that to describe macrocosm processes and to discover their regularities, other, wider categorical apparatus is needed rather than that with which the classical physics had operated with. It became necessary to dialectically link the categories of necessity and contingency, to fill the category of causality with the new content (it was necessary to refuse reduction of causality to Laplacian determinism), to actively use the category of the potentially possible when describing the states of a micro-object.

If the categorical system corresponding to the new type of objects has not formed in a culture, these objects will be interpreted through the inadequate network of categories, which will not allow science to discover their essential characteristics. The categorical structure which is adequate to an object should be elaborated in advance as a precondition and term of cognition and understanding of new types of objects. But then a question arises: how is it formed and how does it appear in science? You know the previous scientific tradition can contain no categorical matrix providing an investigation of the principally new (in comparison with those previously cognized) subjects. As regards the categorical apparatus of ordinary thinking, then, because it is formed under direct influence of a subject environment which has already been created by a human, it is often insufficient to the aims of scientific cognition. That is so because the objects studied by science can be radically distinguish from the fragments of the object world assimilated in production and in ordinary experience.

The task of elaboration of categorical structures maintaining the emergence from the frameworks of traditional modes of objects understanding and comprehension, is solved in many aspects due to philosophical cognition.

Philosophy is able to generate categorical matrices necessary for scientific investigation before the latter starts to assimilate the corresponding types of objects. Developing its categories, philosophy thus prepares a special preliminary program for the future conceptual apparatus of natural science and social sciences. Application of categories, developed in philosophy, in a concrete-scientific search brings about the new enrichment of categories and development of their content. But to fix this new content the philosophical reflexion on science is needed. This reflexion appears as a specific aspect of philosophical comprehension of reality in the course of which the categorical apparatus of philosophy is developing.

But then the question arises about nature and origins of prognostic functions of philosophy regarding special scientific investigation. This question is about how the systematic birth of ideas, principles and categories, which are often redundant for describing the fragments of the world already assimilated by a human, is possible in philosophical cognition of the world. But at the same time they are necessary for the scientific studying and practical assimilation of objects which a civilization faces in the following stages of its development.

Just the simple comparison of history of philosophy and history of natural science gives very persuasive examples of philosophy's prognostic functions regarding special sciences. It is enough to remember that fundamental natural science idea of atomic theory originally emerged in philosophical systems of the ancient world, and it was developing inside of different philosophical schools until natural science and techniques achieved the necessary level which allowed the transformation the prediction of philosophical nature into a natural scientific fact.

One can further show that many features of categorical apparatus, developed in Leibniz's philosophy, appeared retrospectively regarding big systems, though in practice and natural scientific cognition of this historical epoch, the simpler objects were predominantly assimilated. Those objects were the small systems (in natural science of the

17th century the mechanical picture of the world dominated which transferred a scheme of structure and functioning of mechanical systems to a whole nature).

Leibniz, in his monadology, developed the ideas which are in many aspects alternative to mechanical conceptions. These ideas, pertinent to the problem of interaction between a part and a whole, non-forced interactions, links between causality, and potential possibility and reality, discover a surprising consonance with some conceptions and models of modern cosmology and elementary particles physics.

Maximon and planckeon cosmological models enter also the conceptions about correlation between a part and a whole which in many aspects have something in common with the picture of monads' interrelation (each maximon is a particle for an external observer, but the Universe for an internal one). As consonant with Leibniz's ideas one can interpret the conceptions of the branching worlds,[97] which are also developed by H.Everett, J. Wheeler, and B. DeWitt, the modern conceptions about microcosm particles as about those containing all other particles in a potentially possible kind, and the understanding of microscopic objects as those ones representing the mega-world and a series of other modern physical conceptions.

The substantiated opinions that the monadity conception becomes one of the fundamental ones for the modern physics which approached such level of substance investigation, when it revealed fundamental objects turn out to be "elementary" not as unstructured, but in a sense that studying their nature discovers some properties and characteristics of the world as a whole.[98] This, of course, does not mean that when elaborating such conceptions modern physics consciously oriented to Leibniz's philosophy. The rational moments of the latter were fused into the system of the objectively idealistic conception of the world. And one can say only that actual features of complicated systemic objects dialectics were guessed therein. However, Leibniz's conjectures, undoubtedly, made an impact on the subsequent development of philosophical reflection. The new interpretations of philosophical categories content supposed by him made a contribution in their historical development. And in this aspect it is correct to talk about the indirect (through the history of philosophy and all culture) influence of Leibniz's creative work on the present.

Finally, considering the problem of philosophy's prognostic functions in respect to special scientific investigation, one can address the fundamental modern science conceptions about self-developing objects whose categorical network had been elaborating in philosophy long before they became a subject of natural scientific investigation. In philosophy the idea about such objects' existence in nature was originally substantiated and historicism principles were developed. The latter required approaching an object taking into consideration its preliminary development and ability to further evolution.

Natural science only began investigation of objects taking into consideration their evolution in the 19th century. From the external side they were being studied in that period by the formation of paleontology, geology, and biological sciences. But theoretical investigation, directed at studying the laws of historically developing objects, probably, was first given in Darwinism. It is indicative that in philosophical investigations the categorical apparatus necessary for theoretical comprehension of self-developing objects had already been developed. The most weighty contribution into development of this apparatus was made by Hegel.

Hegel had not had at his disposal sufficient scientific material to elaborate the general schemes of development. But he chose the history of human thinking as the initial object of analysis. This history was realized in such forms of culture as philosophy, arts, lawful ideology, morality, etc. This subject of analysis was represented by Hegel as self-development of an absolute idea. He analyzed the development of this object (idea) by the following scheme: object gives birth to "one's another", which then starts to interact with the foundation that gave birth to it, and, shaking it up, formed the new whole.

When he had extended this scheme of developing notion on any objects (because they were interpreted as the other being of an idea), Hegel, though in a speculative form, revealed some specific features of developing systems: their ability, displaying the initial contradiction concluded in their original embryonic state, to increase the new organization levels and to reconstruct the system complex whole when each new level appeared.

The network of categories, developed in Hegel's philosophy based on this understanding, can be rated as formed in the first approximation categorical apparatus which allowed assimilation of the objects regarding type of self-developing systems.

Thus, the comparison of history of philosophy and history of natural science allows us to establish that philosophy possesses prognostic abilities regarding natural scientific search, elaborating in advance the necessary categorical structures.

But then the question arises: what are the mechanisms that provide such elaboration of categories? The answer to this presupposes revealing the functions of philosophy in dynamics of culture, its role in reconstruction of foundations of concrete historical types of culture. These functions related to necessities of assimilation and critical analysis of culture universalities.

Any important changes in human vital activity presuppose changing of culture. Externally it appears as a complicated mix of interacting knowledge, prescriptions, norms, patterns of activity, ideas, problems, beliefs, generalized visions of the world, etc. Elaborated in different cultural spheres (science, ordinary cognition, technical creative work, arts, religious and moral consciousness, etc.), they possess a regulative function regarding different kinds of activity, behavior and contacts between people. In this sense one may talk about culture as a complexly organized set of over-biological programs of human vital activity, the programs in accordance with which the definite kinds of activity, behavior and contact are carried out.[99]

In its turn, reproduction of these kinds maintains the reproduction of the appropriate type of society. Culture preserves, translates, and generates the programs of activity, behavior and contact which consist of the aggregate social-historical experience. It fixes them in a form of different symbolic systems that have sense and meaning. Any components of human activity may appear as such systems (tools, patterns of operations, products of vital activity objectifying its aims, the individuals themselves appearing as carriers of some social norms and patterns of behavior and activity, natural language, different kinds of artificial languages, etc.).

Dynamics of culture is related with appearance of some and disappearance of other over-biological programs of human vital activity. All these programs form a complex developing system in which three major levels may be singled out. The first of these is composed of relic programs. They represent the specific splinters of past programs which have already lost their value for the community of the new historical epoch, but nevertheless reproduce the definite types of people's contact and behavior. Many customs, omens, and superstitions that are still used in our times but that emerged in the culture of primitive society are concerned with them. For example, ethnographers noted that in the beginning of the 20th century the superstition existed in many nations, including Russia,

Estonia, and Ukraine, in accordance to which having sexual contacts before hunting and fishing could result failure. This superstition is the relic of the production-sexual taboos of the primitive epoch.

The second level of cultural formations consists of the programs providing the reproduction of forms and kinds of activity which are vitally important for the given type of society, and determine its specifics. Finally, one can single out one more (the third) level of cultural phenomena in which the programs of future forms and kinds of behavior and activity, corresponding to the last stages of social evolution, are elaborated. Generating in science theoretical knowledge, causing revolutions in techniques and technologies of subsequent epochs, ideals of future social system, moral principles which are being elaborated in the sphere of philosophical ethical doctrines and often passing ahead in their century - all these are patterns of future activity programs leading to changing the existing norms of social life.

Programs such as these appear as a result of the search for ways of solving social contradictions. Their emergence lays the outlines of new types and modes of activity, and their generation functions as a result and expression of personality's creative activity.

In a complex kaleidoscope of cultural phenomena of each historical epoch one can reveal their foundations, deep programs of social vital activity which pierce all other phenomena and cultural elements and organize them into the integral system. Being realized, they provide the reproduction of complex coupling and interaction between different forms and kinds. The foundations of culture determine the type of society on every concrete stage of its historical development; they consist of the worldview of the appropriate historical epoch.

Analysis of foundations of culture and their historical dynamics closely leads to the problem of philosophy functions in society's life. In Russian literature the point of view has already been expressed (M. Mamardashvili) that philosophy is a reflexion on culture foundations. But, here the specification is needed as what the foundations of culture represent. The preceding reflections allow us to make important steps towards this. If the foundations of culture appear as an extremely generalized system of worldview conceptions and attitudes which form an integral image of the human world, the question arises about the structure of these conceptions, ways of their being, and forms in which they are realized.

Such forms are the categories of culture – the worldview universalities systemizing and accumulating the amassed human experience.[100] Exactly in their system the characteristic for historically determined type of culture image of a human and conception of his place in the world, conceptions of social relations and spiritual life, of the environment and structure of its objects, etc. are formed. The worldview universals determine the mode of understanding, comprehension and emotional experience of the world by a human.

The socialization of an individual and forming of a personality presuppose their assimilation, and this means the assimilation of that integral image of the human world which forms a specific matrix for expansion of different concrete patterns of activity, knowledge, prescriptions, and norms and ideals regulating social life in frameworks of given culture type. In this relation the system of culture universals will appear as a specific genome of social life.

In a system of the worldview universals one can single out two main blocks. The first of these is formed by categories in which the most general characteristics of objects, transformed in activity, are fixed: "space", "time", "motion", "thing", "property", "relation", "quantity", "quality", "causality", "contingency", "necessity", etc. The objects which are transformed in activity can be not only natural objects, but also social objects, a human himself and states of his mind. Therefore, the enumerated "subject categories" have a universal adaptability.

The second block of culture universals consists of categories characterizing a human as a subject of activity, structure of his communication, his relation to other people and to society in a whole, and to the aims and values of social life. The following categories are concerned with these: human, society, self, others, labor, consciousness, good, beauty, belief, hope, duty, conscience, justice, freedom, etc.

These categories are concerned only with the sphere of social relations. But in human vital activity they play no less a role than the "object categories". They fix in more general form the historically accumulated experience of introduction of an individual into the system of social relations and communications and they do so with its distinctness as a subject of activity.

Evolution of human activity, and appearance of its new forms and kinds function as a basis for development of both types of categories. The new categories can emerge in their composition, and those ones already composed can be enriched with the new content. In this development the categorical structures, which fix the most general attributes of subject of activity, become interdependent with categorical structures fixing the attributes of the subject world (the world of objects at which an activity is directed).

In different types of cultures which are characteristic for different types and kinds of society, historically replacing each other, one can reveal as general, invariant, as particular, specific features of content of categories. In the consciousness of a human in every epoch all these features are alloyed into a whole, because a consciousness in its real being is not abstract consciousness in general, but the developing social and individual consciousness having its concrete-historical content in every epoch.

From these positions it is purposeful to talk about presence in every type of culture of a categorical system of consciousness specific to them which combines in its content the moments of absolute, imperishable (expressing the deep invariant of the human being and its attributes) and the moments of relative, historically variable (expressing the particularities of culture of historically determined type of society, forms and modes of communication and people's activity, preservation and transmission of social experience, accepted in its scale of values).

Thus, category of being and non-existence appear as the fundamental characteristics of the world in very different cultures. But if one compares, for example, comprehension of these categories in ancient culture and in the culture of ancient China, one can discover a series of essential distinctions. If thinking of the ancient world interpreted non-existence as absence of being, in ancient Chinese cultural tradition another comprehension dominates that non-existence is the source and plenitude of being.

In this system of thinking the world appears as a permanent turnover of transformations of being into non-existence, and moreover, the situations of apparent, real, thing, and moving being as if would emerge dark, rest non-existence and, having exhausted themselves, again become absorbed in it. Non-existence appears as absence of things and forms, but in it all possible richness of the world, all unborn, not emerged and shapeless as if would be hidden.[101]

In ancient Chinese culture the category of emptiness acquires a particular sense which appears as expression of non-existence. And if in the ancient world the category of emptiness meant the absence of things, in Eastern cultures it was understood as a beginning of things determining their nature. Representing the absence of any forms, it appeared at the same time as condition of form of things. In a monument of ancient Chinese culture "Tao te-ching" (the IV-III centuries BC) it was emphasized that just an emptiness contained in a thing between its parts determined a usefulness of thing and its adaptability. The wheel is created owing to particular connection of spokes, but application of the wheel depends on emptiness between them; vessels are created from clay, "but usage of vessels depends on emptiness within them"; "they make the holes for doors and windows to build a house but usage of a house depends on emptiness in it".[102]

Characteristic to Eastern cultures the vision of the world as transitions from being into non-existence and inversely is concretized further in specific meanings of such categories as "causality", "necessity", "chance", "event", "essence", and others. In ancient Chinese and ancient Indian systems of the worldview any situational event was perceived as expression of a thing's or phenomenon's becoming, their "coming up" from non-existence with the subsequent leaving to non-existence. So in any event, in their changing and becoming, in fixation of their originality the truth of the Universe is given. It is uncovered not at the expense of penetration in essence by the way of its separation in a pure analytical form, but at the expense of catching in every ephemeral phenomenon the integrity of being. The world essence is not fixed as much in notions where it is separated from phenomena, as it is expressed in images, when through the individuality and situational nature of phenomena the essences inseparable from them are seen.

All these particularities of categorical separation of the world in the thinking of a human of ancient Eastern societies inseparably linked with comprehension of a human's place in the world specific to their culture. An interpretation of a human being as the active principle, deep-rooted in European thinking and mainly assumed as a basis by ancient culture, which is opposed to thing passivity and proving itself in its actions, extremely differs from comprehension of a human in cultures of the ancient East. Here the ideal of a human being is not so much a self-actualization in subject activity, in changing the external circumstances by a human, as the orientation of human activity on his own inner life.

The ideal of deepening into oneself by way of refusal from active subject work is perceived as the possibility to achieve full harmony with the world, as the exit from a sphere of object causing suffering to the sphere where peace is acquired and suffering is absent. But peace, absence of real subjects and suffering appear as the fundamental attributes of non-existence. The deepening in it is perceived as the necessary condition of training the imperturbability of spirit in situations of complicated everyday collisions, as the way to obtain the truth. Thus "non-existence" appears not as the neutral characteristic of the world on its own, but as value tinted category. Its specific status in the culture of ancient China obtains explanation in the real particularities of mode of living characteristic to Chinese civilization, where the rigid system of social control leaves to personality the right of freedom only in self-knowledge and self-denial. Suppression of personal identity appears here as a condition of demonstrating the creative potentialities of personality (creative work is permissible only in rigidly regulated frameworks of tradition).

A harmony between a human and the cosmos in these cultures was always understood in a way that consonance of human actions to the cosmic order should be related with minimal demonstration of human activity (a human will find the way of truth if he holds to the middle, uses moderateness, and follows the experience of the older ones, etc.). Harmony is achieved by way of dissolving a personality in the cosmic whole. Its actions should be the expressions of the cosmic whole but not a self-actualization.

It is indicative that ancient culture also developed in that epoch the theme of harmony between a human and the world, and the category of harmony, adequacy of things in frameworks of a whole was fundamental for the culture of the ancient Greek city-state. But the semantic substance of this category of culture is already another. Harmony of cosmos is proportionate to harmony of a human himself, but a human is understood here not as dissolving in a mysterious and incomprehensible cosmos, but as a particular separate part appearing as the measure of all things. Beyond this other comprehension of harmony between a human and the world stands the principally different, than in Eastern civilizations, mode of living of Greek city-state, ancient democracy, where the individual activity, aspiration of personality to self-actualization become the condition of reproducing the whole system of its social links.

For a man formed by the appropriate culture the senses of its worldview universals most often appear as something obvious, as presumptions in accordance with which he builds his activity and which he usually does not realize as the deep foundations of his own worldview and disposition. Types of worldview and disposition peculiar to different types of society are determined by different content of categories lying in the foundation of culture.

It is important to emphasize that categories of culture are realized and developed not only in forms of conceptual-cogitative comprehension of objects, but also in other forms of spiritual and practical assimilation of the world by a human. The latter allows characterization of categories as distillation of experience accumulated by mankind including all forms of this experience, but not only a sphere of its theoretical realization. Thus, the categorical structures uncover themselves in all displays of spiritual and material culture of society belonging to this or that historical type (in ordinary language, phenomena of moral consciousness, artistic assimilation of the world, functioning of techniques, etc.).

Universals are not localized in one field of culture but pierce all its spheres. So transformation of the categorical meanings which began under the influence of new social requirements in one or several fields of early or late artistic creative work resonates inevitably in the others.

Thus the universals of culture at the same time make at least three interrelated functions.

First, they provide an original structuring and sorting of diverse, historically variable social experience. This experience is organized by rubrics according to meanings of culture's universals and is gathered into specific clusters. Owing to such "categorical package" it is included into a process of translation and is transmitted from one man to another, from one generation to another.

Second, the universals of culture appear as the basic structure of human consciousness, their meanings determine a categorical order of consciousness in each concrete historical epoch.

Third, interrelation of universals constitutes the generalized picture of the human world that is acceptably called the worldview of epoch. This picture, expressing the general conceptions of a human and the world, introduces the certain scale of values accepted in a given type of culture, and thus determines comprehension as well as the emotional experience of the world by a human.

In all these functions the meanings of culture universals should be adopted by the individual, to become the internal outline of his individual comprehension of the world, his deeds and actions. And this, in its turn, means that in hierarchy of meanings characterizing categorical structures of human consciousness, along with the level of universal which includes the definitions of being that are invariant regarding different concrete historical epochs, and also along with the level of specific, represented by the meanings of culture universals of every epoch, the level of single also exists which corresponds to specifics of group and individual consciousness. At this level the meanings of culture universals are concretized considering the group and individual values. And in stable states of social life the universals of culture may permit a very wide spectrum of concretizing, be supplemented with the values of social groups which are opposed by interest, and do not lose their major meanings.

For example, dominant in medieval culture was the conception of suffering as the permanent attribute of human being which was differently perceived by representatives of ruling classes and by commoners. If the first of them saw the category of the "suffering" as predominantly the official church-religious doctrine of punishing mankind for original sin, the second ones often also put on it a definite heretical sense, supposing the necessity of God's punishment of their oppressors in earth life, for sins and absence of compassion to humbled and aggrieved.

In their turn, the group consciousness stereotypes are specifically refracted in the consciousness of every individual. People always put in culture universals their personal sense according to accumulated vital experience. Resulting from this, in their consciousness the picture of human world acquires the personal coloration appearing as the individual worldview. Form these positions it is appropriate to talk about a huge multiplicity of modifications which are peculiar to each system of worldview attitudes dominating in culture. Basic persuasions and conceptions may combine, and often by the contradictory mode with especially personal orientations and values. And all the complexities of individual persuasions may change during a life.[103] For many Americans in the epoch of slave-ownership the worldview presumption that "people are born as equals" was related with persuasion about the correctness of slave-ownership[104]; famous Russian philosophers N. Berdyaev, S. Bulgakov, and S. Frank in their youth took a great interest in the ideas of Marxism, and then took opposition to it.

Individual variability of worldview attitudes is the important precondition for changing and developing the fundamental meanings of culture universals. But the critical attitude towards them of some personalities does not in itself cause the automatic change of categorical structure of model of the human's world, lying in the foundation of culture. It is necessary but insufficient for changes such as these. The opposition ideas emerge in any epoch but they can find no response in mass consciousness and be seized by it. And only at the definite stages of social evolution do they become a center of remelting of the old meanings by which most of people living in this or that type of society follow.

Transformation of basic meanings of culture universals and correspondingly changing culture type is always related with crucial stages of human history, because it marks the transformation not only of the human world's image, but of the types of personality produced by it, their relation to reality, and their value orientations.

In society evolution the crisis epochs emerge periodically when the previous historically formed and strengthened by tradition "categorical model of the world" stops providing the translation of new experience, coupling and interaction of types of activity necessary to society. In such epochs the traditional meanings of cultures universals lose the function of worldview guidelines for mass consciousness. They begin to be critically revalued and society enters the field of intensive searching of new vital meanings and values appealed to orient a human, reestablish the lost "connection of times", and reconstitute the integrity of his vital world.

In elaboration of these new values and worldview guidelines philosophy plays a particular role.

To modify the previous vital meanings, strengthened by tradition in culture's universals, and thus also in the categorical structures of consciousness of given historical epoch, it is necessary in the beginning to explicate them, to compare with realities of being and to comprehend them critically as an integral system. From unconscious, implicitly functioning categorical structures of human understanding and activity the culture universals must be transformed into specific subjects of critical consideration. They must become the categorical forms at which consciousness is directed. Such reflexion on the foundations of culture constitutes the most important task of philosophical cognition.

The necessity in such reflexion is evoked not by a purely cognitive interest, but by actual needs in searching the new worldview guidelines, in elaborating the new maximally general programs of human vital activity. Philosophy, explicating and analyzing the meanings of culture universals, appears in this activity as the theoretical core of the worldview.

Revealing the worldview universals, philosophy expresses them in conceptual-logical form as philosophical categories. In the process of philosophical explication and analysis some simplification and schematization of culture universals takes place. When they are expressed by means of philosophical categories, in the latter the accent is made on the conceptual-logical mode of comprehension of the world. At that the aspects of emotional experience of the world are eliminated in many respects, the certain personal meaning put in culture's universals remains in a shadow.

The process of philosophical comprehension of worldview structures lying in the foundation of culture contains several levels of reflexion. To each of them its type of knowledge and mode of philosophical categories' arrangement corresponds. Their developing as notions, where in forms of definitions the most general properties, links and relations of objects are reflected, represents the result of sufficiently complicated development of philosophical knowledge. It is as if the highest level of philosophical rationalization of culture's foundations, established, as a rule, in frameworks of professional philosophical activity. But before such forms of philosophical categorical apparatus emerge, philosophical thinking must single out and fix their general meanings in a huge variety of cultural phenomena.

Rational explication of these senses often begins from specific catching of commonality in qualitatively different fields of human culture, from understanding their unity and integrity. So, as the primary forms of philosophical categories being not so much notions, as sense images, metaphors and analogies appear.

In the beginnings of philosophy forming this particularity is tracked very clearly. Even in relatively developed philosophical systems of Antiquity many fundamental categories bear the stamp of symbolical and metaphorical image reflection of the world (Heraclitus' "The fire logos", Anaxagoras' "Νοῦς", etc.). To a larger degree it is characteristic to ancient Indian and ancient Chinese philosophy. Here in categories, as a rule, the conceptual construction is not separated at all from the meaning forming basis. An idea is expressed not so much in conceptual, as in artistic-figurative form, and an image is the major way of comprehending the truth of being. "Nobody can give a definition of Dharma. It is translated as a "law" and as the "elements of being" which are counted from 45 to 100. Each creature has its own universal and solitary Dharmas (an essence is inseparable from phenomenon). You will not find two similar definitions of Tao in Laotzu, or two similar interpretations of Jen or Li in Confucius; he defined Li on dependence of which of his followers asked him a question".[105]

In the process of philosophical reflection all these symbolical and metaphorical meanings of categories played no less a role than conceptual structures. Thus in Heraclites characteristic of soul as the metamorphosis of fire is expressed not only as the idea of secondary nature of spirit relative to the material substance constituting the basis of the Universe, but also a whole series of concrete meanings framing this idea. They allow us to argue about the perfect and imperfect soul as in a different degree expressing the elements of fire. According to Heraclites, the fire component of the soul is its logos, therefore a fire (dry) soul is the wisest, and moistening of the soul leads to losing the logos (the drunk soul is moistened and he loses rationality).[106]

But one should not think that as philosophy is developing symbolical and metaphorical ways of thinking about the world disappear, and everything is reduced to rigid conceptual forms of reflection. And the reason is not only that in any human cognition, including fields of science subordinated to, as it seems, most strict logical standards, a visual figurative component obligatorily presents. But it is also in that the philosophical nature itself as the theoretical core of the worldview requires from it the permanent addressing to the most general worldview carcasses of culture, which must necessarily be caught and revealed to make them a subject of philosophical reflection. It follows the irremovable uncertainty in using philosophical terminology, inclusion into the substance of philosophical reflection of images, metaphors and analogies by means of which the categorical structures piercing all multiplicity of cultural forms are lightened. When, for example, Hegel in *Science of Logic* tried to substantiate the category of "chemism" as a characteristic of a particular type of interaction constituting some stage of the world evolution, he resorted to extremely unusual analogies. He talked about chemism not only as the interaction between chemical elements, but also as the characteristic of atmospheric processes which had "more the nature of physical elements than chemical ones", about interrelations of male and female in the living nature, and about the relations of love and friendship.[107] Hegel in all these appearances tried to discover some general scheme of interaction where the interacting poles acted as equals. And to justify generality and universality of this scheme, to present it in a categorical form, he should reveal its action in the most remote, and at first glance not interrelated fields of reality.

The complex process of philosophical explication of culture universals in primary forms may be implemented not only in professional philosophical activity but also in other spheres of spiritual assimilation of the world. Literature, arts, artistic criticism, political and moral consciousness, and ordinary thinking facing the problem situations of the worldview scale are the fields into which the philosophical reflexion may be fused and in which philosophical explication of culture's universals may emerge in their primary image form. In principle, on this basis the sufficiently complicated and original complexes of philosophical ideas may develop.

In the works of great writers even the integral philosophical system can be elaborated and expressed in the material and language of literary creative work. This system is comparable by its value with conceptions of great creators of philosophy (a famous example in this case is the literary creative work of L. Tolstoy and F. Dostoevsky). But, in spite of all meaning and importance of such kind of primary "philosophemes", rational comprehension of culture foundations in philosophy is not only limited by these forms. On their basis philosophy then elaborates the more strict conceptual apparatus where categories of culture have already been determined in their most general and essential attributes.

In this way culture universals are transformed in the frameworks of philosophical analysis into specific ideal objects (combined into the system) with which one may carry out particular mental experiments. Thus the possibility opens for internal theoretical motion in the field of philosophical problems. The formation of new categorical meanings emerging from the framework of historically formed and typed in substance of present social reality of the worldview foundations of culture, may become the result of this motion.

In this work on two poles, one of them is the pole of immanent theoretical motion and the other of constant explication of real meanings of ultimate culture's foundations the major destination of philosophy in culture is realized: to understand not only what the present human world is like in its deep foundations, but what it can and should be.

It is indicative that the emergence of philosophy itself as a particular mode of cognition of the world falls in the period of one of the most crucial turns in social evolution. This is the transition from pre-class to class society, when the break of traditional kin-tribal links and the crush of appropriate worldview structures, embodied in mythology, required formation of the new worldview guidelines.

Philosophy always actively participates in elaboration of such kind guidelines. In rationalizing the foundations of culture it carries out "forecasting" and "projecting" of possible changes in its foundations. Already the rational comprehension itself of culture foundation, which function in ordinary thinking as the unconscious structures determining vision and emotional experience of the world, is an important enough step. In principle, to live in frameworks of traditionally formed way of living, it is not necessary to analyze the appropriate image of the world represented by categories of culture. It is enough just to assimilate it in the socialization process. The comprehension of this image and its evaluation already sets the problem of its possible modification and that also means the possibility of another image of the world and way of living, i.e. the exit from the formed state of culture in another state.

Philosophy, accomplishing its cognitive work, always offers to mankind some possible variants of its life world and in this sense it possesses prognostic functions. Of course, these functions are accomplished with necessary completeness not in any system of philosophical constructions. It depends on the social orientation of the philosophical system, the type of society which creates preconditions for developing in philosophy the models of "possible" worlds. Such models are formed at the expense of permanent generation of new categorical structures in a system of philosophical knowledge. These structures provide a new vision of the objects transformed in human activity, as of the activity's subject itself, its values and aims. These visions often do not coincide with universals of culture of the appropriate historical epoch and exceed the boundaries of traditional ways of the worldview and the worldview lying in the foundation of a given culture.

Generation of the new categorical models of the world in a system of philosophical cognition is established at the expense of the permanent development of philosophical categories. One can point to two major sources providing this development. First, the reflexion on different phenomena of culture (material and spiritual) and revealing the real changes which occur in categories of culture during the historical development of society. Second, the establishment of informal-logical links between philosophical categories, their interaction as elements of a developing system when changing the one element brings about changing the others.

The first source is related to generalizing the experience of spiritual and practical assimilation of the world. It allows not only the formation of the philosophical categories as the rationalization of human culture's universals (categories of culture), but also to permanently enrich their content at the expense of philosophical analysis of scientific knowledge, natural language, arts, moral problems, political and lawful consciousness, phenomena of the object world assimilated by human activity, and also the reflexion of philosophy on its own history. The second source is based on application of the apparatus of logical operating with philosophical categories as with particular ideal objects. This allows, at the expense of "internal motion" in the field of philosophical problems and the revealing of relations with categories, to work out their new definitions.

Evolution of philosophic knowledge is implemented in interaction of these two sources. Filling of categories with new content by way of reflexion on the foundations of culture appears as the precondition to every next stage of internal theoretical development of philosophical categorical apparatus. Due to this development the forming of nonstandard categorical models of the world is provided in many aspects in philosophy.

Philosophical cognition appears as a particular self-consciousness of culture which actively influences its development. Generating the theoretical core of the new worldview, philosophy thus introduces new conceptions about desirable way of living which it offers to mankind. Justifying these conceptions as values, it functions as ideology. But together with its permanent intention on elaborating the new categorical meanings, setting and solving the problems many of which on the given stage of social development are justified previously by immanent theoretical evolution of philosophy, brings it together with modes of scientific thinking.

Historical development of philosophy permanently introduces mutations into culture, forming new variants, new potentially possible lines of culture's dynamics.

Many ideas generated by philosophy are transmitted in culture as specific "drifting genes" which in definite conditions of social development receive their worldview actualization. In these situations they can stimulate the elaboration of new original philosophical conceptions which can be further concretized in philosophical publicism, essayism, literary criticism, moral doctrines, political and religious teachings, etc. In this way philosophical ideas may obtain the status of worldview foundations of this or that historically formed type of culture.

Generating the categorical models of possible human worlds, philosophy in this process at the same time also elaborates the categorical schemes able to maintain comprehension of objects of the principally new systemic organization, in comparison with those which practice appropriate historical epoch assimilates.

In this way the important preconditions are created for the becoming of science itself and its further historical development.

Thus, in periods of reconstruction of scientific ontology and norms of investigation, philosophical analysis serves as the purpose-orienting methodology of search, and through philosophical justification these new ontologies and norms of science are submitted with accepted and dominating in culture worldview guidelines.

But the coincidence of philosophical heuristics and philosophical justification is not obligatory. It may happen that in a process of forming new conceptions an investigator uses some philosophical ideas and principles and then the conceptions developed by him receive another philosophical interpretation, and only by this way they acquire recognition and are included in culture. Thus philosophical foundations of science are heterogeneous; they allow variations of philosophical ideas and categorical meanings applied in investigation activity.

Philosophical foundations of science should not be identified with the general array of philosophical knowledge. From the large field of philosophical problematic and variants of its solutions emerging in culture of each historical epoch, science uses only some ideas and principles as the substantiating structures.

Forming and transforming of philosophical foundations of science require not only philosophical but also special scientific erudition of the investigator (his understanding of the particularities of the subject of appropriate science, its traditions, its patterns of activity, etc.). It is established by way of selection and further adaptation of ideas elaborated in philosophical analysis to requirements of a definite field of scientific cognition, which leads to concretizing of initial philosophical ideas, their specification, and emergence of new categorical meanings which after the secondary reflexion are explicated as the new content of philosophical categories. All this complex investigation at the turn of philosophy and concrete science is implemented together by philosophers and scientists-specialists in a given science. At the present time this particular layer of investigating activity is marked as philosophy and methodology of science. In historical development of natural science the outstanding naturalists played particular role in elaboration of problems, related to formation and development of philosophical foundations of science. They connected in their activity concrete scientific and philosophical investigations (Descartes, Newton, Leibniz, Einstein, Bohr, and others).

Heterogeneity of philosophical foundations does not eliminate their systemic organization. One can mark out in them at least two interrelated subsystems: first, the ontological, represented by the network of categories which serve as a matrix of comprehension and cognition of investigated objects (comprehension of thing, property, relation, process, state, causality, necessity, contingency, space, time, etc.); second, the epistemological, expressed by categorical schemes. These schemes characterize the cognitive procedures and their result (comprehension of truth, method, knowledge, explanation, proof, theory, fact, etc.).

Both subsystems historically develop depending on the type of objects assimilated by science and on evolution of normative structures providing the assimilation of such objects. Development of philosophical foundation appears as the necessary precondition of expansion of science on the other subject fields.

As a result of conducted analysis the foundations of science appear as a particular link, which at the same time belongs to the internal structure of science and its infrastructure determining the connection of science with culture. The structure of scientific knowledge, determined by links between foundations of science, theories and experience, can be visually depicted in the following scheme (see picture 3).

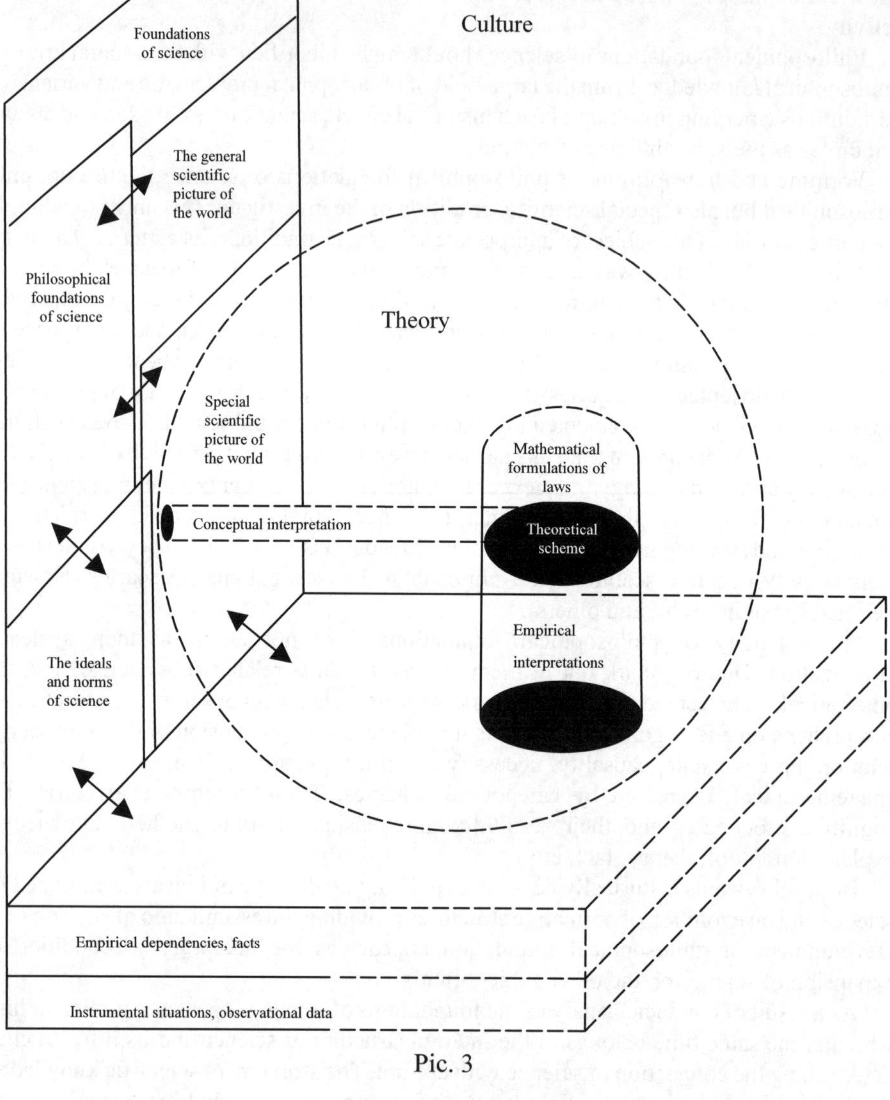
Culture
Foundations of science
The general scientific picture of the world
Philosophical foundations of science
Theory
Special scientific picture of the world
Mathematical formulations of laws
Conceptual interpretation
Theoretical scheme
Empirical interpretations
The ideals and norms of science
Empirical dependencies, facts
Instrumental situations, observational data

Pic. 3

NOTES: CHAPTER 3

[1] On the basic principles of this conception see Suppes (1977). An analysis of the standard conception is given in V.N. Sadovsky's works. See, e.g. Sadovsky (1981, pp. 315-351).
[2] Kuhn (1962).
[3] See, e.g. Gurevich (1972, pp.15-16).
[4] Chanyshev (1982, pp. 38-43).
[5] Holton (1992, pp. 38-39).
[6] Ibid, p. 38.
[7] Ibid, p.41.
[8] Schrödinger (1971, pp.38-42).
[9] Vernadsky (1981, pp.229-232).
[10] See, e.g. Gurevich (1972), (1983); Stepin (1986).
[11] More detailed correlation of philosophical categories and universalities of culture in functions of philosophy in culture will be stated below, in part "Philosophical foundations of science".
[12] Mostepanenko (1969, p.5).
[13] Chernovolenko (1970. p.122).
[14] Planck (1958).
[15] Ibid.
[16] Ibid.
[17] Einstein (1931).
[18] Einstein, Podolsky and Rosen (1935).
[19] Chudinov (1976, p.33).
[20] Einstein (1967, p.40)
[21] Ibid, p.40.
[22] Bohr (1963).
[23] Born (1956).
[24] Ibid.
[25] Bohr (1971, p.505), Einstein (1967a).
[26] Bohr (1963).
[27] Vernadsky (1981, p.237).
[28] Einstein (1945), Wiener (1948), Vernadsky (1977, p.84).
[29] Born (1956).
[30] Dirac (1963), (1963a).
[31] Vernadsky (1978, p.13).
[32] Ibid.
[33] In essence, it is discussed about the ideas of global evolutionism which will find their realization in the modern scientific picture of the world. It will be discussed below.
[34] Vernadsky (1981, p.43).
[35] Wiener (1948).
[36] Friedman (1965, p.5).
[37] Vernadsky(1981, pp.62-63).
[38] Kuhn (1962, p.10).
[39] Ibid.
[40] Lacatos (1970, pp.127-128, 132-133).
[41] Holton (1988).
[42] Ibid.
[43] Laudan (1977, p.24).
[44] Ibid, p.97.

[45] Ibid, pp.24, 61.
[46] In methodology of science in that period several schools emerged. Each of them made its contribution in the elaboration of structure and functions of the scientific picture of the world. The works are of Leningrad philosophers (M. Mostepanenko, A. Mostepanenko, and others); philosophers of the Kiev school (V. Chernovolenko, P. Dishlevy, S. Krimsky, V. Kuznetsov, and others); Moscow philosophers (I. Alexeyev, L. Bazhenov, L. Kosareva, L. Mikeshina, B. Pahomov, V. Shvirev, L. Yatsenko, and others), Minsk methodological school.
[47] Mostepanenko (1969, p.71).
[48] Dishlevy (1973, p.118).
[49] Ibid.
[50] See: Zelenkov and Vodopianov (1987), Stepin (1987), Kuznetsova (1984).
[51] This approach was realized in the investigations of the author and his followers represented in the 1970–80s Minsk school of methodologists. See, for example, Stepin (1976), (1979), (1981).
[52] Lorentz (1953, p.29).
[53] Ibid, p.33.
[54] Ibid, pp.32-33.
[55] Along with determining the attributes of abstract objects of theoretical scheme in terms of picture of the world, notions also include the operational definitions and also the definitions fixing relations between the attributes of abstract objects of theoretical scheme which are revealed through the formulation of appropriate theoretical law. The example of this can be determining a mass as magnitude directly proportional to force and inversely proportional to acceleration. This expresses those major relations between the attributes of a material point, force and spatial–temporal frame of reference, which are expressed in the second law of Newton.
[56] This assumption, however, was found as incorrect when investigating electromagnetic processes. It was needed here to introduce another understanding of experimental-measuring procedures that in the end led to replacement of Newtonian conceptions about space and time by conceptions of relativistic theory (see Mandelstam (1972, pp.160-161, 181-185), Tomilchick and Fyodorov (1987, pp.144-145)).
[57] Einstein (1965-67, vol. 2, p.125).
[58] See Motroshilova (1981, p.91).
[59] Newton-Smith (1981, pp.3-4).
[60] Kuhn (1977).
[61] Ibid.
[62] Ibid.
[63] Ibid.
[64] Ibid.
[65] Popper (1968).
[66] Newton–Smith (1981, p.7).
[67] See Laudan (1984).
[68] Ibid.
[69] Ibid.
[70] Ibid.
[71] Ibid.
[72] Ibid.
[73] Ibid.
[74] See Laudan (1977, pp.16, 25).
[75] Newton-Smith (1981, p.185).
[76] Ibid, pp.185-186.
[77] Ibid, pp.187.
[78] Ibid, p.190.

[79] As it seems to me, thesis about the variety of ideals of scientific character in their sociocultural and historical dimension has already become generally accepted. It entered in different statements into the educational literature on philosophy of science. One example of its qualified explication of the chapters may dedicated to the problem of ideals and norms in the educational manual *Philosophy and methodology of science* (1994) written by A. Ogurtsov, B. Yudin, A. Kezin. I would like to draw attention to the thesis specially oriented in these chapters that victory of this or that ideal of scientific character on the competitors does not eliminate their succession. My point of view is that such succession is implemented on different layers of content of ideals of scientific character and is related to preservation of that layer in which the basic features of scientific rationality distinguishing it from the extra-scientific forms of cognition are fixed.
[80] Here a great number of names can be listed. Along with the investigators mentioned above this problem was successfully analyzed by P. Gaidenko, L. Kosareva, S. Krimsky, M. Rozov and others. Review and estimation of Russian works on the problem of ideals and norms of science can be found in the book: E. Mamchur, N. Ovchinnikov, A. Ogurtsov. Russian philosophy of science: preliminary results. Moscow, 1997.
[81] Results obtained in these investigations were stated in my monograph *Becoming of scientific theory* (1976), and also in parts of books written by me *The nature of scientific cognition* (1979); *The ideals and norms of scientific investigation* (1981).
[82] See Foucault (1966).
[83] Ibid.
[84] See Agassi (1974, pp.506-507).
[85] Ibid, p.513-514.
[86] See Stepin (1981a).
[87] Wartofsky (1976).
[88] Popper (1968).
[89] Popper (1968, p.257).
[90] Kuhn (1962, p.88).
[91] Lacatos (1970, pp.125-127, 132-133).
[92] Holton (1988).
[93] Coyré (1985, pp.14-15).
[94] This type of investigation was wide spread in the Russian philosophy of the mentioned historical period. The most famous works of those years are the investigations of I. Alexeyev, I. Akchurin, L. Bazhenov, V. Bransky, V. Kaziutinsky, S. Krimsky, E. Mamchur, M. Omelyanovsky, R. Karpinskaya, B. Kedrov, V. Kuptsov, Y. Sachkov, I. Frolov, B. Yudin.
[95] See Stepin (1976), (1979), (1986).
[96] See Faraday (1832).
[97] Evolution of the integral Universe is considered in the frameworks of this conception as a specific branching of a multiplicity of worlds from the one trunk. These worlds exist as if parallel to each other and do not interact in energetic and force senses, but together with this they mutually supplement each other as the particular quantum states of the integral Universe (more precisely about physical and philosophical senses of the conception of the "branching worlds" see, for example, Mitskevich (1976, pp.101-104), Krimsky and Kuznetsov (1983, pp.88-120)).
[98] See Krimsky and Kuznetsov (1983, pp.54-55).
[99] In people's vital activity the two types of programs interact: the biological programs (instincts of self-preservation, nutrition, sexual instinct, instinctive predisposition to communication resulting in the adaptation of human ancestry to gregarious way of living, etc.) and the social ones, which have as if been overbuilt on the biological ones in the process of becoming and development of mankind (so they may be called the over-biological programs). If the first ones are transmitted through the hereditary genetic code, the second ones are preserved and transmitted in society as a cultural tradition.

[100] The term the "categories of culture" was widely accepted by A. Gurevich when he investigated the culture of the Middle Ages (see Gurevich (1972)). In further statement terms "categories of culture", "universals of culture", "worldview universals" are used as the convertible terms.
[101] Grigorieva (1979, pp.63-70).
[102] Ancient China Philosophy (1972, Vol. 1, p.118).
[103] Holton (1992, p.38).
[104] Ibid, p.48.
[105] Grigorieva (1979, pp.75-76).
[106] Heraclitus of Ephesus (1954).
[107] See Hegel (1920-1923).

CHAPTER FOUR

GENESIS OF THEORETICAL KNOWLEDGE IN CLASSICAL SCIENCE

An analysis of the structure of theoretical knowledge allows concretizing a problem of their genesis. The key role of theoretical schemes, both in interpretation of the apparatus of the theory and in the process of its contents expansion, makes the most important one in genesis a problem of forming theoretical schemes. It looks like the analysis of theory structure, if to conduct it with an accent on discovering connections between the components of the theory and the represented in it reality, inevitably leads to such setting of a problem. Making an attempt to solve this problem we will be guided by main characteristics of theoretical schemes found during the process of analyzing structure of theory. Knowledge about such characteristics determines a way of analysis of the scientific history material, where are rendered the main methods and operations of the research thought, which leads to a formation of a theory.

The main purpose will be that by reconstructing historical material we will recover these ways and operations and so will find out how the core of theoretical knowledge is created.

As much as structure analysis of the theory showed that there exist two levels of theoretical schemes and according to that two levels of organizing theoretical knowledge, as much as it is purposeful to study genesis of the theory according to these levels. First we should look at how singular theoretical schemes are formed (before their inclusion into a developed theory) and then to proceed to a problem of becoming developed theory.

Engaging in a solution of this problem, we should take into consideration the factors of science evolution which change the ways for constructing theoretical knowledge.

In the history of science classical and non-classical periods are usually distinguished, each one of which has specific ways for creating a theory.

That is why in the beginning it is purposeful to analyze ways of constructing theoretical schemes in classical science, and then to look what has changed in the ways of their construction in the modern phase.

But before we engage in this analysis we should resolve another important problem. It is connected with elucidations of the role of empirical foundations in the genesis of the disciplinary ontologies – special scientific pictures of the world which appear as a specific form of theoretical knowledge. It is important, because in the classical science special pictures of the world always forego theoretical schemes. There are a lot of situations when science starts to investigate a matching object domain, not having any means or possibilities to create concrete theoretical schemes for its explanation. In such situations science studies its field with empirical methods, collecting necessary experimental facts. Principles of the world picture set problems for an investigation, aim observations and experiments and give explanations to them.

Because the world picture belongs to a layer of theoretical knowledge, it has explanatory and predicting functions. By this feature sometimes it is called a theory. Strictly speaking, it is incorrect, since in this case there is no difference between forms of theoretical knowledge. But if we agree with such application of notions (which is spread in the methodologically superficial level of reflection and is used within the frameworks of the so-called "common sense" of science), then we should keep in mind that term "theory" is not strictly applied here, but is applied broadly, like an equivalent of the term "theoretical knowledge".

But in the methodological analysis the differentiate approach is more preferable, distinguishing the picture of the world, which is described in the system of theoretical principles, and concrete theories, including in their composition theoretical schemes and matching them laws formulations. Since theoretical systems gain ontological status only through connections with the picture of the world, then in order to understand the process of their formation it is important to find out how scientific pictures of the world (disciplinary ontology) are created and developed. For this purpose we again should distinguish two situations: development of the world picture under straight influence of experience and the picture's evolution under influence of created theories, which mediate its relationships with the empirical material.

SCIENTIFIC PICTURE OF THE WORLD AND EXPERIENCE

Situation of direct interrelation between the picture of the world and empirical data can be realized in two versions. First of all, on the stage of developing a new field of scientific knowledge (scientific discipline) and, second of all, in theoretically advanced disciplines with empirical detection and investigation of principally new phenomena which go beyond the already existing theories.

First let us look at how the picture of the world interacts with empirical facts on the stage of evolving scientific discipline, which in the beginning goes through a stage of collecting empirical material on studied objects. Under these conditions empirical investigation is purposefully aimed by available ideals of science and by forming a special scientific picture of the world (picture of studied reality). The latter forms that specific layer of theoretical notions which provide a setting of problems of an empirical investigation, a look on situations of detection and experiment and their results' interpretation.[1]

Special pictures of the world as a special form of theoretical knowledge are a product of long-lasting historical development of science. They appeared as relatively independent fragments of the general scientific picture of the world on the stage of disciplinary organized science's formation (end of the 18th – first half of the 19th century). But at early stages of development, during the epoch of becoming natural science, such an organization of science didn't yet exist. This circumstance wasn't always adequately understood in methodological researches. In the 1980s when a question about the status of special pictures of the world was intensively discussed, three points of view were expressed: special pictures of the world do not exist and they shouldn't be distinguished as special forms of theoretical knowledge; special pictures of the world are highly expressed autonomic constructs; their autonomy is highly relative, because they appear as fragments of the general scientific picture of the world. But in the history of science we can find acknowledgements for all three points of view. They only belong to different stages of its development: before disciplinary science of the 17th century, disciplinary organized science of the 19th – first half of the 20th century, modern science with its getting stronger interdisciplinary connections. These stages should be distinguished.

The first of sciences, which had formed a whole picture of the world, based upon results of experimental investigations, was physics. In its evolved forms the appearing physical picture of the world contained multiple natural philosophic layers. But even in this form it purposely aimed at the process of an empirical investigation and collection of new facts.

A characteristic example of such interaction between the picture of the world and experience during the epoch of becoming natural science is the experiment by W. Hilbert, in which the peculiarities of electricity and magnetism were studied.

Hilbert was one of the first scientists, who opposed the worldview ideas of the Middle Ages by a new ideal – experimental study of nature. But the picture of the world which purposefully aimed his experiments included some notions taken from the influential and during the Middle Ages natural philosophy by Aristotle. Even though Hilbert criticized the Peripatetic conception of the four elements (soil, water, air and fire) as a basis of all other solids, he used conceptions of metals as land thickening and about electrified solids as water thickening. Basing on these conceptions Hilbert announced some hypotheses about electrical and magnetic phenomena. These hypotheses didn't go beyond the frameworks of natural philosophic constructs, but they served as an impulse for setting of experiments, which discovered real facts. For example, conceptions of "electric solids" as an embodiment of the "water element" evolved a hypothesis that all electrical phenomena are a result of "fluids" outflow from electrified solids. That is why Hilbert proposed that electrical outflows must be delayed by a barrier of paper and fabric and that fire must destroy electrical operations because it evaporates outflow.[2] That is how appeared an idea of a series of experiments that discovered the facts of screening an electrical field by some types of material bodies and the facts of fire impact on electrified solids (if we use modern terminology, then there was, in effect, discovered that fire has a quality of conductor).

Analogously the notions of a magnet as Earth thickening generated the famous experiments by Hilbert with a ball magnet, which had proved that the Earth is a ball magnet and had discovered the features of the Earth magnetism. Experiment with a ball magnet seems very elegant even by the standards of modern physical experiments. In its basis laid an analogy between a ball magnet (terrella) and the Earth. Hilbert studied the behavior of a miniature magnet arrow, put in different places of terra, and then he compared obtained data with the known from the practice of navigation facts of orientation of a magnet arrow relative to the Earth. From the comparison of this data he concluded that the Earth is a ball magnet.

An assumed analogy between terra and the Earth was hinted by the accepted Hilbert picture of the world, where a magnet as a kind of metal was looked at as an incarnation of the "land nature" Hilbert, even in the name of a ball magnet ("terrella" is small Earth), emphasizes the integrity of the Earth material and the magnet material and the naturalness of an analogy between the globe and the ball magnet.

Aiming observations and experiments, the picture of the world always tests their back influence. We can ascertain that the new facts obtained by W. Hilbert during a process of empirical investigation of the processes of electricity and magnetism, generated a series of fairly important changes in the picture of the world firstly accepted by him. By an analogy with the conceptions about the Earth as a "large magnet" Hilbert includes in the picture of the world concepts of planets as magnet solids. He expresses a brave hypothesis that the planets are delayed on their orbits with the forces of magnetic gravity. Such an interpretation evoked by experiments with magnets radically changed a concept of the nature of forces. During this time power was looked at as a result of solids contact (the power of pressure of one weight onto another, the power of impact).[3] A new interpretation of power was a predecessor of future conceptions of the mechanical picture of the world, where transfer of forces on a distance was looked at as a source of changes in the states of solids in motion.

Facts obtained from observation not only can modify the available picture of the world, but also lead to contradictions in it and demand its reconstruction. Only going through a long stage of development the picture of the world clears from natural philosophic layers and turns into a special picture of the world, which constructs (unlike the natural philosophical schemes) are entered by features that have an empirical justification.

In the history of science such an evolution was firstly performed by physics. At the end of the 16th – first half of the 17th century it reconstructed the natural philosophic scheme of the world, which was the leader in the physics of the Middle Ages, and created a scientific picture of a physical reality – mechanical picture of the world. In its becoming new worldview ideas and new ideals of cognitive activity, which was formed in the culture of the Enlightenment era and the beginning of the New Time played the decisive role. Sensed in philosophy they appeared in a form of principles, which provided new look at the collected – by the previous cognition and practice – facts studied in physics processes and allowed to create a new system of conceptions of these processes. The most important role in constructing a mechanical picture of the world played: principle of the material integrity of the world, excluding scholastic division of the terrestrial and the celestial world; principle of cause and regularity of natural processes; principles of an experimental investigation of nature and the attitude for an integrity of an experimental study of nature with the description of its laws in the language of mathematics.

Provided construction of a mechanical picture of the world, these principles had turned into its philosophical substantiation.

After the appearance of a mechanical picture of the world the process of forming special pictures of the world is flowing already in new conditions. Special pictures of the world that were appearing in other fields of natural science experienced an influence of a physical picture of the world as leader in natural science and, in their turn, had a back influence on physics. In physics itself a construction of every new picture of the world happened not by advancing natural philosophic schemes with their following adaptation to experience, but by changing already established physical pictures of the world, which constructs were actively used in later theoretical synthesis (as an example can serve a transfer of notions of absolute space and time from mechanical into electrodynamic picture of the world in the end of the 19th century).

A situation of interaction between the picture of the world and empirical material characteristic for early stages of forming scientific discipline, is also reproduced in later stages of scientific cognition. Even when science had formed a layer of concrete theories, experiment and observation are capable to discover objects that are not explainable in the boundaries of available theoretical conceptions. Then new objects are studied by empirical means, and the picture of the world starts to regulate a process of such investigation, experiencing a back influence of its results.

A very significant example in this relation is an experimental discovery of cathode rays in the end of the 19th century and the study of their main features.

After that these rays accidentally were discovered in experiments with electrical discharge in gas tubes they had found out that the existing theoretical knowledge does not say anything about the nature of a new physical agent. Then began a fairly long period of studying cathode rays mainly by experimental means. It was stated that a cathode beam is capable to turn around a radiometer (effect of mechanical action of cathode rays), that placed on their way a Maltese cross creates a fluorescent glass a distinct shadow (rectilinearity of propagation of cathode rays), that an approach with a magnet leads to displacement of the fluorescent spot caused by them (an effect of interacting cathode rays with a magnetic field). All these features of cathode rays were revealed in Crookes' experiments. He had stated that cathode rays are a flow of charged particles.

Habitually it is presumed that hypothesis of corpuscular nature of the cathode rays was proposed by Crookes as their conclusion after conducting experiments. But it is not so, because in its general sense this hypothesis preceded Crookes' experiments. They were purposefully planned by a special system of historically established concepts of physical reality, according to which processes in nature were explained as an interaction between the "ray material" (air oscillation) and particles that are carriers of an electrical charge (in their turn capable to make solids both charged and electrically neutral).

A mentioned system of conceptions wasn't a theory in the strict sense of the word, because it didn't contain concrete theoretical models and laws, explaining and predicting the results of experiments. This was a physical picture of the world, accepted in natural science in the late 19th – beginning of the 20th century.

From this picture it was stated that a physical agent, which nature was to be studied, could be either a flow of particles (electrically charged or neutral), or "ray material". From the very beginning Crookes followed the corpuscular hypothesis and his experiments were set with a purpose of its approval. It is characteristic that during this period an experimental verification of an alternative suggestion by other researchers (Lenard, Hertz) was conducted – about wave nature of cathode rays (experiments gave a negative result, showing that cathode rays are not electromagnetic waves).

It is important that in both cases primary hypothesis, according to which the main problem of experimental research was proposed, was generated by the physical picture of the world. Later on as hypothesis was compared to the abilities of an experiment the general aim of researches was concretized and parted into a series of local problems: scientists were elucidating which effects can ascertain corpuscular (correspondingly wave) nature of cathode rays, laid down by which means they can register the indicated effects and so on. From here appeared an intention of every experiment, set by Crookes, Lenard, Hertz and other researchers. A picture of physical reality here determined a strategy of an experimental activity, formulating its problems and showing ways for their resolution.

In their turn obtained facts exerted an active back impact on the established physical picture of the world. A hypothesis of special nature of particles, forming cathode rays, appeared which Crookes understood "particles that lay in the grounds of physics of Universe". "I make bold to suppose, – wrote Crookes – that the main problems of the future will find their solution just in this field and even beyond it. Here, in my point of view, are concentrated the final realities, superfine, determinable, enigmatic".[4]

A later scientific development of physics in many ways had confirmed this hypothesis, had proved that negatively charged particles that make cathode rays are not ions, but electrons (experiments by Thompson and Lenard and the Lorentz's theory).

Function of the scientific picture of the world as a research program of empirical search is discovered in a process of an experimental research and also in sciences, based on observations and not using experimental methods.

Thus, in the modern astronomy, regardless of a fairly developed layer of theoretical models and laws, an important place belongs to researches where the picture of the world directly regulates the process of observation and formation of empirical facts. Astronomical observation very often recovers a new type of objects or new sides of interactions which can't be explained right away in the frameworks of already existing theories. Then the picture of reality is actively aiming all consequent systematical observations, where peculiarities of a new object are gradually uncovering.

A characteristic example in this respect is the discovery and creation of quasars. After discovering the first quasar – a radio source 3C 48 – the question appeared, to which type of cosmic object it belongs to. In the picture of an investigated reality, established by the time when the quasars were discovered, most types of objects "suitable" for this purpose could be stars either very distant galaxies. Both hypotheses were purposefully checked in observations. Exactly during a process of such checking the first features of quasars were revealed. A further investigation of these objects by empirical means was also held with active correction from the side of the reality picture. In particular, we can establish its purposeful role in one of the key moments of this research, precisely in discovery of a big red shift in the quasar spectrum. The essence of this discovery was based on a guess by M. Schmidt who identified emission lines in the spectrum of quasars with regular Balmer series of hydrogen, allowing a big red shift (equals to 0,158). From the outside this guess looks very accidental, because by this time it was assumed everywhere that quasars are stars of our galaxy, and the stars of our galaxy shouldn't have such a shift. That is why in order to this idea of showing identification of lines itself to appear, an extravagant hypothesis was needed to be put forward beforehand. But this hypothesis stops to be so extravagant, if we take into account those general notions of the structure and evolution of the Universe, formed by this period in astronomy, included conceptions of grand explosions happening in the galaxies, which were accompanied by emission of substance with high speeds, and about our Universe expansion. Any of these concepts could generate a starting hypothesis of possibility of a big red shift in the spectrum of quasars.

From these positions behind accidental elements in the reviewed discovery its internal logic already can be seen. Here an important side of regulative function, which was executed by the picture of the world regarding to the process of observation is shown. This picture allowed not only to formulate the primary hypotheses, which purposefully aimed observations, but also helped find the right interpretation of matching data, providing a passage from the data of observations to scientific facts.

So, the primary situation characterizing interaction between the picture of the world with observations and experiments does not die with the appearance in science of concrete theories, but saves its main characteristics as a special case of developing knowledge under conditions when a research empirically finds new objects, for which an adequate theory had not yet been created.

In the methodology of science a study of these heuristic functions of the scientific picture of the world in the first hand were carried through on the material of the history of physico-mathematical natural science. For that there were certain reasons, because physics before the other empirical sciences had reached high stages of theorization and here it was easier to distinguish the scientific picture of the world from a theory as special unit of theoretical knowledge, every one of which has specific interconnections with experience. But after a heuristic role of a physical picture of the world was discovered by the framework of this approach, a problem in the empirical cognition appeared: how universal are the developed methodological notions? Are they confirmed when applied in other sciences? Do forms of knowledge, analogous to the physical picture of the world which exerts a function of a very general research program in science, exist in other scientific disciplines?

A controversy around special scientific pictures of the world (disciplinary ontology) appeared not once in our literature. Two alternative approaches to a problem had formed.

Adherents of the first one assumed that by an analogy with the physical picture of the world matching forms of systematizing knowledge in other sciences can be uncovered and analyzed. Adherents of the second approach denied the available special scientific pictures of the world, thinking that in a methodological analysis structures and dynamics of knowledge can do without a given comprehension. As backing for this position the following argumentation was quoted. First of all criticism was aimed against entering by analogy with the physical picture of the world the terms "biological", "chemical", "technical" and other pictures of the world. These terms are really not very apt, and their critic contained rational moments. The thing is that when applied in fundamental ideas and notions of physics their indication by a term "picture of the world" was acceptable, because an object of a physical investigation are fundamental structures and interactions, which determine the evolution of the Universe and can be followed on all stages of this evolution. But concerning other sciences (biology, chemistry, technical and social sciences) this cannot be said. Studied processes in them are evaluated in the modern system of conceptions about the world like they appeared only on a certain stage of the Universe development. They do not belong to fundamental structures of the universal set which exists on any stage of its development. That is why intuitively the terms "chemical picture of the world", "biological picture of the world" and others provoke aversion.

But a critic of a term is not yet a reason for denying the denoted by its form of knowledge. After all a search for adequate terminology is important, but not deciding in the development of a problem of scientific methodology. By the way, a term "picture of studied reality" (biological, chemical, social and so on) appears fairly acceptable taking into account that an application of matching notions already has a solid tradition (in particular, a notion "biological reality" was analyzed in our literature already in the 1970s in the works by I. Frolov).

Besides objections of terminological character, adversaries of the conception of special pictures of the world had also introduced some general methodological reasons. For example, it was approved that the peculiarities of biological and social sciences make a non-perspective transfer on these fields of those methodological models which were elaborated and reasoned on the material from physics.

But, as the history of science shows, strict prohibitions of such kind are rarely productive. In science itself and in its methodology one of the spread ways of learning a new object field is translation of ideas, concepts, methods and theoretical models from other fields of knowledge. It is clear that application of already developed methodological schemes in a new field presumes their correction, frequently also fairly radical change for an according specificity of this or that scientific discipline. To figure out in advance applicable or non-applicable already worked out methodological means is very hard, and more frequently is just impossible without a concrete analysis of a structure of disciplinary organized knowledge. That is why those not numerous quotations of the results of such analysis, which were conducted by the opponents of a conception of special scientific pictures of the world, deserve special attention.

For example, during the 1980s in the works by R. Karpinskaya, who was deeply involved in philosophical and methodological problems of biology, it was emphasized that an analysis valuable for methodology of physics, yet "has a little concern for biology, because in biology we can't find constructs, about which would have been built the picture of the world".[5] In the given case a thesis was clearly formulated which could have been approved or denied, addressing concrete historical texts of the biological science. An analysis of these texts found out that in biology, like in other sciences, fundamental conceptions of the studied reality (pictures of biological reality) enter a set of basic theoretical constructs, which have an ontological status and are described by a system of ontological postulates (principles) of biology. For example, Cuvier's notions of species which disappear only as a result of environmental catastrophes, entered a typical idealized construct –an unchangeable species. Here an analogy with notions of a non-separable atom is very appropriate which were coming into the physical picture of the world until the end of the 19th – beginning of the 20th century.

Similarly in the picture of biological reality proposed by Darwin there were conceptions of single species as units of evolution, which endowed with an ability to inherit all received features. This was a basic theoretical construct, which was identified with reality, but the scientists had to reject it in the long run, and modified the Darwin's picture of the biological reality.

Various investigations of the latest two decades had approved an assumption about the existence forms of knowledge systematization in different sciences, specifying a general vision of a researched object and analogous by their functions to the physical picture of the world.[6] This opened possibilities for analyzing their heuristic role in the empirical and theoretical cognition, appealing to a wide spectrum of situations of development of different sciences.

Most of these sciences much later than physics entered a stage of theorization, connected to forming concrete theoretical models and laws, explaining facts. That is why when a methodologist analyzes historical dynamics of knowledge in these sciences, he most frequently met with dominating situations of empirical search, where the picture of reality takes functions of theoretical programming of experience and is developed under its influence. With that science can compete at the same time alternative pictures of reality, each one of which executes a role of a research program, proposing their own setting of investigative problems and interpretation of empirical material. In this competition, that research program which better assimilates collected material, provides a transition to constructing first theoretical models and which corresponds with worldview ideas, that had formed in the culture of a certain historical period, usually wins.

This way of empirical cognition is widely spread in science. It can be followed not only in physics, but also in biology. A typical example here is the competition between alternative pictures of the biological world, proposed by Cuvier and Lamarck. Every one of them interacted with experience and set its own aims to empirical search. Cuvier's conceptions of the unchangeable species and geological catastrophes stimulated purposeful collection of facts, which were evidence for that there existed species in the past that radically distinguished from modern and already disappeared ones. A picture of biological reality, proposed by Lamarck, assimilated this empirical material, but gave it another interpretation. Variety of species was evaluated as a result of appearing one sort of species different from another as a result of organisms' adaptation to changing conditions of existence and inheritance of obtained features. In this picture a notion of gradual improvement of the organic world and appearance of more highly developed species was entered.

A new picture of the biological world changed orientations of the empirical search. The main problems now consisted of discovering facts, evidencing about gradual collection of changes and continuous line of evolution (problems opposite to those that were set by the picture of the organic world, maintained by Cuvier and his advocates).[7] It is indicative that by the way of empirical base's broadening Lamarck's picture of the biological reality was specified and concretized. In it appeared a notion of a multigraded ascending ladder of creatures as a result of evolutionary changes and accordingly of gradations of large taxonomic groups of animals and plants. We will emphasize that in the later development of biology classifications and typologies of biological objects, concluding the collected empirical material, most often were realized under direct influence of the picture of the biological world, which functioned as a research program, aiming scientific search.

A role of a picture of studied reality and interpretation of facts and setting of problems of empirical research can be also found in other natural scientific disciplines. For example, that what in chemistry is called phlogiston theory, can't be considered as a theory in the full meaning of this word, since it didn't contain concrete laws and theoretical schemes, explaining facts, but only entered principles of such explanation. By these principles a very general system of conceptions about chemical objects and their connections was fixed. This system of notions formed the picture of chemical reality. Foundations of the picture mentioned were formed in the 17th century in the works by Becher and Stahl. In this picture all chemical compounds were considered as made out of threefold kind of "lands" – special elements, which combine with water and special material substance – phlogiston. "Lands", "water", "caloric" were acting as primary creatures, and all the rest of substances (compounds, "mixed solids") were supposed as built from these essences.

Processes of oxidation and combustion were connected with the activity of phlogiston, and besides that it was considered as "flying substance" that could tell its volatility to the particles of the matter when combined with them. Since during this period Newton's doctrine of world-wide gravitation was only evolving, many successors of Stahl believed that caloric does not gravitate to the center of the Earth, but speeds upward.[8]

This picture of reality accepted by researchers explained chemical reactions as a process of phlogiston's transition from a substance enriched by it to a substance where there is less phlogiston. It allowed to consider the self chemical reactions as interaction of minimum two substances, to integrate processes of combustion with the phenomenon of burning and etc. In other words, it allowed to accumulate empirical facts and interpret them. And what is more, on the basis of this picture were obtained some justified in practice advices for improving processes of melting metals.[9] But with the knowledge development such facts revealed that didn't fit in the reviewed picture of chemical processes. Thus, ascertainment by Ray of an increase in the metals' weight when they are turned into calx came into controversy with the caloric conception, according to which it was considered that in the burning process some part of the inflammables are lost. Nevertheless, one of the founders of the "caloric theory" G. Stahl didn't pay any attention to this fact, and his followers with a purpose to save the available picture of chemical reality, used notions of caloric negative weight (Guyton de Morveau).

Stability of the reality picture concerning anomalies (facts that do not fit in its conceptions) is a characteristic specialty of its functioning as a research program. I. Lacatos emphasized that the kernel of a program (in the given case fundamental principles and notions of a picture of studied reality) is saved at the expense of defending hypotheses which are proposed by the way of anomalous facts appearing.

A hypothesis of "caloric negative weight" is a typical example of an attempt to defend a kernel of a research program.

Along with that collection of anomalies and increase in the number of ad hoc hypotheses in the "protective belt" of the reality picture a critical attitude towards it and proposal of a new picture is stimulated.

In the history of chemistry of a reviewed historical period a new picture of the studied reality was proposed by Lavoisier. For some time it competed with the former notions based on phlogiston conception about chemical processes, and then replaced the out-of-date picture. A new picture of reality, developed by Lavoisier, eliminated concepts about phlogiston and entered a new concept of chemical elements as simple substances that are a limit of substances' dissolubility in chemical analysis, from which due to activity of "chemical forces" complex substances are generated. This picture allowed to give another interpretation of existing facts, and the researchers who accepted it had to face new problems: to study features of chemical elements, to prove experimentally the law of substance's conservation and to analyze nature of "chemical forces" and so on.

A functioning picture of reality as a research program, aiming empirical search, can be also followed by the material of social sciences.

Here a competition of various notions of the reality, each one setting its aims for empirical research can also be found.[10]

Thus, in the historical science of the 20th century pictures of the social reality, proposed by A. Toinby, P. Sorokin, a picture of the society, maintained by the followers of classical Marxism put forward various types of problems when researching concrete historical situations.

Mainly Toinby attended facts, which could have evidenced about peculiarities of each one of detailed by him civilizations and about their cyclical development. He strove to follow the hierarchy of social values and the concept of the meaning of life, which lay in the foundation of every type of civilization and which determine its answers for historical challenges. According to these purposes a selection of facts and their interpretation happened.

The picture of the social and historical reality, proposed by P. Sorokin, also accented the historian's attention to researching fundamental values, which determine the type of culture and the matching type of social connections. Here the main aim was to uncover the facts that substantiate typology of cultures, matching, according to Sorokin, three general types of worldview (sensory, rational and intuitive).

Historians and sociologists who agreed with this system of conceptions concentrated their efforts on analyzing how fundamental values reveal themselves in different stages of religious life, in philosophical and ethical thought, in politics and economical relations.

Concerning Marxist historians, for them the most important in a research of a historical process was the analysis of changing the ways of production, class structure of the society, discovery of dependency of the spiritual life from the powerful production relations.

The picture of the social reality formed by the basic principles of historical materialism demanded to consider all historical events from the point of view of replacement of social and economical formations. Accordingly, all these paradigmatic attitudes of search objectives and interpretation of historical facts were set.[11]

It is significant that when facts were discovered that didn't conform to the initial picture of social reality, they were either left without an explanation, or explained by ad hoc hypotheses. When resistance of the picture of the world to the pressure of "anomalous" facts was stronger the more actively this picture served for ideological purposes. It is known, for example, that Marxist historians had encountered many problems when they analyzed Eastern traditional civilizations, applying to them the concept of five social and economical formations. In particular, there weren't discovered any deciding facts, evidencing that in the history of these societies existed a slave-holding method of production. A model of slave-holding formation at best was applicable to a small number of ancient civilizations of the Mediterranean region. Complexities appeared also when the traditional Eastern societies were studied from the positions of classical Marxist notions of feudal way of production.

All these facts demanded correction of the picture of the social reality worked out by K. Marx and F. Engels. It is indicative that in his own time Marx, having found difficulties in concordance of empirical material that belonged to the history of traditional civilizations his picture of social reality of social typology, tried to modernize this picture a little. He introduced a hypothesis of an Asian way of production as a foundation for Eastern civilizations. Later Marxist historians many times returned to this idea. There were held discussions about the problem of the Asian way of production. But with the strengthening in the USSR ideological control over social sciences and making more dogmatic of Marxism more and more dominated attempts to adjust the facts to conceptions of the five social and economical formations, various, frequently artificial admissions were introduced.

Generally speaking all the attempts to save the kernel of the research program by entering defending hypotheses is a characteristic feature of its functioning.[12] Even more so when such a kernel is represented by fundamental principles in science, certifying accepted in it ontology – a picture of the studied reality.

Revision of the principles of the reality picture under the influence of new facts always presumes addressing to philosophical and worldview ideas. Equally it concerns natural science and social sciences.

In a social and scientific research ideological and political aspects of the worldview play a special role. Their influence can stimulate an elaboration of new conceptions of the studied object domain, but also can increase resistance to new facts, even in those situations when the accepted picture of the social reality less and less provides a positive heuristics of an empirical search.

Thus, an analysis of different scientific disciplines leads us to come to a conclusion about universality of cognitive situations, connected to function of special scientific pictures of the world (pictures of the studied reality) as research programs, directly regulating empirical search, and about their development under influence of empirical facts. Such development in the classical science appears as one of the conditions for building theoretical schemes that istablish the kernel of concrete scientific theories.

GENESIS OF THE PRIMARY THEORETICAL MODELS OF CLASSICAL SCIENCE

Previously it was emphasized that the main peculiarity of theoretical schemes is that they are not results of purely inductive generalization of experience. But analysis of the structure of scientific knowledge had shown that theoretical schemes must represent significant features of the objective side of those experiments and intentions which support the theory.

On the face of it between two shown characteristics is a certain contradiction. But this only seems to be a contradiction. Theoretical schemes are firstly entered as hypothetical constructions, but then they are adapted to a certain integration of experiments and in this process are justified as the conclusion of an experiment.

From here on it seems purposeful to study the theory genesis to single out two stages of theoretical schemes formation: a stage of their introduction as hypotheses and a stage of their justification.

Formation of theoretical scheme as a hypothesis

In developed science theoretical schemes were constructed in the beginning as hypothetical models. Such construction is realized at the expense of using abstract objects, earlier formed in the sphere of theoretical knowledge and applied as the building material when a new model is being created.

Only on the early stages of scientific research, when a transition from predominantly empirical study of objects to their theoretical assimilation is realized, constructs of theoretical models are created by direct schematization of experience. But then they are used in a function of means for yielding new theoretical models, and this starts to dominate in science. The previous method is preserved only in rudimentary form, and its sphere of activity ends up being sharply reductive. It is used mainly in those situations when science meets with objects for theoretical understanding of which enough means had not yet been elaborated. Then objects are started to be studied experimentally, and on this basis gradually form necessary idealizations as means for constructing first theoretical models in a new field of research. An example of such situations are early stages of the theory of electricity development, when physics was formulating the basic notions – "conductor", "isolator", "electric charge" and so on – and by that itself was creating conditions for constructing the first theoretical schemes that explain electrical phenomena.

Most of theoretical schemes in science are constructed not at the expense of straight schematization of experience, but by a method of translation of already created abstract objects. In order to uncover this specific of constructing theoretical models, let us address to concrete material from the history of physics.

One of the most important stages of classical electrodynamics' development was a discovery by Faraday of a phenomenon of electromagnetic induction. Various experiments on researching this phenomenon (experiments with a magnet, which when in motion relatively to a closed wire generated in it inductive current; analogous experiments with solenoids and wires of various configuration, an experiment by Arago, etc.) were explained by Faraday in the framework of the law of induction. According to this law, when a conduction substance, moving about a flow of magnet power lines, crosses it, then electromotive force (*emf*) in the conduction substance appears.

The given law expressed correlations between abstract objects of a theoretical scheme, which characterized electromagnetic induction through a relation of the abstract objects "magnet power lines" and "conducting substance". However, let us peer more attentively where these objects came from. They weren't contained inside empirical schemes of induction, but were transferred from other fields of theoretical knowledge. Faraday took a construct "magnet power lines" from a close field of theoretical knowledge, which was entered for explaining experiments of magnetostatics (research of possible orientations of minute magnet arrows in the action field of constant magnets and currents). Another abstract object – "conducting substance" – was transferred from the field of knowledge about the current of conduction. These objects were "deepened" into a new system of relations, due to which had acquired new features.

A construct "magnet power lines" had obtained a feature to "cause electromotive force (*emf*) in conducting substance" (in comparison, before in the knowledge of magnetostatics, it was determined only by a feature of influence on the testing magnet). A construct "conducting substance" which it earlier represented only features of conductors, connected to current's conductivity action, ended up having a new feature – "arising in a conductor *emf* induction". Giving these constructs new features meant reconstruction of the previous abstract objects, because every one of them was determined only as a carrier of some strictly fixed features. Likewise science had formed a primary variant of theoretical scheme of electromagnetic induction.

Analogous methods of constructing theoretical schemes can be seen in physics practically on every step. For example, in this point of view, let us consider the already mentioned Rutherford's model of an atom. Its basic elements (abstract objects) – "nucleus as a center of potentially repulsive forces" and "electron" – were taken from already established fields of theoretical knowledge. A construct "positively charged center of potentially repulsive forces" was transferred from electrodynamics and determined in relation to an ideal alpha particle and electron as an atomic nucleus. "Electron" was also taken from classical electrodynamics and when it was immersed into new relations, scientists gave it a new feature – to "revolve around a nucleus". At the expense of all these internal theoretical operations was created a hypothesis of planetary building of an atom, meant for explaining an experiment in the atomic field.

Thus, in developed forms of scientific research a theoretical scheme is created by integrating in a new "web" connections of abstract objects, that were taken from other fields of knowledge. But then appears a question: how does a researcher find out, exactly what elements of already in science created theoretical schemes can be used when constructing a new model, and in what relations these elements should be "immersed", to build such a model? An answer to this question leads to understanding important sides of a process of the theoretical scheme appearing on the stage of its birth as a hypothesis.

It seems that a researcher when he chooses abstract objects is fully oriented by those experiments, which much be explained by means of a new model. Thus, in a planetary model of an atom, the results of an experiment themselves (discovering that alpha particles, coming through a substance atoms, disperse on large angles) made natural a conclusion that inside an atom there is a strong positive charge, which behaves itself as a center of potential repulsive forces. From there followed an idea of an atom's nucleus. Its stable existence inside an atom demanded, in its turn, that electrons do not touch the nucleus and do not neutralize its charge. From here naturally appeared an assumption about the turning of electrons around a nucleus, thanks to which they stay away at a certain distance from it.

Principally, exactly likewise, usually, the essence of the Rutherford's hypothesis of an atom's building, is stated. When this statement is used, a problem of forming a hypothetical variant of the future theoretical scheme is solved simply: choice of its abstract objects (positively charged nucleus and electrons) and suggesting the system of their relations an experiment.

But let us proceed with our analysis. As can be seen from the history of physics, long before Rutherford had realized his experiment, in physics such hypothetical models of an atom, according to which positive charges can be concentrated as a nucleus, and electrons must turn around the nucleus were known.

The planetary model of an atom, usually connected with the name of Rutherford, was a hypothesis introduced by Nagaoka in 1904 long before any experiment with alpha particles. Judging by that in the first works, dedicated to the discussion of experiments with alpha particles and ideas of a nuclear building of an atom (1911), Rutherford quoted this work, he seemed to have set his experiments already having at his disposal one of the hypothetical variants, which were subjected for experimental testing, planetary model of an atom.[13] This fact is important for comprehending the logic of introducing a scientific hypothesis. It evidences that a problem of search of abstract objects of the future theoretical model and their relations can't be resolved only by the way of indicating the purposeful role of experiments, which justify a hypothetical model.

Absence of such experiments does not hamper introducing the hypothetical models. It's true that in classical physics situations of such kind are rather anomalies, than a rule. But for analyzing the logic of scientific discoveries they are especially important, because in these situations they reveal exactly in pure look those operations of constructing theoretical schemes on the stage of hypothesis, which are hard to see when there is a developed layer of experiments, which provides a justification of a hypothesis and influences the process of its formation. That is why a special interest gains an introduction of exactly the first versions of a planetary atom's model. They can be looked at as a hypothetical stage of constructing a mentioned model. Rutherford's activity then can be interpreted as a stage of justifying a planetary atom's model.

Of course, such approach means certain reconstruction of the historical material, because Nagaoka's model in its own time wasn't successful and wasn't accepted by the majority of physicists. The idea of an atomic nucleus itself during that period didn't have any approval. What's more, paradoxes of atom instability were discovered (not considered by Nagaoka), to which led a planetary model: revolving around the nucleus an electron must radiate and, losing its energy, fall down on the nucleus.[14] A planetary atom model obtained its second life only after the experiments by Rutherford, having confirmed existence of an atomic nucleus, and so it is by right connected with the name of Rutherford. It is characteristic that during this period all paradoxes of atom's instability were discovered as if for the first time. But now the situation is changing, and physicists, regardless of all these paradoxes, accept a planetary atom's model, thinking that elimination of its contradictions would become possible in the nearest future. But because of all these moments, connected to a problem of accepting a hypothesis by a scientific community, we can abstract from following the logic of theoretical schemes' formation. In definite limits it is acceptable to evaluate an introduction of the first versions of hypothetical models and their later justifying as a continuous process, realized by some "integrated researcher" (in our case Nagaoka-Rutherford). In this case it is indifferent how an introduction of a hypothesis and its justifying is realized – by one scientist or by a group of scientists, every one of whom executes a determined series of cognitive operations, logically necessary for constructing a theoretical scheme.

On this basis we can again turn to a problem of choosing abstract objects and the "web" of their connections, due to which they form a hypothetical model. But now we should approach it from new positions, not appealing to "crucial experiments" like Rutherford's experiments with alpha particles.

First of all it is necessary to find out where the problem of constructing planetary models itself came from, if there didn't exist yet any experiments, approving the idea of atomic nucleus's availability.

Analysis of physics condition during a period of introducing the first hypotheses of an atomic structure shows that a set of such objective was closely connected with the development of the electrodynamics' picture of the world. This picture established due to the success of electrodynamics in the end of the 19th century and developed by way of increasing experimental and theoretical achievements. According to principles of the electrodynamics' picture of the world, all processes in nature must be introduced as an interaction of substance and ether. The forces of nature were presumed to be unified, bringing their different types to the changes in the ether's condition ("One ether for light, warmth and electricity", – wrote Kelvin in the late 19th century).[15] It was thought that even Newton's law of world wide gravitation can be reduced with time to transfer of forces with the final speed in the ether.[16] Interaction between ether and substance's atoms was considered as a source of charges' appearance.[17]

In the beginning according to Maxwell's and his successors' program (for example, Lenard, Hertz) it was assumed that charges can be introduced as special processes of ether disturbance[18] (a reason for that was the key idea of the Maxwell's theory of electromagnetic field about the equality of bias current to conductive current which allowed to imagine density of the charges-currents in a form of a flow of electromagnetic field). But under the influence of ideas of atomistics in physics hypotheses of the possibility of distributing the principle of atomism over to charges were repeatedly expressed. These ideas found empirical and theoretical approval after the discovery of electrons and elaboration of Lorentz's electrodynamics, based on the notion of charges-currents as a system of electrons, interacting with electromagnetic field. After that a new concept of charges finally entered in the picture of the world. Changes initially were considered as special particles of the substance – electrons ("atoms of electricity"), interaction of which with electromagnetic field (ether) was introduced as deep foundation of all physical processes. Then in the physical picture of the world a new element – "atoms of electricity" appeared besides "atoms of substance" and "ether", and also a problem of their relation to atoms of "usual" substance appeared. A great interest to the questions of how a substance is built, appearing in physics in the end of the 19th – beginning of the 20th century, was in many ways generated exactly by this problem.[19] Discussing it, physicists firstly set a question: aren't electrons a part of an atom? Of course, the formulation of such question itself was a courageous step, because it led to new changes in the picture of the world (it was needed to accept a complex construction of a substance's atoms). That is why concretizing a problem of correlation of atoms and electrons was connected to the introduction of a philosophical analysis, which always happens with radical shifts in the picture of the world (for example, J.J. Thomson, who was one of the initiators of setting a question of connection between electrons and substance's atoms, searched for help in the ideas of Bošković's atomistics to prove that reduction in the picture of the world of "substance atoms" to "atoms of electricity" is necessary).[20] But somehow or other we can state that a problem of correlation of electrons and atoms and its analysis from the point of view of the atom's complexity was generated by the development of the physical picture of the world.

With physics development, new experimental data and theoretical notions were appearing (especially after discovering radioactive dissociation and creating of its theory) constructing different models of atomic structure became for physicists a common phenomenon. But a construction of such models began a little before under the influence of the electron's problem, which entered as a special element in the picture of the physical reality.

Thus, we have a right to make a conclusion that an impulse for constructing hypothetical schemes of an atomic structure was given by the electrodynamics picture of the world, which included in its composition new elements under the influence of former development of empirical and theoretical material from physics and participating philosophical ideas.

A physical picture of the world does not only support an introduction of a problem, leading to a search for new hypothetical models in physics, but also shows the ways for its solution, outlines a field of possible means using which we can create hypothetical versions of the future theoretical schemes. In our example of the planetary model of an atom it is not hard to discover that the setting of a problem itself – to reduce atoms of substance to "atoms of electricity" – determined a field of starting abstract objects, which must be used to build a model of an atom. These must be abstract objects of a theory "atoms of electricity", meaning objects of Maxwell-Lorentz electrodynamics – positive and negative charges, interacting through a magnetic field. Relations between these charges were intended for showing an electrically neutral and stable atom.

But to construct a model of an atom, it is not enough just to define its elements. We also need a "network of relations", which should contain these elements. A choice of elements of a future hypothetical scheme of an atom in a way already sets constraints on the character of such "network" (because features of abstract objects must correspond to the character of their relations in the limits of the created model). In particular, unlike charges according to their main feature, by which they had been entered into electrodynamics, charges must have been gravitating in accordance with the Coulomb's law. That means, that the problem was in finding such correlations, in the frameworks of which they, regardless of this gravitation, would be left distantly separated and such configuration would be stable.

One of the first models of an atom, proposing solutions for this problem, was the self model of Nagaoka. Its creator, supported by an idea of Kelvin of a possibility to liken a configuration of charges, of which must be made of, to systems of gravitating masses of the celestial mechanics, had transferred the relations between stable configurations of such masses (for instance, planets and the Sun in the Solar system; planets and their satellites) to charges that formed the atom.

From these positions to imagine a process of constructing an atom's model is possible likewise: the shape of the planetary system was used as a peculiar structure, special network of relations, in which should have been immersed constructs "electron" and "positively charged sphere in the center of an atom". Nagaoka firstly used a model of singular solids, revolving around a center solid, and then, with a purpose to find analogy of many electronic orbits, used analogy between them and rings, revolving around Saturn. Connecting this network of relations, taken from the celestial mechanics, with constructs from electrodynamics (replacing material points, representing central solid and moving around it masses, by charges), Nagaoka obtained a hypothetical model of atomic structure.

A shown procedure of introducing a hypothesis could have been described also in the terms "Gestalt switching", like it is often done in philosophical literature (when a problem of a scientific discovery is discussed).[21] Then stable configurations of gravitating masses of the celestial mechanics (like the Solar system or a planet with satellites) will appear in a role of "Gestalts" (or "examples" by Kuhn), which allowed us to see the problem of atomic structure in a new light. But this approach somewhat darkens the problem of structural separation of theoretical models important for logical comprehension, and also a connection of their forming with processes of transfer abstract objects from other fields of knowledge. Besides, there is another important moment, which in our point of view is not taken into account when the process of discovering in terms switching of "Gestalts".

We are talking about the foundations due to which creation and application in science of analog models happens. Following Kuhn we can talk only about researchers' psychological intuition, which is expressed in the shift of a pattern of seeing a scientific situation. The question about causes of choice of this or that example in Kuhn's works, is taken away.[22]

But a setting of this question exactly leads to an important aspect of theoretical discoveries. Why, for example, researchers creating atom models, all of a sudden turned to a notion of gravitating masses? What stipulated their seeing of an atom as an analogue of the planetary system? After all, to use analogies we need some reasons to assume similarity between two types generally very dissimilar phenomena.

It turns out that there was such a reason, and its source was the electrodynamics' picture of the world. In this picture all types of nature forces, including gravity, are connected with ether. It was thought that activity of gravitating masses principally can be explained by ether's features as a carrier of electromagnetic energy (as a concrete and theoretical reason a similarity between expressions of Newton's law of the world wide gravitation and the Coulomb's law for charges was stated and also a successful reformulation of the latter in the frameworks of the field theory was indicated).[23]

One of the first researchers who started to consider an interaction between charges in an atom in the image and likeness of interacting gravitating masses was Kelvin. In his own time he paid special attention to connections of gravity and electromagnetism and so he was maybe better than the others ready to use analogies between gravitating masses and charges (moments that have to do with psychology of the discovery). But if there was a reason for such step, the step itself can already be considered as logically approved and available for any researcher (logic of discovery). In this concern it is significant that, regardless of unlikeness of the atom's models, proposed in the beginning of our century by Kelvin, J.J. Thomson, Nagaoka and others, the majority of them were guided by analogy between distribution of a charge and distribution of gravitating masses,[24] trying to glean such a configuration of masses in order to, replaced them by charges, get a stable atom.

Thus, the physical picture of the world not only shows field of theoretical constructs, which can be used when new theoretical schemes in science are being constructed, but also helps to find determined relations between such constructs. Constructs and structure, where they must be located, can be taken from different fields of knowledge. But to transfer the structure we need to see analogy between objects of research of already established and only forming field of theoretical knowledge. Such a vision of physical situations provides the picture of the world.

A considered example with the planetary model of an atom, which was created as a hypothesis before appearance of "crucial experiments", which confirmed existence of an atomic nucleus, allows to monitor relatively clearly the main peculiarities of constructing theoretical schemes on the first stage of their development.

The main one of these specialties is the active purposeful influence of the picture of the world on the process of choosing abstract objects and networks of their relations, due to integration of which the first hypothetical versions of theoretical schemes are created. This specialty can be also seen when a theoretical scheme is created with a developed layer of experiments, to explain by which it is introduced. In this case experiments ease the process of forming hypothetical versions of a scheme, but they are not the only factors in the preliminary choice of its abstract objects and their relations.

It is not hard, for instance, to find out that with constructing Faraday's model of induction, which had been created for explaining already accomplished experiments that had established the phenomenon of electromagnetic induction, the picture of the physical reality developed by Faraday played a role in finding their connections. In this picture all electrical and magnetic processes were considered as a display of some integrated essence, and the center of gravity of analyzing these processes was transferred from charges and magnets over to the distance between them, which is considered as "filled by curved lines of electrical and magnetic forces". These primal notions of the picture of the world, elaborated by Faraday, were based on the previous achievements in electrodynamics, considered from the point of view of philosophical ideas of the world's integrity and the integrity of the matter and power.

Depending upon this picture of the physical reality, Faraday, when he constructed a theoretical scheme of electromagnetic induction, transferred to a new field the magnetostatics conception of the movements of magnetic power lines in space. Likewise one of the main relations between the conducting substance and the power lines in an induction model was introduced, more specifically, *emf* appears when a number of power lines, crossing a conductor, changes in time in every unit of its volume.

Through a prism of this notion it would have been possible to understand easily all effects that appear with relative motion of conductors and magnets. But from the knowledge of these self effects a concept of power lines was very hard to deduce, practically impossible. It is enough to remember how surprising for Faraday's contemporaries his explanation was of phenomena of electromagnetic induction, well known from experiments, to find out that the knowledge of such experiments itself didn't hint the idea of a connection between *emf* induction and the change of a number of power lines in a conductor. In this concern particularly characteristic was a surprising explanation by Faraday of an experiment by Arago. The main point of this experiment was in the following: if over a suspended (nonmagnetizing) cupric disc to rotate a magnet, then a disc will also begins to rotate. Everybody knew about this experiment, but only Faraday could explain it: when a magnet rotates in space power lines, which surround it, move, and crossing a conduction substance (cupric disc), evolve in it inductive flows, which makes a disc for some time a source for magnetism (current brings magnetism) and leads it to interaction with rectilinear magnet, causing a disc rotation. Thus, in order to enter such explanation, a picture of moving magnetic power lines in space is needed in advance. But this picture does not follow logically from the self experiments in induction. Faraday elaborated it in magnetostatics, and then extrapolated to a field of new phenomena. A process of such extrapolation became possible only due to a picture of the world elaborated by Faraday, according to which all processes of electromagnetism were to be explained proceeding from a "conflict" of electrical and magnetic forces in space.

A mode of changes in power directions in space as a cause of all electromagnetic phenomena was always before Faraday's internal look. That is why for him it was completely natural to use models from magnetostatics, based on a concept of magnetic power lines as analogies when electromagnetic induction was explained.

We will also emphasize that a transfer of models itself from one field of knowledge about electricity and magnetism to another was possible only because Faraday's picture of the physical world postulated a connection between objects of research in every one of these fields. If we take into account that during this period Faraday had to prove that various types of electricity (frictional electricity, galvanic electricity, magnetoelectricity and so on) are displays of the same electricity, then similar transfers of models seem not so trivial.

Afterwards the picture of the world stimulated a choice of certain types of objects and their relations for creating a hypothetical model; experimental situations correct and specify a hypothesis (for example, a notion of appearance in a conductor *emf* induction is the result of a similar type of correction). But only one type of experiments can't determine a choice of theoretical means for constructing hypothetical models in science.

Thus, we can conclude that a construction of a theoretical scheme on a stage of hypothesis in the classical science started from the picture of the world, which helped to set an objective for research and showed the ways of its solution.

Introducing general conceptions of the structure of natural interactions, the picture of the world shows which fields in science have similar objects of research. Moreover a "prompting" appears, from where abstract objects as a building material for future theoretical schemes should be translated. At the same time the picture of the world also helps to find a preliminary network of relations, a structure, with which must be connected such objects. A means for transfer of a shown structure serves as application of theoretical schemes of one field as analogous models for another research field.

In the cases reviewed above such structure was entered in a form of visual notion of connections which should be satisfied by abstract objects of a new knowledge field, replacing the old elements in analogous model. This is, for instance, a conception of motion of material points around a center solid, entered in celestial mechanics and used in constructing a planetary model of an atom; or a picture of moving magnet power lines, crossing solids, which Faraday had extrapolated from a field of magnetostatics over to a field of electromagnetic induction's phenomena (to create a hypothetical scheme of electromagnetic induction is was enough to substitute in an analogous model, taken from magnetostatics, instead of a construct "solid as a whole" a new abstract object – "conducting substance", where appears an inductive *emf*").

In both stated examples a structure, i.e. "immersed" abstract objects of a created theoretical scheme, were expressed in a form of a vivid mode of correlations between elements of analog model and were fixed by means of informal descriptions like: "material points rotate around a gravitating center", "power lines cross solids" etc.

But principally this same structure ("network of relations") can also be displayed in a form of mathematical dependencies. Then its transfer over to a new field means application in this field of matching mathematical means (equations, which are meant for connecting new theoretical constructs). Such a transfer of equations is realized by already described formula. The picture of the world helps to determine, which theoretical schemes of established fields of knowledge can be used in a function of analog models concerning a new object of research. Then equations connected to such analog models are transferred to a new field of knowledge and integrated there with new abstract objects, which are used for building hypothetical versions of the future theoretical schemes. Thus, equations applied for describing a configuration of celestial bodies in mechanics, were used together with the planetary model when describing and explaining an atomic structure. These equations were used as a means for calculation when a problem of atom stability was solved, together with some equations in electrodynamics (application of the latter was necessary because initial abstract objects of a model were taken from electrodynamics).

In classical physics application of mathematical means in a theoretical research was certainly connected with preliminary constructions of informally expressed theoretical model, even in a form of hypothetical construction. Often a process of integrating such model with an equation could have been separated in time from its primary construction (an example is Faraday's theoretical schemes of electromagnetic and electrostatic induction). In this case a hypothetical model preliminarily went through a stage of empirical justification and turned into a theoretical scheme, which provided an explanation and prediction of facts on the basis of a quality law (a sort of Faraday's law of induction). But then came a stage of searching for quantitative, mathematical formulation of this law. It was ended by introducing a corresponding equation, relating to which a theoretical scheme represented as its interpretation.

It is important, however, to emphasize that an integration of a prepared theoretical scheme with equations most often is accompanied by changes in the system itself. Equations, applied as means for theoretical description, frequently introduce new relations between abstract objects of a theoretical scheme, which demands to give such objects new features. For instance, a series of Faraday's schemes of electrostatic and electromagnetic induction led Maxwell to execute integration of these schemes with Euler's equations (later, when we will talk about constructing theory of electromagnetism developed by Maxwell, we will encounter these changes more thoroughly).

This same specialty of a theoretical search can be found also in other historical examples. To find out more specifically how a change in a theoretical scheme, which already has been justified by experience, does occur under the influence of a mathematical instrument entering a theory, let us investigate as one of such examples a situation that appeared in electrostatics in connection with the formulation of a famous Coulomb's law.

During a period which immediately preceded this discovery, a system of theoretical conceptions of an interaction of charged solids in electrostatics was created. Preliminary these notions were expressed in a vivid theoretical model, explaining processes of attraction and repulsion of electrified solids. Such model was created by the efforts of Aepinus, Cavendish, Priestley and Coulomb himself and demanded to consider solids interaction, which contain electricity, as a process of transfer in space of forces that appear when two types of "electrical fluid" (positive and negative) are influencing each other. Every fluid like this one was assigned a feature to concentrate in solids. Depending from a density of concentration a weakening or strengthening of interactive forces between bodies, containing "electrical fluid" occurred.

Thus, a theoretical scheme of electrostatic solids' interaction, constructed by Aepinus, Cavendish, Priestley and Coulomb, introduced abstract objects of "density of an electrical fluid" and "forces", acting between "electrical fluids".

Correlation between the shown abstract objects was characterized in the following way: it was thought that two identical fluids contained in solids repulse, and two unlike – attract with a force directly proportional to their density in solids and inversely to a distance between the solids. This characteristic matched a qualitative expression of a law of electrostatic interaction of electrified solids. A search for quantitative formulation of a law demanded to find precise mathematical expression of a dependency between the density of fluid in solids, distance between them and the quantity of active forces. Priestley and Cavendish expressed a hypothesis that the character of this dependency is the same as with interaction between point attracting masses of the Newtonian mechanics.[25] Coulomb, having accepted this hypothesis, later ascertained its fairness in experiments with torsion balance.[26] It is worthwhile to emphasize that as soon as this hypothesis was accepted, an integration of an equation for interacting gravitating masses with a model of interaction of "electrical fluids" right away transformed the latter. It can be seen at least in the Coulomb's formulations of laws for charged solids. They are already expressed not in terms of density of electrical fluids in extensive solids, but in terms of "infinitesimal particles" of such a fluid, its densities in points.[27] The latter means that along with a hypothetical equation for interaction between charges a new theoretical scheme was introduced, where such abstract objects, as point charges ("densities of an electrical fluid in a point") appeared. The displayed hypotheses, which correctness was justified by Coulomb, led to a discovery of the famous law of electrostatics.[28]

Thus, the procedure of mathematization of theoretical knowledge not infrequently leads to changes of originally introduced theoretical schemes. But by virtue of such changes they are again transferred from the rank of proven and justified theoretical schemes to the rank of hypothetical constructions, which need justification. That is why in the classical physics we can talk about two stages of constructing singular theoretical schemes as hypotheses: stages of their construction as informal physical models of some field of interactions and stages of possible reconstruction of theoretical models in the process of their connection with mathematical apparatus.

On the highest level of developing theoretical knowledge these two aspects of a hypothesis flow together. But on early stages of physics' evolution, when theoretical knowledge about new fields of phenomena just started to form, these two aspects of constructing hypothetical versions of a theoretical scheme could be separated.

It is important, however, that in both cases a stage of introducing hypotheses runs according to general laws. Even when we are talking about reconstructing a theoretical scheme under the influence of mathematical means, cognitive movement reproduces all main features, peculiar to the process of formation of a hypothetical model. At this stage mathematical means are transferred to a new field with the help of analog models (this whole process of choosing and applying analogies is purposefully aimed by the scientific picture of the world).

A little later we will show how this process went as applied to theoretical schemes of magnetic and electrostatic induction created by Faraday. For the moment we mention that Maxwell's mean for integrating theoretical schemes of electrostatic and electromagnetic induction with equations were analog hydrodynamic models which allowed to transfer equations from hydrodynamics into a new knowledge field. Concerning a reason for analogy between the processes of hydrodynamics itself and the field of electrical and magnetic influences, it was rooted in the accepted Maxwell Faraday's picture of the physical reality. The latter, as was shown above, displayed interaction as a continuous changing of forces in space, and that is why it easily allowed to see analogy between mechanics of continuums and electromagnetism.

The same way a transfer of a Newton's equation for gravitating masses to a field of electrostatic interactions, when the Coulomb's law was derived, was determined by applying an analog model of point masses, connected by the forces of gravity, to a situation with charged solids. The analogy of such kind itself was possible only due to that after Franklin and influenced by experimental and theoretical achievements of the "physics of electricity and magnetism", Aepinus, Simer and Priestley elaborated a picture of the physical reality, which became a modified version of the Newton's picture of the world. It assumed that a quantity of matter which characterizes a mass of Newton's corpuscles, can be combined with some quantity of material of an imponderable electrical fluid and that at the same time with mechanical forces in nature act electrical and magnetic forces, which are instantly transmitted from one solid to another in the absolute space.

Therefore it is clear that physical pictures of the world, participating in formation of theoretical schemes on a level of their introduction as a hypothesis, determine a strategy of a theoretical search. They orient a researcher, from which field of knowledge in physics he can take the initial abstract objects in order to build new theoretical schemes, and help to find a network of connections of such objects, expressed both informal and in a form of mathematical dependencies, which can serve as a mathematical apparatus of a future theory. Therefore if a researcher had chosen a picture of the world, just by that alone he had chosen a program of the future theoretical movement, a global strategy of a theoretical search.

The heuristic role of the pictures of the world in a process of formation of theoretical knowledge not once was specified in philosophical and historico-physical literature. In our point of view, this analysis is performed in Russian investigations most fully. Above-stated argumentation are pretending for proceeding and concretizing such studies regarding a problem of mechanisms that form hypothetical models, underlying a scientific theory.

In foreign logico-philosophical literature dedicated to problems in epistemology, for a long time under the influence of the positivistic tradition when it analyzed processes of forming a theory, excluded the very introduction of a question about a role of the pictures of the world in this process. In many ways exactly with this a refusal of the rational analysis of a scientific discovery's process was connected. It was thought that an act of introducing a hypothesis itself is only a product of a researcher's bold guess (and it is a matter for a psychologist, and not logician to inquire into the reasons of such conjecture).[29]

Some turn around regarding philosophy of science to problems which were earlier qualified as "senseless metaphysics" happened during a post-positivistic period of the Western philosophy of science (T. Kuhn, S. Toulmin, P. Feyerabend, M. Polanyi, I.Lacatos and others).

But the insufficient differentiated description of the theoretical knowledge structure didn't let them clearly distinguish its components, as the picture of the world and theoretical model. As it was stated above that the main notions in concepts by Kuhn, Lacatos, Toulmin and other, more specifically notions "paradigm", "research program", "rational ideas regarding regular order of Nature" are used very ambiguously.

If, for example, Lacatos more differentially considered the structure of the scientific knowledge, then already within the scope of his conception an idea about various types of research programs could have been interpreted, which differ by a breadth of range of phenomena and forms of their generalization. The picture of the world could appear as a kernel of a global research program, relating to which more local research programs form, having their own kernels and their own "safety belt of protective hypotheses". In as much as Lacatos did not elaborate a similar classification of a research program, just a general approach was contemplated and the term "research program" itself, designating heterogeneous components in science, does not clear up their correlation and interaction. In its turn, it does not allow to investigate concrete mechanisms of such interaction and to discover concrete procedures of an introduction of a scientific hypothesis. The same limited nature is peculiar to researches by Kuhn, Toulmin and other representatives of post-positivism.

There is, however, one aspect which is especially emphasized in the works of Lacatos and Kuhn and which should be attributed to strong points of their investigations. They are an idea of struggle between research programs (availability of some paradigms, following Kuhn's terminology), which are characterizing a development of the scientific knowledge. With reference to natural scientific pictures of the world as determinants of a strategy of a theoretical search a given assumption means that on the one and only stage of scientific evolution a number of versions about the picture of the world can compete with each other. Thus, from the history of classical electrodynamics it is clear that about the same period two alternative approaches are constituting to an analysis of electromagnetic interactions: a picture of the world, presuming a description of interactions of nature from the positions of instant transformation of forces on the right line in a vacuum (developed in Aepinus's, Priestley's, Coulomb's, Ampere's and Weber's electrodynamics), and the picture of the world which is based on a notion of "lines of forces", filling the space between solids (Oersted, Wollaston, Faraday and Maxwell). Analogously competing pictures of the world, which determined a struggle of Cartesian and Newtonian schools in mechanics, can be singled out.

Every one of the physical pictures of the world introduced in science was going through a long lasting evolution, changing and detailing under the influence of more new results of the theory and experiments, which it generated.

A researcher proceeding to solution of these of those problems, already by choosing them implicitly also chooses the picture of the world. In this sense Kuhn is right when he mentions that a choice of paradigm determines a choice of scientific problems. A difference in the pictures of the world, accepted in various scientific directions, is capable to evolve also a difference in the problems proposed by them. As Kuhn emphasizes, "a paradigm of one scientific community" can even exclude a setting of problems, which are considered the most important ones in another community. A similar thought, but formulated in terms of "methodology of research programs", can be found in Lacatos's works, who shows that a program's kernel provides positive and negative heuristics, that is to say determines a range of main problems and methods of a research and at the same time can prohibit setting of a series of other problems, which do not have any sense within the scope of the given research program.

There is a grain of truth in this opinion. It is comprehendible that for advocates of the Ampere's line in electrodynamics the main problem of the Faraday's school – to study forms of lines of electrical and magnetic forces in space and the character of their changes in time did not make any sense. Of course, incompatibility of the objectives of two different lines of investigations, in the foundations of which lie different pictures of the physical reality, never is absolute (Kuhn does not consider enough this aspect of the question, excessively exaggerating incompatibility of a setting of research problems within the limits of various paradigms). Taking into account that competing theories must explain some general integrity of experimental facts, even alternative lines of investigations will contain general research problems. But the approach for their solution itself will be different, and in this sense it is very rightful to think, that a kernel of research problematic and the form of setting of theoretical objectives in many respects is determined by the picture of the world. The investigator's choice of the picture of the world allows not only to establish the scope of theoretical objectives but also helps to find specific means for their solution. In theoretical researches as such some types of abstract objects, already collected by previous development of science, and mathematical instruments, formed in definite fields of scientific knowledge appear. The picture of the world orients a researcher for investigating these means, which is a necessary condition of introducing new hypotheses.

For constructing new theoretical schemes on the stage of their formation it is sometimes enough for a researcher to use a picture of the world, which is already established in science. For example, matters that stood with the situation of Coulomb's discovery of a theoretical law, describing interaction of charges. A hypothesis of infinitely small fluids had appeared without any substantial preliminary corrections in the picture of the world, developed by Aepinus, Simer and Priestley. But often when new theories are created scientists have to enter changes in the earlier formed pictures of the physical reality and depending upon a reconstructed picture introduce new hypothetical models of explained phenomena. For instance, Galileo, who elaborated theoretical schemes of uniform rectilinear motion, free fall of solids, movement on inclined plane and so on. Faraday acted analogously when he theoretically explained Oersted's experiments, and later phenomena of electromagnetic and electrostatic induction. In order to create hypothetical schemes, dedicated to explain conforming experiments, Faraday had to introduce preliminarily a new concept of space between solids, having assumed that it is filled with "lines of electrical and magnetic forces". The concept itself already radically changed the physical picture of the world. Changes of such type always are fairly revolutionary steps and demand attracting philosophical ideas, providing a special consideration of the available empirical and theoretical material. An impulse for such changes are usually unexpected from the point of view of former notions about nature experimental facts and theoretical conclusions (for instance, in elaboration of the Faraday's picture of the world an important role played a discovery by Oersted "rotary influence" of current on the magnetic arrow and revealing by Faraday himself rotation under the influence of magnetic forces, evolved by current; this directed him to a thought about the vortical character of the magnetic forces).

But comprehension of such facts and conclusions itself is a very delicate question, in solution of which philosophical ideas and various factors of sociocultural determination of cognition participate. As much as the picture of the world serves as a special bridge between a "population" of theoretical knowledge and culture generated by it, as much it serves in the becoming of the picture of the world whether or not value factors participate. In a certain sense we can think that such factors influence also the process of forming hypothetical models in science.

But if to abstract from the moments of psychology of creative work and look only at the logic of discovery, then an influence of value factors on the introduction of concrete scientific hypotheses always appears as mediated by the picture of the world. The latest circumstance allows to distinguish two aspects in a problem of hypothesis's formation: 1) an analysis of mechanisms of changes in the picture of the world (here we should take into consideration the influence of empirical and theoretical knowledge, from one point of view, and influence of philosophical ideas and a series of sociocultural factors from the other point of view); 2) an analysis of mechanisms of influence of the picture of the world on formation of theoretical models.

In the research practice these aspects are interconnected, and sometimes you can get an impression that all show aspects of the discovery process as if glued to each other. But in an analysis these aspects have to be distinguished.

Frequently such a distinction is troubled when a researcher introducing fairly substantial changes in the established picture of the physical reality, does not describe them, sometimes even does not give a meaning to them as global changes in a strategy of theoretical search (even though they, essentially, are such).

New concepts of systemic organization of the processes in nature studied by science, which directly precede a concrete scientific hypothesis, can be introduced implicitly as a result of introducing a new physical principle. Thus, Faraday's idea of space, filled with changing lines of forces, was proposed in the beginning as a physical principle that only has to do with electrical and magnetic forces. But when we read Faraday's texts from "Experimental researches in electricity" it is not hard to discover that this principle for Faraday himself was thought about through a concept of space, continuously filled with material, where there are solids as centers of forces and where the forces are transferred from point to point. This was a new picture of nature (in the strict sense of this word), introducing a notion of fields of forces as a special reality.[30] It's possible that in the beginning it didn't have such a clear meaning obtained after experimental and theoretical investigations by Faraday that proved real existence of electrical and magnetic forces lines. But in the general sense it definitely was proposed by Faraday from the very beginning (we should take into consideration that in the Oersted's works, on whose ideas Faraday rested upon, already similar conceptions about space as an arena of "forces conflict" can be found).

Causes by which a researcher will not publish the very first forms of the picture of the world, are founded in that these pictures appear in the beginning only as preliminary images of the physical reality, which yet have a small number of endorsing experimental and theoretical results. In some sense they, of course, must be resting upon experimental facts and theoretical generalizations from the former period of developing science. After all, their appearance itself owes consideration from new positions of that empirical and theoretical material, which looked like an anomaly regarding the earlier accepted physical picture of the world or provoked problems when it was coordinated with the indicated picture. But it is not yet enough to assimilate already known facts to settle in science in a status of a new picture of the physical reality. Everything depends on as far as experiments and theoretical hypotheses will give productive results generated by the new picture of the world, and partly from how successful a new picture of the physical reality to obtained within the scope of competing with it research lines.

If the picture of the world comes through all these ordeals, then one from embryonic stage comes to mature state. On this stage it is openly propagandized by researchers and it is accepted as a general scheme of seeing investigative situations or, speaking in Kuhn's terms, becoming a "paradigm" and accepted by a wide scientific community.

But not all preliminarily introduced pictures of the world have such a destiny. Many of them turn out to be unproductive and perish not exiting from embryonic condition. That is why a researcher, having formulated for himself a new system of conceptions of the studied reality, does not rush to introduce it as a picture of the world until when on its base will not be displayed a series of hypotheses which will go through a justification by experience and having turned into a theory will not predict new ones, earlier unknown facts. That is exactly why sometimes it is hard to describe a picture of the world as a basis of definite direction in researches.

But from a principal point of view it is important that it exists and that an introduction of hypotheses on this stage can't do without the picture of the world. Its functions in the beginning phase of investigations is that it purposefully aims at constructing hypothetical models, from which fields of already established knowledge abstract objects and structure in which they must be immersed are suggested.

Justification and transformation of a hypothesis in a theoretical model of an object

Hypothetical models obtain a status of theoretical concepts of some field of interactions only when then they go through procedures of empirical justification. It is a special stage when it is proven that its primary hypothetical version can appear as an idealized representation of the operational structure, which expresses substantial features of exactly those experimental-measuring situations, within the scope of which peculiarities of the studied in theory interactions are revealed.

In our point of view, in researches on scientific methodology people do not pay enough attention to this side of the matter and limit themselves to a simple verification of the fact that an introduced theoretical model is set to a display of a structure of a studied object in that case if the derived within its scope prediction of empirical dependencies are coordinated with dependencies obtained on a base of a real experiment.

A given verification, of course, does not contain anything principally incorrect, but by virtue of its clearly descriptive character it does not show ways to explain predicting functions of a theoretical scheme and does not open objective sources of its contents.

When its hypothetical version was constructed a researcher gave abstract objects, which he used as the raw material for yielding a theoretical scheme, new hypothetical features.

A theoretical law, which expresses a connection between shown hypothetical features of abstract objects, on this stage also is a hypothesis. On the face of it seems that it is easy to be justified, having verified in experiments predictions, obtained on the base of law. In reality such justification is not a simple procedure.

Let us from this point of view consider a concrete situation of justification of a hypothetically introduced law, which took place in the real history of science. We will address to a period of electrostatics' development, when Coulomb verified in an experiment the fairness of the hypothetically introduced equation for interacting electrified solids.

It is well known that in experiments with torsion balance Coulomb obtained empirical dependency which conformed to a hypothetical law for charges (we will once more emphasize that Coulomb did not deduce his law only from experiments; proceeding with an experiment with torsion balance, he already had a hypothesis, which he checked with empirical facts). But a transfer from hypothetical equation to his verification in an experiment was not a simple step.

In an experiment Coulomb operated with volumetric spherical electrified bodies. Coulomb's law, introduced as a hypothesis together with a model of charges' interaction, was formulated not for extensive solids, but for point charges (we will use a modern term "charge" instead of the Coulomb's term "portion of electrical fluid of given density", taking into account that the meaning of these terms is identical). And, strictly speaking, it was unclear, if it is possible to transfer from the value of a point charge to the value of a charge, distributed by the volume of some solid. In other words, in order to verify a hypothetical law, we need to have a compounding of connections between that and the values calculated in experience. Coulomb in the beginning didn't have this compounding. To obtain it, it was needed for him to prove that hypothetical features of a charge "to be point" does not contradict those characteristics of interaction between charged bodies, which were found in real experiments in electrostatics. The proof of such kind consisted in introducing a point charge as an idealization, resting upon real experiments in electrostatics. From experiments it was known that in solids different by volume a charge of an identical density can be concentrated, and in one and only solid – charges of an identical density. Resting upon these features the following mental experiment could have been proposed: mentally decreasing a solid's volume, to save in it a charge of the same density and in extreme case to transfer to an endlessly small volume of a charge.

Thus, a hypothetical model of interacting point charges ended up to be justified as an idealized scheme of real experiments. From this justification exactly followed a method of connection between a value of a point charge and value of a charge, distributed over a solid's volume. It turned out that if to choose a fairly small circular charged solid then it must interact with another charged solid in such way as if their charges are arranged in the solids' center. Thus, in an experiment it was possible to check interactions between solids and to master how electrical power is changing depending from the distance.

From the said can be seen that a procedure of justification a hypothetically introduced model presumes a special verification of features which were provided with its abstract objects. These objects as if afresh are "building" by idealizing real experiments, explaining and predicting for which the model was designed. After that a hypothetical model appears as an idealized scheme of real experimental-measuring situations of that field of interactions, which it pretends to explain. Such justification turns a hypothetical model into a theoretical scheme of the given interactions.

It is possible in the general sense to formulate the main demands which must be satisfied by a justification of a hypothetical model. Having assumed that it is applicable to a new not yet mastered theoretically object field, a researcher thereby supposes that: firstly, hypothetical features of abstract objects of a model can be compared with some relations of objects from experimental situations of just that field, for explaining which pretends a model; secondly, these features are compatible with other determinative characteristics of abstract objects, which were justified by the former development of cognition and practice. A legality of such suppositions should be proven especially. This demonstration is executed through introducing abstract objects as idealizations that are rested upon new experience. Hypothetically introduced features of abstract objects are obtained within the scope of mental experiments, respecting to peculiarities of those real experimental-measuring situations, which are intended for explaining the introduced theoretical model. After that the scientists check if new features of abstract objects concord with those that were justified by the former experience.

In this process of the model justification operational definitions of those main physical magnitudes which figure in the formulation of a theoretical law are automatically created. Operational definitions appear as descriptions of an idealized experiment and change within the scope of which a conforming magnitude is introduced, and a description of ways of constructing respected idealized experiments on the base of those real experiments and changes, which are generalized by a theory. In that way a connection is reached between physical magnitudes that are introduced in equations of the theories, with experience. In the theory appears a compounding of this connections, and the rules of conformity are created.

This whole complex of operations which provides a justification of the features of abstract objects of a theoretical model by experience, we call a constructive introduction of abstract objects, and the theoretical scheme, which satisfies to described theories, is constructively justified.

Since a model is constructed as a hypothesis always an allotment of initial objects with new features take place. A constructive introduction of these objects is necessary even when it seems as if a hypothetical model simply and visually is compared with respecting experimental situations.

External obviousness and evidentness of a model do not guarantee that hypothetical features of its abstract objects have a justification in experience. Obviousness and evidentness can be connected with these features that are associated with experiments from other fields of knowledge, those where from every such object on a hypothetical stage of constructing a model was adopted. But a model is intended for explanation a new field of interactions, and it needs to be justified as an idealized system of such interactions, which have a certain attitude toward a new field of knowledge. That is why even in relatively common situations a researcher has to prove that every hypothetical feature of abstract objects can be obtained at the expense of idealizations that are resting upon the experiments explained by a given model.

From this point of view Faraday's activity on justifying theoretical scheme of electromagnetic induction is a very characteristic example. It seems that this scheme was very simply projected on experiments on studying electromagnetic induction. But an attentive analysis shows that fairly complex problems appeared here.

As it was emphasized before, abstract objects "conducting substance" and "magnetic power lines" in a model of electromagnetic induction were transferred from a field of knowledge about conduction current and magnetostatics. When these objects were connected within the scope of a model of electromagnetic induction, they were subjected for restructuring. A construct "conduction substance" earlier was defined by a series of features, connected with the current's flow in a conductor (the power of current, voltage, resistance). But in an induction's model it must have been defined also by a feature of appearance in it, i.e. *emf* induction. Analogously an object "force line" was defined in magnetostatics by a principle "to orient in a certain way the test magnet". Transfer of it into a model of induction demanded to define this object also through a feature "to evolve in a conductor *emf*". An important moment in this whole activity was that the joining of a new feature to each one of the mentioned abstract objects at the same time presumed a conservation of their former features – features of a conduction substance to "represent the quantity and direction of the magnetic force". But if these features were justified with experiments with conduction current and interaction between magnets in magnetostatics, then in relation to experiments on studying electromagnetic induction they became hypothetical features, of which legality of introduction was obligatory to prove especially.

In Faraday's texts are clearly found traces of such proof. Thus, explaining a phenomenon of electromagnetic induction by the effect of force lines on the conductor, Faraday introduced a new definition of a force line through its relation to a conductor, where current can induce. Magnetic force line is characterized already by that if a placed "across its wire would move along in every direction, there wouldn't be in it an aim for induction, whereas with motion in any other direction this aspiration would have come to pass".[31] This definition represents a description of a special procedure of a mental experiment, resting upon real practice, in the flow of which it was proven that an object "magnetic force line" can be introduced by a feature "to induce current in a conductor" without destroying all else determining it substantial means. It seems that such kind of proof is excessive, because a direct experiment convinces that "magnetic force" does not change its nature when experiments on electromagnetic induction are executed, that it easily installs when a magnet is simply taken from one experimental situation to another. Nevertheless Faraday especially conducts a described demonstration, taking into account that on the level of a theoretical description force lines appear as an idealized object. They are looked at as relatively independent from a character of the magnetism's source "independent substances", marked as carriers of some features, abstracted from reality. On this level a transfer of objects of one model into another quite does not prove their equality. It is necessary to justify this statement. A respective abstract object constructed in a system of a mental experiment, under a necessary condition that the latter would be projected on experimental basis, subjected for generalization within the scope of the created theoretical scheme. Only after transfer from other fields of knowledge the abstract object "magnetic force line" stops to be alien relating to the structure of experimental practice, generalized in a model of induction. Now it is organically included as an element in this model.

Analogously an abstract object "conduction substance" is justified, when it is proven that it is capable to include as one of the determining features "tendency to induction", not destroying its other main characteristics ("ability to be a conductor").

Such demonstration was executed the simplest way, because in experiments on studying electromagnetic induction conductors were used from the very beginning by the feature of "conducting current", which was evoked by a certain influence on the conductor of magnetism's source.

Only after executing all these demonstrations the hypothetically scheme of electromagnetic induction introduced by Faraday is turning into a theoretical model.

Thus, an important regularity in constructing theoretical schemes is uncovered: after their introduction as hypotheses, they are adapted to real experimental-measuring practice, which results must be explained and predicted by a scheme. A means for such adaptation is the constructive justification of a theoretical scheme.

We will mention that the procedure of such justification itself flows as a process of operations with objects of empirical schemes from real experiments and changes. Empirical schemes, replacing real experiments and changes, fix a form of special abstractions (empirical objects) real features and relations of objects that are interacting in an experiment. Using these abstractions, it is possible to operate in thought experiments with features and relations of real objects of experimental-measuring situations. That is why when a construction of abstract objects of a theoretical scheme occurs by way of idealizing real experience, all mental operations are executed with objects of empirical schemes. In this sense singular theoretical schemes adapt to the real experimental practice by means of its empirical schemes.

Constructive justification of a theoretical scheme provides its connection with experience and internal coordination of all determining features of its abstract objects. As it was mentioned above, at the expense of constructive introduction of object "force line"

Faraday had proven that its features "to be a source of *emf*" and "to show the direction of magnetic force" can be combined in one description and do not contradict each other.

But if all objects of a hypothetical model did not go through a procedure of constructive introduction, then there always exists a danger that a model would lead to contradictions in a theory, because objects that have exclusive features can appear. A possibility of such features' appearance is easily explainable. When a theoretical scheme is constructed as a hypothesis its abstract objects are transferred from other fields of knowledge and their former features are provided with additional features which conform to a new network of their relations. In this process a structure can be assigned. It shows a new object of research. But a destruction of abstract objects can also occur when one determing feature will exclude another – also determing – feature.

As an example we will quote well-known facts connected with the application of a non-constructive object in the Rutherford's model of an atom.

Rutherford's model was intended for explaining the results of experiments on scattering of alpha particles on the atom and appeared at the first hand as an accumulation of a structure of these experiments. Along with that it pretended generalization of all other experiments from atomic physics, somehow or another uncovering an atomic structure.

As a preliminary hypothetical model Rutherford used planetary model of an atom introduced earlier. But he gave this model a principally new status at the expense of justification its main hypothetical element: positive charge in the atom's center. Rutherford introduced it constructively, resting upon experiments with alpha particles. He defined a nucleus by the feature "ability to scatter alpha particles" and at the expense of idealizing real experiments he showed that positively charged atomic nucleus is a center of potential repulsive forces. The main progress in the development of theoretical models of atom's construction was realized due to Rutherford's activity. Nevertheless in Rutherford's model a theoretical object without a constructive status was preserved. An electron's feature "to move on an orbit", introduced hypothetically in order to combine this object with other elements of the planetary model, was not justified not in one system of procedures, resting upon real practice of atomic experiments. But this feature exactly was not compatible with other, also determining characteristics of electron. A well-known paradox of an emitting charge showed that an electron can't be determined by the feature "of a stable movement around a nucleus", because it contradicted with its other definition "to be elemental negative charge inside an atom". This paradox, known from the time when Nagaoka's model was created, remained in the Rutherford's model.

It is indicative that an existence in an atom's model of a construct with exclusive features found its expression in contradictions inside a system of theoretical knowledge, concerning a model. Here appeared two logically mutually exclusive statements: "atom is stable" and "atom is unstable".

An appearance of such contradictions is easily explainable, if we take into consideration that a system of theoretical statements is distributing knowledge, connections and relations between abstract objects of a theoretical model. That is why availability in a non-constructive object of exclusive features must sooner or later lead to an appearance in a system of knowledge judgments contradicting each other. Such contradiction serves a peculiar signal of a misfit of a beginning model, on the base of which had grown a given system of knowledge, with features of a real object. In this, obviously, lies a cause for those strong regulative principles, which consists in a demand of consistency of the knowledge system. A discovery of paradoxes always shows that a structure of a studied object is inadequately represented in its theoretical scheme, which in its turn sets an objective of a radical reconstruction of the latter in a new theoretical model. In the reviewed case physics had to rebuild Rutherford's system so that, having preserved an idea of an atomic nucleus, to eliminate a non-constructive element – a charge, moving along the orbit around the nucleus, changed it with a new abstract object (electron which wouldn't have a mentioned feature, but other ones were introduced, providing its existence as an elementary negative charge inside an atom with preservation of the latter stability). This objective in its final look was resolved within the scope of quantum mechanics.

Thus, availability of non-constructive elements in theoretical schemes can lead to paradoxes in theoretical knowledge. With that one more important aspect is brought to light: procedures of constructive introduction of abstract objects. These procedures allow to part constructive and non-constructive elements and models and by that stimulate development of knowledge, showing in which direction a model should be reconstructed.

It is indicative that one of the impulses towards developing quantum-mechanical models of an atom exactly was an aspiration to localize and then to eliminate such element, like an "electronic orbit", having saved with that all other features of objects of the Rutherford's model which have an empirical meaning.

Thus, a process of constructing a theoretical scheme is provided due to an interconnection of two general operations: 1) a transfer of abstract objects from other fields of knowledge and connecting them in a new system of relations within the scope of a hypothetical model; 2) a restructure of a hypothetical model and transforming it into a theoretical scheme at the expense of introducing its abstract objects as idealizations, resting upon the new empirical material (the one that must assimilate a created theory).

All these operations are executed as a conceptual activity of a researcher and represent one of the main cognitive procedures, providing a development of scientific concepts.

Abstract objects are always fixed in corresponding notions. In this sense a translation and restructure within the scope of a new model of abstract objects is equivalent to transformation of concepts from other fields of knowledge and their redefinition in a new field. Due to that a notion includes all new definitions, in which more fully and specifically features and relations of objects in a real world are reflected. In this connection we should again refer to a problem of relationships between a notion and an abstract object.

It was stated above that in logic a concept is looked at as a folded definition and is identified with singular proposition function $P(x)$, which obtains meanings of verity or falsity depending on which objects are substituted for variable x, that is which objects are assigned a predicate P. Influencing proposition function by special operators, for example, lambda-operator (in a concept of lambda-conversion by Church), an abstract object can always be singled out, conforming to this or that concept. There also exists an inverse operation which lets us to pass from an abstract object to a concept. From here we can conclude that enrichment of an abstract object with new features leads to development of notion content. A distinguished specific is revealed even more clearly, if we track how concepts function in a system of developed knowledge.

Let us consider a concrete example, connected with a process of applying concepts in already discussed matter of constructing a model of electromagnetic induction. When an abstract object "magnetic force line" is transferred from a knowledge field of magnetostatics a coordinating notion was used, which fixed a force line as an object of study. A development of this concept can be described in our system of analysis in the following way. In the beginning a theoretical scheme was created, which represented in cognition substantial characteristics of magnetic interaction, found in experiments from magnetostatics (experiments by Oersted on orienting a magnetic arrow by conductors with current, experiments by Coulomb on investigating magnetic interactions, experiments by Faraday on researching orientation of ferrous fillings by magnets and conductors with current and so on). Object structure of all these experiments was represented in an idealized form, as an interaction of ideal magnetic arrow with a source of magnetic force. Within the scope of the given model the following relation was fixed: a magnetic arrow, moving in a direction to the magnetic force, should always be oriented about a tangent to this direction. A mapping of a model on the introduced by Faraday physical picture of the world allowed to look at it as a representation of a special object in research (direction of magnetic forces). Fixed in a model substantial relation represented as a definition of this object. Thus, "direction of the magnetic force" was defined through its substantial feature and was characterized as a line which is described by an ideal magnetic arrow, "if it moves in any way in direction of its length is such way that all the time remains as a tangent to a line of motion".[32] It is very clear that the subject of research is determined through connections fixed in corresponding him theoretical scheme. Since these connections exhaustively characterized a mode of its existence (essential features of an object), then within the scope of a definition, expressed in a form of a respective judgment, a subject *S* turned out to be identical to a predicate *P*. Due to this feature happened a folding of a definition into a notion, which is denoted by some term of the theoretical language (in our case it was a term "magnetic force line").

All these operations introduced "inside a notion" a special "plan" (method) of building an ideal object. A given plan could always have been realized in that sense that a concept could have been developed into a definition and by that to introduce a corresponding ideal object. In this connection a notion can be characterized in double ways: as a discovery of structure of connections in reality, represented in a form of an ideal object, and as an expression of operations (methods) of constructing a given object. Both these sides of a concept – an "objective" content and "operational" function – are inseparably linked with each other (an object of research is reflected in a notion of the activity form). We will mention that in the positivistic and operational interpretation of a concept its second function is parted from the first and contradicts it, which prevent us to comprehend the nature of the notion. Contained in a conception the way of constructing an ideal object allows using the latter in a function of an operational object, applying it as a means for creating new models. If before the appearance of a notion an object could have been distinguished, only pointed at a conforming theoretical scheme, then now this scheme in its substantial characteristics turned to be "folded" in a concept. The main features of the studied object field, fixed in a notion, can be represented in a form of an abstract object – a carrier of given features. By these features a shown object can reestablish by using a notion as a means of studying a new object domain. From this it is obvious that with the help of concepts in a process of a theoretical research turned out to be possible to form abstract objects, turning them from objects of research into ideal operational objects. Due to that objects obtained in one field can be used as the raw material for constructing theoretical models in a new field.

But, as we mentioned before, an inclusion of an abstract object into a network of relations with other objects when hypothetical models are built from them, as a rule, requires its transformation. In its initial form an object is usually not included into a new system of connections, that is why it is reconstructed, "adjusting" it to conditions of a new existence. Constructive introduction of an object accomplishes this reconstruction. Along with that the results of all procedures, that led to appearance in an object new characteristic feature, are formalized in a new definition of this object. Like that was, for example, with introducing a force line by a feature "appearance in a conduction substance *emf* induction". If such an introduction is realized constructively, new object features are included in a concept. Thus, a use of an abstract object, "keeper" and "translator" of which is a notion, in the functions of an operational object leads to redefinition of a concept. A notion includes in itself all new definitions after every operation of constructive rebuilding of an object in new fields of knowledge. In this enrichment of a notion with definitions exactly consists the development of its content.

Operating with concepts in a system of theoretical research flows as an appearance of their connections with each other and their redefinition in a system of these notions, until "categorical structure", reflecting a studied object domain would not be built. Behind all this if we stick to a stated above conception stand operations of translation and constructive introduction of abstract objects that make a theoretical scheme, with subsequent explication of knowledge about the structure of object field, which is reflected in a theoretical scheme.

Thus, in a process of cognitive activity a researcher creates a hypothetical version of a theoretical scheme and realizes its adaptation to empirical material. This adaptation is the one that turns a beginning hypothetical scheme into a theoretical explanation of experimental facts. Even though a theoretical model on the first stage is built as if "from above" in relation to empirical schemes of real practice and isn't deduced directly from experience, in the end it turns out to be an accumulation of real practice and representations of respected natural structures, discovered within the scope of practical activity.

Observe that parting a stage of constructing a model as a hypothetical scheme and a stage of its justification we had executed a simplification in purposes of analysis's convenience. In the real research activity itself during the process of a model's justification often a return to a beginning point of movement happens, when it is found out that a model can't assimilate empirical material, and that means that it is unsatisfactory in some elements. Then it is reconstructed at the expense of transformation of its abstract objects and again is subjected for verification. This motion between a layer of theoretical knowledge and empirical material, when a movement in a theoretical layer is replaced by a movement from it to empirical schemes, and then again from empirism to theoretical objects, repeating until there wouldn't be constructed a model which expresses a structure of the real practice, for generalization of which pretends theoretical knowledge.

All this activity is accomplished with mapping already justified theoretical scheme on the picture of the world. In this process a theoretical scheme, displayed on the stage of its justification as an expression of substantial features of experimental-measuring practice, obtains "ontological" status and appears as a figure of a structure of respective interactions.

In their turn, under the influence of created theoretical schemes a concrete definition and the development of the picture of the world happen. Theoretical schemes due to constructive justification obtain more reach content in comparison with an initial hypothetical model (sometimes such models in general reconstruct in this process, which very radically changes their content in comparison with the initial version). This new content leads to changes in the picture of the world. Thus, under the influence of the theoretical schemes in magnetostatics created by Faraday, electrostatic and electromagnetic inductions the picture of the world, which formerly expressed the ideas of short-range action in a very abstract form, much more concretizes them and, finally, introduces a concept of the power field as a transfer of interactions from point to point in the filling space material surroundings (at that this concept is extended also on the field of gravitation).[33] This was the first openly proposed physical picture of the world, where a notion of a field as a special material began to play the paramount role.

Not less showing facts of the back influence of theoretical schemes on the picture of the world are the consequences of a justification by Rutherford an idea of nuclear atomic structure. This idea fast entered the picture of the physical reality, created in physics in the beginning of the 20th century, evolving a new circle of research problems (a nuclear structure, specialties of "nucleus material" and so on).

CONSTRUCTING A DEVELOPED THEORY IN THE CLASSICAL SCIENCE

An introduction of theoretical schemes as hypotheses with their consequent constructive justification is the main cognitive procedure in the genesis of theoretical knowledge. This procedure determines not only the process of becoming singular theoretical schemes, but also transition from them to a developed theory. In the classical physics a creation of a developed theory usually began before that separate aspects of researched interactions were reflected in some collection of individual theoretical schemes and laws.

The first fundamental developed theory of physics – Newton's mechanics – was created as a generalization of theoretical models and laws of such types of mechanical motion, as oscillation of the pendulum, free fall of solids, movement of solids on the inclined plane, movement of planets (Kepler's laws), etc. Analogous situation is viewed in the history of thermodynamics and classical electrodynamics, where single aspects of the studied processes were expressed in the developed part of individual theoretical schemes and laws long before that the first generalizing theories of these parts of physics were constructed.

A developed theory is constructed on the basis of synthesis of individual theoretical schemes. They are included in the structure of the theory in a transformed form and appear as deduced (constructed) from its fundamental theoretical scheme. Conformably all individual theoretical laws appear as a consequence of fundamental laws of the theory.

In this connection a question arises: which way is a kernel of a developed theory created – its fundamental theoretical scheme and equations, expressing the main laws of the theory, connected with it?

Two assumptions are possible about the method of forming a developed theory. We can concede that a fundamental theoretical scheme and connected with it mathematical apparatus are introduced as hypotheses in already developed form, and then they are justified by those individual theoretical schemes and laws which a theory must include in its content. But another assumption is possible, according to which formation of a fundament of a developed theory happens gradually, by way of consequent synthesis firstly of some close laws, generalized in a theory, and then laws that belong to more far out fields of studied theory interactions.

History of classical physics evidences, more likely, in the favor of the second assumption. It shows that already after constructing a series of individual theoretical schemes begins a process of their extrapolation to the adjacent fields of knowledge, in order to unify laws explaining a certain field of interactions, and to explain all phenomena of this field with a unified point of view. Concerning this we can cite fairly many instances.

A conclusion by Galileo of laws of oscillation as a special case of the law of motion on an inclined plane is characteristic. Very significant is an attempt made by Ampere to represent a basic law of electrodynamics as a law of force interactions of currents and to conclude as a consequence a law by Biot-Savart and the law by Coulomb for magnetic poles.

Assuming that synthesis of such kind is a norm for theory construction, at least for a stage of classical science, let us define a central aspect of this synthesis.

For developed physical theory a relatively high state of mathematization is always characteristic, and that is why a constructing of a mathematical apparatus usually is considered as a main objective of a theoretical generalization. But we shouldn't lose sight of the second, not less important side, more specifically a connection of mathematical formalism with theoretical schemes, which provide its interpretation. Hence when historical material is analyzed we should pay special attention to a connection between these two aspects of the cognitive movement (between constructing mathematical apparatus and constructing its interpretations).

One of the dangers which await a philosopher and methodologist when they look at the complex and scantily explored problem of developing a scientific theory, is the extrapolation on any sort of theoretical synthesis of the methods of theoretical generalization discovered by modern science and widely distributed in it. Such an extrapolation can be accomplished unconsciously, under the influence of the established tradition, but it will get in the way of finding ways for theoretical search, characteristic for every stage of scientific development. In the end this can harden a solution of the main methodological objective: to find out what are changes in the ways of constructing theoretical knowledge in the course of physical evolution and what are the invariants of historically changing forms of cognitive activity, in other words its constant and repeating features, which belong to main regularities of a theoretical investigation.

In order to avoid preconceived notions relatively to the methods of constructing a developed theory at this or that stage of physics evolution, we should refer to an analysis of real historical material. This material should be taken not so much from text books on the history of physics, which objective is to give a short description of main stages of physics evolution (empirical and theoretical discoveries) and to create a main picture of its historical development, but in original texts of the creators of scientific theories in person, in texts that preserved the results of motion of their creative thought. Such texts serve as the general empirical basis, on which a historian of science and a methodologist verify their hypotheses.

With a purpose to find out how the process of constructing a developed theory in the classical physics went, let us look at one of the most important fragments of its history – the development of the Maxwell's theory of electromagnetic field. A choice of this fragment from the history of science for logic-methodological reconstruction represents a special interest because of following causes. First, it is interesting from a point of view of the completeness of "empirical material", very rare for a philosophical analysis of history of science (all main texts, fixing principally important for stages of development of the Maxwell's theory, beginning from the first, sketch versions and ending with its relatively accomplished form, are preserved). Second, the creation of a theory from the beginning until the end was realized here by one researcher, in connection with which logically necessary operations of theoretical construction can be distinguished, without a special temporal reconstruction of the historical facts, because in this case a logic of ideas can be seen in their real historical consequence. That is why on the material of the history of

Maxwell's electrodynamics one can track with a fair fullness the main peculiarities of hypotheses introduction and their transformation into a developed scientific theory.

Analysis of the reviewed historical material is also interesting because an uncritical transfer of modern methods of research relation to Maxwell's theory of electromagnetic field is often realized, which in the Maxwell's creative work in the best case can be found only in an embryonic form. History of Maxwell's electrodynamics most often figures in our literature as an example of effectiveness of a developed theory's apparatus. With that there is an opinion that a fundamental idea of the bias current which led to a formulation of Maxwell's equations, owed to the ideas of symmetry.

It is accepted that having written in the differentiate form the laws by Coulomb, Biot-Savart, a qualitative law by Faraday for electromagnetic induction and having expressed in equations a fact of principal closure of magnetic force lines, Maxwell had discovered an absence of symmetry in the found equations. More specifically, the equation $rot\ \mathbf{E} + \frac{1}{c}\frac{\partial \mathbf{B}}{\partial t} = 0$ (1) includes derivatives as by coordinates, as by time, whereas the equation $rot\ \mathbf{H} = \frac{4\pi}{c}\mathbf{J}$ (2) includes only derivatives by coordinates, and derivatives by time are absent.

For restoring a symmetry Maxwell added into equation (2) a new term – a derivative from a vector of electromagnetic induction (displacement) by time $\frac{\partial \mathbf{D}}{\partial t}$, which (being multiplied by $\frac{1}{4\pi}$) appeared as a mathematical expression for bias current. By that a system of equations of electrodynamics obtained a closed character and turned into a mathematical apparatus of theory. From it naturally followed an equation by d'Alembert for intensity of electrical and magnetic fields, a solution of which straightly led to a prediction of electromagnetic waves.

Such a conception seems fairly persuasive, if taken into account the simplicity of the reproduced course of discussion and its external obviousness (data of experiment which Maxwell could depend on, really didn't allow him to deduce an idea about displacement current and directly didn't say anything about existence of electromagnetic waves). But an acquaintance with real historical material shows that a process of forming Maxwell's theory was going differently from that appears within the scope of a traditional understanding of a mechanism of mathematical hypothesis.

Let us try in a condensed form to analyze the main stages of this process, accenting our attention on its internal logic.

It is well known that the main objective which was achieved by Maxwell during the period of this theory's creation and which was introduced by the whole former flow of scientific development, was reduced to searches of a unified way for describing and explaining different aspects of electricity and magnetism.

By this time single aspects of electromagnetic interactions were fairly well studied and reflected in a whole series of relatively independent systems of theoretical knowledge. They included theoretical models and respective laws of electrostatics (Coulomb law, Faraday law for electrostatic induction), magnetostatics and interactions between stationary currents (Biot-Savart law, Coulomb law for magnetic poles, Ampere law), electromagnetic induction (Faraday law), direct current (laws by Ohm, Joule-Lenz and others) (pic. 4). This knowledge played a role of a peculiar empirical material on which Maxwell depended when he created a theory of electromagnetic field.

The main problem was in reducing all this interaction of laws to some generalizing expressions, from which it would be possible to deduce already available knowledge as conclusions. For this purpose it was needed to find a scheme of synthesis, which would provide a seeing of all material about nature of electricity and magnetism with a unified point of view collected by history. Such synthesizing scheme provided a preliminary picture of electromagnetic interactions, which was accepted by Maxwell and later had developed into electrodynamics picture of the world, established in physics in the end of the 19th century. This picture introduced a notion of electromagnetism as a transfer of electrical and magnetic forces form point to point in accordance with a principle of a short-range action. The introduction of this picture itself was prepared by the preceding history of science and first of all by Faraday's works.

The picture of magnetic and electrical processes, depending on principle of short-ranged action, competed with a contradicting picture of the physical reality, which during this period of history of electrodynamics was developed by Ampere and Weber. Their theoretical research was based on notions of electromagnetic interactions as a momentary transfer of forces between point charges and differentially small elements of current. By this approach, combined with a principle of long-ranged action, Ampere long before Maxwell tried to create a generalized theory of electricity and magnetism. As a whole he and then Weber accomplished to develop a fairly rich theory, though the latter had encountered many troubles, for example when it explained electromagnetic induction.

Maxwell's electrodynamics and Ampere-Weber electrodynamics for a long time competed as two alternative research programs. The victory of the Maxwell's line was gained only after constructing a theory of electromagnetic field and experimental discovery

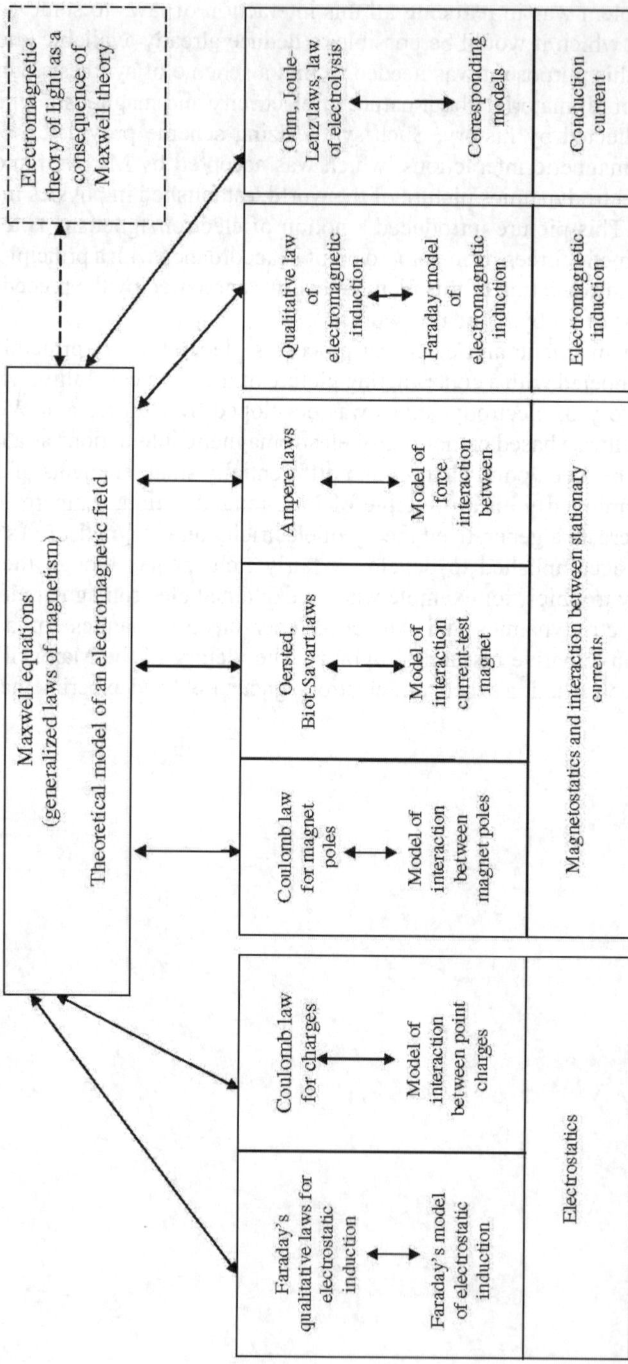
Maxwell equations
(generalized laws of magnetism)
Theoretical model of an electromagnetic field
Electromagnetic theory of light as a consequence of Maxwell theory
Faraday's qualitative laws for electrostatic induction
Faraday's model of electrostatic induction
Coulomb law for charges
Model of interaction between point charges
Electrostatics
Coulomb law for magnet poles
Model of interaction between magnet poles
Oersted, Biot-Savart laws
Model of interaction current-test magnet
Ampere laws
Model of force interaction between
Magnetostatics and interaction between stationary currents
Qualitative law of electromagnetic induction
Faraday model of electromagnetic induction
Electromagnetic induction
Ohm, Joule-Lenz laws, law of electrolysis
Corresponding models
Conduction current

Pic. 4

of the predicted theories of electromagnetic waves. But at the starting point during a period of its formation Maxwell's program did not have any privileges. Moreover, electrodynamics of Ampere-Weber was already developed and by this time had obtained fairly impressive results. Therefore the choice by Maxwell of an alternative picture of electromagnetic processes itself was a fairly courageous research step.

If one analyzes the causes of why Maxwell had accomplished this step, then we would have to state many factors, including of psychological and sociocultural character. But for logic of scientific cognition it is important that this step is prepared by the preceding development of physics. And as long as we are talking about a creation of a theory of electromagnetic field, then an introduction of a picture of the physical reality and based on an idea of short-ranged action, it was a logically necessary preliminary condition of the consequent construction of a mentioned theory. That is why it was necessary that if not Maxwell, then some other researcher had proposed this picture as a basis of a new program of electrodynamics reconstruction.

Faraday's and Maxwell's picture of the physical reality had right away shown concrete ways of synthesis of already known laws of electricity and magnetism. It defined a field of mathematical means with the help of which it was possible to describe very heterogeneous electrical and magnetic phenomena. These means that served as a basis for constructing an apparatus of the future theory of electromagnetic field were taken from mechanics of continua, in particular from hydrodynamics.

Now it is well known that exactly at the expense of transfer of mathematical structures from hydrodynamics over to a field of electromagnetic processes

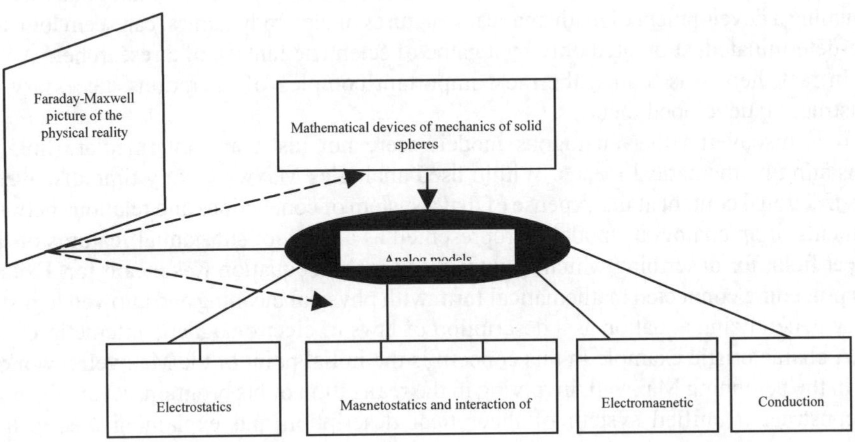

Pic.5

Maxwell had built generalizing equations of an electromagnetic field. But in order to refer to this field in the very beginning of a research, Maxwell had to see the main structural similarity between very far fields (mechanics of continua and electricity and magnetism) beforehand. Such a similarity was found only due to a picture of the physical reality accepted by Maxwell. For comparison we will mention that Ampere and Weber introduced notions of electromagnetic processes as interactions of point charges and currents by instantaneous transfer of forces, which demanded an application in electrodynamics other mathematical structures, taken from Newton's particle mechanics.

A difference in the pictures of the physical reality, lying in the basics of research programs by Maxwell and Ampere-Weber, called forth also a difference in mathematical methods of these programs.

A transfer by Maxwell of equations from hydrodynamics over to a field of electromagnetic phenomena was realized by constructing analogous models of hydrodynamic and quasi-mechanical type (pic. 5).

This aspect of the question is fairly well illustrated in historical physical and philosophical literature. When one characterizes a procedure of creating a mathematical apparatus of Maxwell's electrodynamics, use of analogous models plus a race in strictly speaking mathematical means are always described as a necessary feature of the given procedure.

Outwardly everything looks as if the apparatus of the Maxwell's theory was created as a result of permanent introduction of more perfect equations, describing in a generalized form more new aspects of electromagnetism. But when a more attentive analysis is conducted, we find out that with Maxwell "mathematical analysis interflows with physical content",[34] due to which a theory of electromagnetism is possible to form.

Only with superficial examination a switch and transformation of analogous models, providing a development of mathematical structures of electrodynamics, can seem logically non-determinated, stipulated only by a game of scientific fantasy of a researcher.

In fact, here was hidden the most important complex of operations, necessary for constructing developed theories.

It is discovered that analogous models were not just some intermediate link for translating mathematical means. Within used analogies Maxwell every time discovered constructional content at the expense of that a system of connections and relations between elements of an analogous model he represented as a show of substantial features of that object field, for describing which an introduced by him equation was meant for. Exactly this procedure connected mathematical form with physical meaning and allowed to justify every generalizing equation as a description of laws of electromagnetic interaction.

A characteristic example in this concern is the initial point of the Maxwell's work.

In the beginning Maxwell, involving in the realization of his program, set an objective to construct a unified system of theoretical description and explaining electrostatic phenomena. For that it was necessary to deduce a unified generalized equation of electrostatics. A means for deduction of such an equation served an analogous hydrodynamic model, of which the main element is a single unlocked tube of current of some ideal incompressible liquid. This model allowed to transfer Euler's equation for fluids over to a field of electrostatic phenomena and use it as a hypothetical expression for generalized law of electrostatics. From here expressions for a law of electrostatic induction and the Coulomb law in differentiate form were obtained.[35]

Though the described situation is habitual, from the point of view of logic an absolutely unclear moment appears: why was it possible to interpret hydrodynamic equations in terms of electrostatic values? An answer to this question is provided by the procedure of applying an analog model. It turns out that Maxwell justified it as a display of substantial features of all experimental-measuring situations of electrostatics, after which an assumption about the possibility to interpret hydrodynamic values of Euler's equation in terms of electrostatics obtained a status of a proven hypothesis.

In a logically reconstructed view a procedure of this justification can be described in the following way.[36] In the beginning it was determined that a hydrodynamic analogy shows substantial features of phenomena of electrostatic induction. Already in the process of transfer of Euler's equations over that field a tube with incompressible fluid was liken to a force line, and a series of such tubes – characteristic for induction of a change in strength of force lines in space from point to point. But by that Maxwell had introduced, strictly speaking, a totally strange object – electrical force lines, existing outside of evolving charges. Such object didn't exist in the former theoretical concepts of electrostatics. In a scheme of electrostatic induction by Faraday force lines were shown as appearing in idealized dielectrics, limited by ideal charged plates, and depended on the quantity of a charge on plates (model of an idealized condenser). In Maxwell's model they were introduced by another feature. That is why it was required to specially prove legality of a new hypothetical quality of a force line. A demonstration was conducted by all rules of a constructive justification of an abstract object, which was built as an object of research, in a system of procedures, shown in a model of electrostatic induction.

Thought variations of charges on the facing of an ideal condenser of the Faraday's scheme and verification of the fact that with this energy in dielectrics sometimes decreases, sometimes increases, allowed to make a passage in the limit to a case when the whole electronic energy is concentrated in dielectrics. This was equivalent to a series of force lines, existing when then evolving charges are eliminated. Now the picture of the force lines, estranged from charges, ended up being idealization, reduced to a real experimental material. Exactly due to this within an analog model a content was distinguished, which allowed to represent it as an expression of substantial features of interactions, found in experimental situations on studying electrostatic induction.

In a similar way a hydrodynamic model of tubes was justified by Maxwell and in relation to situations of Coulomb's interaction of charges.

In order to rewrite the basis of hydrodynamic equations the Coulomb law, Maxwell had introduced a display of a charge as a point source (drain), from which evenly flows incompressible liquid, coming in closely connected tubes of variable cross-section. As much as tubes with liquid were already represented as electrical force lines, as much a model of point source became a portrayal of a charge, features of which were characterized through features of force lines. Thereby a hypothetical notion of a charge that is defined through strength of the created by it field was introduced.

But if we take into account that in a Coulomb's model a charge didn't have this characteristic, but was determined through a feature "to change the state of movement of another charge", then it was needed to prove equivalency of the both concepts. Only after that it was possible to consider a new equation for charges equivalent to Coulomb's law.

Constructive justification of a hypothetical notion of a charge was an easy to accomplish operation, if we keep in mind the possibility of a following experiment. In a scheme of Coulomb's interaction of charges, which concentrated in itself substantial features of the respective experimental-measuring procedures of electrostatics, it was possible to vary the value of charges. In the limiting case one charge – the "source of electrostatics" – could be thought as fixed, and the other – arbitrarily small. Then the latter turned into a "test charge", which does not influence at all the value and direction of electrical force, but only allows to characterize density of the force lines, coming through and evolving charge. In the result a charge – source of a field – turns to be defined only through characteristics of a field just as well, as through a feature to influence other identical charge. Accordingly inside a hydrodynamic model of tubes was discovered a constructive content, which relates to situations of the Coulomb's interaction between charges.

After all described operations a given model appeared already as a synthetic show of the most substantial features of experimental-measuring situations in electrostatics as a whole. Behind its external vivid form was conceived a theoretical scheme of electrostatics, which could have been explicated in a form of relations of following abstract objects: a charge, evolving a field, test charge and electrical force lines. Accumulating in itself basic and substantial features of all experiments and calculations in electrostatics, a given theoretical scheme by virtue of this represented an object of study of the distinguished field. It appeared as a display of structural specialties of a stationary electrical field. Such a vision of a theoretical scheme was achieved due to its mapping on the picture of the world, accepted by Maxwell already at a starting of his theoretical research.

A comparison with the picture of the world allowed to objectivize a synthetic model in electrostatics. Force lines, separated from charges, obtained the same status of physical reality, as a charge. In its turn, such ontologization of theoretical constructs, justified by their introduction as idealizations, which are rooted in a real experiment, formed in a physical picture of the world new concepts, specifying and developing beginning shapes of electromagnetic interactions. Physical reality of the "parted" from charges of electrical force lines conformed to notions of electrical force lines, which can be considered as existing in space relatively independent from its sources. Taking into account that an existence of such field itself could be justified only by electrical energy in preserved in polarized dielectrics and in absence of charges, in the picture of the world must have been present (or to appear due to mapping on it a theoretical scheme of electrostatics) counterpart of an ideal dielectric. Such an analogue was ether as a sphere when force lines can exist. Denote the ether, filling space, was introduced in a by Maxwell developed picture of the physical reality by these features, which exactly correspond features of an ideal dielectric.[37]

Thus, due to a creation of generalization theoretical model in electrostatics into the use of physical analysis principally new characteristics of a field, which were unknown before its creation, were introduced. Having obtained the same status of a physical reality that even a charge, electrical field principally could appear already not only as an evolved by a charge, but also as a generative charge (series of electrical force lines, tightened in a point, could be considered as a charge).

Generalizing equations in electrostatics introduced by Maxwell sorted with a new system of physical notions, due to which they obtained a justification as expressions for laws of stationary electrical field.

These equations expressed a deeper physical content than the preceding laws by Coulomb and Faraday, which turned now into singular cases of Maxwell's equations for electrical field.

The system of Maxwell's theoretical activity, connected with deduction of generalized laws of electrostatics, all the time was repeated in the process of accomplished by its grand synthesis of knowledge of electricity and magnetism.

Hypothetically introduced equations every time were justified as laws of this or that field in electromagnetism by finding a constructive meaning of analogous models. These procedures themselves were proceeding as operations with primary theoretical schemes of electrodynamics. In such schemes an invariant content was found, which then was fixed within every analog model. From these positions a construction of a mathematical apparatus of Maxwell's electrodynamics appears not only as a series of mathematical extrapolations. It turns to be tied up with a process of consecutive synthesis of the during the pre-Maxwell's period created primary theoretical models of electromagnetic interaction. It is not hard to see in this synthesis a cognitive movement, directed to the creation of the generalizing theoretical scheme, which should be the fundament of a future theory and appear as an interpretation of its apparatus. Exactly in this process of gradual crystallization of a fundamental theoretical scheme in Maxwell's electrodynamics that internal logic of a research was conceived, which purposefully aimed a selection of analog models. The latter were not just auxiliary means, something like scaffolding, which must be taken away, when the building of a theory is constructed. They served as special carcasses, a part of which became armature for erected walls of theoretical construction, was included in the "body" of a created theory itself, and the second, external part, connected with descriptive-graphic form of a model, were left as scaffolding, which eased the creation of a theory and which were eliminated after its creation.

In the process of becoming a mathematical apparatus of the theory of electromagnetic field a movement in mathematical means was constantly corrected by a motion in the sphere of real objects, constituting theoretical schemes in electrodynamics. That real historical material contains straight illustrations of inseparability of both types of cognitive operations.

In this concern it is extremely significant that when Maxwell couldn't distinguish in an analogous model a constructive content right away was suspended an advance toward mathematical apparatus in electrodynamics.

To all appearances from the field of vision of historians of science, even those who especially engaged in analysis of Maxwell's discovery, falls out the following extremely important fact. It turns out that Maxwell, already fairly far advanced in constructing a mathematical apparatus of a theory, had encountered with insuperable problems exactly in that point where it seemed the most adequate mathematical form of laws of electrodynamics was found. This happened on that stage of theoretical synthesis, when a generalized law of electrostatics was obtained div $\mathbf{D} = 4\pi\rho$,[38] introduced the equation $rot\ \mathbf{H} = \frac{4\pi}{c}\mathbf{J}$, generalizing the laws by Ampere, Biot-Savart and Coulomb's law for magnetic poles,[39] and, finally, was proposed an expression $\mathbf{E} = \frac{\partial \mathbf{A}}{\partial t}$, on the base of which Maxwell tried to obtain a mathematical law of electromagnetic induction.[40] The latest expression Maxwell interpreted as a connection between strength of electrical current of electrical field E and so-called "electrotonic vector" A, which he introduced as a characteristic of potential possibility of appearing *emf* when energy on a magnetic field is changed. In modern understanding the expression $\mathbf{E} = \frac{\partial \mathbf{A}}{\partial t}$ corresponds to a definition of a field E through a vector-potential A. If we look retrospectively, then Maxwell, in effect, already "held in his hands" the mathematical scheme of electrodynamics, which in appearance was very close to its modern formulation (electromagnetic interactions were represented as a relation between electrical, magnetic fields and currents; the relation "current-field" itself was set in energetic form, by introducing a vector-potential,[41] which, as it is known, corresponds to modern, Lagrange formulation of a theory).

From positions of a method of mathematical hypothesis Maxwell was left to do very little: 1) take a rotor from both parts of the obtained equation $\mathbf{E} = \frac{\partial \mathbf{A}}{\partial t}$ and , taking into account introduced by him earlier a definition B = rot A (where B is a vector of electromagnetic induction), to obtain from here a law of electromagnetic induction in a form $rot\ \mathbf{E} = -\frac{1}{c}\frac{\partial \mathbf{B}}{\partial t}$; 2) to compare the latest correlation with obtained earlier equation for Biot-Savart law $rot\ \mathbf{H} = \frac{4\pi}{c}\mathbf{J}$; 3) to find out that in equation $rot\ \mathbf{H} = \frac{4\pi}{c}\mathbf{J}$ for full symmetry is missing a term $\frac{1}{c}\frac{\partial \mathbf{D}}{\partial t}$, matching displacement current.

But in reality exactly in this point, it seemed to be the most perspective, from the point of view of traditionally method assigned to Maxwell, he himself completely renounced from the later development obtained by it formalism, and, in effect, began to build mathematical apparatus of the theory all over again.

This fact, totally unexplainable within the scope of traditional concepts of methods of Maxwell's research, can be easily understood, if we take into account the connection between development of formalism of the theory and procedures of constructive justification of theoretical schemes.

Historically Maxwell approached the explained above formulation of laws of electromagnetism, depending on conception of stationary force lines.

Modeling electrical force lines by means of notion of uniform flow of incompressible liquid in unlocked tube, Maxwell achieved in the beginning generalization of the knowledge in electrostatics. Then the analog "tube model" was extrapolated over to a field of magnetostatics and interaction between stationary currents.

Taking into account a vortical character of magnetic forces, Maxwell altered the initial analogy and introduced a notion of closed tubes of a current of incompressible fluid.[42] By the means of a new model, revealing in it constructive content, he found the generalizing law $rot\ \mathbf{H} = \frac{4\pi}{c}\mathbf{J}$, containing a connection between continuous current and (evolved by it) stationary magnet field. From this law expression for Coulomb laws, Biot-Savart and Ampere as a singular case was deduced.

Thus, two important equations were found at the disposal of the creator of the theory of electromagnetic field, one of which characterized continuous current, and the other – magnetostatic field. Their connection he hoped to find in the process of search for mathematical expression for generalizing law of electromagnetic induction, because in the phenomenon of electromagnetic induction an interaction between electrical and magnetic forces was clearly demonstrated.

In order to find the mathematical law of electromagnetic induction, Maxwell used already earlier applied analogy between circulation of incompressible fluid in a circular tube and stationary magnetic field, evolved by magnets and currents. He transferred this analogy over to a new field of electromagnetic processes, taking into account that in a phenomenon of electromagnetic induction, like in phenomena of magnetic activity of stationary currents, a genetic connection between electrical current and magnetism was traced. By means of a model "circular tube" Maxwell tried to assimilate Faraday's conceptions of electromagnetic induction.

But exactly here did appear deciding problems. An attempt to find in an analogous model a constructive content, corresponding to Faraday's schemes of induction, led to a loss of the most important feature of hydrodynamic analogy – its capability to model magnetic force line. The cause was that a model of "circular tube" principally could substitute and represent in cognition only stationary (continuous in time) magnetic force line, whereas for explaining electromagnetic induction it was substantially important to keep in mind an alternating character of magnetic field (changing at the time of the flow of magnetic force lines, crossing a conduction substance). That is exactly why in a model of continuously flowing in a circular tube incompressible liquid it was impossible to imagine substantial peculiarities of electromagnetic induction, not destroying that content which expressed specialties of the processes of magnetostatics and interaction between stationary currents. Regardless of this equations $\mathbf{E} = \frac{\partial \mathbf{A}}{\partial t}$ and $\mathbf{B} = rot\,\mathbf{A}$ were introduced, from which it was easy to obtain an equation for electromagnetic induction. The absence in analog model of constructive content right away influenced the features of the introduced equations. Their clearly formal characteristics, as it was discovered later, were quite applicable for describing electromagnetic induction. But the physical meaning of the magnitudes that figured in equations was totally unclear.

Maxwell in the beginning tried to interpret them, introducing notions of changing in time of the energy of a magnetic field. But that way he obtained a contradictory definition of a field, because it was given only as stationary (a field was shown as a spatial configuration of stationary force lines). But to talk about changes in time of the field energy, which from the very beginning was determined as continuous in time, was insensible.

In the result equations, introduced on the base of analog model stationary circulating incompressible liquid and meant for describing electromagnetic induction, turn to be without a physical content. They appeared as clearly mathematical formulas, where one vector was defined through another, not having any independent definitions. Because a physical content of a strictly hypothetical character could have been entered in an equation, Maxwell had to leave this very perspective, by itself formalism. He simply did not know how it should be used for characterizing real changes. That is why the whole work, connected with constructing a unified theory of electromagnetism, a researcher had to almost do all over again. This was a kind of deadlock stage of the Maxwell's cognitive process. But it prepared a passage to a productive stage, finished by the creation of a theory of electromagnetic field.

A passage to this stage was connected to changes in strategy of a theoretical search. Maxwell declined the primary attempts to synthesize knowledge about electromagnetic interactions on the base of notions of stationary electrical and magnetic fields and appealed to an idea of time-dependent force lines. In this new point of view he began to look at the former empirical material.

Such a switch of vision was a result of analysis of troubles, which characterized unproductive attempts for knowledge synthesis of magnetostatics and electromagnetic induction. A dead-end situation itself, which appeared on this stage, in so many words showed that abstraction of stationary force line is too harsh for building a generalized scheme of explaining electromagnetic phenomena. That is why it was necessary to find such a concept, from which stationary magnetic power line could "be derived" as a singular case.

A concept of such kind Maxwell introduced with the help of the known model of vortex in incompressible fluid.[43] In this model vortex represented magnetic force in a point, a series of vortexes modeled the magnetic force line. Within the analog model Maxwell found constructive content, conforming to generalized scheme of magnetostatics and interaction between stationary currents, and from generalizing equation, obtained on the base of "vortex's model", deduced the laws by Ampere, Coulomb and Biot-Savart as a singular case.

It can appear that Maxwell did not get anything new, because an equation, generalizing the laws by Ampere, Coulomb and Biot-Savart, already was obtained on the preliminary stages of theoretical synthesis. But if we draw our attention to the physical meaning of such equation, then a situation appears in a new light. Before, writing expressions for general laws of magnetostatics and interaction between stationary currents, Maxwell took stationary magnetic field for main (reference) object, in relation to which the alternating field appeared as some kind of divergence from the reference. In the new version, having rejected the stationary force line as the beginning object of his analogies, Maxwell turns relations. Now stationary field is the one that can in principle be expressed through alternating field.[44]

Later Maxwell had made a consequent synthesis of knowledge about continuous current and electromagnetic induction, every time modernizing the starting analogous construction (in the beginning a corporal element was added to a model of stationary vortex, showing a moving charge;[45] then was introduced a concept of uneven rotation of vortexes and rapid motion connected with corporal elements, which modeled an interconnection between alternating magnetic field and alternating current).

In this process, along with the starting analogy, were reconstructed into a new system of dependencies conforming equations of electromagnetism. It is indicative that obtaining such equations as hypothetical laws for a more widening class of electromagnetic phenomena, Maxwell necessarily proved legality of introduced theoretical notions. The demonstration itself was always held in two plans. On the one hand, from new mathematical expressions for generalized laws every time were deduced all former known laws as a singular case of a new equation. On the other hand, every altered analog model was justified by Maxwell as a show of substantial features of all those experimental situations, which were subjected for theoretical generalization within the scope of the given model. A procedure of such justification was implemented through constructive introduction of theoretical objects in electrodynamics, displayed by an analog model. These objects were entered as idealizations on the base of those primary theoretical schemes, which were assumed to be synthesized within the scope of the existing analogue shape. After the periodical change of the analogue model Maxwell not only discovers that in it can represent substantial features of a new field of interaction, but also checks that its former constructive content did not fall apart.

It is indicative that, for example, assimilated "block" of knowledge of continuous current on the base of vortex model, Maxwell in special way deduces from obtained generalized equations the Biot-Savart law.[46] In the beginning he mentally constructs a magnetic field in a configuration of closed force lines, and then establishes that it corresponds to conduction current of a certain value. This experiment can't be conducted in a real demonstration, but it became necessary for Maxwell to prove that introducing in a model new abstract object – current, marked by "substantial" features (to flow through the conductor, to produce Joule effect and so on), preserves its former feature of current – "ability to generate magnetic field".

Theoretical construct, which represented "substantial" features of current, and construct by the means of which magnetic activity of current was characterized earlier, were different abstract objects, because had different features.

Due to the described contrived experiment the identity of these constructs and the possibility of their replacement by one abstract object, which combined mentioned groups of features, was proven. In that way was discovered the consistency of the two definitions, making a notion "electrical current". In such way, in the process of constructing a more full and enriched by physical content a theoretical scheme of electromagnetic interactions, a conceptual carcass of Maxwell's electrodynamics was gradually formed, which provided interpretation of its mathematical apparatus. In this process an enrichment of contents of earlier established concepts of physics and new elaborated concepts (for instance, a transition to consider force lines in a point led to appearance of notions "electrical" and "magnetic" strength in a point). In formation of a cognitive apparatus in Maxwell's theory an important role was played not only by operations of constructing a theoretical scheme by idealizations, which depended on real peculiarities of experiments, but also a procedure of constant comparison of such scheme with the physical picture of the world. The latter resulted into specification of the more general conceptions of the structure of electromagnetic interactions and provided a development of the most fundamental concepts in electrodynamics. Thus, for example, a transition to analysis of electrical and magnetic force lines, as "forming" in time from one spatial point to another, had formed in the physical picture of the world a notion of electrical and magnetic fields, propagating in space with terminal velocity. Thereby the fundament for later elaboration of the basic concept of electrodynamics – a concept of electromagnetic field was laid.

Taking into account the content-physical aspect of Maxwell's cognitive movement allows in a new way to comprehend many specialties of constructing a theory of electromagnetic field, including its deciding stage, connected with deduction of equation, and including expression for bias current. With this approach it is found out that the famous idea of displacement current which had finished formation of Maxwell's theory, was introduced not at all on the ways of mathematical hypothesis, but was obtained by a much more prosaic method. It appeared as a result of analogy building, constructive content of which corresponded a generalized theoretical scheme of all known to Maxwell fields to electromagnetic interaction. This aspect of a problem we should give a little more attention, because we are talking about a deduction of one of the most important laws of the electromagnetic theory, regarding a method of obtaining accompanied by considerable vagueness and disagreements.

Equation with displacement current was obtained by Maxwell in connection with a necessity to account in already created generalizing laws peculiarities of electrostatic interactions. Before this moment a system of generalizing equations for magnetostatics, conduction current and electromagnetic induction was already built. A similar model, providing an introduction of such equations, served a notion of rotating vortexes of incompressible liquid, between which contacting with them solid elements are arranged. In the given model physical content was specified, it corresponded to generalizing theoretical scheme for magnetostatic processes, phenomena, connected with conduction current, and for electromagnetic induction. An analog model appeared as a symbolical figure for all these processes. In this model a movement of vortexes with steady speed was compared with stationary magnetic field, a movement of vortices with acceleration – with alternating magnetic field; a solid element – with differentially small portion of electricity (charge), a transfer of the solid element – with conduction current; tangential power, influencing solid element, corresponded to a vector of electrical tension. Functioning of the given model (pic. 6 a) expressed the following specialties of the accounted processes of electromagnetism: 1) appearance of conduction current under the influence of electrical field (tangential force leads the vortexes to rotate); 2) appearance of electromotive force under the influence of alternating magnetic field, and 3) appearance of inductive currents (speeded up rotation of vortexes evolves tangential force, which, in its turn, sets in motion the solid elements).

Maxwell was left to account only the phenomena of electrostatics. For that he changed an initial analogy in such way that, preserving its previous physical content, it also reflected the specifics of this new field of interaction. He assumed that vortexes can deform, causing a little bias of the solid element. Such a displacement, in its turn, must had shown polarization of charges with electrostatic induction.

For convenience having switched vortexes by deformed cells of elastic medium, Maxwell then proved that in a new model all essential features of electrostatic interactions are presented (a demonstration was executed on the base of thought experiments with theoretical schemes by Faraday and Coulomb). After that one more demonstration was conducted, establishing that in a new model its former physical content is not lost. Thereby, a theoretical scheme, that generalized all known to Maxwell knowledge about electricity and magnetism, was created.

From the analysis of the given scheme it followed that displacement current and conduction current are introduced by the same series of features, and that means that they are equivalent concepts. In order to show it, let us look at pic. 6.

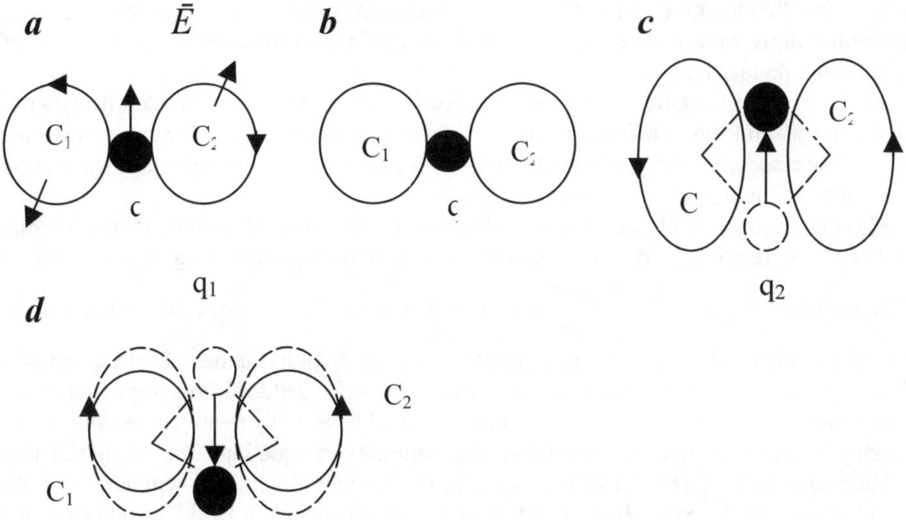

Pic. 6.

a *– rotating vortices C_1 and C_2 set in motion element q and clockwise;* ***b*** *– a solid element, representing charge, is located between deformable, but not yet deformed cells;* ***c*** *and* ***d*** *– deformation of cells C_1 and C_2, matching polarization, and transfer from deformed condition of cells to undistorted condition (removal of polarization) leads to transformation of an element q.*

Let us imagine a beginning of polarization of a dielectric from the moment when external electrical field is turned on (refer to pic. 6 b, c, d). In a model this is shown as an appearance of power which starts to "push through" a solid element q between two cells C_1 and C_2. In the process of transformation of elements between the cells the latter experience deformation of rotation, and appearing with such a deformation elastic force starts to counteract to the motion of an element q until it will not counterbalance the external force. At this moment happens an arrangement of the solid element in position q_1, which corresponds to the setting of polarized condition in the considered differentially small volume of a dielectric.

In the development of the described interaction of a solid element with cells the latter come into rotation with variable speed, which appears in a moment of inclusion of the external force (position q_2) and again becomes equal to zero in the position q_1. As much as the cells' rotation with variable speed means availability of time-dependant magnetic field, as much a described picture of appearing polarization corresponds to the effect of production of magnetic field because of local bias of charges in dielectric. From positions of a dynamic structure of a model it was absolutely indifferent; the rotation of cells is a result of motion of solid elements without deformation of cells or it appears as a result of this deformation. This also meant that it is indifferent is a magnetic field: it is a result of local bias of "elements of electricity" in the process of polarization or it appears as an effect of conduction current. In any case a displacement of charges in the process of induction according to a model was already defined by two features: 1) "to be a series of electrical portions, moving under the influence of electrical field"; 2) "to evolve a magnetic field". If we take into account that in the flow of all earlier executed procedures of justification "conduction current" was introduced as a carrier of same features, then Maxwell simply didn't have any right to draw distinction between it and the bias of charges in a dielectric.

He interpreted the latter as the beginning of current. In this meaning identification of "bias current" and "conduction current" was rather a logical consequence of a motion in a layer of intermediate interpretations, than a genius guess, that out of the blue dawned upon the creator of the theory of electromagnetism.

Having expressed all discovered relations of elements of analog model on the mathematical language, Maxwell obtained his famous equation with bias current of displacement: $rot\,\mathbf{H} = \frac{1}{c}\frac{\partial \mathbf{D}}{\partial t} + \frac{4\pi}{c}\mathbf{J}$, which closed the system of equations in electrodynamics and became the deciding stage of creation of its mathematical apparatus[47]. Here can be seen how much a view that Maxwell deduced this equation from considerations of symmetry, does not match historical facts. Only relatively recently in the work by A. Bork[48] it was mentioned that this symmetry of equations first was noticed by O. Heaviside only in 1885, and any evidence of considerations from symmetry is fully absent in texts by Maxwell himself. Making common cause with Bork's conclusion, we can express a stronger ascertaining: a method of such kind couldn't been used by Maxwell principally, because it does not correspond the logic of that real cognitive movement, the result of which was the theory of electromagnetic field. Such methods of constructing a theory became typical only in the modern epoch of physics' evolution.

After a derivation of a system of Maxwell's equations it was left only to find behind the scaffolding of analog models a created building of electrodynamics. This whole process was over a little later, but already on the final stage of Maxwell's synthesis an obtained system of equations and its interpretation had formed a new view of electromagnetic processes. The latter appeared as an interaction of electrical and magnetic fields tied between themselves, propagating in space with a finite speed. Due to this the theory of electromagnetism itself must be evaluated as a description of the essential characteristics of special physical reality – electromagnetic field.

To sum up let us formulate some conclusions.

1. An analysis of the history of Maxwell's electrodynamics shows that when a developed theory is constructed a researcher does not have to refer directly to experimental data, but can use as empirical material theoretical knowledge of the previous level (Maxwell never operated with immediately data of experiment, but takes them into account indirectly, operating with theoretical schemes by Faraday, Coulomb, Ampere and others). In this aspect the history of electrodynamics illustrates of for constructing a new theory it is not obligatory to have new experiments.

2. One of the crucial points in the creation of Maxwell's theory was a preliminary selection of the picture of the physical reality, which predetermined the character of applied mathematical means and analog models. Availability of a picture of the physical reality serves a necessary link of theoretical synthesis, leading to basic equations of the developed physical theory. A shown picture determines a program of theoretical synthesis, its strategy.

3. A construction of a theory by Maxwell was running as a process of alternate motion in mathematical means and plane of the physical content. Every new step on the way to equations of electromagnetic field was accompanied by intermediate interpretations, which stimulated a new movement in mathematical means. Such transfers from mathematical means to physical content and the other way round happened until the synthesis of all

knowledge, that was generalized within the scope of Maxwell's electrodynamics, didn't accomplish. As an assumption, subjected to a check on concrete historical material, one can formulate a hypothesis, according to which, as it seems, also in modern physics intermediate interpretations are a necessary link for theoretical construction.

4. Ignoring connection between formal-mathematical and informal-physical operations can lead to preconceived interpretation of a history of science. In particular, the opinion that Maxwell obtained an equation with bias current, using conceptions from geometry, is wrong. This equation was obtained on the basis of an informal-physical model.

NOTES: CHAPTER 4

[1] We should note that heuristic functions of the picture of the world in empirical research were fixed and described by myself already in the middle of the 1970s. That is why a little amazing looks a statement by V. Michailovsky and G. Khon, that they are the first to attract their attention to that "the picture of the world as a pre-assigned view allows to investigate objects, for which a developed theory is not yet created. In this case special (individual) pictures of the world and natural scientific picture of the world purposefully aim researches and actively participate in interpretation of the obtained results". (Michailovsky and Khon (1989, pp.11-12)). If we talk about priority in researching these aspects of dynamics of knowledge, then, without a doubt, it belongs to the Minsk methodological school. (See Stepin (1976, pp.71-72), A Nature of Scientific Cognition (1979, pp.163-173, 212-222), The Ideals and Norms of Scientific Investigation (1981, p.15).

[2] See: Gilbert (1900).

[3] See Frankfurt and Frenk (1962, p.192).

[4] Quoted from Liozzi (1970, p.291).

[5] Karpinskaya (1970, p.291).

[6] See Gorokhov (1984), Zelenkov and Vodopianov (1987), Petushkova (1983), Shmakov (1990), Shubas (1982), Smirnova (1991) etc.

[7] More about the structure of the pictures of the biological reality, proposed by Cuvier and Lamarck, and functioning of them as research programs see Kuznetsova (1984, pp.91-94).

[8] Solovyov (1971, pp.35-36).

[9] Ibid, p.35.

[10] The following statement goes after only one purpose – to illustrate universality of functioning of special scientific pictures of the world as research programs in science. What concerns analysis of structure of the social reality as special components of social and scientific knowledge, their historical types, correlation with concrete social theories – is an objective for a special research.

[11] As a consequence of these attitudes in Marxist historical literature descriptions of economical development of various countries, revolutions and rebels of the people's masses prevailed, but a very rare exception were works, dedicated to analysis of deep mentalities and values, determining spiritual climate of this or that historical epoch,

researches of conditions of mass consciousness and the people's way of life, characteristic for this epoch and certain kind of society.
[12] Lacatos (1976).
[13] Rutherford (1972, p.223).
[14] A critical analysis of the Nagaoka's model from these positions can be found, for instance, in the works by V. Wien's works (1905) (see Spassky (1977, p.229)).
[15] New ideas in physics (1913, p.13).
[16] Ibid, pp.23-27.
[17] Ibid, pp.23-27.
[18] Ibid, pp.23-27.
[19] An interest to the problems of substance's structure was drawn also by influence of other factors, for example, by achievements in development of chemistry during this period (here first of all we should keep in mind problems, to which a discovery by Mendeleev of periodical law led to). But if inside physics itself, in its picture of the world the questions of substance's structure wouldn't have been placed especially keen, then, as it seems, this influence would have remained oblique and wouldn't have led to that connection of physical-chemical researches, aimed for atom's study, which was seen from the beginning of the 20th century.
[20] Maxwell (1865).
[21] Kuhn (1970, pp.189-195).
[22] Ibid, pp.192-195.
[23] New Ideas in Physics (1913, pp.18-31).
[24] Liozzi (1970, pp.373-376).
[25] Ibid, pp.184-187.
[26] An opinion that Coulomb obtained his law directly from the measurements without any preliminary theoretical presumptions, is wrong. He had a theoretical hypothesis, which he justified in experiment.
[27] Mattis (1967, p.26).
[28] Theoretical expression for Coulomb's laws, formulated relatively an ideal situation of interaction between point charges in vacuum, shouldn't be mixed up with the mathematical formulation of empirical dependency, which Coulomb had obtained from experiments with real electrified solids, interacting in real air sphere. Logically different status of the similar expressions we already mentioned in Chapter II.
[29] See e.g. Popper (1935), (1968).
[30] Everything that is said in the given part about the pictures of the world, concerns classical period of physics development. In modern physics this situation is reproduced in general outline, but with a series of essential peculiarities, about which we will talk later.
[31] Faraday (1832).
[32] Ibid.
[33] Liozzi (1970, p.276).
[34] Kuznetsov B. G. (1958, p.280).
[35] Maxwell (1865).
[36] An attempt of such reconstruction was firstly made by H. Hertz (see Hertz (1948)).
[37] In the classical electrodynamics of the pre-Einstein period ether, to which connected all demonstrations of electromagnetism, was provided with features of dielectric. This

circumstance specially emphasized L. Mandelstam in his lectures, dedicated to physical basics of the theory of relativity (Mandelstam (1972, pp.118, 135)).

[38] Here and later found by Maxwell expressions for laws of electromagnetism are given in the modern form of their recording.

[39] Maxwell (1865).

[40] Ibid.

[41] Ibid.

[42] Ibid.

[43] Ibid.

[44] Ibid.

[45] Ibid.

[46] Ibid.

[47] Ibid.

[48] See Niven (1890).

CHAPTER FIVE

FORMATION AND DEVELOPMENT OF THEORY IN NON-CLASSICAL SCIENCE

MATHEMATICAL HYPOTHESIS AND ITS EMPIRICAL JUSTIFICATION

The strategies of theoretical investigation do not remain forever given and invariable; they are changing along with the evolution of the science.

Since Bacon and Descartes, philosophy and nature study used to believe that it is possible to find the only true strict way of cognition which could guarantee formation of true theories in any situations and concerning any objects. Foundations of the classical science included this ideal. Changeability and variety of concrete methods were not denied, but the aim of investigator was considered to be a united strategy of theory yielding. It was supposed that first the investigator was to find evident and obvious principles formulated as generalization of experience, and then, on their base, to seek for concrete theoretical laws.

This strategy was believed to be the only true way, the only method which leads to the true theory. As to investigations in physics, they required creation of an integral image of the reality studied, as a preliminary condition for the following employing of mathematical means to describe it.

The development of science in the 20th century has made people to reconsider these methodological attitudes. Even in the late 19th century, when historical changeability of the fundamental principles of science, relativity of their empirical justification accepted by the scientific community (empiriocriticism, conventionalism etc.) was discovered, the first critical observations towards the classical strategy of investigation were made. Certain doubts in the classical methodology as an absolute, reflected in philosophy of that historical period, may be regarded as the preliminary step in the formation of a new paradigm of theoretical cognition. But this paradigm itself was firmly established in science in a great part due to the development of modern quantum-relativisticic physics, the first of sciences that demonstrated non-classical strategies of yielding a theory.

A prominent Soviet physicist, L.I. Mandelstam characterized them in the following way: "Classical physics mostly acted so that determination of links between mathematical magnitudes and real objects preceded equations, i.e., establishing laws. Moreover, the derivation of equations was the main goal because the contents of the magnitudes in advance seemed clear, and scientists sought equations for them... Modern theoretical physics, though not deliberately, but historically it is true, has chosen a different way. It happened by itself. Now first of all we try to guess the mathematical apparatus operating magnitudes meaning of which (at least part) is entirely unclear".[1]

This mode of investigation, which has become domineering in 20th century physics, was connected with a broad application of a special method, which was called mathematical hypothesis or mathematical extrapolation.

General characteristics of this method are as follows. In order to find laws of a new area of phenomena, we take mathematical expressions for laws of a neighboring sphere, which are then transformed and generalized so that we could obtain new correlations between physical magnitudes. The obtained correlations are regarded as hypothetical equations describing new physical processes. After corresponding experimental verification, these equations either get status of theoretical laws, or are rejected as non-fitting to experience.[2]

The characteristic given reflects the most important feature of development of modern physical theories: in contrast to classical patterns, they start as if from top storeys – search for mathematical apparatus, and only when equations of the theory are found, scientists begin to interpret them and look for empirical justification. Though, we probably cannot extract more out of the characteristic of a mathematical hypothesis. Further specification of this characteristic requires that we determine how a mathematical hypothesis is formed in science and what the procedure of its justification is.

Here only first steps have been made yet. First, I should mention S.I. Vavilov's interesting observations about existence of regulative principles (correspondence, simplicity etc.), which give aim and direction to the search of adequate mathematical means.[3] S.I. Vavilov, who introduced the term "mathematical extrapolation", formulated a special group of problems connected with the discussion of the nature of corpuscle-wave dualism. It was said that specificity of mathematical hypothesis as method of today's physical investigation is not the fact that while creating a theory we transfer mathematical means from one field to another (this method has always been used in physics), but mostly in peculiarities of such a transfer itself – in today's mode.

S.I. Vavilov emphasized that mathematical extrapolation (in its modern variation) has appeared due to the fact that visual images, which used to be the basis for creation of mathematical formalism in classical physics, now, in quantum-relativisticic physics have lost their integrity and visuality. The picture of the world taken by modern physics reflects specific features of micro-objects by means of two complimentary representations – corpuscular and wave. Therefore, it looks impossible to work out a unified visual physical model of reality as a preliminary basis for a theory. We have to elaborate a theory concentrating on purely mathematical work connected with reconstruction of equations "dictated" by various analogue images. This is where we can see the unconventionality of mathematical extrapolation of nowadays. "Experience leads to our consciousness reflection on the spheres of the world, which are unfamiliar and alien to a common person. We lack familiar images for visual and model interpretation, but logic... in its mathematical form, still works and introduces order and links in a new, unwonted world".[4]

If we understand mathematical hypothesis this way, we have to ask a question: how does it regard the picture of the world which takes into account the specificity of new objects. It is evident that – in a hidden form – we are dealing with the problem of heuristic picture of the world as a preliminary base for search for adequate mathematical means employing in formulating laws of physics. All these problems need special discussion.

Peculiarities of modern forms of physical picture of the world and their role in putting forward mathematical hypotheses

The specificity of modern pictures of the world may give the impression that they emerge only after a theory has been formed, and so theoretical search nowadays is not directed by their influence.

Though, we may come to conclusions of such kind only after quite prompt consideration of modern investigational situations. More profound analysis discovers that in modern investigation the process of putting forward mathematical hypotheses may also be ruled by ontological principles of the picture of the world.

An example is the establishment of quantum electrodynamics (we'll speak about it in more details in further chapters).

Thus, it is important to emphasize that new strategies of cognition do not cancel the preceding classical models. The latter, though modified, may be reproduced in modern theoretical search as well. Non-classical strategies of investigation may co-exist along with the classical ones, interact with them and appear in a spectrum of variations – from evidently alternative (to the classical models) to hybrid ones, which combine various features of classical and non-classical investigation.

In obviously non-classical situations theories really are created before the new picture of the world appears. And still, the conclusion about disappearance of directional functions of the picture of the world seems hasty. We are to bear in mind two important circumstances.

The first one concerns the process of raising problems, the process which starts construction of fundamental theories. Special relativity theory and quantum mechanics were initiated by discovery of paradoxes in the system of physical knowledge which emerged when scientists tried to correlate new facts and new theoretical conclusions generated under direct influence of a previously formed picture of the world with this image itself. These paradoxes arose in terminological interpretation of corollaries of Lorentz's transformations and corollaries of Planck's law of radiation of absolutely black body. These paradoxes transformed into problems that encouraged theoretical research and led to construction of special relativity theory and quantum mechanics.

Though the new physical picture of the world appeared at the late stage of construction of these theories, its earlier version participated in raising problems. So we may say that certain aspects of directing role of the picture of the world remain also in modern research.

The other circumstance connected with the role of the picture of the world in construction of modern theories may be defined as reinforcement of significance of its operational aspects. We believe that this is the main feature of non-classical strategies of the construction of a new theory. Under modern circumstances, pictures of physical reality are created and reconstructed differently from the way which worked in the era of classical development of physics. They used to be created as visual patterns of structure and interaction of natural objects, i.e. types of measuring procedures, which gave an opportunity to reveal the corresponding objects, were presented in a veiled form. Nowadays the investigation uses a method which can be called – in certain aspects – contrary. The future picture of physical reality is fixed first as the most general pattern of measuring, and objects of a certain type should be inspected within its frame. The new picture of the world is given in its incipiency at this stage, while the structure of the physical reality studied is defined by means of the pattern of measuring: "nature has objective characteristics, recognized within the frame of such and such type of measurements".

By the way, these characteristics first are given as a quite approximate image of structure of the interactions studied, by means of fragmentary ontological ideas which are united in a system due to explication of an operational scheme. Only later does a relatively clear and "quasi-visual" idea appear, the idea of structural features of the physical reality, which is revealed in the type of measurements given and represented by the picture of the world. We can find examples of such way of investigation in the history of modern physics. Let us regard, for instance, Einstein's works of the period when he was working out the main ideas of the special relativity theory. It is well known that formation of this theory started from generalization of the relativity principle and creation of the scheme of spatial and temporal measurements which would consider finite signal propagation velocity necessary for synchronization of watches in inertial frames of reference. First Einstein explicated the scheme of experimental and measuring procedures which was the basis of Newtonian ideas of absolute space and absolute time. He demonstrated that those ideas had been introduced due to a recent postulate: watches, which are in different frames of reference, are correlated by means of instantaneous signal transmission.[5] Since no instantaneous transmission of signals exists, and any interaction propagates at a finite speed, Einstein offered another scheme of measuring space and temporal coordinates in inertial frames of reference, that have watches and rulers. Synchronization of the watches by means of light signals, spreading at a constant speed irrespective of the movement of the light source, was the central point of that scheme. Objective qualities of the nature, which could be also revealed through this type of experimental and measuring actions, were reflected in the ideas of space-temporal continuum, where space and temporal intervals, taken separately, are relative. But these ideas – in their "ontologized" form – were reflected in the physical picture of the world later, only after the special relativity theory had been created. At the early stage of yielding the new picture of the world the features of the physical reality mentioned were presented in direct connection with the operational scheme of investigation.[6]

The same specificity, in certain sense, can be traced in the process of the development of the quantum picture of the world. What is more, here the history of science lets us trace clearly, how the development of atomic physics led us to changes in the classical mode of construction of the picture of the world.

In the history of quantum mechanics we can single out two stages: the first one, which based on the classical methods of investigation, and the second, modern one, which has changed the very strategy of theoretical research.

However unusual the notions of the quanta of electromagnetic energy introduced by M. Planck were, they still did not break the very method of theoretical research. After all, Faraday's ideas of force fields were not less revolutionary than the idea of discreteness of electromagnetic radiation. So, when Planck's works introduced the idea of discreteness of radiation into the electrodynamics picture of the world, it was a revolutionary step, because the old picture of the world was blown up from the inside. But Planck's ideas did not exercise direct influence on the classical methods of yielding the picture of the world, which was created as a visual image of natural interactions. Further development of physics was related to efforts to create a quantum picture of reality in accordance with the ideals of the classical approach. Here de Broglie's investigations are characteristic. De Broglie offered a new picture of the physical reality, which included a statement about specificity of atomic processes, and introduced "visual" image of atomic particles as inseparably connected with the "waves of matter'. According to de Broglie, movement of the atomic particles is tied with some wave spreading in the three-dimensional space (the idea of a pilot-wave). Those ideas played a great role at the initial stages of quantum mechanics development. They gave basis to the natural analogy between the description of photons and electrons and provided transmission of quantum characteristics introduced for photons, to electrons and other elementary particles (de Broglie's picture of the physical reality provided us with the choice of analog models and working out certain theoretical schemes, which were to explain wave qualities of electrons).

Though, de Broglie's picture of the world was "the last of the Mohicans" of visual application of quasi-classical notions to the image of the physical reality. Schrödinger tried to develop this picture, introducing an idea of particles as wave packages in the real three-dimensional space, but failed, because his efforts provoked paradoxes in theoretical explanation of the facts (the problem of stability and reduction of the wave package). After M. Born had found the statistical interpretation of the wave function, it became clear that waves, a "packet" of which should have formed a particle, are "probability waves". Since that time physicists have more and more often regarded the efforts to introduce a visual picture of the world by means of classical models as an anachronism. It is becoming evident that the ideas of a corpuscle and a wave complement each other but are not compatible with each other within the same visual image.

The development of science showed that the object of the new type, studied by quantum physics, is extremely unlike the objects known, and, according to S. I. Vavilov, "we lack familiar images for a visual and model interpretation of its image". But a general image of the reality studied was still necessary, as it defined the strategy of theoretical search, directing the choice of analog models and mathematical means to put forward productive hypotheses.

Under these circumstances a turn to the new method of construction of the picture of the world was happened. Here a great part belongs to N. Bohr. The image of the physical reality was now built as an "operational scheme" of objects studied, and we may say that their characteristic is what is revealed within the scheme. Bohr's approach can be characterized not by introduction of hypothetical ideas of the structure of nature as foundation for new concrete theoretical hypotheses, which are to be verified experimentally, but by analysis of the scheme of measuring, which can help reveal the corresponding structure of the nature.

Niels Bohr was one of the first scientists who clearly formulated the principle of quantum-mechanical measuring, different from the classical pattern. The latter was based on extraction of a self-identical object of the material world. It was believed that the strict demarcation line separating the object from device would be drawn since in measuring it is always possible to take into account all details of influence of the device over the object. But the objects in the quantum sphere are quite specific, and detailing of influence of the device over the object can be accomplished only with precision determined by the existence of action quantum. Therefore the description of quantum phenomena includes description of essential interactions between atomic objects and devices.[7]

General features of a micro-object are defined by means of clear description of characteristics of two complementary types of devices (one is used, for instance, to measure coordinates, the other – to measure impulse). Complementary description is a method to reveal basic and profound features of a quantum object.

All these principles introduced "the operational scheme" which lay in the foundation of the new picture of the world created by quantum physics. Through such a scheme scientists could fix (as activity) essential features of a quantum object. This object, according to the new view, was presented as having a special "two-level" nature: a micro-object in its existence is stipulated by macro conditions, and they are inseparable. D. Bohm wrote that quantum mechanics makes us reject the assumption which lies in the foundation of many common statements and images: that we are able to analyze separate parts of the Universe, and each of them exists independently.[8] But this image of a quantum object has not been differentiated yet and not presented as a system-structural description of interactions in nature. So we can predict further development of quantum-relativistic picture of the world. Probably, it will lead us to notions of the structure of natural objects which include quantum characteristics as natural ones. The decisive part in such development will belong not only to new achievements of quantum physics, but also to philosophical analysis necessary to prepare usage of new system notions for description of the physical reality.

The approach to quantum objects as complicated self-organizing systems seems very fruitful. This problem has already been widely discussed in literature, including Russian literature. As early as in the 1970s authors tried to interpret the specificity of quantum mechanics description in terms of complicated systems. Yu.V. Sachkov, for instance, mentioned the two-level structure of quantum mechanics concepts: there are concepts which, on the one hand, describe the unity of the system, while on the other hand, represent typically random characteristics of the object.[9] The idea of such dismemberment of the theoretical description correlates with the idea of complicated systems, which are characterized by subsystems with stochastic interaction of the elements and, on the other hand, some "controlling" level securing integrity of the system.

The idea that quantum mechanics notions may be correlated with description of the reality in terms of complicated, self-regulating systems has also been postulated by G.N. Povarov[10] and V.I. Arshinov.[11] My works of the 1970s also promoted this idea.[12]

The foreign literature of that time present more or less detailed concepts alike in the works of such physicists as G. Chew, H. Stapp, D. Bohm, B. Hiley, philosopher F. Kapra and others.

In the conception of "bootstrap", which appeared on base of the S-matrix approach, G. Chew offered a picture of the physical reality in which all elementary particles obtain system integrity. They are as though laced together by generating reactions, but no one of them should be regarded as fundamental for others.[13] The American physicist-theorist H. Stapp worked with notions of the physical reality in the same direction. He paid special attention to ideas of non-locality, impossibility to combine requirements of causation and localization of micro-objects in a quantum mechanics description. Such incompatibility is expressed in the complementarity principle (complementarity of causal and spatial description). Correspondingly to these ideas Stapp outlines a new ontology, which states: the physical world is a system unity, irreducible to dynamical connections between its elements. According to Stapp, besides causal connections, the decisive part belongs to non-forced interactions which unite different elements and subsystems into a whole. As a result, we have an image of a weblike global structure of the world, all elements of which are interconsistence. Any localization, any individualization of elements in this global structure is relative, stipulated by general mutual dependence of the elements.[14] Stapp interprets the fundamental probability character of the results of measuring in quantum mechanics from the point of view of correlation of the local and the global.

G. Chew and H. Stapp emphasized the idea of system integrity of the world, but the problem of the level hierarchy of the elements – a very important characteristic of complicated self-regulating systems remained in the shadow. The idea of a web like network, where all elements and substructures are mutually correlated, did not generate enough stimuli for working out notions of their relative fundamentality and complexity of the elements and their connections which are found at different levels of the hierarchy.

Probably, these features of the "bootstrap" conception caused the decay of interest to it among physicists while the quark model of elementary particles has been being elaborated.

But the very idea of relativity of localization and individualization of physical objects and events, their stipulation by the qualities of a whole system became a necessary and important aspect taken into account in most modern efforts to build an integral physical picture of the world which would include quantum and relativistic notions.

This approach has been well presented in the works of D. Bohm, who tried to solve the problem of the quantum mechanics ontology. Bohm stressed the need of the system of notions of the physical world to overcome the classical approach that postulated existence of local elements and events, which are interconnected and may be isolated. The new image of the physical reality, according to Bohm, should be based on the idea of relative locality depending on the integrity of the Universe, on non-dynamic relations that (along with the dynamic ones) define the structure of the nature. Bohm compares the picture of the reality with correlated substructures and elements with a carpet, where parts of the decoration do not form a whole because they interact dynamically.[15] They are individualized through inclusion into the whole and their relation to other parts of the whole. Here Bohm's images of the reality correspond to those offered by Stapp. But Bohm has made a new step. He suggested to consider the world as some kind of order, a hierarchy of different levels. Every level, according to Bohm, is characterized by its own non-locality and non-force interactions. Bohm emphasizes that non-locality and non-force correlations can be revealed not only in the microworld, but also at the macrolevel. In the work written together with B. Hiley, D. Bohm gives an example of the experimental facts of correlation of distant atoms in super-fluid helium. These correlations disappear at high temperature, when the effect of viscid friction arises because of casual collisions of atoms, but they restore when the temperature is lower a certain threshold level.[16]

As to the conception of non-locality in the microworld, it is the most brightly expressed by the reduction of the wave function – which is the corner stone of quantum physics. Even in the 1930, at the time of Bohr's and Einstein's discussions, scientists formulated a so–called paradox of Einstein-Podolsky-Rosen (the EPR-paradox). The point of it is that a wave function is assigned to two interacting particles, and then the distance between particles becomes so considerable that their dynamical interaction can be ignored. But if we measure the magnitudes characterizing the state of one particle (for instance, its impulse or coordinate), we will see reduction of the wave function, and thereby the state of the other particle will automatically change. Einstein regarded this mental experiment as a paradox which proves that quantum mechanics is incomplete. But further discussions of the EPR-paradox, for instance, in the 1970s, showed that it leads to a contradiction if we latently accept the principle of locality, which assumes the possibility to separate system and to measure its spatially separated, distant parts independently.[17]

But if we reject locality as an absolute principle and think that it can be applied only relatively and limitedly, we will come to the probability of non-local interaction. The EPR-paradox may be interpreted as a display of non-locality.

Bohm's picture of the world postulates existence of some hidden order which organized all other types of orders in the Universe; this order is inherent in the net of space interactions. Bohm explains the idea of this hidden order by means of another visual analogy (like the example of a carpet ornament). He uses a metaphor of a hologram in which, if we throw light to any local part, we will be able to see the entire picture, though less detailed than in case of lighting of the whole hologram. Bohm tries to correlate the idea of the hidden order and hierarchy of orders with the notions of the structure of the space. Basing on the general relativity theory and interrelation with gravitating masses and curvature, he believes it possible that these ideas may be widened and generalized within the hypothesis of topological properties of the space correlated with the types of order in the Universe. Hiley and other Bohm's investigation program supporters have also developed these ideas.[18]

This program, as well as G. Chew's and H. Stapp's investigations, can be considered as variations of some general approach to the construction of a physical picture of the world, which would use the ideas of non-locality, non-forced interactions and notions of a complicated self-regulating system, where the features of parts and elements are stipulated by the features of the whole, and the probability causality is a basic characteristic.

The philosophical and methodological basis of such approach is the rejection of methodology of "elementarism", which was domineering in physics for a long time and assumed that the features of physical systems are completely described by characteristics of their elements.

The holistic approach, opposite to elementarism, is based on the idea that the features of the whole cannot be reduced to the features of the elements and their interactions.[19]

This approach was developed mainly in investigations of biological and social objects. Then it was transferred to the system of non-organic nature due to cybernetics, theory of information and the general theory of systems.

The way of investigation chosen (in various forms) by the conceptions of G. Chew, H. Stapp and D. Bohm is based on employment of the "organismical" methodology in the construction of the physical picture of the world. F. Capra says that Bohm's and Chew's conceptions are the two most philosophically sophisticated approaches to describe the physical reality.[20] He denotes their rapprochement – further versions of the "bootstrap" concept tried to consider elements of the S-matrix as types of orders and to link them with the space-time geometry. In Capra's opinion both of these conceptions understand the world as a dynamic network of relations and put the concept of order in the centre; they both use matrices as means of description, and topology – as means to determine categories of order more exactly.[21]

Then Capra emphasizes that Chew's, Stapp's and Bohm's picture of the world present elementary particles not as immutable bricks of the Universe, but as dynamic structures, "energy beams" forming objects which belong to higher levels of organization. According to Capra, for modern physicists matter is not passive and inert, but is always dancing and vibrating, and the rhythmic patterns of the dance and vibrations are determined by molecular, atomic and nuclear structures. The nature is in balance, but dynamic, not static.[22]

Here it would be right to stress that this image of the Universe as dynamics of the physical processes, their mutual correlation and hierarchy of orders, is more likely an image of a self-regulating system, where mass, stochastic interactions are controlled by the whole and reproduce the whole. The classical picture of the world as a simple device, which dominated in classical physics, is now replaced by the image of the Universe as a self-organizing machine.

Though, in this respect we are also to mention narrowness of such approaches to construct a modern physical picture of the world, which are adjoint to the images of a complicated self-organizing system reproducing the basic characteristics of the whole as a hierarchy of orders in dynamics.

Self-organization cannot be brought only to the processes of reproduction of dynamic order and level organization of the system, though this aspect is obligatory. The other aspect is irreversible change and development connected with appearance of new organization levels and transition from one type of self-regulation to another. If we take these aspects into consideration, we are to employ more complicated images of system organization, that is images of complicated, historically developing systems. The notions of such systems include the idea of dynamic balance, but only as one of the states of non-equilibrium processes characterized by changes of the types of dynamic balance, transitions from one type to another.

In the modern science the program, most adequate to such view, is the one connected with working out dynamics of non-equilibrium processes (I. Prigogine) and synergetics (H. Hacken, M. Eigen, G. Nicolis, E. Laszlo, S. Kurdyumov, G. Malinetsky, Yu. Klimantovich etc.). Differently from classical physics – in principle – the synergetic paradigm sees the place of non-equilibrium and irreversible processes and their correlation with equilibrium and reversible processes. While classical physics presented non-equilibrium processes as sort of declination from the standard situation, the new paradigm puts them into the focus of interest, considering them as a way to give birth to stable structures.

Stabilities appear not despite, but due to non-equilibrium states. In these states even small fluctuations, random influences cause attractors leading to new organization; at all levels, either level of macroscopic physics, or level of fluctuations, or microscopic level, the source of order is non-equilibrium. Non-equilibrium is what gives rise to "order from chaos".[23]

When we describe the behaviour of quantum objects in terms of self-organizing systems, we obtain new opportunities to build quantum mechanics ontology.

I. Prigogine emphasizes that we can explain features of quantum mechanics measuring connected with the reduction of the wave function as consequences of instability, immanent to the movement of micro-objects, and measuring – as an irreversible process of causing stabilities in dynamic chaos.

From the point of view of "order from chaos", the basically static character of predictions in quantum mechanics seems not to be the result of activity of the one who is doing the measuring, but to represent the essential characteristics of the nature itself.

Non-localities presented in the behaviour of micro-objects, according to I. Prigogine and C. George, are related to the growth of coherence of quantum ensembles in comparison with classical dynamics.[24] Coherence, in its turn, expresses a special quality of self-organizing systems, related to their non-linearity and ability to cause cooperative effects based on non-force interactions.

I. Prigogine and I. Stengers say: "In our approach, the world follows the same laws, with measuring or without measuring...";[25] "introduction of probabilities, in our approach, is compatible with physical realism, we do not need to identify it with incompleteness of our knowledge. The observer now does not play an active part in the evolution of the nature or, at least, his part is not more active than in classical physics. In both cases, we can put into action the information got from the outer world".[26]

S.P. Kurdyumov has found quite interesting solutions of problems connected with mathematical description of peaking regimes in a nonlinear medium. These regimes are an essential characteristic of behaviour of synergetic systems, and their mathematical description bases on nonlinear links of space and temporal coordinates. The apparatus developed in application to such situations is effective when applied to quantum mechanics problems. It allows to obtain Schrödinger's equation and to explain quantization as expression of the features of a nonlinear medium.[27]

Probably, with the development of all these approaches the quantum picture of the world will one day appear in objectivized form presenting the structure of the nature "by itself".

But in order to consider modern features of the theoretical research it is important that at initial stages of developing pictures of the world in modern physics the "operational aspect" of the vision of reality is accentuated. It is the operational side that mainly determines the search for mathematical hypotheses.

It is quite indicative that the modern theoretical-group approach directly connects the principles of symmetry based on various groups of transformations with characteristics of the measuring devices.[28] An attempt to use a certain mathematical structure in physics in this sense is determined by the choice of a measuring scheme as the "operational aspect" of the corresponding picture of the physical reality.

So far as the starting point of investigation – choice of the picture of the world as operational scheme – often presupposes quite radical changes in the strategy of theoretical research, it requires philosophical regulation. But, unlike classical situations, when introduction of the picture of the world was mainly directed by "philosophical ontology", in modern physical investigations epistemological problems are in the focus of attention. It is significant that in regulation principles, which facilitate the search for mathematical hypotheses, theoretical and cognitive statements (the correspondence principle, simplicity etc.) are evidently represented (in concretizing with reference to a physical research form).

It seems that only by analyzing these problems (while regarding the chain of relations: philosophy – the picture of the world – analog physical model – mathematics – mathematical apparatus of a physical theory) we can reveal at greatest length the mechanisms of developing a mathematical hypothesis.

From this point of view, the discussion of the method of mathematical hypothesis in philosophical and methodological literature has been valuable, not only due to verification that the fact really existed, but – to a greater extent – to the fact that the problems described above were formulated and first attempts to solve them were made.

Still, though we do justice to actuality of the problems raised, when we accentuate heuristic value of the mathematical methods, we should not loose sight of another, not less important aspect of theoretical research: the process of constructing a theoretical scheme which allows us to interpret the mathematical formalism introduced. Inaccurate analysis of this aspect of investigation leads to hidden introduction of a series of simplifying notions, true only in their general formulating. If they are employed without enough specification, it may lead to incorrect ideas. Such notions include:

1. Assumption that experimental verification of a mathematical hypothesis and its transformation into a physical theory is a rather obvious procedure, which is just brought to mere comparison of all corollaries of the hypothesis with experimental data (the hypothesis is accepted if its corollaries correspond to the experiment, and rejected in case of contradicting); 2. Assumption that a mathematical apparatus of a developed theory can be created as a result of advancement in purely mathematical means, by mathematical extrapolation, without constructing any intermediate interpretational models.

We are going to try to demonstrate that such notions of forming of modern theory are not correct enough.

To begin with, we will analyze the situation of construction of local theoretical schemes, and then we will turn to the process of creating a developed theory. As the former we will consider the theoretical scheme which is the foundation of Dirac's relativistic electron theory, the latter – quantum electrodynamics (the theory of interaction of quantized electromagnetic and quantized electron-positron fields).

First we have to denote that the interpretation of Dirac's theory as knowledge corresponding to the level of local theoretical schemes can be employed only in case we take into consideration the fact that it has been assimilated by a developed theory – quantum electrodynamics – and has become a part of it as a fragment which describes one of the aspects of electrodynamics' interactions in the quantum area. In generality the theory of relativistic electron surpasses such classical local theoretical schemes and laws as, say, the system of theoretical knowledge about oscillation of the pendulum (Huygens's model) or Faraday's observations of electromagnetic induction.

But one of the features of the method of mathematical hypothesis is that it raises local theoretical schemes and laws to a new stage of generalization; it lets us start constructing a developed theory from synthesis of theoretical knowledge of a higher degree of generality – compared to classical examples.

The problem of empirical verification of a mathematical hypothesis

In classical physics the pattern of investigation was the following: a theoretical model was created (it was introduced as a hypothetical construction, then scientists proved that it included essential features of the generalized experimental situations) and only after that were mathematical expressions for the laws of the theory derived. The latter appeared as the result of revealing connections of abstract objects of the theoretical model and expressing them in the language of mathematics. So, introduced equations immediately got an adequate interpretation and connection with experience.

In such structure scientists had no difficulties in empirical justification of the equations. But in modern physics the situation is different. Using the method of mathematical hypothesis, physics began to create equations to construct rules of correspondence, which link magnitudes of the equations with the object of experience, and then emerge certain difficulties connected with the search for interpretation of the equations.[29]

We would like to emphasize that these difficulties essentially consist not in the fact that first a mathematical hypothesis is introduced without any interpretation at all. In that case the hypothetical equations could not be regarded as expressions for the laws of physics and would be only formulae of pure mathematics. Since certain symbols in the equations are considered as physical magnitudes, interpretation of the equations is indirectly assumed. But the problem is that, at the first stage, we, as a rule, inadequately interpret the hypothetical equations. The reason is the following. Formulating a mathematical hypothesis, we reconstruct the equations which used to express physical laws of some area. Such expressions were tied with a corresponding theoretical model, or scheme, which provides their interpretation. The magnitudes, tied in them, fixed attributes of abstract objects of the model given. But when the initial equation was reconstructed, it gave birth to new connections of the physical magnitudes and, consequently, new definitions within the new equations. Nevertheless, in a physicist's mind these magnitudes are still combined with the ideas of abstract objects of the old theoretical model. So, having carried out mathematical extrapolation along with physical magnitudes whose links are postulated in the equation, he borrows such objects and tries to use them to interpret the equations obtained. In correspondence with the new links of physical quantities in equations he sets down new attributes to the abstract objects and determines their correlations. This is the way to get a hypothetical model which then is showed as a representation of essential features of the new sphere of interactions. But in this quality it is not proved. We have not checked if it is possible to derive the element objects (with their new attributes) by means of idealization and passage to the limit from real object relations of the new area. Therefore it is very likely that the hypothetical interpretation of the new equations will be wrong. In this case, if we check the equations at once, comparing them with the experimental data, the results of the checking may lead to mismatch between the equations and experiment, even if the equations are productive.

To consider this aspect of the question at length, let us take an example which has already become classical: the justification of Dirac's relativistic equation. We know that Dirac had constructed – in complete correspondence with the canons of the method of mathematical hypothesis – a system of four linear differential equations of the first order for four independent wave functions and obtained, as one of the basic mathematical results, solutions which corresponded to negative value of rest-mass (complete energy) for a free particle.

It is usually believed that these results, when compared to the experience, led at once to prediction of positron. But reality was far more complicated. The initial comparison of Dirac's equations with the experience caused such predictions, after which it seemed impossible to save the equation.

The most extravagant and obviously contradicting to the experience were conclusions about possibility of spontaneous collapse of electrons and, as a consequence, about instability of hydrogen atoms.

It is easy to see that such conclusions contradicted to all experimental treasures of atomic physics so brusquely, that they were enough to reject Dirac's equation as an unsuccessful mathematical extrapolation. But the point is that the results mentioned were pressed not by the qualities of Dirac's equation, but by its initial interpretation. Since the equation was obtained out of classical correlation between mass and energy for one particle, and contained an ordinary expression for quantum mechanical operator of the impulse of that particle, then it seemed very natural that Dirac's equation described the behaviour of a separate quantum mechanical particle under condition that non-relativisticic restrictions are removed.[30] In other words, the transformation of traditional quantum mechanical equations into a relativistic equation for electron took place along with the introduction of a new system of abstract objects, which were taken from theoretical models of non-relativistic quantum mechanics and provided with new qualities. Solution of Dirac's equations indicated existence of areas with positive and negative energy, separated by the energy barrier $2mc^2$. Thus, the equations introduced the following system of abstract, theoretical objects: "particle" (in the sense of quantum mechanics, unable to move at relativistic speeds), "area of positive energies" and "area of negative energies". In accordance with the general principles of quantum mechanics, a particle with charge e and mass m got ability to move through the barriers between the areas mentioned under indefinitely small electromagnetic influence, and get into the area of negative energies. As Dirac's equation did not contain any "lowest" limitation of the possible quantity of negative energy ($-\infty < E \leq mc^2$), it followed that any particle, once in the area of negative energy and tending to the state with the lowest energy (the system stability principle), has to fall down into a bottomless energetic hole with a zero probability to return to the area of positive energies. It is easy to understand, that the above indicated paradoxical conclusions from Dirac's equation were somehow or other connected with the effect of particles' (electrons) disappearance "without trace" from the observed area when they got into the negative energy zone.

O. Klein was the first to find out these paradoxical corollaries soon after the theory of relativisticic electron had been published. They made many prominent physicists of the time be skeptical about Dirac's theory. For instance, W. Pauli stated that Klein's paradox (according to which, electrons are capable of overcoming barriers mc^2-order high and get from the area of positive energies into the area of negative energies) is a corner stone difficulty of Dirac's theory.[31]

Pauli wrote that states with negative energy have no physical sense. Nevertheless, unlike classical relativisticic mechanics, in Dirac's theory it is impossible to eliminate in general case states of negative energy for free electrons.[32]

The example of Dirac's quantum-relativisticic equation is quite instructive methodologically. It shows that the initial theoretical model introduced together with mathematical extrapolation may result false and dangerous even for productive equations. Hence we may draw an important feature of justification of a mathematical hypothesis. At the first stage verification of the equations by the experimental data does not let us determine whether the equations are fit for description of a new sphere of phenomena. Even if the conclusions from the equations do not agree with the experiment, it does not necessarily mean that they should be rejected as a fruitless hypothesis. Mismatch with the experiment is just a sign that in the integrity "equations plus interpretation" some part is inadequate to the new sphere of phenomena. The investigator does not know in advance, which part (we may speak about productivity of equations only in retrospective, when we already know their role in the history of physics, as, for instance in the case of Dirac's equation).

Nevertheless, as the initial interpretation of the equations is hypothetical, it is quite probable that it bears responsibility for contradictions between corollaries of the equations and experimental data. So, if we discover mismatch of the equations and experiment, it is the start of the second stage of empirical justification of a mathematical hypothesis. Here the initial interpretation is being changed; the initial hypothetical model, which used to serve the equations, is transformed into a new model. To illustrate characteristic features of this process, let us return to the example of Dirac's equation.

After mismatch of the equation and experiment had been discovered, Dirac reconstructed its initial interpretation. He refused to treat the equation as description of one particle's behavior. The theoretical model, due to which Dirac's mathematical formalism turned into an effective apparatus, was connected with the idea of many-particle systems. In this model the area of negative energies was forbidden for free particles, though presence of two signs for energies was a direct mathematical corollary of the strict solution of the equation. Such exclusion was obtained thanks to Pauli's principle formulated, as we know, for a system of electrons. Within the new interpretation, all negative energy states were considered as totally filled by electrons. Such "quasi-continuum" of electrons, according to Pauli's principle, could never manifest itself externally, because the electrons transfer (moving) inside the continuum, as an indispensable condition of its experimental discovery, stipulates change of the electrons energy, which is impossible because all energetic levels are already full.[33] The only possibility to find out at least one particle of the continuum was to transfer the particle to the positive energy zone, where there were free levels. It was possible to reach under energetic effect not weaker than $2mc^2$ (volume of the energy barrier). But when an electron is extracted in such a way from the continuum, there appears a "free place" (a hole) which behaves as a state with positive charge and positive energy (since to eliminate this state we have to, by definition, place an electron, negatively charged, there). This "non-filled state" can already be experimentally revealed. The "hole" in the electron continuum may be filled by an electron from the neighboring cell of the continuum where an electron from another cell can "jump" etc. Efficiently this process should be appeared as basically observable motion of a positive charge with positive energy. So, the very qualities of the new model naturally caused prediction of the positron.

Though, interpretation of the "hole" also required some creative efforts. At the early stage Dirac associated the "hole" with proton. But soon R. Oppenheimer proved that if the "hole" was interpreted as proton, this would preserve the conclusion from Klein's paradox of instability of hydrogen atoms (according to which the lifetime of a hydrogen atom was to make about 10^{-10} sec). To find a solution of the contradiction, Oppenheimer suggested that we should consider the "holes" as positive electrons, different from protons. It was Oppenheimer who introduced the term "positron".[34] H. Weil proved that the mass of the holes has to coincide with the mass of electron. About three years after Dirac's new interpretation of the quantum mechanics equation for electron, in 1932 C. Anderson discovered positron experimentally.

According to the new interpretation of Dirac's equation, any "hole" (positron) which appeared in the continuum, may be destroyed when an electron from the zone of positive energies enters it. Such transition of electron must cause discharge of quanta of energy (no less than $2mc^2$), in the same way as energy is discharged when an atom, which has lost an electron from one of the internal shells, captures a free electron. It is easy to notice that the properties of the new theoretical model directly led to the idea of annihilation.

Dirac's reinterpretation of his equation removed mismatch of the latter with the experiment. The equation was not only put in concord with experiments, but also enables scientists to predict most unexpected phenomena: positrons and annihilation and pair creation effect.

The new theoretical scheme providing an adequate link of quantum relativistic equation for elecron and experiment in correlation with physical picture of the world introduced basically new ideas of electromagnetic interactions. In the physical picture of the world new notions of electron-positron vacuum as a specific state of the physical world were appeared, actively reflected in interactions of electrons, positrons and photons.

The new interpretation of Dirac's equation, after all details of its physical sense had been clarified, was recognized by the scientific world quite soon. The physicists who had been skeptical to Dirac's theory first, reconsidered their positions. A characteristic example here is W. Pauli. He paid attention to Dirac's grace in his new interpretative scheme of prohibition principle and recognized perspectives opened by the notions of physical vacuum as potential generator of particles.

In his Nobel Prize lecture delivered December 13, 1946, Pauli, considering Dirac's discovery from historical distance, said: "P. Dirac's response led to what could really happen if we employed the prohibition principle". In his Stockholm lecture, Dirac himself spoke of his proposal of new interpretation of his theory, according to which in the true vacuum all negative energy states are to be filled, and we can consider as observable only deviations from this minimal energy state, i.e., holes in the sea of the filled states. The prohibition principle is what guarantees stability of the vacuum in which all negative energy states are filled. What is more, the holes possess all the qualities of particles with positive energy and positive charge, since they can be born and destroyed in pairs in external electromagnetic fields. Indeed, thus predicted positrons, exact mirror images of electrons, were found experimentally.

It is evident that in principle the new interpretation does reject the point of view proper for one-particle problem, and from the very beginning it considers the problem of many particles".[35]

This example, to our mind, allows us to distinguish a few peculiarities of experimental justification of a mathematical hypothesis connected with construction of new interpretation of equations. In general sense, it is well known that when an experiment does not confirm a mathematical hypothesis, the investigator starts searching a new interpretation. But we would like to draw our reader's attention to the following mechanisms of the search.

The first important thing is that the initial material for new interpretation consists of abstract objects of the model initially introduced. Constructing the new model, Dirac used abstract objects "particle", "area of positive energies" and "area of negative energies" which already existed, and only the last object was changed (the feature "to have free energy levels" was eliminated).

The investigator does not yield his new interpretation "out of nothing", but uses abstract objects introduced before, while constructing the mathematical hypothesis, as his building material.

The second important factor directing construction of the new interpretation is the following requirement: the theoretical model should be justified as an idealized scheme of interactions which are observed in real experimental situations. That is what makes the investigator reconstruct abstract objects, finding correlatives of their features in real interactions observed in experiments. As early as in primary experimental verification of mathematical hypotheses it becomes clear which of the abstract objects do not meet this requirement. This is how non-constructive elements in the primary interpretation are discovered, and the ways of its changing are indicated. So, when the primary model in which Dirac's equation was held, was mapped on experimental situations in the atomic area, such mapping showed that its contradicting to the experiment was caused by the notions of the negative energy zone.

But, just as the equations required that such abstract object should be introduced, so there remained only one way: to provide "the area with negative energies" with features which would prohibit electrons to enter this area. This is probably the source of the right conjecture on electron continuum, which allowed to shape a productive interpretation of the equations.

It is characteristic that, introducing a new system of abstract objects (continuum of electrons filling all states with negative energy and free electrons in the positive energy zone) instead of the previous model, Dirac justified this system as an idealized scheme of experimental measuring situations of the atomic area. He found reason for features of the abstract objects in experimentally observable situations. Such abstract objects as "electron" and "area of positive energy" were justified easily enough (in principle all preceding development of atomic physics proved lawfulness of their introduction). The task was harder in case of "electron continuum". Nevertheless, this abstract object also got a correlative in real interactions fixed by experiments in the atomic area. The idea of continuum was a result of analysis of all theoretical and experimental material of physics connected with studies of electron shells of atoms. Dirac introduced continuum of electrons as an analogy to filled shells of an atom which also could lose electrons at external shells. Having imagined such shells in extremely idealized form, Dirac interpreted them as a sort of system of fermi-particles in general. After that the electron continuum turned justified by all experimental measuring situations in which investigations of many-electron systems were held. Then such justification allowed to use effectively Pauli's exclusion principle in constructing a new theoretical model.

So the process of empirical justification of a mathematical hypothesis includes a number of procedures, complicated enough. We may point out the following: 1) explication of a hypothetical model introduced initially along with new equations; 2) mapping of this model on experimentally observable interactions of natural objects; 3) comparison of the "equation plus model" system with the experimental data; 4) reconstruction of the primary model in case of mismatch with experiment; 5) constructive justification of the new model; 6) new experimental verification of the system "equations plus their new interpretation".

Only when all these operations are completed, one may decide whether the equations (introduced by method of mathematical hypothesis) are fit for description of the sphere of interactions. As to the statement that the judgment about the hypothetically introduced equations is passed by means of their comparison with the experiment, it is true only in case we take into consideration all peculiarities of the empirical justification of the equations. But if we simplify it – "equations are rejected if they are not confirmed by the experiment, and are accepted if they coincide with the experimental data" – it may turn out false: mismatch with experiment at the first stage of empirical justification of a mathematical hypothesis is not a sufficient reason to reject the equations.

From all said above we may conclude that the main difficulties in creation of a non-contradictory system of theoretical knowledge are not over when equations are found. What is more, here a theorist faces the hardest and most important stage of his work.

P. Dirac wrote: "It is easier to discover mathematical form necessary for some fundamental physical theory than find its interpretation. It is true because the number of objects we deal with while discovering formalism is strictly limited, but, dealing with physical interpretation, we may find strikingly unexpected things".[36] We do not think it would be an exaggeration if we postulate: at the current stage of development of theoretical knowledge, when the investigator's first steps are connected with mathematical hypothesis, construction of a theoretical scheme which provides interpretation of the equation and their comparison with experiment still remains the key stage of the investigation.

HOW A DEVELOPED THEORY IS FORMED IN MODERN SCIENCE

Considering genesis of a theory in modern physics, it is important that one should not forget about differences in levels of theoretical organization of knowledge. Plain extrapolation of construction methods of a local theoretical scheme to all cases of theoretical research may lead to erroneous notions of ways of today's theoretical research. Such extrapolation makes one think that mathematical apparatus of a developed fundamental theory can be obtained thanks to continuous series of mathematical hypotheses, like the way Dirac, for instance, got his equation for relativistic electron.

Even if we agree with the statement of universality, assume that means of construction of mathematical apparatus for all primary theoretical schemes of modern physics (such as

Dirac's scheme) are the same, we still cannot conclude that mathematical apparatus of a developed theory should be obtained in the same manner.

This apparatus is a more complicated system of mathematical means, first of all because it allows to get – due to certain methods – the regularities characterizing local theoretical laws from the basic correlations. There are no reasons to believe that such apparatus can be worked out merely by means of continuous series of mathematical hypotheses. The contrary is more likely. If every stage of creation of apparatus of a developed theory ends at putting forward a hypothetical equation, consequently, the investigator has to justify the legitimacy of this equation before taking it for initial base for putting forward the next mathematical hypothesis. Philosophical literature has always taken somehow or so this circumstance into consideration discussing the problem of mathematical extrapolation. It is evident enough that only a hypothesis which has been verified empirically has got "the right to live due to dictate of experiment" and gets the role of "starting point for a new hypothesis which will inevitably replace it".[37] Though we have seen that the procedure of comparing mathematical hypothesis with experiment turns a complicated system of operations aimed at constructing a theoretical scheme, which provides interpretation of the equations.

If we take this circumstance into account, we will come to a non-trivial conclusion: forming of mathematical apparatus of a developed theory should be interrupted by intermediate interpretations, which would direct every new series of mathematical hypotheses. Naturally, this conclusion is to be checked. But if we accept it as a preliminary assumption, we will see a parallel between process of theoretical synthesis in classical physics (which has already been discussed) and situations of construction of a developed theory in modern physics. We should not be surprised by such analogy, because the process of evolution provides succession between higher and lower levels of development.

The very idea of evolution in scientific thinking claims for seeking not only specific, but also repeating, invariant contents in historically changing methods of construction of theory. However greatly the past is transformed in the present, their genetic link always lead to reproduction in compact of the main features and specificities of their historical development. That is why history of scientific cognition should be analyzed in two aspects: revealing of specific features of the investigation characterizing the modern stage of evolution of physics, and search for invariant contents inherent in both classical and modern forms.

Now let us consider modern situation in construction of a developed theory from this point of view. To reach this goal, we reconstruct logically the process of settling of quantum electrodynamics. Even cursory comparison of classical and modern situations of theoretical search show up several characteristic features of theoretical activity nowadays.[38]

One of these features is the fact that developed theories of high community degree now are elaborated by research groups, and the duties are distributed among them clearly enough. For instance, we of course can regard the creators of quantum electrodynamics W. Heisenberg, W. Pauli, P. Dirac, P. Jordan, N. Bohr, L. Rosenfeld, L. Landau, R. Peierles, V. Fok, S. Tomanaga, J. Schwinger, R. Feynman, F. Dyson and others as a "collective creative subject" who executed all logically necessary operations which led to construction of a new theory. Just for comparison, we would like to remind the reader that for classical theory of electromagnetic field all operations of the kind were carried out by one investigator – J. C. Maxwell. For classical physics it was more a rule than exception; of its three most important theories – mechanics, electrodynamics, thermodynamics – only the latter can be looked at as production of a "collective creative subject".[39]

In quantum relativistic physics, after creation of general relativity theory, we cannot find a situation when a developed theory was constructed by creative efforts of one investigator. The objects studied became far more complicated; construction of a theory now requires far greater quantities of information, so each of the investigators carries out only some of the logically necessary procedures which provide construction of a new theoretical system.

In this respect the following example is characteristic. N. Bohr, who, together with L. Rosenfeld, did the main work on interpretation of the mathematical apparatus of quantum electrodynamics, joined the creative group working on the new theory, when its mathematical formalism had already been basically built. According to Rosenfeld, Bohr not only had taken no part in creating this formalism, but even did not know its basic principles at the early stage. Rosenfeld recollected: "Bohr's state of mind when he attacked the problem reminded me of an anecdote about Pasteur. When the latter set about investigating the silkworm sickness, he went to Avignon to consult Fabre. "I should like to see cocoons," he said, "I have never seen any, I know them only by name." Fabre gave him a handful: he took one, turned it between his fingers, examined it curiously as we would some singular object brought from the other end of the world. He shook it near his ear. "It rattles," he said, much surprised, "there is something inside".[40]

L. Rosenfeld continued: "My first task was to lecture Bohr on the fundamentals of field quantization; the mathematical structure of the communication relations and the underlying physical assumptions of the theory were subjected to unrelenting scrutiny. After a very short time, needless to say, the roles were inverted and he was pointing out to me essential features to which nobody had as yet paid sufficient attention".[41]

Another important specificity of modern theoretical-cognitive situation is that fundamental theories more and more often are created without a well-developed layer of primary theoretical schemes and laws, which could characterize certain aspects of the new area. In this respect it is significant, for instance, that quantum electrodynamics, as preliminary knowledge of microstructure of electromagnetic interactions, had only fragmentary theoretical laws and models which characterized quantum properties of radiation and absorption of light by the matter. The other intermediatory links, necessary for construction of the theory, were created in the course of theoretical synthesis.

Last but not least, the third specificity of construction of modern physical theories is application of the method of mathematical hypothesis considered above. This method allows to pass in compact the stage of forming primary theoretical schemes and laws, finding at once equations of some vast object domain and then getting on their base the corollaries – theoretical laws which characterize particular aspects of this area.

In order to imagine visually the peculiarities of this way of theoretical investigation, let us consider the following hypothetical situation. Suppose Maxwell, while working on the electromagnetic field theory, did not have laws of electromagnetic and electrostatic induction or Coulomb's interaction of charges. Imagine then, Maxwell's theory was being created through introduction – by method of mathematical extrapolation – of generalizing equations for blocks of electromagnetic induction, electrostatics and others, which were derived out of Coulomb's, Faraday's and other laws, i.e. laws experimentally verified. In this case synthesis, leading to Maxwell's equations of electromagnetic field, would have been carried out on base of the mentioned generalizing laws.

Something of this kind is happening in construction of modern physical theories, and quantum electrodynamics is a typical example. It was formed in complete accordance with the requirements of mathematical hypothesis, and "intermediate" theoretical knowledge necessary to construct the new theory was created in the course of theoretical synthesis, which led to the system of its fundamental equations.

The main stages of development of the mathematical apparatus of quantum electrodynamics

The process of creation of the mathematical apparatus of modern quantum electrodynamics can be conventionally divided into four stages.

The first stage: apparatus of quantized electromagnetic field of radiation (field not interacting with the sources). The second stage: mathematical theory of quantized electron-positron field (quantization of sources of the field). The third stage: description of the interactions of the said fields within framework of the disturbance theory in first approximation. The fourth stage: apparatus characterizing interaction of quantized electromagnetic and electron-positron fields and taking into account the second and further approximations of the disturbance theory (development of renormalization method which allowed to describe the interacting fields in highest orders of the disturbance theory).

Each of these stages also consisted of several logically necessary steps which led to the corresponding equations of quantum electrodynamics. From this point of view, for instance, the first stage – construction of the apparatus of free quantized electromagnetic field – could be executed only due to preliminary investigation of quantum properties of radiation.[42] On this base scientists formed the notion of electromagnetic field of radiation as a specific quantum system which, on the one hand, has continual characteristics (frequency, wave vector), and, on the other hand, can be presented as set of photons in different quantum states. In the aspect of wave properties the field traditionally has been described by Maxwell's equations. Thus, there emerged the problem to transform the equations so that to take into account corpuscular properties of free electromagnetic field as well.

In order to do this, the magnitudes bound in Maxwell's equations, by analogy with now customary quantum mechanical approach, were regarded as operators subordinated to transposition correlations. So Maxwell's equations were transformed into equations of quantized electromagnetic field. Taken together with the commutation rules for operators (transposition correlations), they formed mathematical apparatus describing this field.[43]

The next step in investigation of the microstructure of electromagnetic processes stipulated an account of interaction of the radiation field with quantized sources (densities of charge-current). It required development of mathematical formalism describing quantum qualities of electron system in relativistic area. The solution of such a problem led to notions of electron-positron field. Finally the initial problem of quantization of sources of electromagnetic field was reformulated as problem of mathematical description of quantum properties of electron-positron field. Its solution marked the second stage of working out the apparatus of quantum electrodynamics.

From the point of view of logic of cognitive motion, the initial point of this stage is Dirac's relativisticic quantum mechanics of electron. We would like to emphasize again the fact that Dirac's theory, which opened for physics the area of electron-positron interactions, served as a kind of intermediate "pack" of knowledge for construction of modern quantum electrodynamics. It was a typical example – how, in the course of theoretical synthesis, investigators introduced missing links (local theoretical schemes and laws), which provide successful progress toward future fundamental equations of the theory.

Generalization of Dirac's equations was connected with quantization of the electron-positron field. This object, introduced within the scope of electron relativistic mechanics, was considered in the same way as previously had been considered the electromagnetic field of radiation subject to quantization. It was presented as some integral dynamic system having both wave and corpuscular qualities. Quantum nature of this system was described by introducing operators that influenced the wave function (state vector) of the system, which had been defined as a function in the space of filling numbers (particles corresponding to numbers – electrons and positrons, which were in certain quantum states and formed electron-positron fields). Wave functions $\psi(x)$ and $\overline{\psi}(x)$, which characterized states of electrons and positrons in Dirac's equations, were considered as the main operators of the field. Influence of these operators upon the field state vector changed the filling numbers; that corresponded to description of the field in terms of creation and annihilation of electrons and positrons in certain quantum states.[44]

Thus scientists created the mathematical theory of free quantized electron-positron field. The notion of such field made them reformulate the problem of theoretical description of quantized electromagnetic field interacting with the sources. Now it emerged as the problem of interaction of corresponding quantized fields.

Foundations of the mathematical apparatus describing this interaction were found at the third stage of forming quantum electrodynamics. The said apparatus consisted in a system of equations which united equations for quantized electromagnetic and electron-positron fields (correspondingly Maxwell's and Dirac's equations for operators of the fields). Besides, it included methods of their approximate solution by means of the perturbation theory which had been developed within non-relativisticic quantum mechanics and then transposed to the sphere of interaction of quantized fields. In quantum electrodynamics such interaction is presented as scattering of corresponding particles (electrons, positrons and photons) connected with their mutual transformations.[45] First the processes of dispersion were described only in first approximation of the perturbation theory. This became foundation for the theory of interaction of quantized electromagnetic field with charges. The theory allowed to describe and explain two types of processes: 1) transition of electron (or positron) from one state into another with emitting a photon and 2) formation or absorption of electron – positron pairs accompanied by absorption or emitting of photons.

Attempts to explore interaction of quantized electromagnetic and electron-positron fields in other approximations of the perturbation theory not only failed to make the results more precise, but even led to mathematically meaningless expressions. Observable magnitudes for characteristics of electrons and positrons, i.e. charge, mass and other connected magnitudes got infinite expressions in the form of divergent integrals.

The problem of construction of mathematical apparatus, which would take into account higher approximations of the perturbation theory, was solved only at the fourth, final stage of evolution of quantum dynamics. S. Tomanaga, J. Schwinger, R. Feynman, F. Dyson in their works developed the perturbation theory in relativisticic invariant form and suggested the renormalization method, which eliminated deviations by replacing formally computed infinite values of physical magnitudes by finite values known from experiments.

In the issue the sphere of processes, described and explained by quantum electrodynamics, considerably widened. It became possible to solve problems of scattering electron by electron, photon by electron, predict interaction of electron and vacuum, scattering of photon on photons etc.

This is the history of quantum electrodynamics taken in the aspect of forming its mathematical apparatus. It is easy to trace clearly expressed internal logic of its construction: first formalism, describing free quantized fields, was created, then on base of it the apparatus characterizing interaction of fields was constructed.

Outwardly the whole process (in its main part, at least) looks like a series of mathematical extrapolations leading to a system of equations for interacting quantized fields and methods of solving such equations. Wonderful achievements of quantum electrodynamics can be interpreted as one more evidence of efficiency of the modern method of constructing a theory. It is enough to say that the equations preceded such unexpected predictions as the one of electromagnetic vacuum (the state of electromagnetic field with the lowest energy which, despite absence of photons, influences upon charges behaviour, for instance, electron in atom). Predicted effects of vacuum polarization (effects connected with formation – due to an electromagnetic field – of virtual pairs, which cause certain distribution of charges in space, like polarization of dielectric, and have opposite action upon the external field, screening the primary charge creating this field) were quite unusual.

Nevertheless, speaking about heuristic functions of the method of mathematical hypothesis, we cannot stop at a trivial statement that in modern physics construction of a theory starts with attempts to "guess" its future mathematical apparatus.

Reflection of creators of new theories evokes a lot of judgments of this kind.[46] But this is only the first step toward understanding genesis of the theory. The main goal is to see logically necessary operations, leading to construction of new systems of theoretical knowledge, behind external features of modern investigation. In this respect we would like to pay attention to two important factors which refer to the process of becoming of quantum electrodynamics: 1) stipulation of putting forward mathematical hypotheses by the picture of physical reality preliminarily accepted by the investigators and 2) correlation between construction of the apparatus of the new theory and creation of a theoretical scheme which provides interpretation of this apparatus.

Quantum mechanical picture of the world and its role in forming the mathematical apparatus of quantum electrodynamics

Tracing the shifts of mathematical extrapolations in the history of quantum electrodynamics, we inevitably face the problem of initial ideas, bases for this or that extrapolation. Here it becomes clear that the putting of theoretical problems and indication of the ways of their solving were generated (at starting point, at least) by physical picture of the world grown out of the development of quantum mechanics. In that image the physical reality was depicted as two linked layers: macro and microlevels, and microlevel physical systems were considered as objects included in certain macroconditions and expressing their wave-corpuscular nature. In "operational" aspect the idea of wave-corpuscular features of microobjects was revealed by means of the complementarity principle. An object was regarded as a physical system which essential aspects, expressing in macrocircumstances strictly fixed by certain devices, could turn out mutually eliminating.

But that they were regarded as some kind of projections of an integral whole, united within one and the same method of description as complementary characteristics, discovered the specificity of the microobject.

The investigator who accepted this picture of physical reality had to take into account two possible aspects of considering physical systems: from the directions of their macro and microstructure. Correspondingly, he should apply a certain method of description of the system (classical or quantum mechanical). The connection between macro and microlevels of physical reality stipulated the connection between mentioned description methods within the correspondence principle[47].

We may find the decisive role of such picture of the world in putting initial problems of quantum electrodynamics, if we take into consideration the following. The program of quantizing fields was based on extrapolation of methods of quantum mechanics of points to a new sphere – fields and their interactions. But, in order to realize such extrapolation, scientists first had to see resemblance of fields with already studied quantum mechanical systems. Such view of fields was not at all evident because known and familiar quantum systems, physics had dealt with before quantum electrodynamics was constructed, in classical limit could be regarded as systems of a finite number of particles (systems with a finite number of degrees of freedom). Here, in a quantizing field, a classical analog was a continuum medium which could be compared with a dynamic system with an infinite number of degrees of freedom. That is why extrapolation of quantum mechanical description to the new area required certain justification. It could be provided by the quantum mechanical picture of the world which fixed the most general features of discernment of quantum objects. Previously collected empirical and theoretical knowledge of microstructure of electromagnetic interactions revealed such features of electromagnetic field (dualism of wave-corpuscular qualities). On this basing electromagnetic field was considered as an integral system which had quantum nature. Then this type of consideration was extended to electron-positron field. But such transfer was as well connected with functioning of quantum mechanical picture of physical reality, as consideration of an electron system in the image of electromagnetic field stipulated non-standard vision of it. The electron system now acts not as a mere multitude of quantum mechanical particles, but as an integral object – field whose separate quanta are particles belonging to the system.

Such vision was unusual since there was no classical analog for such an object (unlike quantized electromagnetic field which has a classical analog, the idea of electron field is meaningless in classical physics: in classical language electrons are particle with a finite – in principle – number of degrees of freedom).

We may follow T. Kuhn and characterize such approach to new consideration of electron system as a sort of gestalt-switching caused by change of model of vision in investigational situations. It is important that the latter was prepared and happened due to an already formed picture of the physical reality.[48]

Just as the picture of the world identified field and set of quantum mechanical particles as objects of the same nature, having the same combination of qualities (wave-corpuscular dualism), so it was possible to choose any of these objects as a model for considering the other (possibility to consider field as a system of particles, or to define a system of quantum particles as field).

Thus, the picture of the world in physics contributed to the idea of fields as special quantum objects which are to be theoretically described. This was the foundation for formulating initial investigational problem, which led to creation of quantum electrodynamics. The picture of the world served as stimulus to put forward such a problem, and it also pointed out the ways to solve it. These ways were founded in transfer of mathematical structure of quantum mechanics of points to the new area (fields and their interactions). Field was to be quantized in the same way as non-relativistic quantum mechanics did with systems of particles. On this base the method of secondary quantizing was developed. It provided transition from equations describing classical electromagnetic fields, and the ones describing quantum mechanical particles, to equations of quantized fields. Taking into consideration what was said about the role of physical picture of the world in constructing mathematical apparatus of quantum electrodynamics, it would be interesting to compare the modern way of investigation and models of theoretical investigation in classical physics, for instance, method of constructing a theory used by Maxwell (described above). The comparison shows that, at least in initial points, there is no sharp rupture between traditional and modern ways constructing a theory, despite the fact that in 20th century physics theories are constructed by the method of mathematical extrapolation. In both cases the investigator first "guesses" new equations due to directing influence of the picture of the world, which defines the putting of theoretical problems and points at the sphere of mathematical means, which would provide construction of a theory. The new element in modern investigation, along with explication of operational aspects of the picture of the world, is more active reverse influence of even early studies of mathematical synthesis upon the picture of the world. In the history of quantum electrodynamics we can see examples when the mathematical apparatus being created made scientists correct the quantum mechanical picture of the world from the point of view of relativistic ideas. The need in such correcting was caused by the requirement of Lorentz-invariance of the equation created (Lorentz-invariance of classical electrodynamics equations, when synthesized with the formalism of quantum mechanics, should be transferred to the equations of quantized field). But after the general relativity theory had emerged, to require Lorentz-invariance meant to accept relativistic notions of space-time. Consequently, such notions were to enter the quantum picture of physical reality in hidden form. Though the program of joining of quantum and relativistic notions within the framework of an integral physical picture of the world was accepted by all investigators after quantum mechanics had been completed, the first real steps toward its realization were made only in the process of constructing relativistic quantum mechanics and the quantized fields theory. In any case, it was stipulated by the very character of the mathematical formalism of the new theory, and that is why creation of the latter may be regarded as a considerable contribution to construction of the quantum-relativistic picture of physical reality.[49]

Paradoxes of the theory created and the problem of interpretation

The second important aspect of modern investigation is connection between mathematical hypotheses and procedure of construction of theoretical schemes.

In analysis of modern theoretical activities this side is usually lost sight of, because search for mathematical structures, especially at the early stages of formation of a theory, becomes the cognitive task number one. The problem of interpretation emerges only when the mathematical apparatus is already quite developed.

So we come to an impression that mathematical formalism of a developed theory is created independently from its interpretation, by a series of mathematical hypotheses realized in succession. Apparently the history of quantum electrodynamics proves it is right. But deeper analysis reveals that if we agree with it, we will have to make a very strained statement.

As we have emphasized above, equations of physics cannot exist outside of connection with theoretical schemes. Otherwise they would be purely mathematical statements but not expressions for laws of physics.

Since the process of reconstruction of equations taken from already formed spheres of theoretical knowledge into a new sphere always stipulates translation and redefinition of the corresponding abstract objects, then any mathematical hypothesis inevitably introduces a model which is supposed to be the theoretical scheme of the new sphere of physical processes. This model is reflected in the picture of the world and obtains ontological sense. It determines the initial semantic interpretation of the created formalism of the theory. At this stage, usually there is no empirical justification, so empirical sense of many magnitudes linked in the equations may be unclear. But their semantic interpretation doubtlessly should exist. Until some moment this interpretation encourages development of the mathematical formalism of the theory. The process of working out the mathematical apparatus of quantum electrodynamics is a good illustration. Let us take, say, the first stage of development of the apparatus. In the course of quantizing of electromagnetic field, the quantities of Maxwell's equations were tied in a new network of relations, in accordance with the principles of quantum mechanical description. Correspondingly, abstract objects transferred from classical electrodynamics and quantum mechanics to the new area of theoretical knowledge, also get new features. This was how, along with mathematical formalism, a preliminary theoretical scheme characterizing microstructure of electromagnetic field was created. Its authors introduced fundamental theoretical constructs: states of electromagnetic field and classical observables, whose probabilities of numeric values are correlated with the state of field. It was supposed that field described by the wave function (state vector) ψ_{nk} can be defined through superposition of some elementary states k', k'' etc., and to each of them these correspond photons (quanta of field) which are in the given state ($n'_{k'}$ photons in state k', $n''_{k''}$ photons in state k'' etc.). The field state vector allows to fix probability of emergence of photons in every "elementary" state.

In ontological aspect, which corresponds to reflection of this scheme in the picture of the world, it corresponded to the idea of electromagnetic field as a system with a varying number of photons, which appear in certain state with certain probability.

At the same time the theoretical scheme expected that the field state vector should be connected with some probability of observation of classical field components in a point. It followed from the basic principles of quantum mechanical description, in accordance with which the apparatus of quantized electromagnetic field was composed. According to these principles, operators of the field should be juxtaposed with physical quantities whose numeric values can be determined exactly at macroscopic registration level by a device set for measuring the corresponding value. The probability of these quantities is determined by the field state vector (or, squared modulus of the wave function). For example, field could be characterized by field strength operators $\hat{\mathbf{E}}$ and $\hat{\mathbf{H}}$, so the experiment should give values $\mathbf{E}$ and $\mathbf{H}$, corresponding to mathematical expectations of these operators.

The considered theoretical scheme at the early stages was accepted without the procedure of its empirical justification. For instance, there was no special verification of the question: how legitimate is to transfer such idealizations (abstract objects) as field in point, taken from classical electrodynamics, to a new area of interactions. The ideas of classical electrodynamics that the states of field can be characterized by strengths **E** and **H** in a space-time point were conserved within the framework of quantum mechanical description of electromagnetic field. Such description introduced only one evident change to the classical notions: it claimed to use classical observables in order to characterize the field state taking into account basically the statistic character of expectation of the concrete values, but put no limitations for possibility to determine exactly each of the values for each separately taken value measured. Therefore at the early stages the preliminary theoretical model of quantized field of radiation, as its determined semantic interpretation of corresponding equations, was accepted as legitimate from the point of view of empirical sense as well. In any case, empirical interpretation of the magnitudes linked in the equations seemed then obvious and easy to realize according to patterns of standard quantum mechanical description.

The conviction that the introduced theoretical models are quite reliable for some time encouraged development of mathematical formalism of quantum electrodynamics. It is enough to say that immediately after quantizing electromagnetic field attempts were made to build a similar apparatus for description of electron field.

But successful progress toward generalizing equations of quantum electrodynamics was stopped, when in the very foundation of the theory paradoxes were found. It became clear that classical field strengths in a point cannot have exact value. If the field consists of separate quanta, appearing and disappearing with certain probability in different quantum states, chaotic fluctuations of every component of the field in a point are always possible.

Thus, two equally fundamental principles of the quantum field radiation theory, which seemed completed – 1) state of the field can be characterized by classical values of components of the field in a point and 2) the field is a system with a varying number of photons filling certain "elementary" states whose superposition characterizes the field – resulted contradictory. Emergence of such contradictions destroyed the primary theoretical scheme and made the corresponding mathematical apparatus physically meaningless.

For methodological analysis this circumstance is of paramount importance. It leads us to the conclusion that at a certain stage of constructing apparatus of modern theory it is mandatory that mathematical hypotheses should be supported by analysis of theoretical schemes and their constructive justification. In other words, progress in the plane of mathematical formalism can be relatively free only until some stage, and then it can continue only in case it is correlated with movement in the plain of physical contents.

Paradoxes discovered in the primary variant of the theory of quantized electromagnetic field were one of very characteristic moments of modern theoretical investigation. Mathematical hypothesis, altering connections between theoretical constructs of preceding equations, provide such construct with new features, and one feature may rule out another. Exactly that happened when constructing the apparatus of quantized radiation field, when scientists tried to synthesize equations of Einstein-Lorentz electrodynamics with a quantum mechanical method of description.

The paradoxes of quantized radiation field were a signal of emergence of constructs with mutually eliminating features in the theory.

This situation was similar to already discussed paradoxes of Rutherford's model of atom and Dirac's relativisticic electron theory. As to history of classical electrodynamics, we analyzed a similar situation speaking about the period in Maxwell's work when he tried to introduce an equation for electromagnetic induction basing on the model of stationary force lines.

Naturally the first efforts to eliminate paradoxes had to be aimed at finding non-constructive elements inside of the theoretical scheme, introduced together with the apparatus of quantized radiation field at the stage of mathematical hypothesis. It was necessary to do some selecting among theoretical objects, to discover the ones "responsible" for paradoxes and replace them by new abstract objects which would be fit to the procedure of empirical justification.

The first part of this problem was partly solved in the work by V. Fok and P. Jordan,[50] and more completely by L. Landau and P. Peierles.[51]

Strictly speaking, the paradoxes could be caused either by definition of the field state vector as (in contrast to familiar quantum mechanical approach) superposition of states with a varying number of particles (photons), or by a hidden assumption that observable magnitudes are strengths in a point.

Since the idea of field as a system with a varying number of photons allowed to explain well known dependencies of absorption and emission of light quanta by atoms, then the corresponding characteristics of vector-state were based empirically and got constructive meaning. Then it was to be checked whether classical observable fields possess such meaning in a point. Certain intellectual experiments we held to understand if we could, introducing the said observables into the new area, conserve their main quality – fundamental measurability (i.e., possibility to get exact values of every observable magnitude using a classical device). Intellectual experiments made by Fok-Jordan and Landau-Peierles revealed that if both quantum and relativistic effects are taken into consideration, measuring strengths of quantized field in a point is impossible.

They came to this conclusion by the following reasoning. According to the approach typical for classical theory, intensities of **E** and **H** are determined through influence of the field upon a charged test body. In case of component **E** that influence is measured through impulse passed to the experimental charge, in case of component of **H** – through moment of magnet or some distribution of charge-current. Just as we are to measure the field in a point, so the experimental body should also be a point. Suppose, the task is to determine component E_x. For this we need a point charge. Thought experiments by Fok-Jordan took an electron accelerated by the field, while experiments by Landau-Peierles admitted a point particle of any nature (it could have, a particle with, for instance, greater mass than electron).

Measuring the field component means that the impulse by the experimental particle from the field should be registered by a classical device. In this case the value of this impulse will let us determine the value of the corresponding field component exactly.

Thus, the procedure of thought measuring the field components in a point in moment t stipulated two conditions: 1) localization of the experimental particle in the given point of the field at moment t, where the particle gets an impulse, 2) exact registering this particle by a classical device.

Since the experimental particle submitted quantum laws, both conditions were basically impracticable. The first was impossible because of indeterminacy relationship: localization of the particle in a point led to its fundamental indeterminacy Δp in the value of its impulse. Consequently, the value of the field strength can be accurate up to no more than Δp.

The second condition is impracticable because of two reasons. First, it was impossible to register exactly the impulse of the point experimental particle because of quantum laws of energy-impulse exchange of the particle with the device. Since there is indeterminacy relationship $\Delta\varepsilon\Delta t{\sim}\hbar$ (ε – energy, t – time), so collision of the particle with the device when during time Δt the former gives its energy to the latter causes indeterminacy $\Delta\varepsilon$ in the value of this energy. The connection between energy and impulse causes corresponding relation between time Δt and impulse measured P_x. This relation is expressed by formula $|v''_x - v'_x|\Delta P_x \Delta t \sim \hbar$ (1),[52] where v'_x and v''_x – velocities of the particle before and after measuring, Δt – time of measuring, ΔP_x – indeterminacy in the value of the impulse of the particle.

Accounting of relativistic effects stipulate that $|v''_x - v'_x|$ should not exceed light speed c. Consequently, on the base of (1) there emerges relation $\Delta P_x \Delta t \geq \frac{\hbar}{c}$, according to which, the less time measuring the impulse of the particle takes, the greater indeterminacy in the value of the impulse is measured.

In measuring component E_x in a space-time point it is stipulated that the impulse of the experimental particle should be registered practically immediately. We are to reduce the period of measuring $\Delta t \to 0$ infinitely to avoid side effect upon the impulse of test particle. But in this case ΔP_x will increase infinitely. So observance of one necessary condition, which would provide exact measuring field strengths in a point (practically immediate registering the impulse of the experimental particle) leads to fundamental impracticability of the other condition, as much again necessary condition (exact measuring of that impulse by a classical device).

Secondly, exact registering the impulse of the experimental particle is impossible because the particle is radiating at the moment of collision with the device and starts interacting with its own radiation. The influence of the particle's own radiation can be taken into account only with a basically irremovable error.[53]

Thus, measuring the field component by a point experimental particle we face three irremovable types of indeterminacy: because of its localization in a point of the field; because of its interaction with the device during time Δt; because of its interaction with its own radiation.

In its turn, indeterminacy of impulse of an experimental particle means fundamental impossibility to measure every component of quantized radiation field strengths in a space-time point. Consequently, the theoretical constructs (of field in a point) are meaningless when extended to the area of quantum processes. From the point of view of methodology, it is important to pay attention to the structure of the intellectual experiments which led to this conclusion. It is significant that they took into account not only quantum, but also relativistic effects which were expressed when the field components changed, and because of this they expressed – in idealized form – characteristic features of possible experiments and measurements in the new area. Analysis of measurability of field in a point shows whether we can introduce the mentioned abstract objects as idealizations basing on real specificities of experimental-measuring activity in quantum field studies. Here we can easily see characteristic features of constructive introduction of abstract objects.

The negative result meant that the objects mentioned are non-constructive elements in the preliminary theoretical scheme. Discovery of such elements was the first necessary step toward rebuilding the theoretical scheme on a constructive base. Further task was to change it so, on the one hand, to conserve the constructed apparatus of the theory, at least in its basic characteristics, on the other hand, to justify the theoretical scheme introduced by idealization of experiments and measurements related to the new area of interactions. In the history of quantum electrodynamics this problem was solved due to cognitive activity known as Bohr-Rosenfeld measuring procedures.

Idealized procedures of field measuring and interpretation of the apparatus of quantum electrodynamics (the initial idea of Bohr-Rosenfeld procedures)

Bohr-Rosenfeld measuring procedures occupy a special place in settling quantum electrodynamics, because it was thanks to them that a non-contradictory interpretation of its mathematical apparatus was developed. At first Bohr and Rosenfeld interpreted the apparatus of quantized radiation field, and then revealed the physical meaning of the formalism which described interaction of the field with quantized sources. We will try to show that Bohr-Rosenfeld procedures are a typical example of stage-by-stage shaping of a constructively justified theoretical scheme in the modern epoch of theoretical investigation.

First we would like to describe the historical situation in which the cognitive activity took place. After Landau and Peierles had proved that it was meaningless to apply the idea of field in a point for description of quantum processes, quantum electrodynamics entered a period of crisis of its foundations.

First, it was entirely unclear, how to change the theory in order to get non-contradictory interpretation of the mathematical apparatus introduced. What is more, nobody knew if it was possible in principle. Only retrospectively (we retold Landau's and Peierles's work mainly from the point of view of its logically necessary contribution to construction of the new theory) can we see that the only right position in those circumstances was the desire to reconstruct the initial theoretical scheme so that it could allow only to reject use of field quantities in a point but conserve the idea of classical observables (field strengths).

But this step was not at all easy. In any case, the investigators who had discovered paradoxes of impossibility to measure the field components failed to do the necessary work themselves.

At that stage of development of electrodynamics Landau and Peierles regarded their results not as a proof of limitedness of the initial interpretation of the mathematical apparatus of the theory, but as evidence that this apparatus was worthless and basically could not bear any physical meaning. It seemed their point of view had solid ground. The state of electromagnetic field in classical theory was characterized by strengths **E** and **H**. As to quantum mechanical description, it contained a well known principle: quantizing of a system limits simultaneous measurability of complementary (in Bohr's sense) pairs of quantities, but puts no limitations to measurability of a separate magnitude (classical observable). So, Landau and Peierles believed it was impossible to get the exact value of strengths **E** and **H** taken separately, it meant that there are no ways to apply quantizing methods to such an object as radiation electromagnetic field.

Later Landau and Peierles extended this conclusion to quantizing field sources. They showed that determination of state of electrons, provided that they are measured by means of a point experimental particle during a very short period of time, led to irremovable indeterminacies of each of the separate quantities characterizing the state of electron.[54] It could be automatically concluded that it was impossible to create a quantum mechanical description of the field sources, or, what is equivalent, to construct a quantized electron field theory.[55]

Last, Landau and Peierles appealed to numerous difficulties which had emerged in quantum electrodynamics with efforts to find the physical meaning of its apparatus, extended through a series of mathematical extrapolations. They meant difficulties with interpretation of Dirac's equations (they included solutions with negative energy values) and difficulties in search for sense of so called zero fluctuations of electromagnetic field. The former have already been discussed. We are only to remind the reader that though Dirac had already proposed an interpretation of his equations, a lot of investigators who worked on the quantum theory of field first took his model of "holes" as quite artificial[56] (especially since at the early stages there existed a tendency to connect the "holes" with presence of proton, which led to contradictory conclusions in calculations of mass-energy of particles; only later there appeared the hypothesis of positron, empirically proved only in 1932). Under those circumstances Landau's and Peierles's thesis that quantum mechanical methods cannot be applied in the relativistic area did not at all seem unconvincing nor illogical.

Besides, there were more difficulties connected with paradoxical corollaries of the mathematical apparatus describing quantized radiation field. According to them, the energy of zero energy level of the field was infinite.[57]

Landau and Peierles linked those corollaries with the idea of fundamental incommensurability of the field components in a space-time point. They indicated that it follows from the expression for indeterminacy of each of the components **E** and **H** $\Delta E \geq \frac{\sqrt{\hbar c}}{(c\Delta t)^2}$ and $\Delta H \geq \frac{\sqrt{\hbar c}}{(c\Delta t)^2}$ (where ΔE – indeterminacy in the value of electrical intensity, ΔH – indeterminacy in the value of magnetic intensity, Δt – indeterminacy in the time of measuring, c – light speed, $\hbar$ -Planck's constant), that if we decrease the time of measuring Δt to zero (to realize measurement of the field in time point t_1) correspondingly ΔE and ΔH will tend to infinity. From this position the conclusion of infinite values zero energy level of the quantized field was presented as a special type of incommensurability paradoxes.[58]

Taking all this into consideration, we may understand why there appeared a tendency to preserve quantum mechanics methods only within the sphere of non-relativistic processes.[59]

The crisis of the early 1930s in quantum electrodynamics provides one more proof that fundamental theories of higher degree of generality are constructed differently from the way it seems when we use a simplified approach to mathematical extrapolation. Usually for such theories it is impossible to build mathematical apparatus at once by means of a continuous series of mathematical hypotheses and then find interpretation of the ready formalism. Quite long progress in mathematical means enlarges the danger of hidden introduction and accumulation of non-constructive objects in the theory. So it is urgent that we should use special analysis of physical sense of already constructed links of the mathematical apparatus and their interpretation as early as at intermediate stages of forming of the theory's fundamental laws.

In such periods the central point of the research passes to the area of search of theoretical models which could provide interpretation of the equations introduced.

Let us consider the logic of this search at the period of struggle against crisis in quantum electrodynamics.

First of all, to provide progress in the development of the theory, it was necessary to formulate the theorem correctly. To do this, the investigators had to see in the incommensurability paradoxes only limitations for classical idealizations of the field strengths, but not prohibition to use quantum mechanics methods for description of relativistic processes.

Correspondingly, the investigation task was to be formulated as search of classical observable, which would be fit for characterizing wave properties of quantized electromagnetic field (without using field strengths in a point). But after Landau's and Peierles's work many investigators would have regarded such formula as inherently contradictory.

Here we have come to a very important aspect in evaluation of the crisis caused by incommensurability paradoxes. The fact is that Landau and Peierles, speculating on unsuitability of quantum mechanical description in relativistic area, unwittingly accepted one ill-founded assumption which caused their too categorical conclusion. We mean the supposition that the test particle, used for measuring field quantities, is always a point particle and is of quantum mechanical nature. Such idealization of the test particle was legitimate when the problem dealt with measuring instantaneous value of **E** and **H** because of the problem itself. Indeed, if we measure the force which is to influence upon the test particle in a point of the field, it means that the particle should be located in that very point at the moment given. But for this the particle itself should be regarded as point. Naturally, in measurements in very small areas only microparticles which submitted quantum mechanics laws, could satisfy these requirements.

But then the idea of quantum mechanical test particle was automatically transferred to any situation of idealized measuring field magnitude in quantum area. Landau and Peierles concentrated on its interaction with the device and discovered that here that increasing indeterminacy of the impulse of quantum test particle inevitably appears, if measurements take short periods of time.

For determination of the magnitudes characterizing state of quantum system in relativistic area only short periods are necessary, because here the state of the system can change rapidly enough during the time of measuring. So, it would be easy to conclude: it is impossible to register the corresponding parameters of the test particle exactly, and, consequently, to determine classical observables characterizing quantum system in relativistic area.

This conclusion would be logically immaculate only in one case: if we assume that the means of measurement is a point quantum test particle.

It just never occurred to the majority of scientists to throw doubt on that assumption. But its critical analysis led to decisive clearing on the situation. It was N. Bohr who carried out this analysis. Bohr put forward an idea which provided overcoming the crisis: he proposed to use in the intellectual experiments testing measurability of field quantities a classical experimental body instead of point quantum mechanical particle. Historians of quantum electrodynamics, including Bohr's co-author L. Rosenfeld who brilliantly depicted that "heroic" (Rosenfeld's word) period of the development of quantum physics, usually emphasize great productivity of Bohr's idea, but they rarely reflect the logic of its emergence. Though, from a methodological point of view, understanding of this logic is of extreme importance, because here Bohr's idea is presented not only as a product of highly gifted intuition and a "spontaneous guess" but also as a logically necessary step of theoretical investigation. Probably, the main condition for this step was analysis of the notion of experimental body in the aspect of specificities of quantum mechanical measuring. Let us examine this point in more detailed way.

It is well known that the most part of investigations connected with experiment stipulate use of a special physical agent – a means to transfer the information about the state of the object measured to the observer. The role of such agent may be played by, for instance, a well charged body in experimental measuring of electrical field strengths, some volume of liquid in experimental measuring of temperature, a polarized beam in experiments with crystals etc. All agents of this kind are concrete variations of experimental bodies.

The construction of correspondence rules (operational definitions) is based on thought experiments which are just idealizations of real experimental-measuring activity. In this connection theoretical discourses of physics start using a special idealized object – experimental body. Its general features are derived from analysis of functions of concrete variations of experimental bodies in experiment. Such analysis lets us distinguish three basic and necessary features of experimental body: 1) it should interact with the physical system studied, changing its state in correlation with the state of this system: 2) it should translate the accepted state until interaction with the register device;[60] 3) its interaction with the register device should give the observer enough information about the state of the experimental body, that he could judge the state of the physical system studied (in this case the observer comes to conclusions about values of physical quantities characterizing the state of the system measured, basing on the data from the device).

The features mentioned of experimental bodies can be easily illustrated by simple examples. Suppose, we are measuring temperature with a mercury thermometer. The role of experimental body belongs to a volume of mercury in a glass vessel. The possibility to use it as an experimental body is conditioned by the following: 1) change of the volume of mercury (state of the experimental body) is correlated with the change of temperature of the bodies observed: 2) within certain limits we always can fulfill the requirement than, until observation of the scale (register device) which fixes the height of the mercury column, either the height (volume of mercury) will not change at all under external influences, or, if such change still takes place, it can be taken into account using corresponding equations (for instance, the heat balance equation); 3) when the height of the mercury column is registered by the observer, this act by itself does not change the state of the experimental body so, that it could prevent the body from transferring information about the temperature measured (this condition is practicable because we can, for instance, ignore the influence of light upon the mercury column, take into consideration in the very construction of the thermometer in graduation the change of the volume of mercury caused by its heat exchange with the scale etc.). In other words, really we can use a container with mercury as a means of temperature measuring, because the criteria of correlation, translation and possibility to register the state of this experimental body as the result of interaction with the object measured are observed. It is easy to see that requirements of this kind are observed in any experiment concerning any experimental bodies. They are common and significant features of the whole class of experimental bodies, which is why they form the sense of the corresponding idea.

In experimental-measuring situations of classical, quantum and quantum relativistic physics the indicated features are specified in several special assumptions.

For example, classical physics assumed that, first, the experimental body does not influence upon the state of the object studied during their interaction; second, that perturbing influences upon the experimental body from the register device at the moment of registering can be ignored. Of course, both assumptions are idealizations, but they take into account circumstances of real experiments and measuring in classical area. No doubt, perturbations caused by the experimental body always exist, and the experimental body itself also is object of influences from the register device during the period of time which is needed for measuring (it starts with interaction of the experimental body with the device and ends with finish of the device's indication). But in such experimental-measuring situations, where elements of the system – experimental body and register device – are classical objects, it is always possible either to provide such conditions of experiment that these perturbations would be negligible, or to take these disturbances into account by means of calculations and corrections.

In the measurements of quantum objects all these assumptions lose their legitimacy. In such measurements the physical system whose state is measured is always a microsystem, while the device registering quantities, which characterize the state of that system always belongs to macrolevel. The experimental body, as mediator between the microsystem measured and the experimental body, should interact with the former as a microsystem. Existence of quantum of action prevents us from ignoring the reverse influence of the experimental body upon the object measured, so in quantum area we should avoid an idealized image of a register device, which does not influence upon the object of measuring. This rejection means that in quantum mechanical measurements, unlike classical situations, we cannot identify the state of the system before and after measuring. Reproducing the same conditions and repeating the same measuring of the "prepared" state of the system, we will get different results every time. But each of them can be expected with a certain probability, if we characterize the state of the system before measuring by some wave function. Such connection between mathematical expectation of the results of measuring and characteristics of the state of the system measured allows us to predict (as we know the wave function) the results of measuring (measurements of quantum systems are not repeatable but predictable).[61]

Thus, quantum mechanical character of interaction of the experimental body with the object measured does not prevent the observer from receiving information about the state of the object. The experimental body takes part in quantum interactions and changes its state in correlation with the state of the system studied (though the characteristics of the state are different from those in classical physics). In this sense the first feature characterizing experimental bodies is still valid, when their interactions with the object measured submit quantum laws.

But there exists one more interaction: the experimental body transfers information about the object to the register device. If the experimental body interacts with the device also in accordance with quantum laws, how can it influence the functions of the experimental body? Can it, being a quantum particle, first, translate its state in interaction with the system measured until interaction with the registering device, and, second, transfer without errors the information about the system measured to the device?

In non-relativistic area, when the state of the quantum system is constant during a period of time comparable with the period of measuring, it is possible to fit both conditions.[62] But in relativistic area the situation is entirely different, as Landau – Peierles investigation proved. Here the function of experimental bodies belongs to quantum particles, and observation of one condition automatically excludes the other. The test particle enters interactions in which the state of systems changes during period of time comparable with the period of measuring. After interaction with the system measured the test particle – before it transfers information to the register device – may undergo a new type of influence from the system, since interaction in relativistic area is related to creation of new particles, generated by both the system measured and the experimental point particle itself. The longer is the time of measuring, the harder is the influence of the particles mentioned upon the experimental particle whose state is being transformed. Hence, it is necessary to register the state of the experimental particle as soon as possible after its interaction with the system measured. But, as we have already mentioned, observance of this condition leads to irremovable increasing errors in determining magnitudes characterizing the state of the test particle. Thus, requirements of translation of the state transferring information about the system measures, and requirement of registration of this information without errors are mutually eliminating for a point quantum mechanical particle used as experimental body in measurements in relativistic area. Measurements made through such particles resulted unpredictable.

Investigators saw that a point particle, when used in relativistic area as experimental body, loses its features which could make it belong to the class of experimental bodies. This was the key moment in transition from Landau-Peierles analysis to Bohr-Rosenfeld procedures. From Landau-Peierles intellectual experiments the only conclusion could be drawn: a quantum mechanical particle cannot be experimental body in measuring quantized field, but from this it did not follow that methods of quantum mechanics are inapplicable in relativistic area. Such conclusion considerably changed the situation. Now the task was put into practice idealized procedures of measuring in quantum relativistic area without quantum mechanical experimental bodies.

There was only one way to reach this goal: to return to classical experimental bodies. This approach automatically eliminated all problems connected with translation of state of the experimental particle and its interaction with classical device. If the experimental body is a classical object, in description of its interaction with the register device, it is absolutely correct to apply classical idealizations, which allow either to ignore the perturbing influence of the device or to take it into account by means of corresponding corrections. The only question to solve was that of interaction of the experimental body with the quantum object.

Evidently, such interaction should proceed in accordance with quantum laws. How can it be, when the experimental body is not a microparticle, but a classical object? The answer was simple: quantum systems always include description in terms of macroscopic parameters, and quantum interactions by definition should have in their last stage interaction with a classical device. The latter can be accomplished as early as at the first step (Mandelshtam's words), when we deal with direct measurements, and through a series of further links, where the measurements are indirect.

Application of classical experimental bodies as means of obtaining information on quantum systems in relativistic area may be carried out in two variations: 1) investigator abstracts himself from detailed examination and calculation of atomic structure of experimental bodies, considering the latter as a special part of a classical meter unit adjusted to measuring corresponding field quantities and 2) the said structure is taken into account, i.e., the experimental body is considered as a kind of aggregate of microparticles (for instance, distributions of electrons in certain volume forming experimental charge), which is set for interaction with the object and then interacts with the device, presenting itself as a classical object.

In the first case the measurements are direct, but, unlike direct measurements in non-relativistic area, here we should bear in mind the measured quantum objects' ability to change their state during period of time comparable with the period of measuring. Because of this there are restrictions first marked by Landau and Peierles (but these restrictions now concern not experimental bodies, but the objects measured and are their immanent characteristic). These restrictions consist in the following: to measure a separate classical quantity determining the state of the system, we need time, not longer than the period during which the state described by the quantity measured can be disturbed. If this is beyond our possibilities, measuring not pairs but a separate quantity will give a certain indeterminacy (for instance, for coordinate q and impulse p of a point particle in relativistic area there emerge indeterminacies $|\Delta p \sim \frac{\hbar}{c\Delta t}|$ and $|\Delta q \sim \frac{\hbar}{mc}|$).

In the second case, when atomic structure of the experimental bodies is taken into account, measurements are more like indirect ones. Here we can trace quantum effects of interaction of the object measured and the experimental body, say, some distribution of charge accounting microstructure of this distribution. Such interaction in relativity area causes creation of new particles, and that makes certain contribution to macro–effects fixed by the register device.

So, a classical experimental body used in quantum measurement has dual nature: at microlevel it interacts with the object measured, at macrolevel with the register device. Thanks to this it transfers information about the object measured to the observer and works as means of measuring quantum systems.

The given analysis may be regarded as logical reconstruction of the cognitive activity which secured transition from Landau-Peierles conclusions to Bohr's fundamental idea.

We would like to draw the reader's attention to the fact that analysis of functions of experimental bodies in idealized measurements is a special investigation, which uses metatheoretical language as the language of quantum electrodynamics (or any other concrete physical discipline: classical mechanics, non-relativistic quantum mechanics etc.). This is the language of logical-methodological analysis, an instrument of analysis of common features of experimental bodies and understanding of the very idea of "experimental body".

This peculiarity is important because it discovers exit (characteristic for investigation) to the area of methodological problems every time when science comes across seemingly unsolvable paradoxes. Solution of the paradoxes (or justification of impossibility to solve them with further reconstruction of previously suggested investigational program) is provided by metatheoretical investigations connected with analysis of the most general features of objects studied and comprehension of methods of theoretical cognition.

In this respect let us mark that analysis of function of experimental body was purposeful, on the one hand, by general methodological condition to link basic quantities of the equations with experiment by means of corresponding idealized measuring, on the other hand, by specificity of quantum mechanical objects, which require that for their description classical idealizations should be applied. The fact that it was Niels Bohr who succeeded in this analysis has profound foundation. We should take into account Bohr's decisive part in revelation of conceptual foundations of quantum mechanics, his permanent attention to the key problems of quantum mechanics measurement theory, his methodological erudition which let him grasp the very core of such problems and find solutions. All this gave Bohr the opportunity to be the first who overcame the psychological obstacle which had appeared due to blind using a point quantum object as experimental particle[63]. But these factors refer more to psychology of scientific creative work. In respect of logic of investigation, it is important that there existed logically necessary transition from Landau-

Peierles thought experiments to the fundamental idea of Bohr-Rosenfeld procedures. From this point of view we may say that once the problem of quantizing of fields had been raised and difficulties in interpretation of the introduced equations were found, so if not Bohr, then somebody else had to make the described steps toward the program of idealized measurements by means of classical experimental bodies.[64]

Reconstruction of the theoretical model of quantized electromagnetic field and justification of its consistency

After N. Bohr's program had been put forward, scientists started its accomplishment. The work was done in several stages.

First of all it was necessary to interpret apparatus of quantized radiation field within the framework of idealized measurements with classical experimental bodies. In case of success of this part of the program, it should be extended to the area of quantizing of sources of the field and then – to the area of interaction of the quantized field with quantized sources.

Naturally, no one could guarantee that Bohr's program of interpretation of the quantum electrodynamics equations will successfully solve all problems of the new theory. Only concrete investigation could demonstrate it. But still there was progress, as it became clear how to overcome contradictions of the previous period of development of quantum electrodynamics.

The very formulation of Bohr's basic idea showed concrete ways to positive reconstruction of previously introduced theoretical scheme of quantized radiation field.

First, it became clear, what observables were to be introduced into the scheme instead of field strengths in a point. Measurements of field components should be performed by means of classical experimental body which always occupies certain volume V, while displacement of the experimental body measuring the field strength takes certain period of time τ. So the field strengths could be determined exactly in thought experiments with classical experimental bodies only in area $V\tau$, but not in a point. The conclusion suggests itself: these magnitudes should be observables characterizing the state of the quantum field.

Introduction of such observables meant decisive change of the previous scheme (there appeared a new abstract object, and correspondingly all connections among all other elements transformed). Naturally, the new scheme gave new semantic interpretations of the equations of the theory: it meant that only quantum field strengths averaged in some space-time area (not in a point!) should have physical sense.

Clearly, such interpretation still remained a hypothesis. It could turn out that it is inconsistent with the structure of already built formalism, or requires such corrections which contradict to general foundations of quantizing fields. Instead of the past paradoxes of the theoretical scheme there might emerge new ones, and interpretation might be logically impossible. The possibility of such paradoxes and mismatches at the stage of reconstruction of the initial theoretical scheme is easy to explain, if we take into consideration the basic specificities of structure and functioning of such schemes.

First, a new element introduced to the scheme always changes correlations among all other elements. Just as such correlations are described in the equations, so first of all one has to check whether the offered modernization of the theoretical scheme will be fit for already shaped mathematical formalism or the latter should be transformed.

Second, the change of correlations among abstract objects which form the theoretical scheme can hiddenly provide the objects with such new features, which would be incompatible with the previous ones, those undergone constructive justification. So it is necessary to clear up, whether the new object destroys that constructive and heuristic meaning which was loaded into the theoretical scheme by previous development of the theory.

Clear, successful execution of this operation does not guarantee correctness of the new (reconstructed) scheme.

Even if it is proved that it corresponds to the theory apparatus and is inherently consistent, still the scheme remains a hypothetical construction. It will leave this status behind only by procedures of constructive introducing abstract objects, when the scheme is validated as a generalized model of corresponding experiments and measurements.

In this sense the final semantic interpretation of the theory apparatus appears only when its empirical interpretation is built. Their separation and consideration out of mutual influence is possible only up to certain limits. But, just as the procedures of constructive justification require a lot of work, so, before starting them, it is necessary to make sure the way of their realization is expected to be fruitful. That is why we verify correspondence between the theoretical scheme and the theory apparatus, and verify its object's inherent consistency. We will call such verification "potential interpretation", since final ("actual") semantic interpretation is formed only due to finding empirical sense of the basic quantities linked in the theory equations.

Analysis of the history of quantum electrodynamics shows that the first steps toward realization of Bohr's program of idealized measurements really were connected with potential interpretation of quantized electromagnetic field equations. Having proposed to reconstruct the initially introduced theoretical scheme into new one, in which the place of the observable components of field in a point was occupied by other observables (field components averaged on finite space-time area), Bohr first of all checked how such scheme conforms with the mathematical formalism of the theory, and then, together with Rosenfeld, justified inherent consistence of the new scheme.

Verification of the first type showed that there is complete agreement between the main idea of the new interpretation and character of the mathematical apparatus of quantized electromagnetic field.

Analyzing this apparatus, Bohr proved that there idealizations of field in a point are used only as a formal auxiliary construct and does not have real physical meaning, while field components averaged on some finite space-time area do have such meaning. It followed from the very character of commutation relations for field operators $\hat{\mathbf{E}}$ and $\hat{\mathbf{H}}$. The fact is that the commutation relations were expressed through generalized functions of the kind of δ-function introduced by Dirac in construction of commutation relations in a continuous spectrum. The fundamental feature of this function is its ability to be reduced to zero in all points except one, where it equals to infinity. The field quantities in a point should have behaved correspondingly. But δ-function has one more remarkable property: being integrated over all values of the variables, it turns into one. In commutation relations the role of arguments of generalized functions expressed through derivatives of δ-function belonged to space and time coordinates. Hence, integration with respect to some part of space-time area gave finite values for the right side of commutators of field values and corresponding uncertainty relations for these values. In other words, integrals of field components taken over finite space-time area got unambiguous meaning.

As followed from the structure of the mathematical formalism of quantum electrodynamics, that physically meaningful statements are not those of fields in a point, but those of average values of field components taken over finite space-time areas. This was the first sign of fruitfulness of the reconstructed theoretical scheme and, correspondingly, of Bohr's program of idealized measurements of quantized field components by means of classical experimental bodies.

We have to mention that the described above period of Bohr's cognitive activity is usually related "topsy-turvy" in the history of physics. It is believed that first Bohr discovered that in the mathematical apparatus only averaged field strengths are meaningful, and only then, on base of these specificities of the apparatus of the theory, came to conclusion about application of classical experimental bodies. Such statements can be found, for example, in L. Rosenfeld memoirs of his work with N. Bohr. What is more, the original text written by Bohr and Rosenfeld, dedicated to analysis of measurability of electromagnetic field, offers us a similar version[65]. No surprise that authors of historical essays follow the same way; describing development of quantum electrodynamics, they usually retell recollections of the investigators who built interpretation of the equations of quantized electromagnetic field. Yet when we regard some reproduction of a theory made by its creators, we are to keep in mind that logic of rendering results of the investigation and logic of obtaining these results do not usually coincide. Deductive posing usually starts with statements which were final results in the investigation itself. Therefore real historical progress of thought leading to certain result rarely is rendered without swerves in a scientific text rendering the obtained result. As to retrospective historical analysis of a discovery made by its authors, we should never forget that numerous publications of the obtained results, which searched for the easiest and most compact logic of rendering, are capable of quite considerable deformations of notion of ways to the desired result. We have to be very careful with historical testimonies made by creators of a theory. In this respect A. Einstein said: "If you want to learn something about their methods from physicist-theorists, judge deeds not words". Of course, it does not mean that the creators' reflection cannot give us any more or less valuable historical evidences. We only mean that not every such evidence should be treated as an undisputable historical fact, moreover, retrospective analysis in memoirs normally reproduces only key moments of the creative work, but not the progress of thought which led to them. The latter remains "behind the scenes" of empirical history of science and needs special reconstruction. No doubts, the discovery that only field averages, and not fields in a point, have physical meaning in the structure of mathematical formalism of quantum electrodynamics, was one of the key moments in construction of adequate interpretation of this formalism. But to fix the said circumstance which, by the way, was missed by almost all investigators, it was necessary to approach analysis of the mathematical apparatus from very special positions. We may say N. Bohr possessed brilliant intuition, but this is not enough to explain why other investigators (including such theorist of highest rank as W. Pauli and W. Heisenberg) who paid close attention to the discussion dedicated to problems of measurability of field, did not notice this circumstance. The reason, probably, is that Bohr's intuition was attributed to a special point of view which allowed him to see what remained unseen by other physicists. Above we tried to show that that special point was formed by preliminary analysis of the idea of experimental body in the aspect of the corner stone of quantum mechanics description – the question of relation of quantum object and classical device. This analysis reached the top intensity probably in February 1931 in Copenhagen in discussions between Bohr, on the one hand, Landau and Peierles, on the other hand. In L. Rosenfeld's mentioned works dedicated to the history of quantum electrodynamics we can find a bright description of the emotional atmosphere of those discussions[66]. Rosenfeld's text clearly shows that discussions of the foundations of measuring procedures in quantum electrodynamics and talks on status of experimental bodies preceded Bohr's decisive statement that field components in space-time points are used in the formalism of the theory as an auxiliary idealization which has no direct physical meaning. Analysis of the idea of experimental body showed that quantum particle used in thought experiments on measurability of quantized fields are not fit for the basic definitions of experimental body. From this the hypothesis of classical experimental bodies followed. In its turn, it logically led to the hypothesis of averaged field components which were to replace field in a point. The latter was what stimulated the corresponding analysis of the mathematical formalism of the theory.

Bohr postulated consistency between mathematical apparatus and reconstructed theoretical scheme of quantum electrodynamics; that allowed transition to the second stage of verification of such scheme within framework of potential interpretation. This stage consisted in fixation of inherent mutual consistency of objects forming the theoretical scheme. In particular, it was necessary to find out whether the idea of field as a system with a variable number of particles does not contradict to the idea of field averages. Both characteristics were equally indispensable for description of quantum fields, because one of them appointed corpuscular qualities (field as a system of particles able to appear and disappear in corresponding quantum states with certain probability), and the other – wave qualities (field as an integral system, described by classical wave quantities, observable values of which form spectrum of values of corresponding field operator).

Preliminary analysis showed that field strengths averaged on area $V\tau$ must undergo fluctuations because of effects of creation and annihilation of photons in this area and, consequently, cannot have exact values. Landau and Peierles had also paid attention to this peculiarity, emphasizing that fundamental indeterminacy of field components in a point extends over the averages on some area field components. Landau and Peierles saw here confirmation of their thesis of fundamental inapplicability of the term "electromagnetic field" in quantum sphere.

It might seem that the new theoretical scheme reproduced paradoxes of the old one: the idea of field as a system with a variable number of particles and the idea of field as a system characterized by classical components of strengths averaged on some space-time area are incompatible.

Yet N. Bohr and L. Rosenfeld proved that the situation with field strengths in a point and the situation with averaged field strengths were radically different. As opposed to the first situation, the second one does not lead to logical contradictions, even if we accept the idea of fluctuations. After scrupulous analysis of the apparatus of the theory, Bohr and Rosenfeld showed that in measurements of averaged field components we are to distinguish two cases: 1) when time interval of averaging τ multiplied by speed of spread of electromagnetic wave c is large enough in comparison with linear sizes L of volume V, over which averaging is carried out (i.e., $L \leq c\tau$), and 2) the opposite case, when time τ multiplied by c is small in comparison with L (i.e., $L > c\tau$). In the first case we cannot ignore fluctuations while determining averaged over area $V\tau$ field strengths. It occurs through the fact that during the time of measuring photons, emerging due to radiation, can spread from other areas to the space area V, on which strengths are averaged. Abstraction from fluctuations is possible in this case only if we accept degeneracy of quantum electrodynamics into classical electromagnetism theory.[67]

Entirely different is the situation when field quantities are averaged over area where $L > c\tau$. In this case the averaging area is not connected with neighboring areas by light signals, that is why it contains only photons which have penetrated there before (the light wave covers distance smaller than L during the time of measuring). This allows us to ignore fluctuations while determining averaged field components without losing the field's quantum features. Values of such fluctuations every time will be included in values of strengths determined in area $V\tau$ and with $L > c\tau$ they can be minimized.

Availability of such version is the decisive circumstance which makes the old and new situations of measurability of field components entirely different. It is easy to see that at consideration of the field quantities in a point the described variant ($L > c\tau$) disappears by definition (since $L \to 0$). Therefore paradoxes of incommensurability here are fundamentally irremovable.

Inherent consistency of the objects of the reconstructed theoretical scheme was the second signal of efficacy of Bohr's program. Now, after verification the theoretical scheme of quantized radiation field from the point of view of its consistency and correspondence to the character of the mathematical formalism, there were all opportunities to start the decisive moment of interpretation: the procedure of constructive introduction of abstract objects forming the mentioned theoretical scheme.

The proof of measurability of quantized radiation fields

Let us consider more thoroughly the main features of the procedures of constructive justification of Bohr's scheme of quantized radiation field. Starting their work, N. Bohr and L. Rosenfeld stipulated those initial features of abstract objects which were introduced as their definitions within the framework of the theoretical scheme of quantized radiation field and which now were to be obtained as result of idealized measurements. Such features corresponded to the main correlations of abstract objects inside the theoretical scheme and could be settled through analysis of fundamental dependences of the mathematical apparatus.

After in the theoretical scheme field strengths had been replaced by strengths averaged on a space-time area, the basic mathematical dependences of the theory which had direct physical sense were commutation rules for operators $\overline{\hat{\mathbf{E}}}$ and $\overline{\hat{\mathbf{H}}}$ of averaged fields. They occupied the places of commutation relations for the field operators in a point and formally were easily derived from them by means of integration with respect to corresponding areas of space-time. Then, from the commutation rules for $\overline{\hat{\mathbf{E}}}$ and $\overline{\hat{\mathbf{H}}}$ it was easy to get correlations of indeterminacy for averaged field components. From here followed:

1) one can always determine exactly the value of separate components of the field strengths, averaged on some space-time area (it was supposed that in measuring one can always get exact value of each separate component $\overline{\hat{\mathbf{E}}}$ and $\overline{\hat{\mathbf{H}}}$, a value which belongs to the spectrum of its operator's eigenvalues and, consequently, should be expected in the experiment with certain probability);

2) field components of the same name, for example, $\overline{E'_x}$ and $\overline{E''_x}$, averaged on two different, not coinciding areas of space-time, can be determined together only up to $\hbar$;

3) two field components of different names $\overline{E'_x}$ and $\overline{E''_x}$, averaged over two different, not coinciding areas, can be determined within any accuracy.

Constructive justification of the theoretical scheme meant that observables, having the features listed, should be introduced as idealizations based on real specificities of the experiments and measurements in quantum relativistic area.

The first step on that way consisted in checking measurability of a separate averaged field component. This goal was reached by a theoretical experiment, where an experimental body, which volume V coincided with the boundaries of the averaging area of the field measured, was put into this area and, during the time equal to the averaging time, got an impulse from the field. The impulse should be registered by the device. According to conditions previously set for verification of the theory ($L > c\tau$), it was supposed that linear sizes of the experimental body are greater than time of measuring τ multiplied by light speed.

Bohr and Rosenfeld were to prove that all fundamental difficulties which had emerged in Landau-Peierles thought experiments with point experimental bodies are eliminated in the idealized measuring procedures of the new type.

If we study more or less thoroughly Bohr's and Rosenfeld's arguments, we will see that the proof was presented by means of scrupulous analysis of details of the intellectual experiment on base of continuous juxtaposing of theoretical corollaries and real possibilities of the experimentation. The arguments which helped solve this problem leave a deep impression (emphasized by many historians of science) of graciousness of investigators' thought, which found the exit from seemingly unsolvable paradoxes, but, what is more, can serve models which provide adequate interpretation of mathematical apparatus of modern theory.

In these judgments we can trace, how operational definitions, or (in Mandelshtam's terms) connection receipts of physical quantities of mathematical apparatus with experiment are created, and how in the process of creation of such receipts conceptual structure of a modern physical theory is formed.

The knots of the matter in the proof of fundamental measurability of a separate averaged field component were:

1) analysis of possibilities to locate the experimental body in area V during time of measuring τ;

2) analysis of the process of transmission of impulse from the experimental body to the register device;

3) exact account of fields radiated by the experimental body when measuring the field component.[68]

The typical method of Bohr's and Rosenfeld's judgment at this stage of analysis was the following: first they fixed difficulties and seemingly paradoxical corollaries revealed by theoretical analysis of field measurability, based on abstract notions of experimental bodies, then they showed how to overcome the difficulties if the notions of experimental bodies are adjusted and, correspondingly, conditions of idealized measurements are made specific, i.e. real specificities of physical experiments and measurements in quantum relativistic area are taken into consideration.

Moving this way – from general and abstract scheme of idealized measuring procedure to its detailed and concrete pattern, Bohr and Rosenfeld solved emerging questions of measurability of fields step by step.

It is indicative, for instance, the solution of the problem of localization of experimental body in space-time measuring area $V\tau$. In accordance with the main idea of the measuring procedure, the task was to determine exactly that particular impulse that the experimental body obtained in the area.

To do this, it was necessary that the experimental body should be isolated – as entirely as possible – from influences of neighboring areas during time and strictly fix the interval of measuring (otherwise the borders of the measured area would be blurred). To gain observance of this condition, the investigators had to determine impulse of the experimental body twice: one time just before its interaction with the field in area V, in the very beginning of period, the second time – at the end of that period, after the experimental body had interacted with the field in area V. Then the difference of impulse values p'_x and p''_x in the beginning and at the end of τ they could determine the value of the measured field strength. At the same time, to preserve the strictly determined period of time of averaging, the process of registration of the experimental body impulse p'_x and p''_x should take time Δt, which should be much smaller than general time of measuring τ.

Nevertheless, this refinement of the measuring procedure, though a necessary condition of localization of experimental body in space-time measuring area, by itself did not eliminate the main obstacles on this way.

For instance, there were difficulties connected with displacement of the experimental body in the course of measuring. The problem consisted in the following. Interacting with the field, and then with the register device, the experimental body every time should have received some recoil. Due to this, the body which initially occupied a certain spatial averaging area V, then it crosses the borders of this area and, if its displacement is considerable enough, it experiences perturbing influence from the field in neighboring areas.

In this case the difference p'_x and p''_x and impulses of the experimental body in the beginning and at the end of period τ one cannot determine the measured field strength in area $V\tau$. To avoid that, it was necessary to provide infinitesimal little displacements of the experimental body during the time of measuring. Bohr and Rosenfeld solved the problem by refinement of features of the experimental body and conditions of the measuring procedure. They supposed that the experimental body should have large mass which minimized its recoil[69]. For classical experimental bodies it is easy to gain (unlike point charges). We can easily see that this feature of experimental bodies is justified by idealization of real experiments in which the investigator can vary mass of the experimental body over a wide range.

In solving the problem of localization of experimental body in area there emerged more complicate difficulties, for instance, connected with correlation of indeterminacies between impulse and coordinate of the experimental body.

Just as the impulse of the experimental body should be measured exactly, there appears growing indeterminacy in its coordinate and, consequently, it becomes impossible to locate the experimental body exactly in a specified spatial field measuring area. Any exact measurement of impulses p'_x and p''_x of the experimental body in the beginning and at the end of the measuring period τ and, correspondingly, exact determination of impulse p'_x and p''_x means growth of indeterminacy Δx of the spatial measuring area. From here follows that a field component, say, $\overline{E}_x$, averaged on area $V\tau$ cannot be measured. Remembering that value of the component $\overline{E}_x$ is determined by the formula $\overline{E}_x = \frac{p''_x - p'_x}{\rho V \tau}$ (where ρ - density of charge of the experimental body),[70] and taking into consideration uncertainty relation, $\Delta x \Delta p_x \sim \hbar$ we can derive the expression for the indeterminacy to which the averaged component $\overline{E}_x$ will be measured every time, i.e. the expression $\Delta \overline{E}_x \sim \frac{\hbar}{\rho \Delta x V \tau}$.

At a glance it seems that we have returned to Landau's and Peierles's thesis of fundamental incommensurability of field. But Bohr and Rosenfeld demonstrated that classical nature of experimental bodies allows to overcome the paradox. Accepting indeterminacy Δx in position of the experimental body, one may make it much smaller than displacement of a heavy experimental body caused by change of its impulse. Then, at quite small Δx, one may increase density of the charge distributed over the volume of the experimental body. Then, as follows from the formula $\Delta \overline{E}x \sim \frac{\hbar}{\rho \Delta x V \tau}$, $\Delta \overline{E}_x$ will diminish. Basically, this way can always lead to change of a field component within the accuracy necessary for verification of the theory.[71] Thus, the problem is solved by introduction of a large charge experimental body compensating errors in field measurements caused by uncertainty relation. This feature of the experimental body, as a necessary condition for exact field measurements, like requirement of relatively large masses of the experimental bodies, was easily justified by real possibilities of physical experiment.

For classical experimental bodies varying of their charge basically can be realized (unlike point quantum particles used by Landau and Peierles in their thought experiments). But, admitting any densities of charge at their even distribution of charge over the volume of the experimental body, Bohr and Rosenfeld at once came across new difficulties. To assume even distribution of charge of any density over the volume of the experimental body is legitimate only in case when atomic structure of the experimental bodies is ignored. So a question arises: may one ignore this aspect of quantum area? Is it legitimate not to take into account quantum properties of the experimental charge in thought experiments on field measuring? This was a problem of principle, as it was evident that in interaction with field experimental bodies are added to the field sources, and their atomic structure should influence the quantum processes characterizing the field. Therefore there was a necessity to prove especially the legitimacy of abstraction from atomic structure of experimental bodies in measuring components of quantized field.

Bohr presented such proof on base of scrupulous analysis of specificities of the apparatus of quantum electrodynamics and reference to general principles of quantizing fields.

Mathematical apparatus of the theory of quantized radiation field did not introduce any universal scale of space-time sizes: formalism of the theory contained only two constants – $\hbar$ and c, "which could not form a characteristic length or interval".[72]

This exactly meant that in frameworks of quantized radiation field theory while describing field interactions with experimental bodies, the latter are to be considered just as classical charge distribution ignoring quantum specificities of such distribution.

Yet the reference to specificities of the theory was necessary but not sufficient for complete solution of the problem because, as Rosenfeld indicates in his review of quantum electrodynamics, "mathematical consistency of the formalism in particular was doubted at that time"[73]. Moreover, when Bohr-Rosenfeld procedures were performed, there already existed apparatus describing interactions of field with quantized sources. That is why it was unclear whether it was correct to execute field measuring procedures ignoring the more complicated process of quantum interactions in relativistic area.

Taking account of all this, Bohr supported the projected solution with methodological analysis of specificities and principles of quantizing electromagnetic field.

Rosenfeld emphasizes in his historical review that Bohr first paid attention to the basic side of the question – the fact that any characteristic of quantum system claimed on usage of classical idealization.[74]

Howsoever many intermediate links interaction of quantum systems included, the last link would require classical objects used as devices. Atomic structure of such objects is not taken into account by definition, they are described only by the language of classical physics. Hence there appeared fundamental possibility to abstract from atomic structure of experimental bodies, regarding them as a part of a classical aggregate.[75]

That such abstraction was necessary, followed from the very logic of the construction of quantum electrodynamics. Its mathematical apparatus was built so that quantizing of free fields preceded introduction of description of their interaction within the framework of the perturbation theory.

Construction of the interpretation should follow the same way. While equations for free quantized fields were not validated, it made no sense to interpret more complicated cases referring to interaction of such fields. But free fields require direct measurements (otherwise, if we introduce mediatory interactions between field measured and a classical device, the field is no longer free by definition). As we have already mentioned, observance of such conditions requires that we should consider the experimental body only as a fragment of a classical aggregate. Atomic structure of the experimental bodies should be taken into consideration only at the next stage of interpretation, in considering interacting quantized fields.

Thus, it was proved that at changes of free fields there is no need to include quantum qualities of the experimental charge. As a result, idealization of experimental charge of any, however high density was two-fold validated: from the point of view of real possibilities of the experiment and from the point of view of theoretical reasons of quantized field measurements.

Such two-fold justification of features of the experimental body is one of the important aspects of Bohr-Rosenfeld procedures.

Providing the experimental bodies with various ideal qualities, Bohr and Rosenfeld not only prove that such qualities can be obtained through idealization of real experiment, but also check whether the new idealization could destroy the fundamental conditions of measurement dictated by the principles of the theory.

For the most part, such verification did not require more or less complicated deductions and hence was not reproduced in the rendering of the results of the investigation. But cases with features of even distribution of charge of high density prove fundamental necessity and importance of such justification. The latter can guarantee correct synthesis of specificities of the theory verified with those of experimental practice.

When features of the experimental body had been specified, the thought experiment, which was to provide verification of measurability of the field, was seemingly developed enough to consider the problem of localization of the experimental body in measuring area as solved. Nevertheless, Bohr and Rosenfeld present one more proof: they demonstrate fundamental practicability of such experiment in real practice. In this respect it is very much characteristic that, having introduced such idealized assumptions as large mass and high density of the charge, Bohr and Rosenfeld at once work out a concrete scheme of an experiment including description of details of the device unit, which could provide localization of experimental body in measuring area $V\tau$.

It was supposed that the experimental body can be linked to a rigid carcass representing a spatial frame of reference. This allows to fix strictly its position before interaction with the field during time τ. In the beginning of this period tie with the framework is broken, and the experimental body experiences recoil under influence of the field. Then, at the end of the period τ, impulse is changed and tie with the framework is restored.[76]

Here we face with one more important peculiarity of idealized measuring procedures which throws light on their epistemological nature the method of their construction.

It is well known that concrete description of device units, which help study experimentally interactions in the nature, is usually inherent only to empirical schemes. In theoretical models it either is eliminated (as in quantum theory), or is replaced by abstract characteristic of the type of device, in correlation with which the vector of state of the quantum system is determined.

Introducing description of concrete details of structure and functioning of the measuring unit in discussion of the problem of localization of experimental body in measuring area, Bohr and Rosenfeld resort to notions which are characteristic for empirical schemes. But in return they get a guarantee that thought experiments with classical experimental bodies by their structure correspond to real specificities of physical experiments in the new interaction area. And this is that same condition which provides constructive justification of theoretical objects.

Thus, even the analysis of the first stage of proving measurability of a separate field component lets us conclude that in the process of idealized measurements Bohr and Rosenfeld perform many times repeated motion from the most general theoretical principles to concrete specificities of a physical experiment, and then, basing on these specificities, again turn to solving theoretical problems. In the course of this cognitive movement they reach such specification of the measuring procedure that it guarantees that specific features of quantum interactions in relativistic area are taken into account in the very process of idealized measuring.

The described process of analysis let Bohr and Rosenfeld successfully solve not only the problem of localization of the experimental body, but also two other key problems of field measurability: transfer of the impulse of the experimental body to the register device and account of disturbing influence of fields radiated by the experimental body.

In their analysis of the process of registering impulse of the experimental body, Bohr and Rosenfeld first of all demonstrate that using classical experimental bodies allows to avoid difficulties discovered by Landau and Peierles in their theoretical experiments with point particles. For classical bodies the impulse could be measured within such accuracy which would guarantee determination of the given averaged field component even with very little period of measuring Δt (much smaller than general time τ of measuring the field component).[77]

Though here emerge difficulties unknown in the theoretical experiments with point particles. For prolonged experimental bodies it becomes important that the speed of interaction is finite and it does not exceed the light speed. Because of this separate parts of the experimental body can transfer their impulse to the device not simultaneously but with certain retardation. If we accept an idealizing assumption that the experimental body is absolutely hard, the minimum time Δt, during which all its parts will pass their impulse (say, by means of collision with some membrane of the device), cannot be smaller than L/c (where L – linear size of the experimental body).

But, according to the conditions of measuring ($L > c\tau$), even general time of measuring, during which the experimental body interacts with the field and its impulse is measured, should be smaller than L/c, as to the period Δt of measuring impulse of the experimental body, it should be far smaller than τ. So, a new paradox emerges: measuring impulse of the experimental body requires time far exceeding permissible periods of measuring averaged field strengths.

Bohr and Rosenfeld found solution of this paradox using the method already tried. They reconsidered and refined characteristics of the experimental bodies and correspondingly specified the measuring procedure.

The first refinement which allowed to eliminate the contradiction consisted in rejection of idealization of absolutely hard experimental body. Instead Bohr and Rosenfeld introduced notion of experimental body as a system of small charged bodies which, interacting with the field and the register device, undergo approximately same recoil. To observe the last condition, it was supposed that the whole charge of the experimental body is evenly distributed among the elementary parts-components, and the density of the charge of each of them is also evenly distributed in its volume.[78]

Correspondence of such construction of experimental bodies to possibilities of the experimental practice was quite evident, if we take into account that systems of charges moving as a whole under influence of the field had been used in electrodynamics many times. As to idealization of even distribution of charge of high density in each of the elements of the experimental body, it could be easily justified by both empirical and theoretical reasons (for instance, by already proved possibility to ignore atomic structure of the experimental bodies). The described construction of the experimental bodies by itself allowed to transfer the whole impulse to the device unit during time Δt, given for registering impulse. To prove real practicability of this process, Bohr and Rosenfeld consider two possible ways to measure impulse: by collision of the experimental body with the diaphragm of the register device and on base of Doppler effect.[79] They introduce empirical schemes of possible experiments – corresponding to each of the ways, and then the idealized procedure of measuring the field component is justified as invariant contents of both types or real experiment and measuring. Having solved the problem of measuring impulse of the experimental body, Bohr and Rosenfeld proceed with the final stage of proving quantized electromagnetic field measurability – account of radiations generated by experimental bodies in the process of measuring.

At this stage the thought experiment, which provided measuring averaged field component, reaches highest possible completeness of development and justification – both from the point of view of theoretical reasons and from the point of view of real possibilities of physical experiments. To minimize radiations of the experimental body, which has reverse action upon its impulse, Bohr and Rosenfeld make new corrections in the measuring procedure. They propose to settle it so, that for the most part of the time τ of interacting with the field the experimental body remained stationary. To gain this, it would be enough to join the system of charges forming the experimental body with another set of charges, of the same distribution density but opposite in sign. Then, after a push from the field in the beginning of the period τ, during time Δt the experimental body would run some distance D_x (much smaller than linear sizes L of the field measuring area) and remain in this position during most part of the measuring time under influence of compensating charge, which could be fixed with hard framework of space-time frame of reference.[80] The displacement of the experimental body by itself would represent polarization of charges of a neutral (as a whole) distribution, combined from system of charges of the experimental body and compensating charges of the opposite sign. The value of such polarization would inform us of the field strength in the given averaging area. To measure this value, it would be enough either to determine value of displacement D_x, or register the impulse of the experimental body at the end of the time period τ, when polarization was removed, and the experimental body returned to its initial position under influence of neutralizing charge. The impulse could be measured during a very short time period Δt.[81]

The described scheme of experiment conserved all previously justified features of experimental bodies and conditions of measuring. In particular, it was easy to gain localization of the experimental body in the measuring area, for instance, necessary for this purpose joint and disjoint of the experimental body with rigid carcass of frame of reference were performed as if automatically – first under influence of the field displacing the experimental charge, then under influence of neutralizing charge which, being fixed with the framework, would attract the experimental body. It was easy enough to register its impulse during time Δt (the measuring could be carried out, for example, on base of the Doppler effect, lighting the experimental body with a beam at moments of its deviation and returning to the initial position).

At the same time the described thought experiment obtained minimal radiation of the experimental body in the measuring area. To determine the field component measured, the investigators only had to account this radiation and find a method of its compensating.

The main of the fields, radiated by the experimental body during time period τ and perturbing its impulse, referred to displacement, caused by interaction of the experimental body with the field measured. Bohr and Rosenfeld found a very simple way to compensate this radiation. For this they took a mechanical spring which elasticity corresponded in value to the force with which radiation caused by displacement D_x, will have reverse action upon the experimental body. Having fixed it at the experimental body and joined it with the rigid carcass, they could obtain such measuring unit that its construction had taken into account the perturbing influences of the radiation of the experimental body caused by its displacement under influence of the field measured.

It is clear that such compensation was possible only in respect to classical experimental bodies (it is impossible to fix mechanical springs to a point quantum particle).

By selection of compensatory springs they could also take account of fields appeared as a result of recoil of the experimental body during time Δt of its interaction with the device registering impulse. All these fields were computed by methods of classical mechanics, because the experimental body which radiated them was a classical charge distribution.

But there was one more radiation which was connected with account of quantum characteristics of measuring (relation of uncertainty between coordinate and impulse of the experimental body). This radiation was caused by displacement of the experimental body by distance Δx, which was to be accepted as minimal error in determination of position of the experimental body because of need to measure its impulse as exactly as possible.

It was obvious that this radiation could not be exactly taken into account in measurements because Δx is an indeterminate magnitude. But then there emerged an irremovable indeterminacy which cancelled all previously obtained proofs.

There appeared a situation which was characterized by Bohr and Rosenfeld as one of the most critical moments in physical justification of the apparatus of quantum electrodynamics[82]. It resulted that radiations of the experimental body can be compensated while they are considered within the framework of classical electrodynamics. But after transition to the area of quantum processes, where decisive role belongs to uncertainty relations, there appeared uncontrolled radiations of the experimental body which prevented investigators from exact determination of the field component,

If this problem had not been solved, Landau's and Peierles's conclusions about impracticability of quantum mechanical approaches to field description would have proved right.

But the solution, simple enough though unexpected, was found.

Bohr and Rosenfeld, estimating the highest possible value of disturbances, which radiation connected with Δx into the field measured, found out that in order of value the data of disturbance correspond to fluctuations of the field in the area of measuring emerging due to creation and annihilation of photons.

Processes of such creation and annihilation are the main and integral characteristic of quantized fields. One of remarkable features of these processes – the fact that they are statistically independent events (creation of photon in one of possible states, superposition of which form the radiation field, does not influence the probability of creation of other photons in other states and does not depend on the number of photons filling the said states). So in every possible state the number of photons, emitted by classical field sources and passed to measuring area $V\tau$, should oscillate about some average number in accordance with Poisson distribution. Such variations will cause small changes of energy in the measuring area, and, as energy is correlative to intensity, this will mean emergence of fluctuations in values of classical components of the field strength.

But if the condition of measuring $L > c\tau$ is observed, such fluctuations do not prevent exact measuring of quantized field component but, moreover, they are prerequisite for this measuring.

In this case the investigator will deal with statistical distribution of photons for various states characterizing the field only in area V. Fluctuations connected with statistically independent emission of photons by classical sources should, by definition, be included in the characteristic of the field. Their existence, on the whole, causes a statistical character of predictions of field quantities in quantum electrodynamics (unlike classical theory, what is predicted here is the probability of occurrence in the experiment of some value of the magnitude which belongs to the spectrum of the eigenvalues of the corresponding operator). This including of fluctuations into values of the field strengths measured in area $V\tau$ allows to obtain a certain set of values, and each of them can be expected in the experiment with a certain probability). But this would be that very set which coincides with the spectrum of the eigenvalues of the field strength operator.

In this respect fluctuations do not prevent exact determination of field magnitudes measured by a classical device. Quite another matter is that these magnitudes will somewhat differ from the values predicted by classical electrodynamics (field strength, predicted by the classical theory, is correlative to average energy of all photons which are in the measuring area, without account of fluctuations about average number of photons in each state; account of the photons leads to deviation from values predicted by classical electrodynamics). But this difference is that important property which allows to discover quantum characteristics in measurements of field magnitudes.

Thus, including fluctuations in area $V\tau$ (when $L > c\tau$) into observable field magnitudes not only conserves predictability of measurements, but also allows to discover those features of the field which are proper to it as a quantum system with a variable number of particles. From these positions it was easy to solve the problem of radiation connected with indeterminacy Δx in position of the experimental body.

Since disturbing influences of such radiation do not exceed fluctuations connected with the process of creation and annihilation of photons, then including of these disturbing influences into the measuring process makes us discover in the experiment those very values of the field component, whose mathematical expectation is predicted by the theory.

Perturbations caused by indeterminacy in position of the experimental body may lead only to difference between values calculated on base of the quantum field theory and those calculated by methods of classical electrodynamics. But this difference is the one which allows to predict measurable magnitudes with account of quantum specificities of the field – creation and annihilation of photons.

This way the last problem in the proof of measurability of a separate field component was solved. As a result, the main feature of Bohr's theoretical model of quantized radiation got constructive meaning.

It is characteristic that here, at this final stage of the proof the idealized measuring procedure obtained such degree of specification that there could be no doubt, neither in its correspondence to the basic principles of the theory verified, nor in the fact that it accumulated essential features of practicable experiments and measurements in quantum relativistic area. But it was then that it was discovered that all essential features of the object studied (quantum radiation field) are automatically taken into account within the scope of the given procedure. All influences which prevented exact determination of averaged field strengths were eliminated or compensated in an experiment, planned in details, with classical experimental bodies. As to perturbations which could not be controlled or compensated, they were necessary conditions for determination of field quantities measured with account of quantum effects.

Emphasizing the importance of the last circumstance, L. Rosenfeld wrote that the impossibility to compensate or erase completely the perturbations introduced by the experimental body is not at all the result of imperfectness of an idealized measuring unit, used in the intellectual experiments with classical experimental bodies. On the contrary, "the impossibility of compensating or controlling them in any way, far from being an imperfection of the measuring device, is a property which it must necessarily possess to ensure that all the consequences of the theory are in principle verifiable by measurement".[83] "The fact that the zero-field fluctuations are superposed on to the classical field distribution is indeed a well-defined theoretical prediction, and we see that we are able to suppress the perturbations arising from the manipulation of the test-bodies to an extent which just leaves scope for the test of this prediction".[84]

The proof of fundamental measurability of field components, averaged on some space-time area, was a key moment in constructive justification of Bohr's theoretical scheme of quantized radiation field. The main argument of Landau and Peierles against employment of quantum methods in description of electromagnetic fields was eliminated.

But for complete justification of the quantum theory of electromagnetic field the investigators also had to verify the work of connection between the field strengths in two different space-time areas interacting by lights signals. This connection characterizes spread of the field in space in the course of time, so analysis of correlations between pairs of averaged field components, taken in different space-time areas, was necessary for investigation of dynamic characteristics of electromagnetic field in quantum area.

As we have emphasized, the theoretical scheme postulated impossibility to measure pairs of components of the same name together (for instance, $\overline{E'_x}$ and $\overline{E''_x}$ averaged over space-time areas $V_1\tau_1$ and $V_2\tau_2$ with accuracy exceeding Planck's constant by an order of value).

Bohr and Rosenfeld proved validity of this statement by their thought experiments with classical experimental bodies. Then they proved that features of joint measurability of two components of different types (for instance, $\overline{E''_x}$ and $\overline{H''_x}$ or $\overline{E'_x}$ and $\overline{H'_x}$ derived from thought experiments with classical experimental bodies, also coincide with the features postulated in the theoretical scheme of quantized radiation field).

Thus, the main abstract objects of this theoretical scheme (observable fields, averaged on finite space-time area) were introduced as idealizations based on real specificities of physical experiments. After all these procedures the theoretical scheme of quantized radiation field got its constructive meanings.

At the final stage of justification of Bohr's theoretical scheme the authors offered new development of details of thought experiment on measuring the field strengths. That was characteristic. It was required, in particular, due to the fact that at verification of measurability of pairs of field components of the same name, uncertainty relation for these components, derived from the measuring procedure, initially did not coincide with analogous relation given by mathematical formalism of the theory. But, remembering that between areas of field measuring there exists light signal exchange, basically it was possible to make corrections in the measuring procedure. These corrections should be connected with using such signals as "messages" automatically transferred by experimental bodies and carrying information of their mutual positions. Refinement of the thought experiment in this respect led to coincidence of predictions of the theoretical formalism verified and the results of idealized measurements.[85]

Later L. Rosenfeld described this stage of interpretation of the apparatus of quantum electrodynamics: "It is very striking indeed to see how the greatest accuracy compatible with the commutation law can only be achieved by exploiting to the utmost the possibilities, afforded by the physical situation of controlling the course of the measuring process".[86]

From these positions we can evaluate again heuristic function of the method used by Bohr and Rosenfeld in justification of their theoretical scheme of quantized electromagnetic field.

Consistently moving from the most general shape of the thought experiment, dictated by the mathematical apparatus and hypothetical model of its interpretation, to empirical schemes of a possible experiment, Bohr and Rosenfeld gained that idealized field measurements gradually accumulated essential features of real experimental measuring activity. In the framework of such measurements they traced the process of interaction of device units (including experimental bodies) with the field measured and discovered its characteristics. The latter were compared with the characteristics postulated by the previously accepted theoretical scheme. Coincidence of the field tokens obtained in two described ways proved that the given scheme was an adequate reflection of quantum specificities of electromagnetic radiation.

Thus they solved the main problem of theoretical search at the stage of interpretation of the theory's mathematical formalism: features of the abstract objects got their empirical justification.

We would like to pay attention to one important feature of the described method of investigation: its application no longer requires those real experiments, which provides verification of constructive meaning of the theoretical scheme, should be realized in practice. Enough if they are basically possible and practicable. The investigator can make sure that the latter is true when he develops analysis of measurability of theoretical quantities to concrete empirical schemes of real experiment, when possibility to realize one or another device unit and its interaction with the object measured becomes evident at least because similar device units and methods of their functioning are familiar by previous practice.

So, Bohr's and Rosenfeld's procedure of measuring a field component did not leave place to doubts in fundamental practicability of the corresponding experiment, because in previous physical experiments similar measuring devices and methods of measuring had been used many times. There was no sense to especially prove that the measuring unit might contain, besides experimental charge, a body carrying compensating charge; that the field would cause polarization of charges in a neutral (as a whole) charge distribution: that it was possible to settle rigid connection between the carcass of the frame of reference and compensating charge etc. – similar device units and methods of their functioning could easily be found in previous practice.

Taking into consideration the fact that in creation of a theory by method of mathematical hypothesis the layer of real experiments, where specificity of new interactions is seen, may be developed insufficiently (sometimes there can be no such experiments at all), we may say that the described way of investigation is probably the only possible way of justification of the theory at the modern stage of evolution of physics. Using it, the investigator as if shortens the way of development of the theory. He does not have to wait until a vast enough set of local theoretical schemes and laws justified by real experiments is created. He reproduces in thought empirical schemes of basically practicable intellectual experiments and develops analysis to the foundations where the possibility to realize experiment of the given type is quite evident. The latter only means that such and such type of device unit and the principle of its interaction with the object studied has already been realized in previous practice, so it would be redundant to repeat what has been done.

The necessity to develop and refine procedures of idealized measuring until they accumulate essential specificities of real experiments, which provide studies of corresponding object, Bohr often expresses as a requirement of fundamental controllability of interactions of object and device.

Rationally this requirement can be reduced to the following: any real measuring indeed stipulates a special set of conditions under which the investigator could eliminate (or take into account) perturbing external influences which distort real values of the magnitude measured. The possibility to eliminate such influences or to take them into account introducing corresponding corrections means that the investigator controls the condition of measuring.

Since thought experiments and measurements should be idealization of real experimental measuring activity, then the investigator also should completely discover in them the controllable conditions of measuring. From these positions he has to scrupulously check (basing on already known theoretical laws) consequences of every new detail in the mental scheme of the device unit and, at the same time, correlates the scheme with real possibilities of the experiment. Constructing idealized measuring procedures, the investigator step by step discovers those mentally fixed interactions of the object with the devices which could cause indeterminacies in values of magnitudes characterizing the object. Having revealed such interactions, he checks whether they refer to disturbing influences of the device unit which can be eliminated by its new refinement and application of compensatory devices.

Exhausting possibilities to control the conditions of measuring, the investigator makes sure that the idealized measuring corresponds as much as possible to the possibilities of real experimental measuring activity. If indeterminacies of magnitudes characterizing the object remain, it means that such indeterminacies should be considered as essential characteristics of the object itself.

In this respect everything what is fundamentally uncontrollable within the scope of idealized measuring, justified as scheme of a real experiment, should be included in the specificities of the object measured, since the measuring procedure itself is constructed in such a way that it reveals objective characteristics of the reality studied. Hence we cannot, of course, conclude that quantum characteristics appear due to uncontrolled interaction of the device and the microobject measured. The real structure of Bohr's cognitive activity and his method of construction of idealized measuring were not connected with the idea of uncontrollability in the sense above. They were based on an entirely opposite approach, according to which idealized measurements, structured in concordance with real specificities of quantum mechanical and quantum relativistic experiments, should reveal objective characteristics of the processes in atomic area.

Bohr's requirements of control over conditions of interaction of the object measured and the device were identical to requirements to construct idealized measuring drawing it as close as possible to real specificities of physical experiment. Then characteristics of a quantum object, which could be discovered within real experimental practice, undoubtedly should find expression in the results of idealized measurements.

Intermediate interpretations of apparatus of modern physical theory as a condition of its development

Constructive justification of the theoretical scheme of quantized radiation field automatically provided empirical interpretation of the formalism of the theory. Bohr-Rosenfeld procedures allowed to correlate field strengths from the equations of quantum electrodynamics with experiment indicating mechanism of such connection. This mechanism could be involved by means of description of Bohr-Rosenfeld procedures of thought experiments. The description itself formed a system of operational definitions for corresponding physical quantities.

In this respect the process of construction of idealized measurements in quantum electrodynamics can be taken as some model of activity which provides introduction of operational definitions at today's stage of development of physical theories. But Bohr-Rosenfeld procedures not only formed empirical interpretation of the equations of quantum electrodynamics. They discovered new aspects in characteristic of such field and urged to introduce corresponding corrections also in the semantic interpretation of the formalism of the theory.

The idea of field resulted to be applicable only to finite space-time areas and inapplicable to a point. Thus the idea of quantized field as transfer of electric and magnetic forces from point to point was destroyed. Such idea, acceptable within classical electrodynamics, was inapplicable in quantum area.

Then it became clear that, because of field fluctuations caused by creation and annihilation of photons, the connection between the field and its sources is more complicated than classical theory used to believe. The latter ties sources and fields in a strictly determinate way. At the same time in quantum theory Laplace's determinism of classical electrodynamics is replaced by a wider form of statistical causality. Fields are causally connected with sources only from the point of view of statistical predictability of field magnitudes measured in the experiment. Strictly determined connection, characteristic for classical physics, restores only when the field in the measuring area "consists" of a large number of photons, which, in accordance with Poisson distribution, oscillate about some average number in every of the possible states forming the field. As the average number of photons is large enough, we can ignore their fluctuations and turn to classical description of the field. All these field characteristics were revealed due to measurability procedures, because it was here where investigators determined the physical sense of influence of fluctuation field upon the magnitudes measured. The said fluctuations transformed traditional idea of radiation field determination by its sources.

Finally, in the process of idealized measurements the unbreakable link between radiation field and vacuum was justified. This is probably the most important consequence of Bohr-Rosenfeld procedures.

It may seem at the first glance that the idea of connection between quantized radiation field and vacuum was born due to mathematical apparatus of the theory and did not depend on the proof of the field measurability, as application of methods of quantizing to electromagnetic field automatically led to notions of infinite field energy in absence of photons.

But the matter of fact is that before justification of the field measurability it was entirely unclear whether it was possible to provide vacuum with real physical meaning or it should be accepted only as an auxiliary theoretical construct lacking such direct meaning.

Paradoxes with infinities push physicists to the latter conclusion. They supported opinion that for non-contradictory interpretation of quantum electrodynamics in general it was necessary to exclude somehow "zero field" from the "body" of the theory. We should remember, then, that Landau and Peierles linked the idea of vacuum with paradoxes of incommensurability, and in their analysis energy was presented as one of the evidences of fundamental inapplicability of quantum methods to description of electromagnetic field. Productively criticizing conclusions of Landau and Peierles, Bohr eliminated the last objection, but the question of physical sense of vacuum states still was not solved.

Only in the course of Bohr-Rosenfeld procedures was the problem clarified and connected with the discussion on the role of fluctuation of the field components in the measuring process. But there was one more aspect of the problem, which we have not yet touched for the sake of easiness of the account. Let us consider this aspect now.

Besides fluctuations connected with the presence of photons, there is one more variation of field fluctuations predicted by the apparatus of the theory. It is zero fluctuations which appear in absence of photons and connected with the zero energy level of the field. From the apparatus of the theory it followed that these fluctuations have finite positive value (nothing to do with the infinite energy of the field in zero state!).

As we have mentioned, Bohr and Rosenfeld proved that fluctuations connected with creation of photons should be included in values of the field components. They are discovered due to declinations of values of the field quantities predicted by the quantum theory, from the values calculated by methods of classical electrodynamics.

The empirical sense of fluctuations connected with creation of photons followed from the structure of idealized measuring of the field, since only taking them into account could the investigators determine exactly the averaged field component. But in that case fluctuations of the zero field also obtained empirical justification, as they were fundamentally inseparable from fluctuations connected with presence of photons.

As zero fluctuations were display of "zero field", the latter as well got real physical sense. It resulted that, if vacuum and zero fluctuations caused by it were removed, the very idea of quantized radiation field would become physically empty, because the averaged field component could not be measured exactly.[87]

As a result, Bohr-Rosenfeld idealized measuring procedures led to conclusion about real connection between the radiation field and vacuum and impossibility to obtain description of quantized radiation field without taking vacuum states into account.

In principle, the new vision of electromagnetic field caused by realization of the procedures of measurability is not something unusual or extraordinary in the development of theoretical knowledge. On the contrary, here we can see a certain pattern of epistemological nature; its manifestation we have already seen in the history of science (for instance, in analysis of the history of classical electrodynamics). The essence of it is the following: realizing constructive introduction of abstract objects of a previously accepted theoretical model, the investigator fills this model with new physical contents, because he organizes real experimental research activity, revealing characteristics of the reality studied.

The obtained content is objectified due to mapping of the theoretical model on the picture of the world, and the result is a new vision of the object under study, which fixes its essential properties and relations. The last procedure finishes construction of interpretation of the corresponding phenomena of the corresponding equations of the theory, which are presented now as description of new essential characteristics of the physical reality studied. At this stage the theory obtains new physical notions, and its conceptual apparatus gets further development. Due to this the preliminary accepted semantic interpretation is refined and developed. Thus, constructive justification of the theoretical scheme leads to decisive development of the contents of the scientific theory. This is an accomplishment of the process of formation of its conceptual structure, started at the stage of mathematical hypothesis. Bohr-Rosenfeld procedures can present us a characteristic example of the process developing at modern stage of evolution of theoretical knowledge. After measurability of quantized radiation field had been proved, fundamental possibility to apply quantum mechanical methods in description of relativistic processes provoked no further doubts (unlike initial conclusions made by Landau and Peierles).

The foundation of quantum electrodynamics – the theory of free quantized electromagnetic field – became now a non-contradictory and experimentally justified system of knowledge.

Now the researches only had to interpret the fragments of quantum electrodynamics which described interaction of quantized radiation field with quantized sources (measurability of electron-positron field).

Bohr and Rosenfeld solved this problem was by at the second stage of realization of their research program. It was connected with construction of idealized measurements for sources (distributions of charge-current) interacting with quantized radiation field.[88]

First, they proved measurability of classical sources interacting with quantized electromagnetic field, and then presented a proof of measurability of field sources with account of creation of electron-positron pairs. Thus they completed the interpretation of mathematical apparatus of quantum electrodynamics describing free quantized fields and their interactions in the first approximation of the perturbation theory.

At this stage they not only formulated the correspondence rules, which connected all physical magnitudes of the equations of quantum electrodynamics with experiment, but also discovered early unknown characteristics of quantized fields. In particular, the procedures of quantized measuring allowed to raise the question of space-time boundaries beyond which the field approach to description of quantum properties of charge-current loosing its force.

From the mathematical apparatus of quantum electrodynamics it followed that, unlike fluctuations of electromagnetic field, the fluctuations of charge and current within any strictly limited space-time area are to be infinite. But the analysis of the situation of idealized measuring revealed new field specificities. It was discovered that in areas related to shell of finite depth consisting of experimental bodies (which served to measure the field sources), averaged on the same areas fluctuations became finite. If we infinitely reduce the depth of the shell, the fluctuations infinitely grew tending to infinity. When they are equal to mathematical expectation of the field quantities predicted by the apparatus of the theory, it indicates the limits of applicability of quantum electrodynamics.[89]

Thus, constructive justification of the theoretical scheme of interaction of quantized radiation field with quantum sources, providing empirical interpretation of the formalism of quantum electrodynamics, introduced new aspects into its semantic interpretation as well.

To sum up, we can now once more evaluate the way made by Bohr and Rosenfeld in construction of this interpretation.

Gradually justifying features of free quantized electromagnetic field, then interactions of this field with classical sources, and, lastly, with quantum sources, by means of idealized measurements, Bohr and Rosenfeld were creating a richer and richer theoretical model, which took into account new aspects of electromagnetic interactions in atomic area. This way of construction of interpretation reproduced the basic steps of historical development of the mathematical apparatus of quantum electrodynamics at the level of conceptual analysis.

No essential stage of its development was missed – the logic of construction of the interpretation mainly coincided with the logic of historical development of the mathematical apparatus of the theory.

In this respect, it is interesting to compare interactions of the mathematical apparatus and theoretical models in modern and classical situations in yielding of a scientific theory.

As we have shown above, in construction of classical electrodynamics every step toward the generalizing field equations (Maxwell's equations) was supported by a corresponding theoretical model, which was constructively validated even at the intermediate stages of the theoretical synthesis.

While quantum electrodynamics was being formed, the situation changed. Here for a quite long time mathematical apparatus was built without constructive justification of the theoretical models; there were only hypothetical schemes which introduced preliminary semantic interpretation of the equations. As to procedures of their constructive justification, which provided empirical interpretation of the formalism created, and then its final semantic interpretation, they were carried out later and were separated in time from construction of the formalism as such. Nevertheless, in those procedures investigation repeated all the main stages of development of the apparatus of the theory in brief. Step by step does it reconstruct the developed hypothetical models and, through their constructive justification, introduces intermediate interpretation which correspond to the most important stages of development of the apparatus. The accomplishment of this way consisted in clearing of the physical meaning of the generalizing system of equations of quantum electrodynamics.

So, the method of mathematical hypothesis does not at all reject the necessity of content-physical analysis at intermediate stages of forming the mathematical apparatus of the theory. The specificity of modern investigations is not that intermediate interpretations become redundant, but that the activity aimed at their construction becomes a continuous transition from one intermediate interpretation to another in accordance with the logic of development of the apparatus, which reproduces the history of its development in brief. Classical theory was constructed according to scheme: equation$_1$ $\rightarrow$ intermediate interpretation$_1$, equation$_2$ $\rightarrow$ intermediate interpretation$_2$ ... , generalizing system of equations $\rightarrow$ generalizing interpretation; in modern physics theory is constructed in a different manner: first equation$_1$ $\rightarrow$ equation$_2$ $\rightarrow$ etc, then interpretation$_1$ $\rightarrow$ interpretation$_2$ $\rightarrow$ etc. (but not equation$_1$ $\rightarrow$ equation$_2$ $\rightarrow$ generalizing system of equation and immediately accomplishing interpretation!). Clear, the shift of interpretations in modern physics does not entirely reproduce analogue processes of the classical period. We should not believe that we have only discrete transition from one intermediate interpretation to another replaced by continuous transition, only the number of intermediate links is changed. In modern physics it is as if packed, and therefore the process of construction of interpretation and development of conceptual apparatus of the theory takes cumulative form. There are at least two reasons for that.

First, as we have already emphasized, the process of constructing theoretical models reproduces the history of development of mathematical formalism not entirely, but in brief. Search for adequate interpretation requires verification only of those links of its historical development, which were accomplished by creation of equations included in the theory (for example, Bohr and Rosenfeld in their procedures of measurability of quantized radiation field, investigated the mathematical formalism created by Heisenberg, Jordan and Pauli on base of the initial variant, suggested by Dirac; this variant as such was not considered because it had been put away from a further, more perfect mathematical apparatus).

Second, the mathematical hypothesis by itself reduces the number of intermediate links on the way to generalizing equations of the theory (since at once there are introduced equations of generalization of great enough level – as basic dependences subject to further synthesis and generalization). In its turn, it leads to reducing of the number of intermediate stages on the way to the final interpretation of the theory formalism.

All said lets us conclude that, in comparison with classical models, in modern theoretical investigation the procedures of constructive justification of theoretical models and construction of operational definitions, which connect the formalism of the theory with experiment, are somehow packed. So we may state that at the modern stage of evolution of physics some features of theoretical synthesis, distinctive only of the classical period, are reproduced, but in a packed and pressed form.

In principle, that should be this way – if we take into consideration dialectical way of development: in self-developing systems (and scientific cognition is one of them) higher stages of evolution always repeat in their functioning some features of historically preceding forms. It is important to remember that such features can be both transformed enough or reproduced comparatively purely. The latter variant allows finding new aspects of interaction of mathematical apparatus and interpretation in development of modern theory. As we understand, at some stages of this development it is possible to see sort of return to classical scheme of theoretical synthesis, according to which advance in mathematical formalism should not happen before its exhaustive interpretation is created.

But such return is not the same as absolute repetition of classical methods. It goes on new basis and requires usage of modern methods of theoretical search.

Breakthrough in mathematical extrapolations usually takes place, when they have already helped to build quite rich formalism able to be base of the future apparatus of the theory. But the theory itself is not accomplished yet. The necessity of its further development at this stage may be evident enough, at least because necessary problems are solved only partly (there are theories which should be solved, according to requirements of the theory, but which are unsolvable by means which exist).

But not at all always it is clear, how to find new mathematical means. Moreover, there are doubts if such search is possible on previous basis, as existence of unsolvable problems can be evidence of inner contradictions in the formalism already created. Then we need content analysis of the foundations of the theory, proofs of consistency of the created apparatus and construction of its interpretation.

Development of mathematical formalism is relatively independent from its interpretation (including empirical aspects) only to certain extent. In modern physics there always are periods when further perfection of mathematical apparatus of the fundamental theory created entirely depends on construction of its consistent interpretation, which gives a new impact for further mathematical synthesis and accomplishing of the theory.

In this respect the history of quantum electrodynamics can be a most eloquent example.

Between the third and the forth stages of forming of its apparatus there emerged crisis of its foundation, caused by discovering of incommensurability paradoxes. Further generalization and elaboration of the formalism of quantum electrodynamics would have been impossible as the very principles of quantizing fields were doubted, if that crisis had not been overcome.

Bohr and Rosenfeld laid the way out of the crisis when they constructed a consistent interpretation of the created apparatus, which described processes of interaction of quantized electromagnetic and electron-positron fields in the first approximation of the perturbation theory. Only after that did it become possible for quantum electrodynamics to recover in the 1950s. That recovery was connected with construction of renormalization theory. Firm belief in fundamental applicability of quantum electrodynamics methods of description in the relativity area (shaken because of the crisis and restored thanks to success in solving the problem of measurability of quantized fields) was a necessary condition for search for theory of interaction of quantized fields with account of higher orders of the perturbation theory. The very setting of the problem was correct due to Bohr-Rosenfeld procedures, which had previously proved that the description of interaction of quantized fields in the first approximation of the perturbation theory was consistent.

But Bohr-Rosenfeld procedures gave an impact to further development of quantum electrodynamics not in this generally theoretical aspect only. They exercised concrete influence upon further evolution of the theory, as they revealed such new characteristics of electromagnetic interactions, the information about which made it considerably easier to elaborate the basic physical idea of renormalization.

We usually pay little attention to this circumstance, but still it is extremely important for understanding patterns of evolution of theoretical knowledge.

The general idea of renormalization appeared, as it is well known, due to understanding the limited nature of idealization of a free particle in respect to quantum relativistic area. Any particle is not free, in a strict sense of the word, because it interacts with vacuum, which corresponds to the lowest energy state of quantized fields. The result of such interaction is change of charge and mass of the particle, and then charge and mass of the particle observable in experiment become a summary of this interaction. For instance, if there are mass m_0 and charge e_0 of an electron not interacting with vacuum, in the experiment we observe other mass and charge which are equal to $m = m_0 + \Delta m$ and $e = e_0 + \Delta e$. The magnitudes Δm and Δe express changes introduced in charge and mass of the electron by vacuum.

It seems possible to calculate charge and mass of the electron (observable in the experiment) by means of determining corrections Δm and Δe for interactions with vacuum. But such corrections turned out infinite expressions having the form of divergent integrals. All this caused enormous difficulties in description of interaction of particles (considered as quanta of the field) by methods of the perturbation theory.

Renormalizations, which allowed to eliminate these difficulties, were based on a quite simple physical idea. Magnitudes m_0 and e_0 representing mass and charge of non-interacting (or "bare" in modern physical terminology) electron, as well as corrections, were considered as auxiliary theoretical constructs which had no real physical meaning, because a real electron always is in interaction with vacuum and never exists beyond such interactions. Then mass and charge of a free electron was identified with expressions $m = m_0 + \Delta m$ and $e = e_0 + \Delta e$ which are really observed in experiment. But since these magnitudes have finite values, finite values m and e were to be acquired through special selection of divergent values for Δm and Δe. The method of such selection formed the essence of the renormalization method.

It means that the renormalization method was based on the idea of observable magnitudes characterizing particles, which are considered as quanta of some field, as display of total result of interaction of these particles with vacuum.

But this very idea firmly occupied its place in physics due to the procedures of idealized measuring.

Let us recall that Bohr and Rosenfeld justified measurability of quantized radiation field, and this fact lead to a conclusion: there is a contribution of vacuum in the field observable magnitudes characterizing the state with presence of particles (photons). Further analysis spread this conclusion also on magnitudes describing electron-positron fields (for instance, on such dynamic variables of the field as charge and mass).

Beyond the measurability procedures the initial idea of observables having a contribution of vacuum looked no more than a hypothesis. But idealized measurements got the status of a validated theoretical statement for that hypothesis.

Since works of Bohr and Rosenfeld containing the results mentioned above were well known among the physicists-theorists of the 1940s,[90] we may quite naturally conclude that they prepared the necessary base for development of the idea of renormalization. In any case, we are to remember that the approach to observables, which became a necessary condition for the idea of renormalization, was prepared by Bohr-Rosenfeld procedures.[91]

It is characteristic that this stage coincided with new development of mathematical apparatus of quantum electrodynamics. Here we can see the reverse influence of Bohr-Rosenfeld theoretical model upon the search for new mathematical structures characterizing quantized fields. By the way, such influence can be seen even at quite late stages of development of quantum relativistic ideas. So, we would like to draw the reader's attention to the following important circumstance.

In axiomatic quantum field theory the mathematical apparatus from the very beginning is constructed, meaning that physical sense can belong not to fields in a point, but to magnitudes of fields averaged on some finite space-time area. The modern theory characterizes field not by operator functions (as it was at the earliest stage of development of quantum electrodynamics), but operator functionals, whose description openly contains the operation of averaging on finite space-time area. Such apparatus allows describing easily and briefly quantum processes in relativistic area. For reaching this goal, it uses mathematical structures of higher "information capacity" than those which were in foundation of the mathematical formalism of quantum electrodynamics of the 1930s–1940s.

It is obvious that the physical foundation for the application of new mathematical means were the specificities of fields uncovered by Bohr-Rosenfeld procedures. It means that the interpretation procedures prepare new development of the theory apparatus, encouraging search for more perfect mathematical structures.

To summarize all said above, we may formulate the following epistemological and methodological conclusions.

1. In modern physics the process of construction of a theory is even more autonomous in relation to new experimental data, than in classical physics. Mathematical hypothesis lets us move toward fundamental equations of developed theory even if the local theoretical laws, which are to be synthesized and which are based on real experiments, are presented scarcely enough.

2. Still an important directing role in theoretical investigation belongs to the picture of physical reality. It provides base for choice of principles of mathematical description of new area of physical processes. But, unlike classical models, its operational structure is accentuated.

3. A mathematical hypothesis is able to provide working out a quite developed apparatus, but only to certain extent, because equations manipulation is linked with corresponding transformation of abstract objects of theoretical schemes. If a series of mathematical extrapolations is quite long, it can cause accumulation of non-constructive objects with mutually eliminating features. So, for the development of non-contradictory

theoretical system of knowledge interpretation of mathematical formalism at intermediate stages of construction of the theory is required.

Creation of a theory keeps going on as an alternate correlated movement in mathematical means and plain of physical contents. But, in comparison with classical models, the relatively independent "run" at each of this level grows, and the movement from equations to interpretation and vice versa goes on in larger steps.

4. Construction of intermediate interpretations in modern physics goes on as procedures of idealized measuring and often without preliminary real experiments. Nevertheless, due to consistent development of details of the thought experiment – up to reproduction of empirical schemes of possible future experiment – the very idealized measuring procedures can be justified as schematized and idealized real experimental-measuring activity in the field of interactions. That is why they are capable of bringing to light objective characteristics of such interactions.

5. The idealized measurements not only verify characteristics hypothetically introduced in base of the specificities of the theory apparatus, but also discover new, unknown features of the physical processes studied. Hence the mathematical apparatus obtains new physical meaning, and the notion structure of the physical theory is reconstructed and presented as a deeper and more adequate reflection of the object area investigated. In turn, it raises foundation for search for new, more perfect means of its mathematical description.

6. Stages of development of idealized measurements, which end at construction of an adequate scheme of new area of interactions, reproduce the main stages of construction of the mathematical apparatus, as if repeating its history, but in brief. At the same time, idealized measurements of modern physics shorten the way of constructing the theory as well because they do not require long forming of preliminary theoretical models and laws based on real experiment. In the very process of construction of idealized measurements the investigation briefly passes the stage of forming of such models.

Thus, the evolution of physics at modern stage conserves some basic operations of construction of the theory characteristic for its past forms (classical physics). But is develops the operations, partly modifying them, partly repeating – on a new base – some features of construction of mathematical apparatus and theoretical models, appropriate to the classical models.

In modern investigation the process of theoretical search characteristic for classical physics is reproduced in transformed and pressed form – as it should be at higher stages of the evolution in relation to the historically passed stages.

MUTUAL CONNECTION OF GENESIS AND FUNCTIONING OF A THEORY.
THE CONSTRUCTIBILITY PRINCIPLE

If we compare specificities of development of a theory in classical and non-classical science, some common laws of the process of their development can be revealed.

Analysis of content aspects of the structure and genesis of a scientific theory demonstrate that in formation of its conceptual apparatus the key role belongs to procedures of constructing a theoretical scheme. Such construction is done as interaction between foundations of the science, mathematical apparatus, empirical and theoretical material generalized in the theory. First it stipulates transition from foundations of the science to a hypothetical variant of the theoretical scheme, and then – to empirical material.

This is the first cycle of the process of constructing the theory, connected with the hypothesis put forward. But then we face reverse movement – from generalized empirical and theoretical material to theoretical scheme and again to the foundations of the science. This is the second stage connected with justification of the hypothesis. Here the initially introduced theoretical schemes are reconstructed, saturated with new contents and actively influence upon the foundations of the science, preparing new changes in them.

The hypothesis suggested marks only the most general framework of the conceptual structure of the theory, which is formed – in its main features – with justification of the hypothesis.

Methodological literature usually characterizes the very process of suggesting hypotheses in terms of "discovery context". It is urgent to emphasize that transition from the foundation of the science to analog model and then to a hypothetical scheme of the interaction area studied makes a certain rational outline of this process. It is often described in terms of the discovery psychology and creative intuition. But such description, if it is supposed to be constructive, should, for sure, be linked with clearing of the intuition "mechanisms". It is characteristic that here investigators at once came across the so-called mechanism of gestalt-switching which lies at the base of intellectual intuition.[92]

Detailed analysis of this process shows that the intellectual intuition is considerably characterized by usage of some model ideas through which we examine the new situations. The model ideas stipulate the image of the structure (gestalt) which is transferred to new object area and organizes, in a new way, the before collected elements of knowledge of that sphere (notions, idealizations etc.).[93]

The result of such work of creative imagination is a hypothesis which allows to solve the problem offered.

Further consideration of mechanisms of intellectual intuition has marked clearly enough that the new vision of reality, corresponding to gestalt-switching, is formed due to substituting new elements – ideal objects – into the initial model-idea (gestalt), and it allows to construct a new model shaping new vision of the processes studied.[94]

Here gestalt is a kind of "mold" according to which the "model is molded".[95]

Such description of the procedures of generation of hypothesis corresponds to investigations of the discovery psychology. But the process of putting forward scientific hypotheses can be also described in terms of logical-methodological analysis. In this case its new important aspects will be uncovered.

First, let us emphasize once more the fact that the search for hypothesis cannot be reduced only to the method of trials and mistakes. In forming a hypothesis, a considerable role belongs to the investigator's foundations (ideals of cognition and the picture of the world) which aim the creative search, generating investigation problems and indicating the field of the solution means.

Second, the operation of forming a hypothesis cannot be entirely transferred to the sphere of individual creative work of a scientist. They are obtained by an individual, just as his thinking and imagination are formed in the cultural context absorbing samples of scientific knowledge and samples of their production activity. The search for a hypothesis, including choice of analogies and substituting new abstract objects, determined not only by historically developed means of theoretical investigation, into the analog model. This choice is also determined by translation in the culture of certain samples of the investigation activity (operations, procedures) which provide solution of the new problems.

T. Kuhn is right when he mentions that such samples are included into scientific knowledge and mastered in the process of learning.

Translation of theoretical knowledge in the culture means also translation of samples of the problem solution activity. Such samples reflect procedures and operations of generating new hypotheses (foundations of the science – analog model – substitution of new abstract objects into the model). That is why in the process of adoption of already obtained knowledge (formation of a scientist as a specialist) also some quite general schemes of intellectual work, providing generation of new hypotheses, are mastered.

Translation of schemes of intellectual work in the culture, which provide solution of the problems, allows considering the procedures of such generation, abstracting from personal qualities and abilities of a concrete investigator. From this point of view we can talk about the logic of forming hypothetical models as a part of the logic of forming a scientific theory.

Finally, summarizing specificities of the process of forming hypothetical models of science, it is important to emphasize that the base of this process is the combination of abstract objects from one field of knowledge with the structure ("network of relations") taken from another field. In the new system of relations the abstract objects are provided with new features, which makes appear, in the hypothetical model, new contents, which can correspond to not yet studied connections and relations of the object area, for description and explanation of which the hypothesis put forward is dedicated.

This feature of hypothesis is universal. It can be marked at the stage of formation of local theoretical schemes, as well as in construction of a developed theory.

As to procedures of justification of the hypothesis, they also have a quite complicated structure and internal logic. As it follows from reconstructions of development of classical and quantum electrodynamics, traced above, empirical justification of a hypothesis is not reduced to comparison of its corollaries with the results of experiments and observations. It includes procedures of constructive justification which is a condition and a premise of comparison of hypothetical models with experimental facts. Only after these procedures, does the theory get receipts of connections of its fundamental magnitudes with experiment – operational definitions, which guarantee efficiency of empirical verification of the theory. Further justification of hypothetical models and turning them into a theoretical scheme is connected with procedures of their correlation with disciplinary ontology (scientific picture of the world) and philosophical foundations of the science. When these procedures are completed, the ontological status of theoretical schemes as the core of the new theory is justified.

The process of justification of the hypothesis contributes to the construction of conceptual apparatus of the theory not less than the process of generation of the hypothesis. In the course of justification the contents of the basic notions of the theory are being developed. In turn, it creates premises for future theoretical search, as every new hypothesis stipulates usage of already developed notions and models as material for its construction.

If we take into consideration this specificity of development of scientific knowledge, it will be clear how incorrect the positivists were who strictly separated "the discovery context" and "the theory verification context".[96] The logic of discovery and the logic of verification are two aspects of one and the same process of the theory becoming, and there exists close mutual connection between them.

Historical approach to the problem of structure and genesis of the theory requires that we take into consideration not only mutual connections between different aspects of the theory genesis, but also the connection between the process of becoming and peculiarities of functioning of the theory.

Anti-historicism of the positivist analysis of scientific knowledge consists, for instance, in the fact that theory was considered only as given knowledge, without the peculiarities of its genesis. The result of such type of analysis was a quite poor idea of the process of functioning of a formed theory. Positivism could mark only some formally logical aspects of deductive development of theory and the process of theoretical explanation and prediction of events. Informal aspects of theoretical investigation were lost by the positivist history of science.

The interest to these aspects of theory emerged in the Western philosophy of science in connection with formation of post-positivist branches, whose representatives referred to analysis of the history of science. Studying informal aspects of theoretical investigation, they came across the connection between functioning of theory and its genesis. Probably the most interesting results, revealing this connection, were contained in Kuhn's conception of "model" problem solutions. Kuhn noted that operating models in the process of theoretical description and explanation of concrete events is analogous to the way of forming of new knowledge in the history of science.[97] In his analysis, Kuhn closely approached the question of reproduction of the peculiarities of theory's genesis in its structure and functioning. Still, he failed to determine clearly this problem and logical-methodological approaches to its solution. He tried to answer, how the first model problem solutions are created in a theory, appealing to the psychology of perception of the investigator included into the scientific community. At the same time objective origins and premises of formation of the "models" remained outside Kuhn's analysis.

Just as the problem of the "models" can be formulated as the problem of way of reduction of a fundamental theoretical scheme to local ones and transition from basic equations of the theory to their corollaries, so its solution is of greatest importance for understanding the laws of functioning of a theory. The key to the solution of this problem is to be sought in the logic of historical development of scientific knowledge.

Interaction of the operations of putting forward a hypothesis and its constructive justification is that key moment which allows to get the answer, how paradigmatic models of problem solutions appear in the theory.

Having raised the problem of getting models, the Western philosophy of science failed to find corresponding means to solve it, because it did not reveal and analyze, even in the first approximation, the procedure of constructive justification of hypotheses.

Discussing the problem of models, T. Kuhn and his followers emphasize only one side of the question: the role of analogies as basis of problem solving. The operations of forming and justification of meanwhile appearing theoretical schemes remain outside their analysis.

It is quite indicative that within such approach there emerge fundamental difficulties in trials to elucidate, what is the role of the correspondence rules and their origin. For instance, Kuhn believes that in the activity of scientific community these rules do not play such an important role as methodologists usually attribute to them. He especially emphasizes that the most important thing in solving problems is search for analogies between various physical situations and application of already found formulae on this basis. As to the correspondence rules, they, according to Kuhn, are a result of further methodological retrospective, when methodologist tries to ascertain criteria used by the scientific community in application of different analogies.[98] Kuhn is consistent in his views, because the question of procedures of constructive justification of theoretical models is not brought up in his concept. To detect this procedure, we need a special approach to investigation of structure and dynamics of scientific knowledge. It is necessary that we should consider theoretical models included into the theory as reflection of object in the shape of activity. Referring to a concrete investigation of nature and genesis of theoretical models of physics, such approach orients us to a special vision of them: theoretical models are considered as ontological scheme, which reflects essential characteristics of the reality studied, and at the same time as some kind of "closure" of object-practical procedures, within which in principle we can disclose the characteristics. That vision allows to discover and describe operations of constructive justification of theoretical schemes.

With other theoretical-cognitive basis the mentioned operations remain outside methodologist's field of investigation.

But, as it is constructive justification that provides appearance of the correspondence rules in theory, defining their contents and meaning, it cannot surprise us that Kuhn came across difficulties in determining the ways of forming and functioning of these rules.

It is characteristic that in discussion of the problems of samples Kuhn refers to the history of Maxwell's electrodynamics. Analyzing it only in the plane of application of analog models, he believes that the main results of Maxwell's investigation were gained without any construction of correspondence rules.[99] But, as we have seen, this conclusion lies far from real facts of the history of science.

We think that the above described analysis of procedures of construction of a theory allows getting answer to the question: where do model situations appear from in theory. Such model situations (examples of solution of theoretical problems) demonstrate methods of construction of local theoretical schemes on base of a fundamental one, and ways of transition from basic laws of theory to local theoretical ones. Forming and including such model situations into the theory take place in the course of its becoming.

In construction of a developed theory its fundamental theoretical scheme is created by means of consequent generalization of those theoretical schemes which either preceded the theory, or were constructed in the course of theoretical synthesis. This generalization is carried out by means of creation of several intermediate models, and each of them is aimed at representation of new, not considered before, characteristics of interactions studied, in the theory.

First the investigator introduces each of such models as a hypothesis and then gives its constructive justification. In the course of constructive justification of the model he works out two main proofs.

The first one determines that the model is able to express essential characteristics of situations being generalized. Such characteristics previously could be represented in cognition by local theoretical schemes. Now, when constructive justification of the model is done, the content of the mentioned schemes is included in the generalizing model.

During the second proof the investigator makes sure that in course of new generalization of the model its previous constructive content is not destroyed. This content corresponded to the local theoretical schemes which were assimilated by the generalizing model at previous stages of theoretical synthesis. To make sure this content is preserved, the investigator explicates it. From the generalizing model he derives corresponding local theoretical schemes which, in their content, are equivalent to the theoretical schemes assimilated in the theory.

Thus, in the course of the process of construction of a theory the investigator reduces the fundamental theoretical scheme being created to local theoretical schemes. The methods of such reduction reproduce, in their main features, the methods used for including of essential characteristics of concrete physical situations reflected in the theory, into generalizing model. Such including was executed by means of intellectual experiments based on real possibilities and peculiarities of the experiment. In the course of such experiments the investigator's thought traveled from model to experiment and from experiment to model, studying all main intermediate links between model and experiment. The same thought experiments in their main features are repeated in explication of constructive contents included into the model, when the latter is reduced to some local theoretical scheme. As in the process of justification of the model by new experiment, the investigator first considers concrete specificities of physical situations, and then imposes restricting conditions on the model and constructs a local theoretical scheme.

It is characteristic that at the final stage of theoretical synthesis, when the main equations of the theory are introduced and constructive justification of the fundamental theoretical scheme is accomplished, the investigator executes the last proof of correctness of the equation introduced and their interpretation: from the main equations he gets, in a new form, all generalized local theoretical laws, and then, on base of the fundamental theoretical scheme, he constructs local theoretical schemes corresponding to the said laws. A typical example of such justification is the final stage of formation of Maxwell's theory of electromagnetic field, when it was proved that on base of the theoretical model of electromagnetic field it is possible to obtain, as particular cases, theoretical schemes of direct current electrostatics, electromagnetic induction etc. and from equations of electromagnetic field – to deduce Coulomb's, Ampere's, Biot-Savart's laws, laws of electrostatic and electromagnetic induction discovered by Faraday, etc.

Final justification of the main equations of the theory and the fundamental theoretical scheme at the same time present as account of the "ready" theory. The process of its becoming is reproduced now in reverse order, in shape of deductive development of the theory, deriving corresponding theoretical corollaries from the main equations. Each conclusion here can be considered as account of some method and result of solution of a theoretical task.

Thus, the very process of constructing a theory forms and includes model situations of solving theoretical tasks.

Further functioning of the theory and expansion of its application area creates new examples of solving problems. They are included into the theory, along with those introduced in the beginning of its formation. With development of scientific knowledge and changes of previous form of the theory, the initial models are also modified. But, in their modified shape, they are normally preserved in all further accounts of the theory. Even the latest formulations of classical electrodynamics demonstrate methods of application of Maxwell's equations to concrete physical situation; the example used is derivation Coulomb's, Ampere's, Biot-Savart's, Faraday's laws from these equations. The theory, we may say, preserves in itself traces of its past history, reproduces – as typical problems and ways of their solution – the main specificities of the process of its forming.

Genesis of the theory is imprinted in its organization and determines its further existence. If we define genesis of the theory as intensive way of knowledge development, and functioning of the theory – as extensive way of such development, we will see that both ways are closely linked. Reproduction in a logic of unfolding the theory formed of the main specificities of its becoming is one of the sides of such mutual connection. But there is another side: active influence of the process of functioning of the formed theory upon future shapes of intensive development of theoretical knowledge.

After the theory is constructed, it enters the stage of explanation and prediction of new phenomena. Here the empirical basis of the theory is extended, it being known that the new empirical material is not only mechanically absorbed by the theory, but has active reverse action. The theory is now changing in the course of application to new situations.

One of the main reasons of such changes is difficulties emerging with solving new problems by old methods. To work out methods which would provide solution of wide range of such problems, we have to change mathematical means and develop new theoretical models of the reality studied. As the result we have reformulating of the existent theory: new mathematical apparatus is created, and its conceptual structure is developed.

The history of science presents us a lot of evidences of such development of a theory already settled. For instance, Newton's mechanics first was reformulated, on base of application of analytical methods, by Euler, and then reconstructed into Lagrange mechanics and Hamilton-Jacobi mechanics. Any such reconstruction was connected with application of mechanics to new physical situations and desire to work out general methods of solving various problems. Euler developed the analytical apparatus of mechanics to obtain universal methods of determining states of a material point or a system of such points under action of forces. The new methods enabled him to work out an absolutely new part of mechanics: solid body dynamics. Lagrange's, and later Hamilton-Jacobi's reformulating of mechanics were – to a considerable degree – caused by needs in description and explanation of complicated mechanical systems. Analytical methods, based on the accelerating forces principle, could not be applied in the process of solving quite a number of problems, as the value of forces applied to each body, which was a part of a complicated system, is normally unknown in advance. Lagrange's mechanics, and then Hamilton-Jacobi mechanics solved such problems successfully. In this process of development of mechanics its new mathematical apparatus were formed, its new principles (for instance, the smallest effect principle) were introduced, its new fundamental notions (effect, energy etc.) were formulated.

This kind of specificities of development of already settled theory can be traced also in other historical examples. Thus, predictions of electromagnetic waves and further application of Maxwell's theory to explanation of optical phenomena led to development of conceptual apparatus of electrodynamics (there appeared ideas of electromagnetic wave, electromagnetic radiation etc.). At the same time, as the sphere of empirical application of Maxwell's equations expanded, so it required that the mathematical shape of the theory should be improved. In H. Hertz's and O. Heaviside's works Maxwell's equations were expressed in a form close to a modern one, and then electrodynamics was accounted with help of modern methods of vector analysis.

Finally, we can refer to one more example of reconstruction of a settled theory: historical development of quantum mechanics. After it had been created in its initial version (by W. Heisenberg, E. Schrödinger, N. Bohr and M. Born), its application for explanation and prediction of a wider and wider set of processes in atomic sphere was accompanied by development of the apparatus and the conceptual structure of the theory. The stages of such development are, for example, Dirac's strict operational formulation of the theory in terms of q-numbers, von Neumann's axiomatic model of the quantum theory, Feynman's formulation of quantum mechanics (path integrals).

Reconstruction of a theory in the process of its functioning not only generates new methods of solving problems but also creates means for building new fundamental theories. Mathematical apparatus and conceptual structures, which are developing in the process of application of the settled theory to new physical situation, might be precisely those means needed whose employment in a new area of theoretical search provide intensive development of scientific knowledge.

Electrodynamics could not have been worked out, if mechanics had not formed the mathematical apparatus which provided solution of hydrodynamic problems. The development of quantum physics was carried out, in a great part, due to mathematical structures and notions formed in Lagrange's and Hamilton-Jacobi mechanics. The number of such examples can be increased.

Thus, means for future theoretical search and construction of new theories are created not only at the stage of becoming of the theory, but also, at even a greater extent, at the stage of functioning of a developed theory. This side of mutual connection of genesis and functioning of a theory was missed in Kuhn's analysis. In his conception of development of science the stage of extensive increase of knowledge is sharply opposed to its intensive development. In the real history of science these two sides are closely connected: genesis of the theory determines its functioning, while functioning of developed theories prepares basis for new theoretical structures.

Forming the conceptual structure of a new theory is the result of interaction of mathematical apparatus, theoretical schemes and experiment. Dynamics of such interaction is mostly determined by procedures of constructive justification of theoretical scheme. These procedures have practically never been analyzed in methodological and philosophic literature.[100] Meantime, their disclosure opens new perspectives for getting concrete methodological conclusions and recommendations. First of all, we can present the idea of "constructability" as a methodological rule, which indicates ways of construction of adequate interpretation of mathematical apparatus of the theory. This rule can be formulated in the following way: after a hypothetical model of explanation of empirical facts is introduced, new, hypothetical features of the abstract objects of the model are to be introduced as idealization based on a new layer of experiments and measurements, the layer which was intended to be explained with help of the model. Moreover, we have to make sure that the new features do not contradict to the features of the abstract objects justified by previous experience.

This rule does not mean the same as the requirement to verify theoretical knowledge by experiment. According to analysis of the historical material, verification of this kind stipulates (especially at modern stage) complicated activity connected with construction of adequate interpretation of the equations introduced. The core of such interpretation is constructive introduction of abstract objects. That is why the rule of "constructability" not only says that empirical justification of a theory is necessary, but also indicates how, in what manner such justification is done.

From the requirement of constructive introduction of abstract objects follow quite nontrivial methodological conclusions. One of them has already been discussed. It refers to connection between existence of non-constructive objects in the "body of the theory" and paradoxes emerging there. Since the presence of non-constructive objects can lead to paradoxes in a theoretical system (though not necessarily), then application of the "constructability" rule allows uncovering contradictions inside knowledge before they are uncovered in the spontaneous course of the investigation. This, in turn, can be a means to reconstruct the theory effectively, and to form a conceptual structure which would adequately reflect the new object. To find such a criterion is especially important in respect to modern knowledge, which is quite complicated in its system organization and where it is not always easily to find inconsistency.

The model of such activity aimed at analysis of inconsistency of knowledge by means of constructive justification of theoretical schemes may be Bohr-Rosenfeld procedures in quantum electrodynamics.

When we find non-constructive elements in a theoretical model, we can see weak points of the theory, which are – sooner or later – excluded through replacement of corresponding elements of the theoretical model and its constructive reorganization. This problem should be analyzed especially, as the requirement of elimination of non-constructive objects is close to the requirement of the "observability" principle. Here we are to discuss the question of relationship of ideas of "constructability" and "observability".

As we know, the "observability" principle meant that in construction of a theory the investigation should apply only magnitudes that have operational meaning, while ideas which cannot be verified in experiment should be eliminated from the theory.

Vast philosophical and physical literature gives us quite exhaustive analysis of the ideas of fundamental "observability". It shows that the "observability" principle, applied along with other principles of physics, had quite an important heuristic role in its development, but its usage took place differently in different investigational situations. The strict requirement to eliminate non-observable quantities from the theory has never been applied in physics. This requirement, if understood literally, prohibits us at all to use non-observable magnitudes, while without them we fundamentally cannot construct any hypothesis, because at the stage of such construction the investigator uses mostly non-observable objects (when he supplies the objects of the model with hypothetical features, he, usually, does not know which of them would be justified by experiment, and which of them not). Besides, in a theory already developed there always can exist auxiliary constructs (like "bare electron" in quantum electrodynamics) which are important for the development of the theoretical contents but which are fundamentally non-observable.

At the same time, in some investigational situations the ideas of "observability" unexpectedly turned out quite heuristic. For instance, in the period of construction of quantum mechanics elimination of non-observable electron orbits was a powerful impulse to development of the theory. A situation like this can be found in the period of construction of the special relativity theory, when elimination of non-observable absolute space allowed to develop new images of space and time.

All this is an evidence of certain part of rationality in the ideas of "observability", but, at the same time, of inadequacy of the very formulation of the "observability" principle, which does not include concrete directions: where and when it can be applied in the investigation, how we can tell observable quantities from non-observable ones, and at what stage of construction of the theory we are to eliminate non-observable objects.

Consequently, the regulative role of the "observability" principle was reduced to a trivial claim: to construct the foundation of the theory on magnitudes, tested by experiment, and to base on the intuition of the investigator who should find out, which magnitudes are to be considered as observable, and which ones are to be rejected as fundamentally non-observable.

The inadequacy of the very formulation of the "observability" principle was, in a major part, connected with its genetic, theoretical-cognitive origins. One of the first formulations, given by E. Mach, proceeded from false statements of his philosophy, that theory does not reflect the objective world, but experience and is not more than a brief reproduction of the facts observed. Later logical positivism tried to revive that idea in the form of the method of logical analysis. Positivism required that theory should eliminate all metaphysical ideas which have not been verified (checked up on base of reduction of the concepts to the data of observation). But a theory cannot be reduced to a brief summary of observations, and its notions cannot be treated as just fixation of phenomena observable in the area described by the theory: the theory reflects not the events, but the essence of processes in the real world, while scientific concepts have meaning not only within a certain theory, but they accumulate all preceding history of cognition which uncovers – step-by-step – new characteristics of the objective world.

The positivist interpretation of theory and following "linear prescriptions" of elimination of all non-observable concepts from science led to conclusion that no scientific theory could survive if "purified" in accordance with prescriptions of methodology of logical analysis.

No surprise that inadequacy of such statements to real specificities of scientific cognition led to a deep crisis in positivist philosophy of science.

At the end, the very positivist interpretation of the "observability" principle was put away. But at the same time there emerged an urgent problem of correct understanding of methods of empirical verification of a theory and discovering rational part of the "observability" principle, falsely interpreted by positivism.

In the course of this process investigators started to gradually understand that the abnormal hardness of the observability principle followed from the fact that theory is presented as result of purely inductive generalization of the facts observed. Understanding real methods of construction of a theory caused efforts to make a less hard formulation of the observability principle. We were to indicate, at what particular stage of development of the theory it could play the role of a methodological regulator.

A great part in the right formulation of this goal belonged to methodological investigation of the problem of "observability" made by classics of modern natural science A. Einstein, M. Born et al. What is especially interesting is the analysis of A. Einstein's comments of 1926 concerning W. Heisenberg's understanding of the "observability" principle. Einstein indicated that the very idea of "observability" depends on the theory. Only the theory determines what is observable, and what is not.[101] Einstein's criticism exercised influence upon Heisenberg's works of the 1930s, where the latter postulated that a considerable number of new conceptions should be introduced into a theory, and only then the nature will decide, whether to revise them or not – in every point. In this respect M.E. Omelyanovsky told a truth saying that for concretization of the ideas of "observability" we are to add: introduction of new concepts into a theory should take place at the stage of creation of the theory, and their verification should be done basing on new experience.[102]

Further investigation of the "observability" principle required analysis of the structure of the theory, methods of organizations of concepts inside the theory, distinguishing main and auxiliary abstract objects. Such analysis leads to ideas of constructive justification of the abstract objects of the theory.

After all above we can formulate the difference between requirements of "constructability" and the "observability" principle.

1. "Observability" stipulated inductive construction of the theory, while the "constructability" ideas are based on the opposite vision of genesis of the theory (from the very beginning they take into account that theoretical models are introduced from above, in respect to experiment, as hypotheses and only then are justified constructively).

2. The "observability" principle, at the best, only marks that at the stage of putting hypotheses forward we can use various notions, and only at the stage of justification of the hypothesis verify their empirical sense. The requirement of "constructability" clearly differs these to stages from the very beginning, meaning that constructive introduction of abstract objects into "the body" of the theory starts only after introduction of the supposed hypothetical model.

3. In the "observability" principle there is no differentiation of ideal objects of the theory, so it is not clear which of them are to be considered as observable, and which are non-observable. Criteria of such differentiation are transferred to the sphere of the investigator's intuition. In the requirement of "constructability" we have an effort to introduce such differentiation (at least, in the first approximation). It is supposed that what should be constructively justified (i.e., introduced as an idealizations based on new experience) is abstract objects of the theoretical model which lies in foundation of the theory. Such model is pretty clearly indicated in any theory (so we can agree with Einstein that concrete structure of a concrete theory indicates what there should be observable and non-observable). Then, taking into account that a concrete theoretical scheme (model) and picture of the world should be distinguished, we may divide the problem into two parts: constructive justification of the theoretical scheme and constructive justification of the picture of the world. The latter can as well include non-constructive elements (visual auxiliary images which let us inscribe the created scientific knowledge into the culture of a certain period). These elements are eliminated from the picture of the world only in the long course of historical development. At the best, they can be fixed as non-observable essences, but "criticism of the pictures of the world" takes place only on the eve of their breach. As to abstract objects of concrete theoretical schemes, they are mandatory to be introduced constructively.

4. The "observability" principle, in its strict formulation, required that non-observable objects should be eliminated from the theory immediately after they are discovered. According to the ideas of "constructability", the process of replacement of such objects can be executed as a long search for a new constructive meaning of the theoretical model. But the very fact that a non-constructive object has been found allows us to develop a consistent investigation. In this case the process of construction of theoretical knowledge can be run not by means of immediate elimination of the non-constructive object from the theoretical scheme, but by its localization and use of the theoretical scheme in further cognitive movement so that it could "work" only with its constructive elements. A characteristic example of such investigation is the process of development of knowledge based on the atom model, offered by Bohr and developed by Sommerfeld. That model included electron orbit (a non-constructive element), but Bohr, knowing that it is a "non-observable" object, constructed the system of postulates describing basic relations among main elements of the model, so that they "localized" the main paradoxical corollaries of employing electron orbits (it was supposed that electron, in its stationary state, does not radiate).

Considering the chance of this way of development of knowledge, we may come to conclusion that the very fact of discovering non-constructive objects provides progress of the theory, even if they are eliminated much later than they are discovered.

Thus, the method of constructive justification of theoretical schemes, indicating a concrete procedure of discovering non-constructive objects in "the body" of the theory, can make it easier to solve many investigation problems.

NOTES: CHAPTER 5

[1] Mandelstam (1972, p.329).
[2] Vavilov (1956, pp.156-157, 282-285). Mandelstam. (1972, pp.326-329), Kuznetsov (1975, pp.140-155).
[3] Vavilov (1972, pp.79-80).
[4] Ibid, p.80.
[5] Einstein (1965-1967, vol.2, pp.23-25).
[6] The process of becoming of the relativity theory and forming relativistic ideas of space and time in the physical picture of the world will be analyzed in more detail in chapter 6 "Scientific Revolution".
[7] Bohr (1970-1971, vol.2, p.510).
[8] Bohm (1952),
[9] Sachkov (1974, pp.71-72).
[10] Povarov (1972).
[11] Arshinov (1974), (1973).
[12] Stepin (1976, pp.290-300), (1982, pp.169-172).
[13] Chew (1966), Chew, Gell-Mann and Rosenfeld (1965), Chew (1968).
[14] Stapp (1971).
[15] Bohm (1971, p.28).
[16] See Bohm and Hiley (1977, pp.207-209).
[17] Nordin (1979, p.72).
[18] See Bohm and Hiley (1977), Philippidis, Dewdney and Hiley (1979).
[19] Concerning differences between those two strategies see Blauberg, Sadovsky and Yudin (1969, p.49).
[20] Capra (1994, p.298).
[21] Ibid.
[22] Ibid, p.174.
[23] Prigogine and Stengers (1984).
[24] George and Prigogine (1979, p.380).
[25] Prigogine and Stengers (1994, p.214).
[26] Ibid, p.215.
[27] Kurdyumov (1982, pp.235-236).
[28] Konopleva and Sokolik (1972, p.119), Vizguin (1975, pp.95-96).
[29] Mandelstam (1972, pp.329-337).
[30] It means that a particle is able to move at any speed — from zero to light speed (or, what is the same, energy of its movement is not necessarily small in comparison with the energy at rest).
[31] See Pauli (1964).
[32] Ibid.

[33] We would like to remind that, according to Pauli's principle, each energy state can include not more than one electron. In accordance with the identity principle, exchange effects are non-observable in such system.
[34] See van der Warden (1962, p.282).
[35] Pauli (1956, p.373).
[36] Quoted from Vavilov (1956, p.80).
[37] Kuznetsov (1975, pp.153-154).
[38] The term "modern situation" is used here in broad sense, in application to physics of the 20th century where quantum and relativistic theories prevail. In a narrower sense and within more differentiated approach it is desired that we should distinguish non-classical physics of the first and the second halves of the 20th century. In relation to the latter, special relativity and general relativity are sort of "quantum-relativistic classics". The image of physics of the last third of the 20th century is determined by investigation programs represented by quantum chromodynamics, the Grand Unification theory, theories of supergravitation and superstrings, in which investigators make efforts to come to synthesis of quantum and relativistic ideas, and create a whole theoretical description of strong, electroweak and gravitational interactions.
[39] Here we bear in mind construction of a developed theory in its first version (for instance, Newton's mechanics, Maxwell's electrodynamics), but not previous knowledge of some aspects of the object domain studied in theory, nor development and perfection of foundations of the theory already created (such as reformulation of Newton's mechanics made by Lagrange and then by Hamilton).
[40] Pauli (ed.) (1955, p.71).
[41] Ibid.
[42] The idea of quantum properties of radiation was historically the first fact which was integrated into foundation for development of quantum mechanics. But quantum mechanics for electromagnetic radiation (the theory of free quantized electromagnetic field) was created later than quantum mechanics for atom and atomic particles (electrons, nuclei etc.). This is due to the fact that atomic particles have different from zero rest mass, so for them there is energy area where it is possible not to take into account the effects of the relativity theory. As to photon, its rest mass is zero, so it does not have non-relativistic area. That is why the idea of electromagnetic field as a system of photons could be theoretically expressed in corresponding apparatus only after the quantum theory for non-relativistic particles had been created.
[43] The logic of construction of mathematical apparatus of quantized radiation field can also be traced in its "historical realization". P. Dirac started building this apparatus, then P. Jordan, W. Pauli and W. Heisenberg kept on working at it. In 1926–1927 Dirac suggested the first version of the quantum theory of electromagnetic radiation, which already included the method of quantizing of free field typical for modern physics. The base for transition to the quantum theory was a special method of classical description of the field. The classical radiation field was considered as a set of plane transverse waves, encased in a large but finite space volume. Correspondingly, the classical field equations were expressed through Fourier's transformation and then written down in a shape analogue to canonical equations of mechanics (Hamilton's equations). The expression for energy (Hamilton's functions) of each of the waves whose superposition as the field of study coincided with Hamilton's function for oscillator, which allowed to correlate a set of waves and an analogue set of oscillators. This method of description of electromagnetic field was known even to classical physics. Using it and then applying the quantizing rule for oscillator, Dirac carried out quantizing of the radiation field; its Fourier-components, preliminarily presented as canonical variables (generalized coordinates and impulses) were considered as operators which obey commutation relations. The influence of these operators upon the wave function characterized processes of creation and annihilation of photons in various quantum states. Formally it was expressed in the following way. The wave function (state vector) was defined as a function in space of filling numbers, i. e. photon-particles in various quantum states. Acting on it, the operators, corresponding to the field's Fourier-

components, either increase, or reduce by one the filling numbers, which means either creation, or annihilation of a photon in the given quantum state (correspondingly, these operators are called creation and annihilation operators).

This apparatus, mainly worked out by Dirac, allows us to explain many facts of interaction of electromagnetic radiation and matter (in particular, investigators got corollaries from it: well known rules describing emitting and absorption of electrons).

Jordan, Pauli and Heisenberg perfectioned Dirac's theory of electromagnetic radiation. They constructed the apparatus of the theory in the form satisfying Lorentz's transformations. Here we are to say that Dirac, who developed a perspective method of field quantizing, still failed to create equations relativistically covariant. Jordan and Pauli were the first to overcome this obstacle when they found a Lorentz-invariant expression of commutation relationships for the field operators (see Jordan and Pauli (1928)). It became possible, within the new formalism, on base of initial creation and annihilation operators, to create other operators which would correspond to various field quantities, answering the requirements of the theory's relativistic invariance.

[44] In construction of the mathematical apparatus of quantized electron-positron field Dirac's equations played a role similar to the one of Maxwell's equations in construction of the apparatus of quantized electromagnetic radiation field. The wave functions for electron and positron, in Dirac's equations, were presented as magnitudes characterizing electron-positron field and then regarded as operators satisfying anticommuting transposition relationships (this method, based on presentation of wave functions as operators, was then called the secondary quantization method).

[45] To find probabilities of the quantum effects characterizing dispersion of particles, which form electromagnetic and electron-positron field, we build a so-called dispersion matrix, or *S*-matrix. Squares of modules of this matrix's elements characterize probabilities of transition of the system described from some initial state to some final state. To find the *S*-matrix, we solve a connected system of operator equations which describe interacting quantized fields. The exact solution of this system is unknown, but we may find an approximate solution by means of the perturbation theory. In the framework of this theory interaction is considered as perturbation of the state of one free field by another in some area of interaction. Such visualization corresponds to consideration of particles, which interact only in the process of collision, while before and after collision they move independently. The states of non-disturbed system (of non-interacting photons and electrons, in this case) represent some basic integrity of quantum states. Perturbation (interaction of fields) leads to quantum transitions between these states (to changes of the number of the particles, their energies, impulses etc.). In the perturbation theory the dispersion matrix is presented through operators of free quantized fields and is computated in expanded form under the interaction constant which, in case of electromagnetic interactions, has the form of a dimensionless quantity $\alpha = \frac{e^2}{\hbar c} = \frac{1}{137}$ where α – electromagnetic interaction constant (or thin structure constant), e – charge of electron, $\hbar$ – Planck's constant, c – light speed.

[46] Feynman (1968, p.180).

[47] The correspondence principle has two aspects. The first one can be defined as generally methodological. Here the correspondence principle plays a specific form of connection between old and new theories (see Kuznetsov (1948)). The other aspect of the correspondence principle marks peculiarities of quantum mechanical description: the quantum object theory cannot be constructed without the language of classical mechanics. This aspect, though tied with the first one, cannot be reduced to it. It expresses the special nature of quantum objects: their physical being, characterized by physical magnitudes, is determined by macroconditions, the way of interaction of a quantum object with a classical body (see Kuznetsov (ed.) (1967, p.105-109)).

[48] We would like to remind the reader that, according to T. Kuhn's views, the change of vision of investigation situations is always stipulated by changes of some models, as "patterns", which indicate how to consider the said situations. From this point of view, the transition from vision of the system

of electrons as of a set of particles with quantum nature; their vision as of a field could be explained by choice of a new "pattern". The latter is understood as quantized electromagnetic radiation field, through which the investigator sees also other objects, for instance, he evaluates the system of electrons as a set of quanta of some field. Still, this approach, correct to some extent, leaves some important sides of the investigation process in the dark. It does not take into consideration the above mentioned difficulty of transfer of ideas about the system of photons as a field to a system of electrons (presence of a classical pattern in the first case and its absence in the second one). To carry out such transfer, we, previously, are to refer them to some general class and only then consider one object in the image, after the likeness of another. In other words, to compare, we are to have a base for comparison; to assimilate one image to another, we need a scheme of image distinguish. In this case the role of such a scheme belonged to the picture of physical reality which introduced an extremely general notion of the nature of quantum objects. Correlation of electromagnetic field and system of electrons with it was a base for further representation of one of the objects as a model of the other.

[49] The modern stage of quantum relativistic picture of the world is connected with elaborating the program of Grand Unification which is aimed at synthesis of the four main types of interaction: strong, weak, electromagnetic and gravitational. A considerable success of this program was construction of the electroweak interactions theory.

[50] Fok and Jordan (1931, S.206).

[51] Landau and Peierles (1965).

[52] The given relationship was first obtained by N. Bohr in 1928. Landau and Peierles give a derivation of that relationship (Landau and Peierles (1965, pp.59-61)). Energy and impulse exchange between the particle and the device should follow the conservation laws of impulse and energy. The impulse conservation law provides the following dependence between change of the particle impulse P and the device impulse p before and after measuring: $p'' + P'' - p' - P' = 0$ (1) where p' and p'' – the state of the device before and after the exchange of impulse with the particle, P' and P'' – corresponding states of the particles. The energy conservation law requires the same dependence for energy exchange between the particle and the device during measuring time Δt. Considering the relationship $\Delta\varepsilon\Delta p \sim \hbar$, this dependence looks $\varepsilon'' + E'' - \varepsilon' \sim \hbar/\Delta t$ (2), where ε' and ε'' – energy of the device before and after measuring, E' and E'' – corresponding values for energy of the particle. Values of p' and p'' and ε' and ε'' as related to the device, are always known within any accuracy. So, the equations (1), (2) give relationships $\Delta P' = \Delta P''$ and $\Delta E'' - \Delta E' \sim \hbar/\Delta t$ (3) for impulse and energy of the test particle. In accordance with the correlation between energy and impulse, $\Delta E' = \frac{\partial E'}{\partial P'}\Delta P' = v'\Delta P'$ (v' – speed of the particle before collision), and $\Delta E'' = \frac{\partial E''}{\partial P''}\Delta P'' = v'\Delta P''$.

Substituting these values to (3), we get $|v''-v'|\,\Delta P \sim \frac{\hbar}{\Delta t}$. Thus, this correlation appears because any measuring takes some period of time Δt, during which there appears indeterminacy in the exchange of energy-impulse between the measured quantum particle and the classical device.

[53] It is connected with the need to control the change of speeds of the particle at the moment of its collision with the device — to compute the disturbing influence of its own radiation field upon its impulse. But such control, in turn, stipulates new measuring (determining velocities v' and v'' before and after collision of the particle with the device), measuring during infinitesimal time period. The situation is repeated also due to $\Delta P \Delta t \sim \frac{\hbar}{\Delta t}$, if $\Delta t \to 0$, then $\Delta P \to \infty$, i.e., any control over the disturbing influence of the field radiated by the particle, upon its impulse, leads to increase, not decrease, of indeterminacy of this impulse.

[54] To avoid analysis of disturbing effect of charged experimental particles on the electron, Landau and Peierles, treating photons as such particles, constructed their thought experiments in accordance with the scheme of experiments based on Compton's effect. In that case it was important that the impulse of photon, colliding with electron and transferring information of its state to the device, can be measured during time period Δt only with indeterminacy ΔP which cannot be made smaller than $\frac{\hbar}{c\Delta t}$ (according to the relationship $\Delta P\Delta t \geq \frac{\hbar}{\Delta t}$). If we take this circumstance into account, it means that a classical device can fix the magnitude, characterizing the state of the electron, with the corresponding indeterminacy.

[55] The quantum mechanical description of densities of the charge-current stipulates their representation as a set of separate electrons. The latter can be interpreted as quanta of electron field. According to a postulate of quantum mechanical description, classical quantities characterizing the system should be used also as observables in description of its quantum properties. Sources of the field were characterized in classical electrodynamics by vector of density of charge-current in a space-time point. When we determine this magnitude in the process of measuring it is taken that the time period, required for measuring, should be infinitely small. But in this case, quantum effects taken into account, it is impossible to get the exact value of this fundamental quantity, which contradicts to the quantum mechanical description postulate, which sets no limitations to exact measuring of one observable.

[56] The evidence is W. Pauli's skepticism expressed in 1932 (in. Collected Scientific Papers by W.Pauli, in Two Volumes. Ed. by R. Kronig and V.Weisskopf. New York-Sydney, Interscience Publishing, 1964. (.P. 284 — 286)).

[57] We would like to remind to the reader that the initial model for quantizing the field was the idea of it as of an infinite set of oscillators, each of them is subject to quantizing. The field energy was written down as sum of expression for energy of each oscillator. These expressions meant that the energy values of zero oscillations of all field oscillators are different from zero. At the same time, the said expressions showed that the state studied cannot include photons, i.e., physically it should be pure vacuum. As the number of the field oscillators was infinite (according to the number of the degrees of freedom), we had that, without photons, instead of the expected zero energy there emerged infinite energy which should be attributed to vacuum. That conclusion was so unexpected that initially it could well be regarded as evidence of profound defects of the theory created.

[58] Landau and Peierles (1965, p.69).

[59] Ibid.

[60] Here the term "translation" means that the state of the experimental body during time $t_1 - t_2$ between interactions with the object measured, on the one hand, and the register device, on the other hand, either does not change, or changes in time in accordance with the known law, on base of which the observer can determine the initial state of the experimental body, which is an indicator of the studied state of the measured object.

[61] Landau and Peierles (1965, p.57).

[62] In this case we can always operate so that the experimental body, once having interacted with the measured quantum system, would move as a free particle, without any more influences (translation of its state would follow Schrödinger's equation, and at any moment we could receive information about this state on base of the said equation). As to perturbing influence of the register device upon the state of the experimental particle during time Δt (time of registering this state), we can minimize emerging indeterminacies by means of corresponding choice of Δt. If we bear in mind values of energy ε or impulse P of the experimental particle as characteristics of its state, indeterminacies $\Delta\varepsilon$ and ΔP (caused by quantum effects which emerge with transmission of energy-impulse of the experimental particle to the device) can be reduced by increase in measuring time Δt (in accordance with correlations $\Delta\varepsilon\Delta t \geq \hbar$ and $|v''_x - v'_x|\Delta P_x\Delta t \geq \hbar$). All this makes measurements in the area of non-relativistic quantum interactions quite predictable, even if the experimental particle interacts with the

register device as a quantum object. Analysis of such measurements, when we get information about state of quantum systems not through their immediate interaction with the device (direct measurements), but through a number of intermediate links – quantum mechanical particles (indirect measurements), and justification of fundamental possibility of such measurement in non-relativistic area can be found, for instance, in L. Mandelstam's lectures on quantum mechanics (Mandelstam (1972)).

[63] By the way, the discussions of incommensurability are very close in time to two Solvay congresses of 1927 and 1930, where the famous disputes on foundations of the quantum theory between Bohr and Einstein took place. The corner stone of these disputes was specificity of quantum mechanical measuring and clearing of special role of classical device in determination of states of the quantum system measured.

[64] Appearing psychological barrier and overcoming it is one of the characteristic features of the psychology of discovery in science. A detailed discussion of this aspect of scientific creative work can be found in B. M. Kedrov's writings.

[65] Bohr (1970-1971, vol.1, pp.125-131).

[66] Rosenfeld's memories of his work together with Bohr start from a later period (late February, 1931), when Bohr's discussions with Landau and Peierles were over. We believe, at that time Bohr had already come to the general idea that it is mandatory that classical experimental bodies should be employed in idealized measuring procedures. Describing the corresponding period of the history of electrodynamics, Rosenfeld intended to reproduce the main stages of the measuring procedures which led to justification of fundamental measurability of the components of quantized field. Naturally, he pays closer attention to the procedures, and not to the preliminary period. That period is mentioned by Rosenfeld without any specific details. No wonder that the logic of thoughts, which led Bohr to his remark about the field averages, remains at the background. Reconstruction of that speculation never was among the aims of Rosenfeld's essay.

[67] Bohr (1970-1971, vol.2, p.130).

[68] It is easy to understand that here we see that very set of questions that made Landau and Peierles to come to the conclusion of fundamental impossibility to measure field. Bohr and Rosenfeld return to discussing these questions, but on a fundamentally different base: analysis of the measurability problem within thought experiments with classical experimental bodies.

[69] Bohr (1970-1971, vol.2, p.132).

[70] This formula is easily deduced from Lorentz's equation $F_x = \frac{1}{\rho}\frac{dP_x}{dt}$ for the force of the field acting on the charge ρ at moment *t* in direction of *x*-axis. Turning to integral form of this expression for force component, affecting a charged body of volume *V* during time *t* averaged over area *V*τ, and taking into account that the force of the field action upon a charged body, by definition, gives the value of the field strength, we get the formula $\overline{E}_x = \frac{P_x}{\rho V \tau}$, where $P_x = p''_x - p'_x$.

[71] Bohr (1970-1971, vol.2, pp.132-133).

[72] Bolzman (1929, p.121).

[73] Pauli (ed.) (1955).

[74] Ibid.

[75] The conclusion that it is possible to present the experimental body as a part of the device was, probably, prepared by analysis of the functions of experimental bodies. N. Bohr carried out this analysis while constructing his program of idealized measurements.

[76] Bohr (1970-1971, vol.2, pp.141-142).

[77] Certainly, here we will have some error in the impulse ΔP_x due to the relation $\Delta P_x \Delta t \sim \frac{\hbar}{c}$. But, with fixed Δt, such error has a certain order of magnitude. As Bohr and Rosenfeld have demonstrated, it exactly corresponds to the value ΔP_x which appears with fixed indeterminacy Δx in position of the experimental body along with its displacement caused by interaction with the register device. The presence of indeterminacy ΔP_x, when Δx is fixed, does not prevent us from exact measuring the averaged on $V\tau$ field component, since, as we have proved, that error can be compensated by increase of the density of the charge of the experimental body (for more details see Bohr (1970-1971, vol.2, pp.137-138)).
[78] Bohr (1970-1971, vol.2, p.137).
[79] Ibid, pp.139-140.
[80] Ibid, pp.142-143.
[81] Ibid, pp.142-143.
[82] Bohr (1970-1971, vol.2, p.149), Pauli (ed.) (1955).
[83] Pauli (ed.) (1955, p.78).
[84] Ibid.
[85] Bohr (1970-1971, vol.2, pp.153-158).
[86] Pauli (ed.) (1955, p.76).
[87] In this case we would have to consider the radiation, caused by displacement of the experimental body by Δx at measuring its impulse, and which cannot be compensate, as that perturbing influence, which basically prevents us from exact determination of the field component.
[88] Bohr (1970-1971, vol.2, pp.434-445).
[89] Pauli (ed.) (1955).
[90] The first Bohr's and Rosenfeld's publication dedicated to the problems of measurability of quantized electromagnetic field was made in 1934. The work referring to measurability of densities of current charge was published, in its final version, in 1952, but its first edition, as a review, was prepared in the mid 1930s and was quite well known for the majority of theorists who worked at the problem of field quantizing (see. L. Rosenfeld's memories in Kuznetsov (ed.) (1967, p.76)).
[91] In modern exposition, the need to consider the observables as summary of interaction of a bare charged particle with vacuum is often corroborated by references to vacuum polarization (interacting with vacuum, electron gets polarization "cover" made of virtual electrons and positrons, which an outside observer perceives as effective reduction of the electron charge). But we are to remember that the very discovery of vacuum polarization was a quite late achievement (compared to Bohr-Rosenfeld procedures) and, by itself, needed the preliminary idea of physical reality of vacuum and possibility to observe effects of its interaction with charged particles in an experiment. Such ideas were formed due to idealized measurements of quantized fields.
[92] In Kuhn's conception of paradigmatic models of solutions of problems, new non-standard solutions, leading to perspective hypotheses, are described in terms of gestalt-switching (see Kuhn (1962)).
[93] See Karmin and Khaikin (1971, pp.36-39).
[94] See Bransky (1978, pp.40-41, 36-39).
[95] Ibid, p.40.
[96] Reichenbach (1961, p.6-7).
[97] Kuhn (1962).
[98] See Kuhn (1974).
[99] Ibid.
[100] They were discovered and first described in Stepin and Tomilchik (1970), Stepin (1972, 1976).
[101] Heisenberg (1969, S.91-92).
[102] Omelyanovsky (1973, p.99).

CHAPTER SIX

SCIENTIFIC REVOLUTIONS

SCIENCE IN THE TECHNOGENIC CIVILIZATION CULTURE

In the dynamics of our scientific knowledge, a special role belongs to development stages connected with reconstruction of investigation strategies, required by foundations of science. These stages were called scientific revolutions. Foundations of science provide growth of knowledge, till common features of system organization of the objects studied are included in the picture of the world, and methods of their cognition correspond to the existing investigation ideals and norms.

But developing science may come across basically new types of objects, which require other vision of reality, different from that suggested by already developed picture of the world. The new objects may require that the scheme of method of cognitive activity, represented by a system of investigation ideals and norms, should be changed. In this situation growth of scientific knowledge stipulates reconstruction of foundation of science. Such reconstruction can be realized in two variations: a) revolution connected with transformation of the special picture of the world without important changes in investigation ideals and norms; b) revolution which causes radical changes not only in the picture of the world, but also in scientific ideals and norms, as well as in philosophical foundation of science.

In the history of natural science we can find samples of both situations of intensive knowledge growth. An example of the former: transition from mechanistic to electrodynamics picture of the world in physics of the last quarter of the 19th century due to construction of the classical theory of electromagnetic field. This transition, though followed by quite radical transformation of vision of the physical reality, did not change essentially cognitional attitudes of classical physics. It conserved understanding of explanation as search for substantial foundation for phenomena explained and strictly determined links among the phenomena; any indications to observation means and operational structures, which uncover essence of the objects studied etc., are eliminated from the principles of explanation and justification.

An example of the second situation is the history of quantum-relativistic physics, characterized by reconstruction of not only the scientific picture of the world, but also the classical ideals of explanation, description, justification and knowledge organization, as well as corresponding philosophical foundation of science.

The new picture of the reality studied and new norms of cognitional activity, while settling in a concrete science, then can have a revolutionary influence on other sciences. In this aspect we can mention two ways of reconstruction of investigation foundations: first, due to intradisciplinary development of knowledge; second, due to interdisciplinary connections, "grafting" of paradigmatic statements of one science to another.

In real history of science both ways superpose, so in most cases it would be more correct to speak about domineering of one of them in each science at either stage of their historical development.

INTRADISCIPLINARY REVOLUTIONS

Paradoxes and problem situations as premises of a scientific revolution

Most often science includes new objects into investigation unconsciously, through empirical studies of new phenomena or in process of solving special theoretical problems.

To analyze the peculiarities of this process in details, let us consider the historical situation immediately preceding construction of the special relativity and became one of the premises of the revolution in the 20th century physics.[1] That situation was linked with discovery of paradoxes in classical electrodynamics of moving bodies.

When Lorentz developed Maxwell's electrodynamics, and the electron theory was built, it became possible to solve the class of problems considering interaction of moving charges and bodies with electromagnetic field. In the process of solution investigators were to formulate Maxwell's equations in different frames of reference, and then it became clear that the equations were no longer covariant, when using Galilean transformations. Introduction of new transformations offered the way. The transformations were first offered by Vogt, and then by Lorentz, who has given his name to them for the history of science.

The coordinate transformations (space and time) in transition from one inertial system to another are an important characteristic of such system. Inertial frame of reference is one of the fundamental theoretical objects of any physical theory. In Maxwell-Lorentz's electrodynamics it played the role of a component of the theoretical scheme, which lay in the foundation of the theory. That scheme presented electromagnetic processes through relations of abstract objects: electric and magnetic fields in a point, elementary point charge (electron, and inertial frame of reference. The scheme was objectified through mapping to the electrodynamic picture of the world: elementary point charge correlated with the image of electron as a charged spherical body of very small size, immersed in ether; space-time characteristics of the frame of reference were connected with features of absolute space and absolute time. This connection was established thanks to the fact that space and time intervals of the frame of reference were seen as unchangeable in transition from one frame of reference to another. Stability of the intervals allowed us to consider them as independent from motion of the body (frames of reference) and, consequently, to present them as absolute space and absolute time. Galilean transformations (which automatically inferred this quality of inertial frames of reference) acquired this way their physical interpretation.

But when new transformations were introduced into the theory, the frame of reference, in a hidden manner, gained new features: from Lorentz's transformations it was inferred that separately space and separately time intervals are not conserved in transition from one frame of reference to another. In mapping to the picture of the world these frame of reference features were objectified, which raised inconsistent definitions of space and time. Relativity of space and time intervals was incompatible with the principle of absolute space and time.[2]

Paradoxes are symptoms indicating that science draws into sphere of its investigations a new type of processes, whose essential characteristics have not been reflected in the picture of the world. Formed in mechanics notions of absolute space and time allow consistently processes taking place at speeds low compared to the light speed. At the same time, in electrodynamics investigators dealt with fundamentally different processes characterized by light speed or close speeds. If old notions were implied here, it would have caused contradictions in the very foundation of physical knowledge.

Thus, a special theoretical task became a problem: the system of knowledge could not remain inconsistent (a theory should be consistent, which is a norm of its organization). But to eliminate paradoxes, it was necessary to change the physical picture of the world perceived by investigators as adequate reproduction of reality.

Such situations are quite characteristic for science entering the stage of a scientific revolution. Scientific problems emerging at this period appear due to solving special tasks. From our point of view, a task grows to a problem the following way: theoretical schemes and laws, generated by already formed foundations of science, are rebuilt in the process of their empirical justification, are correlated with new facts and so include new meaning. In reverse mapping to foundations (to the picture in the world, in particular) this meaning can cause mismatch with notions of reality introduced into the picture of the world. If the picture of the world does not take into account specificity of new objects, then theoretical scheme, considering some essential peculiarities of such objects, may lead to paradoxes in the system of knowledge.[3]

Science solves paradoxes by means of reconstruction of foundations previously formed. Such reconstruction without fail leads to change of the picture of the world. Though, revision of the picture of the world is not at all easy, as at the previous period stimulated theoretical and empirical investigations and was perceived as adequate image of the essence of processes studied.

For instance, it is characteristic that Lorentz, who prepared breakdown of the electrodynamic picture of the world, failed to make a decisive step himself.

He interpreted changes of space and time intervals as fictitious, “local” space and time. What was true, he believed, was absolute space and time of the picture of the world accepted by the late 19th century physics.

As early as deducing his transformations, Lorentz was eager to provide them with physical sense introducing into the picture of the world a number of assumptions, which would preserve ether and absolute space and time. He supposed that electron, moving past ether and interacting with it, could change its own space configuration. This was how Lorentz interpreted change of space and time intervals as a by-effect of electron’s dynamics, but not as a real property of space and time. From the same positions Lorentz interprets the results of Michelson’s experiment.

It was Einstein who radically transformed the electrodynamic picture of the world. The transformation was connected with rejection of the conception of ether and revision of the ideas of absolute space and time.

Characterizing Einstein’s transfer to a new vision of the physical reality, we could follow Kuhn and use terms of discovery psychology as Gestalt-switching. But such approach would conceal the logic of cognitive movement, which lay at the foundation of Einstein’s works, and which characterizes foundations of the mechanism of reconstruction of science foundation at the period of scientific revolution.

When Einstein’s predecessors tried to preserve the previous picture of the world, they did not eliminate paradoxes, but only transferred them to a deeper layer of science foundations.

In this case there usually emerge contradictions between the system of knowledge being created and science ideals, while a theory should be constructed according to the latter.

Additional principles, introduced into the picture of the world to explain new phenomena, appear as ad hoc postulates. If we permanently use such postulates when we discover new phenomena, we face a chaotic increase of initial principles of theoretical investigation. In the extreme, such increase may lead to the situation when the number of principles may start equalizing with the number of empirical facts explained through these principles, which would destroy the very idea of theoretical explanation.

Einstein's criticism of the notions of classical physics was, to a considerable extent, stimulated by understanding of the mentioned paradox. In turn, such understanding stipulated the investigator's specific position. He had to leave the limits of especially scientific problems and consider them in the aspect of regularities of cognition process, i.e. turn to the language of philosophical and methodological analysis. Cognitive activity, aimed at reconstruction of science foundations, always stipulates change of investigator's position and turn to philosophical and methodological means (see pic. 7).

Einstein proceeded from the following methodological postulate: a theory should not only fit to normative experimental justification, but also, in the ideal case, it should be organized so that diversity of very different phenomena should be explained and predicted on base of relatively small number of principles which would define the essence of reality studied.

At later stages of his activity (after the special relativity had been created), Einstein pointed at those methodological criteria, according to which a physical theory should be created, as at requirements of its experimental verification and internal perfection.[4] He justified both requirements as profound characteristics of scientific investigation; and, in effect, he regarded them as explication of invariant contents of ideals of science, which controls creative search at all stages of development of natural science.

Justification of the indicated requirements as universally important characteristics of ideal of natural science theory stipulated analysis of the nature of theoretical cognition. Einstein returned to this analysis many times, at different stages of his career, improving and developing notion of ways of formation of a scientific theory. Theoretical reproduction of essential aspects of reality, according to Einstein, is realized by means of creative search for a moderate number of principles, on base of which all the rest conceptual construction of theory is unfolded. The principles themselves can only be "blown" by experiment, they are not deduced directly from experimental facts by induction. They are the result of active reconstruction of historically collected conceptual means, which are developed in the very cognition process and, to large extent, determine the character of theory created. To be true, a theory should rest on experiment. But one and the same experimental sphere can be described by different theories, and each of them offers its own vision of facts.

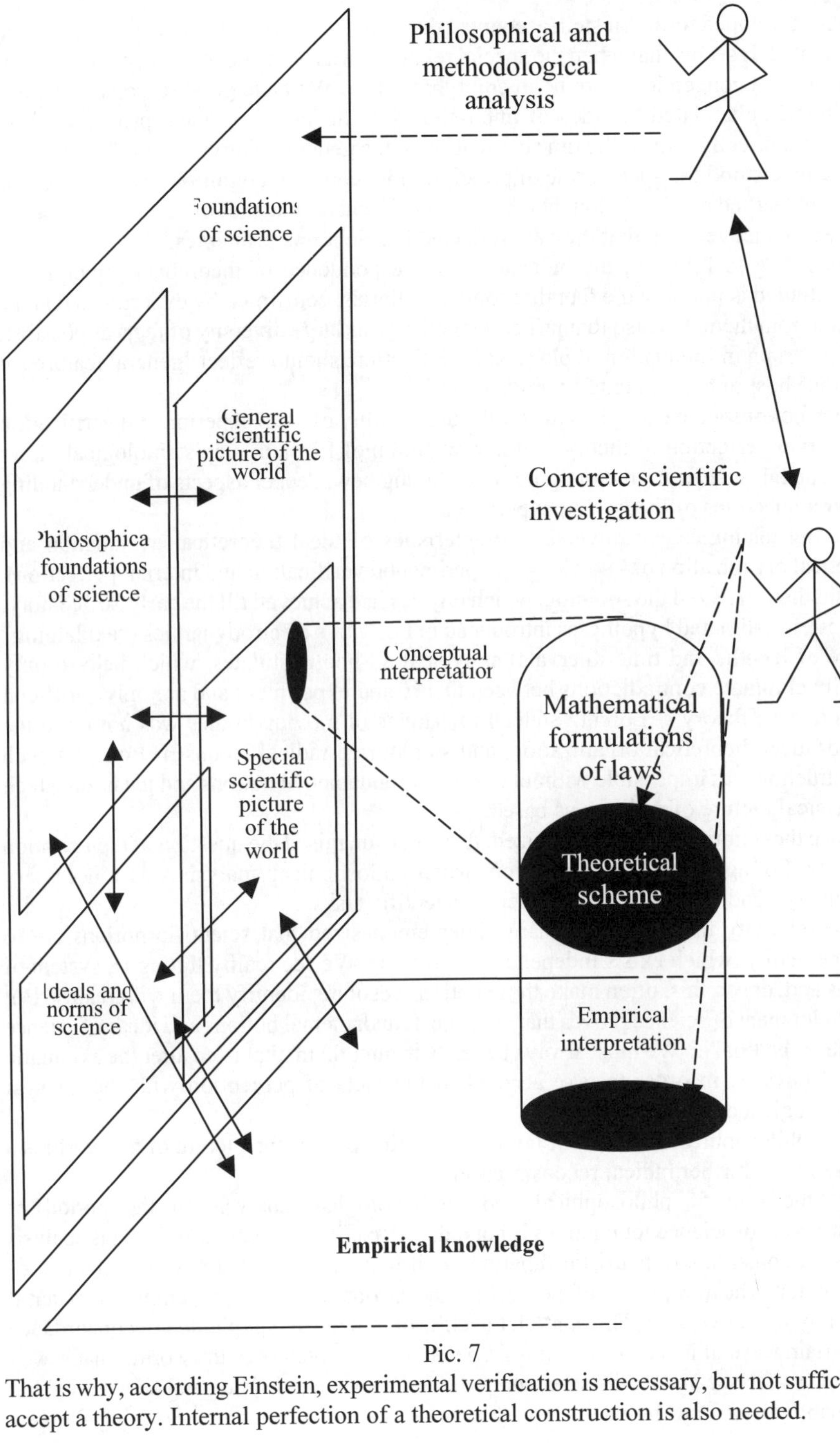

Pic. 7

That is why, according Einstein, experimental verification is necessary, but not sufficient to accept a theory. Internal perfection of a theoretical construction is also needed.

In its developed form, Einstein accounted that conception after the special relativity had been created. It seems that when the special relativity was being developed, many ideas of the mentioned conception were in an embryonic state. We have good reasons to believe that Einstein elucidated the idea of impossibility to deduce theoretical principles from experiment directly only at the time of creation of the general relativity.[5] But Einstein had always understood the special role of principles in theoretical cognition. All stages of his works are marked by conviction that there are profound regularities of nature, that science is called to uncover, and that they are reflected in science as principles.

According to Einstein, the indicator of correspondence of theoretical principles to reality studied is not only the fact that some corollaries, confirmed by experiment, can be deduced from them, but also that principles embrace as large diversity of facts as possible. Principles, laid in foundation of physical investigation, should reflect "general features of enormous host of experimentally proved facts".[6]

Such notions were enough to justify the universality of ideal experimental verification and internal perfection of theory. Further evolution of Einstein's epistemological views only made this justification more precise, including new, deeper aspects of understanding of interconnections of theory and experiment.

Having distinguished universal characteristics of ideal theoretical explanation and theoretical organization of knowledge (experimental verification and internal perfection), Einstein then analyzed the situation, which physics had achieved till the early 20th century.

Einstein estimated hypotheses introduced in Lorentz's electrodynamics ("explaining" change of lengths and time intervals) as typical ad hoc postulates, which help to only formally eliminate contradictions between theory and experiment and are only "artificial means to save theory".[7] Lorentz's electrodynamics of moving bodies was not up to the mark of ideal theoretical organization, and so required radical reconstruction. But such reconstruction was impossible without change of fundamental notions and ideas, on which the physical picture of reality was based.

Since these notions were ontologized, their revision raised the question of their relation to reality. So, again we had a situation, when philosophical analysis was a necessary preliminary condition for solving concrete scientific tasks.

The relativity theory creator many times emphasized that scientific notions are to describe reality, which exists independently from us. We see reality through a system of notions and, due to this, often make these notions absolute, identify them with reality. But the development of science proves that even most fundamental notions and ideas of science "can never be final". "We must always be ready to alter them, that is, to alter the axiomatic basis of physics, in order to take account of the facts of perception with the greatest possible logical completeness".[8]

Such philosophical criticism of notions and principles of the picture of the world is a premise for its further radical reconstruction.

But the role of philosophical and methodological analysis in the period of reconstruction of science foundations is not reduced to critical functions only. This analysis has also a constructive, heuristic function, as it helps to work out new foundations of investigation. The new picture of the world cannot be obtained from new empirical material in a purely inductive way. This material itself is organized and explained in concordance with certain ways of its vision, and this is specified by the picture of the world. That is why empirical material can only discover contradiction between old vision and new reality, but it is unable to indicate by itself, how this vision should be changed. Forming the new picture of the world claims special ideas, which would let us regroup elements of the old ideas of reality, eliminate a part of them, include new elements, so that we could solve paradoxes existing and assimilate collected facts. Such ideas are formed in the sphere of philosophical and methodological analysis of cognitive situations in science and play the role of quite general heuristics, which provides intensive development of investigations.

Heuristic role of methodological schemes

It is well known that the creation of the relativity theory was linked to application of several methodological principles, which played a heuristic role in developing new ideas in physics.[9] These principles (simplicity, observability, invariance etc.) were the result of philosophical analysis of scientific investigation processes and procedures of formation of physical notions. They can be regarded as methodological rules, which are certain concretizations of philosophical ideas in application to demand and need of the corresponding sphere of natural science. The system of such regulations in explicit form expresses certain norms of cognitive activity and directs reconstruction of the picture of the world already formed in science.

In his retrospective estimation of the process of creation of the special relativity, Einstein emphasized that the fundamental role in its construction belonged to an epistemological postulate: "notions and judgments have meaning only as we can correlate facts observed with them unambiguously".[10] (Requirement of meaningfulness of notions and judgments.) It would be just to regard this postulate as one of formulations of the "observability" principle.

We know that E. Mach widely advocated the "observability" principle. In this principle he saw expression of his conception of theory and experiment (in Mach's views, theory is a condensed summary of data, which, in turn, were interpreted as cognition subject perceptions).

Einstein's interpretation of the "observability" principle differed from that of Mach, as it followed from a different conception of scientific cognition and grasped some real moments of formation of theory and its relation to experiment.

First, Einstein required that theoretical notions should be justified by observed facts, as he understood the nature of fact differently from Mach. Unlike Mach, he did not reduce facts to observer's perceptions, but regarded them as phenomena of the physical world, fixed by the observer. These phenomena are revealed in procedures of experiment and measuring. In his first works dedicated to account of the special relativity, Einstein often referred to observable fact by the term "event". Mach widely used the same term. But Mach understands event as subject's perception, while Einstein – as a physical phenomenon registered in experiment and observation.

Second, Einstein never reduced theory to "a condensed summary of data". His epistemological base was recognition of objective existence of nature and independence of the physical world's laws from cognition subject. Search for principles expressing these laws was, to his mind, the main goal of physical investigation. Characterizing the starting stage of formation of the special relativity, Einstein emphasized in his "Autobiographic Notes"[11] that motivation, which led him to this theory, was desire to "come to true laws by means of constructive generalizations of facts. It is important that constructive generalization was understood as "discovery of general formal principle" that "can lead us to reliable results". The idea of "observability", from this point of view, meant discovering of correlation between principles, which made up the core of the theory, and experimental measuring procedures, in the system of which an experimental fact is formed.

Later, having developed the conception of impossibility to deduce inductively theoretical principles directly from experimental data, Einstein introduced new corrections. He emphasized that the core of the theory by itself determines which sphere of experiment we are to apply to justify its notions.

The "observability" principle did not mean that every notion of theory, in all its definitions, should from the very beginning be introduced as schematization of experiment. At the stage of hypothesis a theoretical construction is created on base of conceptual means, elaborated by science, by means of transformation of preciously formed notions into new ones. At early stages a part of these new notions may not be up to the mark of the "observability" principle. But when the core of the theory is already circumscribed, notions, which make up this core, should be introduced in correspondence with "observability" requirements.

The "observability" principle was a methodological normative, expressing ideal experimental justification of theory. At the same time, it was connected with ideal theoretical explanation and organization of knowledge, which Einstein characterized as internal perfection of theory. Ordering to eliminate from the theory core notions, which are not up to the mark of operational criteria, the "observability" principle indicated ways to minimize fundamental notions, by means of which experimental facts are explained.

We know that the tendency to minimize fundamental theoretical notions, explaining facts, is formulated as the simplicity principle. This principle is a normative directly expressing the ideal of "internal perfection of theory". So, there is correlation between "observability" and simplicity principles, which proves a certain system organization of methodological rules explicating norms of scientific cognition.

In a system of such rules separate elements play different roles at different stages of theoretical search. In modern physics at the stage of shaping the conceptual core of a theory, of search for mathematical apparatus and primary hypothetical models called to provide its interpretation, the simplicity principle is often domineering, and the "observability" principle is subordinate. At the stage of justification of hypothetical core of the theory created, when the experimental sphere which would be base for this core, is already shaped, the domineering role transfers to the "observability principle". It provides refinement and reconstruction of fundamental theoretical notions and directs formation of the new theory in its complete form.

Philosophical and methodological literature has already noted that in Einstein's construction of the general relativity the "observability" principle did not play that decisive role, which belonged to it in construction of the special relativity.[12] This fact can be explained in the following way. In construction of the general relativity the main task was to develop mathematical apparatus and to create a primary conceptual core of the theory. Unlike this, Einstein started working on construction of the special relativity, when the base of mathematical apparatus of the future theory (Lorentz's transformations) and initial interpretation of that apparatus had already been formed, and the way to a new theory required reconstruction of that initial interpretation which caused paradoxes.

The "observability" and simplicity principles are principles of not only modern, but also classical physics. Inside them we can distinguish some universal meaning, characterizing universal, constantly reproduced features of cognition tendencies of physics, as well as concretizing layer of statements, where historical stages of development of science are distinguished: this layer expresses the style of physical thought prevailing at each stage.

Transition from classical to modern physics was accompanied by reconstruction of the mentioned concretizing layer, which corresponded to reconstruction of norms of physical investigation, formation of new cognition tendencies which provided progress of science.

Methodological investigations have already indicated that concrete meaning of the simplicity principle changed in the history of science.[13] As we know, the simplicity principle was formulated as early as in the 14th century by W. Ockham as requirement not to increase entities over necessity explaining phenomena ("Ockham's razor"). Classical natural science conserved this requirement, but it was joined with a special system of interpreting statements: the idea of minimization of theoretical principles was deduced from the postulate of "ontological simplicity of nature", and criteria of correspondence of logical simplicity of a theory to the simplicity of nature were proclaimed not only by the possibility to verify experiment and wide grasp of explained and predicted phenomena by the principles, but also by visualization of the principles.

Modern natural science does not take the latter criterion for decisive. At the same time, mathematization of modern physics and wide application of the mathematical hypothesis method introduced a new layer of concretizing statements into the simplicity principle, connecting with the invariance and symmetry principle.

Concrete meaning of the "observability" principle also changed in the process of historical development of physics. In the period of creation of the special relativity reconstruction of this meaning corresponded to forming a new ideal justification of theory, which, in turn, meant transition from classical to non-classical style of thought. This transition can be traced even in early versions of Einstein's interpretation of "observability". It was connected with genesis of a special method of construction and justification of the conceptual core of physical theory.

The mentioned core can be defined (on base of the above described analysis of structure of theoretical knowledge) as a fundamental theoretical scheme mapped to the picture of the world. Notions, making up the core of theory, include definitions reflecting connection between features of ideal objects of theoretical scheme and objects of the picture of the world. Consequently, analysis of fundamental notions of theory from positions of the "observability" principle is inseparable from revealing of experimental foundations of the physical picture of the world, explication of operational foundation of those features which are given to its ideal objects with ontological status. Classical physics also justified the picture of the world by experiment, but such justification was understood as experimental verification and measuring consequences deduced from the principles of the picture of the world.

The new approach used by Einstein, when he started constructing the relativity system, was based on requirement of selective operational control over notions and principles of the physical picture of the world. It was not reduced to indication to concrete experiments and measurements, which verified the picture, but stipulated discovering of essential features of the whole experimental and measuring practice, within the scope of which characteristics of the reality studied, postulated by the picture of the world, should be revealed. Though Einstein did not clearly formulate in his methodological explications the described understanding of 'observability", his investigation practice speaks for such understanding. It was oriented at analysis of profound premises and foundations of experimental and measuring procedures, which constitute empirical basis of the physical picture of the world.

Let us consider this aspect of the question in more details. Experimental and measuring procedures of physics are always based on evident or hidden assumptions concerning characteristic properties of the investigation taking place. These assumptions have a complicated structure. They include statements about those perturbing influences that can be ignored (or taken into account) in a concrete measuring situation, so that it would be possible to reproduce the studied states of the object (and to fix its corresponding parameters). Such assumptions are based on employment of concrete physical laws and are usually distinctly explicated by the investigator. For example, in measuring temperature by thermometer we take into account possible changes of the thermometer scale at its contact with heated body, and, on base of the linear expansion law, determine corrections considered in scale graduation.

But assumptions, which form the base of measuring procedures, also include quite general postulates, which are perceived by the investigator, in most cases, as something evident, and are not formulated directly. For instance, these are profound foundations of physical measuring expressing their very nature, the things common for different concrete types of experimental measuring procedures.

Explication of the mentioned foundations and their analysis are held in a system of philosophical and methodological means, at the interfaces between physics and philosophy.

At each stage of development of science profound foundations of measuring appear as some kind of investigation "presumptions". Such presumptions are, for instance, postulate of objective reproducibility of experiment, and law conformity of phenomena, researched in experiment and observation (subjection of these phenomena to the natural laws).

Physical investigation stipulates fundamental possibility of "dissection" of complicated "superposition" of the natural laws in experiment by means of selection of such conditions, that action of side and obscuring factors is minimum, and action of the laws studied is exhibited in the best form for observation. This postulate of fragmentation and localization of studied processes in physics is complemented by one more presumption: the natural laws, controlled by observable processes, can be expressed in the language of mathematics. What is more, diversity of observed phenomena can be described and explained by means of relatively simple mathematical formulations of physical laws.

Postulates of this kind express norms that lie in the very foundation of physical cognition. Between them, as profound layer of principles of experimental and measuring activity, and layer of assumptions, based on application of concrete physical laws, intermediate layers are presented.

In particular, we may distinguish the layer of physical principles, which have more general character than physical laws, but, concerning fundamental postulates of physical investigation, are their concretizations. For example, the postulate of reproducibility of experiment is concretized by means of principles, according to which the same experiments can be repeated in various points of space and at various moments of time. Statements of the kind seem evident: in Paris and in Moscow the same experiment will give the same results; Huygens's experiments with collisions of elastic bodies or pendulum oscillations can be reproduced nowadays, more than 300 years after they were first executed.

But behind external evident character of such statements quite the powerful assumptions concerning the nature of the physical world are hidden. For instance, the statement of fundamental reproducibility of experiment at various moments of time means that all time points are equivalent, i.e. physical laws act in the same way in all these points. Thus we introduce the ontological principle of time homogeneousness, connected with the postulate of unchangeability of physical laws. But it means that, investigating natural processes, physics abstracts from the idea of evolution and considers physical world as out of its historical development (development stipulates formation of qualitatively different levels of the world organization and corresponding laws in time, so that every new level appears on base of those previously formed, then exercises reverse influence upon them, transforms them. Thus in the process of development not only new laws of functioning of objects appear, but also previously formed laws might be transformed under new ties over them).

Here we come across one of the most important specificities of the principles of measuring. Their system introduces an idealized and quite general scheme of experimental and measuring procedures, by means of which essential features of studied reality are revealed. But along with this scheme, or, more accurately, in accordance with it, notions of the physical picture of the world are created.

Principles and postulates of measuring play the role of concretization (as applied to specificity of physical investigation) of method singularities reflected in ideals and norms of science.

Fundamental postulates of measuring express the most general and profound foundations of this method, which constitute physics as science. The layer of principle concretizing them expresses those singularities of the method, which are characteristic for certain stages of development of physics and can be reconsidered at other stages of development, while new types of objects enter the sphere of physical cognition.

Since correlatively to the method scheme, physics introduces notions of physical processes expressed in the picture the world, then the physicist, accepting such and such principles of measuring, tacitly accepts a set of ontological postulates. From these positions it is clear that analysis and reconsideration of the measuring principles sooner or later cause revision of ontological schemes, accepted by physics at a corresponding stage of its historical evolution.

Usually classical physics did not carry out such analysis in an explicit form. Changes introduced into the picture of the world played the role of hypotheses, which then underwent long experimental verification. The process of such verification and development of the picture of the world on this base, gradually and concealed from the investigator, correlated ontological postulates with the measuring scheme which, tacitly lay in foundation of corresponding experiments.

Modern physics has developed a different way of theoretical assimilation of objects. Principles of the picture of the world are introduced so, that their correlation with the method scheme is fixed in an explicit form. Such approach means that reconstruction of the picture of the world starts with explication and analysis of experimental and measuring activity.

The new approach not at once strengthened itself in physics. In the period of formation of the special relativity it had its initial, yet immature form. Only after the special relativity and the general relativity had been constructed, and especially in the period of formation of quantum mechanics, when Bohr had developed the conception of complementarity, the new approach to construction of conceptual core of physical theory got its clear shape.

The mentioned approach became firmly established as the new ideal justification of theory which, conserving the requirement of empirical justification of its fundamental notions, stipulated special (different from the one of classical physics) interpretation of that requirement.

Becoming and development of the new ideal was accompanied by formation of corresponding norms and their explication as methodological principles. The "observability" principle was just such explication. As the modern ideal justification of theory developed, the contents of the "observability" principle developed as well. First it appeared as "semi-classical" principle, since it could be coordinated with traditional interpretation of classical notions as taken directly from experiment. Later, when construction of the general relativity revealed specificity of becoming the conceptual core of physical theory, Einstein corrected the "observability" principle taking into account the idea of impossibility to deduce theory inductively directly from experimental data.

Development of the "observability" principle was connected not only with refinement of the sphere and features of its action at various stages of theoretical search, but also with detection of its connections with other normative principles of physics. In particular, application of the observability principle in investigation practice stipulates analysis of presumptions and principles of measuring, according to which fundamental constructs should be introduced into theory. In the course of such analysis the meaning of the observability principle displayed itself through explication of a whole system of normative principles, through which measuring postulates were formulated.

Thus, working out methodological principles, expressing new norms of scientific cognition, is not a separate act, but a quite complicated process, developing and concretizing the initial contents of methodological principles. In the starting phase they may not act as an alternative to traditional investigation method; only in the course of development they – more and more evidently – are presented as opposition to the old style of thought.

The condition of working out new normative principles, which change the strategy of theoretical search, is origin of new scientific ideals inside the old method. These ideals express new understanding of the investigation goals, while norms, formed on their base, indicate ways to reach such goals.

Just as investigation ideals and norms are included in culture thanks to philosophical justification, so forming of new ideals and norms often stipulates reconstruction of old and elaborating new philosophical justifications of science.

In this respect it is indicative that the non-classical way of theoretical investigation and modern ideal empirical justification of theory to a large extent were the result of comprehension and philosophical analysis of processes which were revealed in classical natural science due to acceleration of scientific development.

Philosophical premises of reconstruction of foundation of science

Practically one generation of scientists of the 19th century carried out a quite radical reconstruction of the natural scientific picture of the world. First, due to rejection of conception of weightless substances, such as caloric, electric and magnetic fluids, the prevailing in physics mechanistic picture of the world was altered. Then it was transformed into electrodynamic one. Not only notions of "substrate" of physical processes changed (from vast family of weightless substances there was only world ether left). Also views on the nature of physical interaction changed: the short-range action principle gradually replaced old ideas of momentary transmission of forces in vacuum, different kind of forces became now regarded as turning one into another.

Analogous processes of reconstruction of vision of reality took place in other sciences close to physics. Ideas of phlogiston and various biological fluids as special substances – carriers of "chemical and biological forces" – were eliminated from the scientific picture of the world. Connections between physics and chemistry were built on atomistic views. Gradually chemical processes started being regarded as foundation of biological phenomena. In biology the picture of evolution of living organisms was forming; once and for all was it established after Darwin's theory had been created and caused radical displacements in natural scientific picture of the world.

All the processes of revision of "ontological postulates" of natural science, which took place within a relatively short period of science evolution, revealed a number of important features of forming of scientific theory. It became clear that the same laws can be expressed through different notions, and alternative systems of theoretical postulates can (till a certain moment) rest upon the same experimental facts and serve as base to formulating laws explaining these facts. For instance, phenomenological thermodynamics, based on the conception of thermogen, successfully explained and predicted lots of empirically fixed phenomena. Transition to a molecular-kinetic heat theory provided a different explanation of the same phenomena. Mathematical expressions of laws in many cases were preserved and passed to the new theory, though there they got new interpretation.

In electrodynamics long competition of alternative investigation programs (Ampere-Weber's on the one hand, Faraday-Maxwell's, on the other hand) demonstrated that different formulations of electricity and magnetism laws are possible. Victory of Faraday-Maxwell's field conception did not mean that laws formulated grounded on ontological postulates of Ampere-Weber's program (Coulomb's, Ampere's, Biot-Savart's laws) were inadequate to the studied regularities of the physical world.

Thus, natural science put forward the problem of selection and justification of ontological postulates, on base of which the investigation is carried out. One of the most important of its aspects was the question of ontological status of fundamental abstractions, embedded in the foundation of the picture of the world. Many of such abstractions, which previously had been considered as adequate copies of fragments of the objective world, lost their ontological status and appeared as hypostased objects. The fate of phlogiston, thermogen, electric and magnetic fluids was quite significant evidence of this process. It is characteristic that rejection of substantialization of various "force types" gave rise to a quite radical program of reconstruction of the physical picture of the world. Many physicists of the late 19th century (including most respectable ones) start expressing doubts in ontologization of the notion of force, traditionally included into the physical picture of the world as one of its most important components. Kirchhoff suggested to exclude force from fundamental notions of physics, keeping it only as a derived, auxiliary notion. Hertz deliberately patterned his work on this program constructing his mechanics.

The discussion of the problem of choice of theoretical postulates and justification of the basic notion of science was stimulated not only by revolutions in the 19th century natural science, but also by progress in mathematics of that historical period. Discovery of non-Euclidean geometries and further application of axiomatic method in its formal and formalized variations in mathematics revealed that visualization criteria were not enough for selection of axioms of theory. There emerged the urgent problem of existence of postulated mathematical objects.

Science of the 19th century considerably accelerated its development in comparison with previous epoch, so it more and more often faced situations, where ideals of classical natural science revealed their scantiness. These ideals were formed in the culture of the 16th–17th centuries and remained domineering during Enlightenment, expressing orientation of cognition at active comprehension of the world. That comprehension was interpreted as once and for all the given natural ability of human mind to reproduce essence of things, basing on experience, and see action of natural laws in the experimental data.

Relatively slow development of technogenic civilization at early stages of its history (in comparison with latest stages) in classical science did not cause often reconstructions of its foundations[14] (foundations of scientific search, formed in the epoch of scientific revolution of the 17th century, were permanently translated until the 19th century).

Because of this, it was natural that cognition of formed ideals and norms were perceived as expression of the very nature of human thinking activity, which was regarded as basically invariable. Till dynamics of social development revealed dependence of thinking on social conditions in which it is taking place, science and philosophy had not especially considered social determinations of human cognition, had not uncovered those social premises which outline features of intellectual assumption of the world in every historical epoch. Active-practical nature of scientific cognition, dependence of notions (worked out by science) of objects on operational structures, worldview factors and values of the corresponding epoch rested in shade and were not object to reflection in science and philosophy of the classical period.

From that followed interpretation of ideals and norms of explanation and justification of knowledge, characteristic for that period. Ideals and norms, constituting science as a specific form of cognition (such as requirement of objectivity and concreteness of knowledge, empirical verification etc.), were tied with their unique interpretation. It was believed that objectivity and concreteness of theory are reached in case that in "firmly proved" facts speculation reveals essential characteristics and eliminates from the explanation all things referring to the subject and procedures of its activity. Justification of theoretical knowledge was regarded as ability to explain and predict facts and base on self-evident principles, taken from experience.

Classical philosophy of the 17th – first half of the 19th centuries set these natural science statements for expression of the very nature of existence and thinking. A special role in the system of justification of these statements as "the only possible" scientific ideals and norms belonged to mechanistic materialism, whose principles were methodological base for natural science investigation up to the beginning of the 19th century. Only accelerated development of science (especially after the first industrial revolution) drew investigators to new evaluation of ideals and norms of classical natural science. The role of hypothesis in theoretical investigation was distinctly shaped, more and more often there appeared situations when various theoretical explanations referred to the same sphere of experimental facts; it became clear that criteria of experimental verification and self-evidence were insufficient for justification of created theory postulates.

Need of critical look at ideals and norms, established in classical natural science, was first grasped and thought over in philosophy. Even within classical philosophical tradition, on the eve of the 19th century, Kant brought up the problem of premises of cognitive activity and foundations of natural science. More and more clearly philosophers understood active nature of cognition and historical development of its categorical structures (Fichte, Hegel). Finally, transition from the classical stage of philosophy development to the modern one and revision (begun in the middle of the 19th century) of attitudes of "classical philosophy", prevailing from the 17th to the 19th century,[15] gradually uncovered involvement of the cognizing mind into historically formed and historically changing structures of social life. Philosophers raised the problems of social determination of cognition and, as one of its aspects, the question of historicity of profound foundations and premises of scientific investigation.

That was the epoch of becoming of non-classical rationality, when, beside the classical paradigm of sovereign reason, as if from the outside apprehending the world, there emerged alternative approach to understanding of the cognizing subject. In the new paradigm he is regarded as submerged into the world, acting inside it and comprehending objects depending on how historically determined states of human life provide inclusion of objects into human cognitive activity.

Understanding of the roots of consciousness in the structures of human existence and its dependence on these structures was expressed in many philosophical ideas of the second half of the 19th – early 20th centuries (Marx, Cassirer, Rickert, Windelband, Weber, Nietzsche, Freud et al.).

In the philosophy of science these ideas were first of all expressed in intensive discussion of the problems of scientific ontology. Traditional identification of fundamental abstractions of science and reality was now opposed by criticism of such identification, based on experience of science historical analysis. E. Mach, P. Duhem, H. Poincaré quite clearly fixed historical relativity of the applied in science principles and notions of reality and presence of hypostased objects in the system of notions – abstractions like phlogiston or thermogen, which illegitimately had the status of really existing substances.

The central place in elaboration of philosophical questions of science in the last third of the 19th century belonged to search for methods of justification of fundamental scientific abstractions and criteria, according to which they are to be included in the structure of scientific knowledge.

Some important aspects of these problems were developed by conventionalism and empirical criticism, which exerted direct influence upon Einstein's works. Rational moments of conventionalism were connected with raising the problem of out-of-science criteria of acceptance of various ontological postulates. Though, conventionalism only brought out this problem. Emphasizing relative character of ontological postulates of science, it paid little attention to succession in development of their contents, and did not carry out its analysis to investigation of mechanisms, through which various fundamental scientific notions enter the culture and, consequently, the scientific community can reach agreement on their ontological status.

Empirical criticism concentrated on another idea: empirical justification of scientific ontology. It believed that reduction of fundamental scientific abstractions to observation could serve as a criterion for separating constructive scientific abstractions and hypostased objects.

A. Einstein, in his search for solution of electrodynamics paradoxes, used some of these ideas and approaches. But he did not borrow them as they were; he distinguished their constructive moments and reconsidered them in accordance with the new situations in the development of physics.[16]

It is a matter of principle to stress once more, that up to the late 19th century in the sphere of philosophical cognition investigators had worked out necessary means, which let them realize critical analysis of the situation in science, when its further progress stipulated revision of its "classical" ideals and norms of investigation.

It was the transgression to the sphere of philosophical means and their application in problem situations of natural science, that let scientists transform ideal explanations and justification of knowledge, affirming new method of construction of the picture of the world and fundamental scientific theories connected with it.

Epistemological platform, where Einstein solved methodological problems of physics, appeared as result of creative comprehension of vast historical-scientific and historical-philosophical material (analysis of the history of Copernicus's, Galileo's, Descartes', Newton's discoveries, critical comprehension of Kant's, Mach's et al. conceptions).

We are to emphasize Einstein's critical considerations of Mach's philosophical views. Einstein, as well as many natural scientists, had been under influence of Mach's criticism of philosophical foundations of classical natural science, philosophy of metaphysical materialism in particular, which, in the late 19th century, more and more clearly demonstrated its disparity to requirements and needs of science. Critical impulse of Mach's works, aimed against methodology of mechanistic materialism, included rational moments, for instance, criticism of hypostased objects and requirement to eliminate them from foundation of physical theories, as they are not based on experiment.

But Mach's interpretation of this requirement in the tradition of understanding experiment as observer's sensations,[17] and phenomenalist interpretation of theory led to throwing off, along with ideals of classical science, its aim at working out object knowledge. And that meant elimination of the contents, which formed stable core of ideals of scientific approach in all historical ages and essentially characterized the very specificity of scientific cognition.

Also we are to mention the fact that Mach, in his criticism of ideals of classical natural science, failed to overcome some essential unilateralities of classical conceptions. In particular: interpretation of notions and principles of physics, traditional for classical style of thinking, was not only preserved in Mach's philosophy, but obtained hypertrophied features; theoretical notions were considered as fundamentally reducible to the data of observation.

Such approach (expressed in Mach's interpretation of the "observability" principle) quite soon displays its destructive function in science; from these positions Mach opposed to ideas of atomism, finding it possible to exclude the idea of atom from the physical picture of the world.

Einstein, like the majority of other natural scientists, rejected Mach's views of the kind. In the positivist philosophy he distinguished only moments which could be used in the process of reconstruction of cognitive attitudes of classical natural science and working out new investigation normatives. In Mach's criticism of the ideals of classical natural science, Einstein found a rational for the idea of operational control over foundations of theory. But, unlike Mach, he separated historically transient contents of ideals of classical natural science from a part of those ideals – characteristics expressing specificity of science and distinguishing it among other forms of cognitive activity (objectiveness, system approach and justifiability of scientific knowledge, intention of theoretical investigation to reproduce regularities of the reality studied). Therefore operational control over theoretical foundations, in Einstein's interpretation, appeared as one of the conditions of objectivity of theory, as method providing adequate selection of its notions and principles and reproduction of essential characteristics of the reality studied in theory.

In the light of all said above we would estimate as simplifies Holton's opinion: Einstein, in the early period of his work (including the stage of creation of the special relativity) stood on "positivist ground", and only when the general relativity was being created, he drifted away from Mach's positions and started criticizing more and more radically the positivist methodology.[18]

Though Holton mentions the facts contradicting his conception, but he regards them as exceptions to the rule. But, to our mind, these facts are so important for characterizing Einstein's methodology, that they should be regarded more as a rule than as an exception. Holton gives extracts from articles "On Electrodynamics of Moving Bodies", "Autobiographic Notes" and Einstein's letter to M. Grossman of April 14, 1901 as evidences of the fact that even in the period of creation of the special relativity Einstein was not an "orthodox machist", but proceeded from acceptation of objective reality, not reducible to physical events (phenomena discovered in experiment), but including physical laws which "themselves will be seen to be built into the event-world as the undergirding structure "governing" the pattern of events".[19]

We can also add one more quite important evidence in favour of radically different (from Mach's) Einstein's understanding of ways of construction of physical theory and ideals of its justification. In his "Autobiographic Notes" Einstein emphasized that just before creation of the special relativity he had paid attention to the construction method of classical electrodynamics, which included energy conservation and entropy growth principles as equivalent to statements of impossibility of perpetual motion of the first and the second type. Evaluating the situation in electrodynamics of moving bodies, Einstein saw the solution of the difficulties raised in application of method analogous to thermodynamics construction method, i.e. in finding a generalizing principle like the one given in the statement: the laws of nature say that it is impossible to create perpetual motion machine (of the first and the second type).[20] Here we can easily see embryonic forms of that ideal of theory justification and construction method which has been settled in modern physics and stipulates that fundamental ontological postulates of theory should be introduced correlatively to the scheme of practice, which allows us to find characteristics of the reality studied defined in postulates.

Einstein moved to refine this ideal through selection of rational moments, which were included in the known variations of philosophical criticism of the ideals of classical natural science. But that selection itself took place from the positions of philosophical directions, which formed stable basis of natural scientific investigation (from positions of conviction in objective existence of the nature and its laws and in ability of theoretical investigation to express these laws).

It was those aspects of Einstein's philosophical and methodological orientation that provided his success in search for new ideal justification of theory and working out of a normative which would correspond to this ideal (adequate interpretation of the observability principle). In Einstein's works, this principle was closely connected with another one, the invariance principle, which became a most important methodological rule in science of the 20th century.

Invariance in a general sense is the property of a system to conserve some relations, essential for it, in certain transformations. Transformations (operations), carried out by the cognizing subject upon the system studied, are expression of the connection between subject and object through activity. In this sense the invariance principle, orienting us at discovering essential relations and connections which conserve in transformations of the system (as well as adequately understood "observability" principle), appears as expression of non-classical approach to cognition. This approach rejects the idea of parallelism between being and thinking, which in the classical era was believed to be a condition of adequate comprehension of the world. It is based on alternative idea that between reason and objects cognized there is always a special mediator – human activity, and development of its means and methods determines the character of things, uncovered and initially understood by humanity in the surrounding world.

Close connection between the "observability" and invariance principles is stipulated by common activity attitude: to discover regular connections and essential relations through clear stipulation of the operation system in which they are expressed.

The ideas of invariance first were developed in mathematics and then spread to other spheres of scientific investigation. In 1882 F. Klein, a well known mathematician, formulated the famous Ehrlangen program (at that time F. Klein worked at the university of the German town Ehrlangen), aimed at construction of generalized geometry. The program induced the investigation strategy to search for invariants in a certain group of transformations of mathematical objects.

The success of the invariant methods in mathematics stimulated its translation to other disciplines. It seems most interesting that one of the disciplines which first assimilated was not sciences, but humanities: linguistics.

In the late 19th century the so-called linguistic avant-garde (I. A. Baidouin de Courtenay, N. Kruszewski, F. de Saussure) considered language as an integral, variable system, and concentrated on search for invariant entities in language variations.[21] One of the first works, which realized that principle, was the investigation, made by the Swiss linguist I. Winteller. He considered language as a system of elements, where we should distinguish variable and invariant (stable) qualities. The method of search for essential characteristics through uncovering invariants in the language system of variation properties was named by Winteller "configurational relativity" principle.[22]

Winteller's ideas had direct influence upon A. Einstein's works. While studying in Switzerland, young Einstein met Winteller and visited his seminars, which became an important fragment of his biography.

Later, when Einstein joined the works on problems of electrodynamics of moving bodies, he used the ideas of invariance as a basic principle in construction of his theory.

"Observability" and invariance determined new features of the ideal of theory and its ontological postulates. From the point of view of this ideal, new light was shed on the process of forming the physical picture of the world as a disciplinary ontology. Justification of its notions now stipulated explication of its operational structures, the system of which should reveal fundamental essential characteristics of the nature.

This was the way to outline the "method network" which allowed us to justified characteristics of objects studied introduced in fundamental notions of physics. That was the way, which led to creation of the special relativity.

From methodological ideas to theory and a new picture of the world

The first step to the special relativity was fixing the relativity principle as one of the most important operational foundations, correlatively to which we should introduce ontological notions into the foundation of physical cognition.

This interpretation of the relativity principle was first outlined by Poincaré, but Einstein shaped it in the brightest way.

Einstein regarded the relativity principle in two aspects.

The first aspect of regarding the relativity principle characterizes it as methodological regulative of theoretical description of the reality. In the language of such description a physical laboratory, which is moving uniformly and rectilinearly, is marked as an inertial frame of reference, and according to the relativity principle, the laws of nature do not depend on movement of the frame of reference.[23] In theoretical description, physics uses the language of mathematics. In this language, the frame of reference is characterized as a system of coordinates, and the laws of nature are expressed as equations, in which physical magnitudes are connected in a certain way. Independence of the natural laws from motion of the frame of reference can be formulated as requirement of covariance of the corresponding equations relative to transformation of the coordinate systems (in transition from one inertial system to another).

The other aspect presented the relativity principle as profound postulate of experimental and measuring activity. In this aspect the formulation of the relativity states that physical processes are going on similarly in all laboratories moving uniformly and rectilinearly, and so no experiments executed inside a physical laboratory can discover its inertial motion.

The very existence of physics as science stipulates that experiments and measurements can be reproduced. This presumption of physical investigation is concretized not only by principles of reproducibility of experiments in various "points" of space and at various "moments" of time (as we have indicated above), but also by principles which fixed the influence of the laboratory's motion upon physical processes.

Physical laboratories are always connected with moving bodies, and the problem of reproducibility of experiments and measurements requires that this condition should be taken into account. If there are situations where motion of the laboratory introduces perturbations into the process, we are to find a way to take these perturbing influences into consideration. For this we have to select t some standard situation, where relative motion of two laboratories would not change the picture of the process studied. Deviations from this situation can be regarded as perturbations which fundamentally can be revealed and taken into account (fixation of and control over such perturbations are possible only in case when we know the situation free of them). Classical physics, since the very beginning of its existence, considered inertial motion as a standard situation.

Such approach has quite deep foundations (though they were not always understood in classical natural science). The problem is that experimental investigation of a physical process requires that it should be considered in the "purest" possible form. To do this, we are either to isolate the laboratory from outer influences, which can affect the process studied, distort or darken it, or to compensate such influences. In the extreme case, when our laboratory is completely isolated from outside influence, we have an idealized laboratory, which, by definition, is an inertial frame of reference (not influenced by outer forces).

Experimental and measuring activity in physics stipulates that it should always be possible to find a situation when motion of a real laboratory may be, to some tolerance, considered as inertial. In each of such (local-inertial) laboratories, with other conditions being equal, all processes would go on in the same way (no experiment would reveal its relative movement), and so the results of the experiments would be reproducible. As natural processes occur in accordance with objective laws, the possibility to reproduce one process in various inertial moving laboratories means that laws of nature do not depend on inertial motion of the frame of reference.

The relativity principle reflects that very meaning and, correspondingly, appears as the formulation of a quite important assumption, which lays in the foundation of physical experimental and measuring procedures. These assumptions let us concretize profound postulates (presumptions) of physical investigation: reproducibility of physical experiment, its subordination to laws of the nature and possibility to separate expression of the laws by means of different experiments.

The formulation of the relativity principle as a regulative of theoretical investigation (requirements of covariance of equations) plays the role of the invariance idea, expressing operational sense of the principle as applied to peculiarities of theoretical description of the natural laws. Exactly this understanding of the relativity principle, when it represents expression of fundamental features of physical investigation method, providing adequacy of cognition of its objects, was characteristic for Einstein's approach to analyze physical problems.[24]

Interpreting the relativity principle as the most important component of the method scheme revealing characteristics of the physical world, Einstein formulates the problem of ontological postulates of physics in an unusual form (from the classical point of view): how physical reality will look like (what physical picture of the world we will get), if the relativity principle spreads to description of any interactions (including electromagnetic ones).[25]

Accomplishing that program, Einstein analyzed the ontological postulates of the late 19th century physics, which constituted the electrodynamic picture of the world. That was the second step toward the special relativity.

The analysis showed that the postulate of the world ether, filling the absolute space, is incompatible with the relativity principle, as it leads to a different description of electromagnetic processes in different inertial frames of reference. It meant that the world ether was a fundamentally unobservable object, as it could not keep within the scheme of experimental and measuring procedures of physics.

Let us emphasize this important fact especially. Eliminating notions of the world ether as a substance passing electromagnetic interactions, from the physical picture of the world, is usually associated with Michelson's, Fizeau's and other experiments, which did not discover the Earth's movement relative to ether. In his numerous accounts of the special relativity, Einstein also uses these arguments. But in his first work "On Electrodynamics of Moving Bodies" which contains all basic ideas of the new theory, Einstein only casually mentions of failed attempts to "reveal the Earth's movement" relative to "light-carrying medium", but never mentions Michelson's experiment.[26] What is more, in one of his letters he said that in the construction of the special relativity Michelson's experiment did not have a decisive role (this fact was accurately analyzed by Holton, whose analysis confirmed Einstein's correctness in the mentioned statement).[27]

Within the framework of our reconstruction of settling the special relativity, we can easily find a natural explanation to the fact mentioned. To qualify the postulate of the world ether as not corresponding to the "observability" principle, it was not necessary to refer to results of concrete experiments, like Michelson's (though those experiments themselves could confirm "unobservability" of the ether). It was important that investigators should disclose the structure of experimental and measuring practice and show that it fundamentally cannot fix such hypothetical object as the world ether. The relativity principle characterized most essential aspects of that structure. That is why contradiction of postulates of the picture of the world and the relativity principle meant that the postulates have no operational justification and are to be revised.

In Einstein's analysis the "observability" and relativity principles were raised not separate, but connected with each other. The former determined general strategy of the investigation, aiming at disclosure and elimination from the picture of the world of those abstractions which did not correspond to postulates of measuring, the latter concretized that strategy. Interpreted as a measuring postulate, it played the role of one of concrete criteria of "observability" of the objects introduced in the picture of the world.

From these positions Einstein criticized not only the notion of ether, but also the postulate of existence of absolute space and time. The latter determined a laboratory, which was in rest relative to absolute space, as a privileged frame of reference different from moving laboratories.

The famous article "On Electrodynamics of Moving Bodies" starts with indication that such approach, based on the idea of "absolutely resting space", gives birth to asymmetry in description of electrodynamic events improper to such events as they are. Einstein mentioned that such asymmetry contradicts to the relativity principle. Only after that did he point at unsuccessful attempts to find experimentally the Earth's movement relative to the ether, and interpreted these facts as inefficiency of the absolute space conception.[28]The first stimulus to revision of that conception was the desire to eliminate its inadequacy to the relativity principle.

So, new ideals of the theory justification and corresponding new normatives of physical investigation (the "observability" principle and the relativity principle concretizing it) directed reconstruction of the physical picture of the world and stimulated construction of new fundamental physical theory.

After the "weak points" of the electrodynamic picture of the world had been revealed, there appeared new problems. Elimination of the ideas of ether and absolute space destroyed the previous picture of physical reality, which served as base for Maxwell–Lorentz's electrodynamics. Consequently, it became necessary to understand how it would influence the electrodynamics of moving bodies and whether the change in the signs of ether, absolute space and absolute time will lead to destruction of constructs of the theoretical scheme lying in the foundation of classical electrodynamics (vectors of electric and magnetic fields, vector of density of charge-current, inertial frame of reference), as signs of these constructs were connected with signs of objects of the electrodynamic picture of the world.

The kind of analysis generated the foundation of the second (after the relativity principle) fundamental principle of the special relativity – the postulate of constancy of the light speed.

In Lorentz's theory ether had one important physical quality: regardless to motion or rest of the radiating body, the light ray is spread in a system, resting relative to ether, with universal speed c.[29] So if elimination of ether could not destroy classical electrodynamics, it had to be postulated that the light ray is spread in vacuum with universal speed, regardless motion of the source and in absence of ether. The fact of independence of the light speed from the speed of the source was well known in optics. Einstein gave this fact status of a fundamental principle of the new theory. Since according to the relativity principle all inertial frames of reference are physically equivalent, it follows that the constancy of the light speed principle is true for any frame of reference.[30] This postulate included specific contents and in this respect did not depend on the relativity principle. But the latter allowed justifying universality of the constancy of the light speed principle, which appeared the third important step in construction of the special relativity.

The fourth, decisive step consisted in analysis of measuring procedures justifying properties of space and time. In accordance with ideal operational justification of postulates of the theory, Einstein thoroughly analyzed the procedures of measuring space and time intervals. He disclosed the scheme of these procedures; he demonstrated that they are based on operations with a hard core of inertial frame of reference and its clock, synchronized by means of light signals.[31] The role of these procedures in construction of the relativity theory is analyzed completely enough in the literature dedicated to methodology and history of physics. But authors not always emphasize the fact that analysis of the scheme of measuring space and time intervals induced Einstein Lorentz's transformations (this conclusion is expressed in Einstein's work "On Electrodynamics of Moving Bodies").

Such conclusion gave real physical sense to Lorentz's transformations and their consequences. Just as characteristics of space and time intervals, which follow from Lorentz's transformations, are justified by the scheme of measuring which disclosed space and time properties and relations of natural objects, so these characteristics were to be considered as reflection of features of space-time of the nature itself.

This key moment of construction of the relativity theory is worth special attention. In the 1960s – 1980s methodological and historical scientific literature discussed: was A. Einstein the only creator of the relativity theory, or it was created by, at least, H. Lorentz, H. Poincaré and A. Einstein. Historians of science are still arguing on this matter. Really, analyzing Lorentz's and Poincaré's works, we can see that it was Lorentz who discovered the main mathematical contents of the theory: transformation of space-time coordinates, which provided covariance of laws of both mechanics and electrodynamics. Poincaré was the first one who clearly formulated the relativity principle for inertial motion,[32] having expressed the idea of relativity of spatial location and relativity of simultaneity.[33] His works also contain the principle of constancy of the light speed as a basic principle of physical measuring.

Thus, we may conclude that axiomatic base of the relativity theory was created before Einstein and independently from his works. A.A. Tyapkin in the afterword to a book of collected articles "The Relativity Principle" (1973) says: "Poincaré, the most prominent mathematician of that time, made a decisive contribution into discovery of physical principles of the relativistic theory".[34] We could agree with all this, if it had not been for quite a considerable objection: any hypotheses and justifications of the basic postulates still do not give a theory, and its main meaning is not simply deduced from the axioms. We may accept that Poincaré was really close to creation of the special relativity, but still he did not make the last, probably, the most important step. He did not prove that the corollary of Lorentz's transformations of relativity of space and time intervals and Lorentz's transformations themselves have real physical meaning, they are characteristics of real, not fictitious space and time. In Poincaré's works there is no justification of physical interpretation of Lorentz's transformations, out of which his hypothetical ideas referring to their semantic interpretation remained nothing more than hypothetical ideas.

Also ill-founded are statements that Lorentz can be regarded as co-author of the relativity theory. The discussion, whether Einstein knew Lorentz's article of 1904, where the latter demonstrated new transformations of space-time coordinates, does not solve the question. The main thing is that Lorentz regarded his transformations as mathematical form, which does not require radical change of classical notions of space and time, but preserves them. “Local time”, in Lorentz’s works, is fictitious, not real physical time. What Lorentz believed to be fictitious space and time, Einstein presented as real physical space and time, as it followed as a corollary from analysis of idealized measurements accumulating essential features of real physical experiment.

If we describe all these cognitive procedures in terms of modern methodological analysis, it can be said that Einstein carried out an operation of constructive justification of new hypothetical properties of space-time intervals, properties followed from Lorentz’s transformations. And that operation, which connected corresponding magnitudes with experiment and thus introduced Lorentz’s transformations as having empirical interpretation – that cognitive operation was executed by Einstein. It was that lacking link, which tied all separate inlaid suggestions, principles and mathematical expressions in an integral system of a new physical theory. Only when Lorentz’s transformations were connected with experiment, it was right to regard all main consequences physically correct (the law of composition of speeds, the law of mass change with speed change, relation between mass and energy etc.).These consequences were also inferred and justified by A. Einstein.

That is why A. Einstein is named the creator of the special relativity. G.Holton is absolutely right saying that “Lorentz’s article, essentially, does not interpret the relativity theory in the way we understand it after Einstein”.

Thus, the decisive step, which determined the contents of the theory, was made only by Einstein and not by his contemporaries who examined the problems of electrodynamics of moving bodies. It was Einstein who deduced Lorentz’s transformation not from requirement of covariance of the equations, but on base of analysis of local procedure of synchronizing clocks. Poincaré pointed at the importance of such procedure but did not demonstrate how Lorentz’s transformation could be deduced from that.

In the respect of methodology it is especially important to emphasize that Einstein’s approach to justification of hypotheses connected with new space-time transformations was that very method which determined a certain watershed between classical and non-classical construction of a physical theory.

In its evident form, the procedure of constructive verification of new abstract objects, appearing at the stage of hypothesis, was applied only in non-classical investigations.

It can be found, for instance, in the history of quantum mechanics, when the famous uncertainty relations, deducible as a corollary, in principle, from commutativity relations, applied in the mathematical apparatus of the theory, were obtained by Heisenberg on base of the famous theoretical experiment on observation positions of electrons by means of ideal microscope (Heisenberg demonstrated that electron interaction with a light quantum does not allow us to determine its coordinate and impulse at the same time to any accuracy).The same strategy founded Bohr-Rosenfeld procedures in quantum electrodynamics.

Only when magnitudes and their main features, introduced “from above” based on mathematical hypothesis, are confirmed in a system of thought experiments, accumulating real peculiarities of the experiment, they can be supplied with real physical meaning. Then it would be correct to compare their new properties to constructs of the physical picture of the world and, correspondingly, to return a verdict on truthfulness of certain traditional ideas of physical reality. This was the way of development of the relativity theory, after Einstein had introduced the new interpretation of Lorentz’s transformations. Notions of world ether and absolute space and time were eliminated from the physical picture of the world. They were replaced by relativistic notions. Though there still was no integral idea of space-time, transition to it could already be seen. Though the new understanding of space and time, included into the physical picture of the world, contradicted to stereotypes of everyday common sense, it quite soon was accepted by the scientific community and echoed in other spheres of culture.

The European culture of the late 19th – early 20th centuries was prepared – by all its preceding development – to acceptance of the new ideas in the course of non-classical type of rationality. We may point out not only certain call-over between ideas of Einstein's relativity theory and conceptions of "linguistical avant-garde" of the 1870s–1880s (J. Winteller et al.), but also their resonance with the forming of a new artistic concept of the world in impressionism and post-impressionism, as well as new (for the literature of the last third of the 19th century) ways of description and comprehension of human situations (for instance, in Dostoevsky's works), when the author's consciousness, his spiritual world and his worldview do not oppose spiritual worlds of his heroes, describing them as if from outside, from some absolute frame of reference, but co-exist with these worlds and enter an equal dialogue with them.[35]

This specific resonance of ideas, developed in different spheres of cultural creative work in the late 19th – early 20th centuries, uncovered profound worldview foundations, which were ground for non-classical science and which were developed with its active assistance. New worldview meanings, gradually striking roots in the culture of technogenic civilization at that period, in a considerable part provided ontologization of notions of space and time, unusual for common sense, which had been introduced by Einstein into the physical picture of the world.

Further development of these notions was connected with works of H. Minkowski, who completed a new mathematical form for the special relativity and supplied the physical picture of the world with an integral image of space-time continuum, characterized by absolute space-time intervals along with relativity of their division into space and time intervals in every inertial frame of reference.

In physics settling of the new picture of the reality studied was accompanied by philosophical and methodological discussions, where new notions of space and time and new methods of construction of a theory were thought over and justified. In the course of such analysis investigators corrected and developed philosophical premises, which provided reconstruction of classical ideals and norms of investigation and the electrodynamic picture of the world. So they were being turned into philosophical foundations of relativistic physics, considerably contributing to its integration into the texture of modern culture.

Thus, reconstruction of the foundations of science was not an act of immediate change of paradigm (as T. Kuhn wrote), but a process which started long before direct transformation of investigation norms and the scientific picture of the world. The initial phase of this process was philosophical comprehension of tendencies of scientific development, reflection over the foundations of culture and movement in the field of the very philosophical problems, which allowed philosophy to see shapes of the future ideals of scientific cognition and define categorical structures which would found the construction of new scientific pictures of the world.

All these premises and "sketches" of future foundations of scientific search are concretized and then finished off in the process of methodological analysis of problem situations in science. In the course of such analysis investigators correct justification of the new ideals of science and form corresponding normatives, which direct construction of the core of new theory and the new scientific picture of the world.

But even at this stage forming of new foundations of scientific search is not over yet. Real investigation practice can introduce corrections into methodological attitudes worked out preliminarily. It is indicative, for instance, that the idea of operational control over notions and principles of theory Einstein understood at early stages more in the spirit of classical ideas of theoretical investigation ways (when principles are regarded as the result of direct generalization of experiment). But after the special relativity had been constructed, it became clear that such understanding is inadequate, since in construction of the special relativity there already existed previously created mathematical structure (Lorentz's transformations) and its hypothetical interpretation (relativity of space and time intervals), which highlighted the area of experiment necessary for its justification and so oriented investigators to application of corresponding operational structures in their theoretical search.

Reflection over theory already constructed usually leads to correction and development of methodological attitudes, to more adequate understanding of new ideals and norms, reflected in corresponding theoretical models. That is why reconstruction of science includes not only initial, but also final stage of becoming of new fundamental theory, stipulating numerous transitions from the sphere of special scientific analysis to the sphere of philosophical and methodological speculation.

SCIENTIFIC REVOLUTIONS AND INTERDISCIPLINARY INTERACTIONS

Scientific revolutions are possible not only as the result of intradisciplinary development, when the investigation sphere absorbs new types of objects, assimilation of which requires that foundations of the scientific discipline should be changed. They are also possible due to interdisciplinary interactions, based on "paradigmatic grafting" – transfer of notions of the special scientific picture of the world, as well as investigation ideals and norms, from one scientific discipline to another. Such transplantations are able to cause transformation of the foundation of science without paradoxes and crises connected with its inner development. The new picture of the reality studied (disciplinary ontology) and new investigation norms, emerging as the result of paradigmatic grafting, discover another field of scientific problems, different from the one which existed previously, stimulate discoveries of phenomena and laws, which were completely out of the sphere of scientific search before "paradigmatic grafting".

Generally speaking, this way of scientific revolutions has not been analyzed deeply enough neither by T. Kuhn, nor by other investigators in the Western philosophy of science.

Still it is the key for understanding the processes of appearance and development of many scientific disciplines. Moreover, without taking into account features, based on paradigmatic transplantations, we cannot understand that great scientific revolution which was connected with forming of disciplinarily organized science.

The majority of sciences, which are now considered as classical disciplines – biology, chemistry, technical and social studies – date back to ancient times. Historical development of knowledge accumulated facts about separate features of objects studied. But for a long time facts were systematized and explained through natural philosophic schemes.

After the first theoretically formed sphere of scientific knowledge appeared – physics, and the mechanistic picture of the world received a status of universal scientific ontology, a special stage of history of sciences began. In most of them investigators made efforts to apply principles and ideas of the mechanistic picture of the world to explain facts.

The mechanistic picture of the world, though formed within physical investigation, functioned as both natural scientific and general scientific picture of the world at that historical period. Justified by philosophical attitudes of mechanist materialism, it set reference points not only for physicists, but also for scientists who worked in other spheres of scientific cognition. No surprise that investigation strategies in those spheres were created under direct influence of the ideas of the mechanic picture of the world.

In this respect a quite characteristic example is development of chemistry of that historical period (the 17th – 18th centuries).[36]

In the middle of the 17th century, when chemistry was not yet constituted as independent science, it was either included in the system of alchemic notions, or was presented as set of knowledge used in medicine. The first steps of becoming chemistry as science was, to a great extent, connected with atomic-corpuscular ideas entering chemistry. In the second half of the 17th century R.Boyle put forward a program translating principles and models of explanation, formed in mechanics, into chemistry. Boyle suggested that all chemical phenomena should be explained through notions of movement of "minute particles of matter" (corpuscles). According to Boyle, this way chemistry could allow to separate itself from alchemy and medicine, and transform into independent science. Proceeding from universality of laws of mechanics, Boyle concluded that the principles of mechanics should be also applied the hidden processes taking place between the smallest particles of bodies.[37]

Functioning of the mechanistic picture of the world as an investigation program can be traced not only in interaction of chemistry and physics. Analogous mechanism of development of scientific knowledge can be found also in analysis of relations of physics and biology at the stage of predisciplinary natural science (the 17th –18th centuries).

On the face of it, biology had no such close contacts with physics as chemistry had. But still the mechanistic picture of the world in several situations quite strongly influenced the strategy of biological investigations. In this respect it is interesting to consider investigations made by Lamarck, one of the founders of the idea of biological evolution.

Trying to find natural reasons of development of organisms, Lamarck, to considerable extent, was guided by the principles of explanation taken from mechanics. He based on the 18th century's variation of the mechanist picture of the world, which included the idea of "imponderable" fluids as carriers of various types of forces, and believed that imponderable fluids were sources of organic movements and changes in architectonics of living beings.

The nature, in Lamarck's vision, was a field of permanent motion, transfer and circulation of innumerable fluids, among which the main "stimuli of life" are electric fluid and thermogen.[38]

The development of life, from his point of view, rose as "growing influence of motion of fluids", which made organisms more and more complicated. He wrote: "Who cannot see here the historical motion of organization phenomena, observed in animals considered, who cannot see it in this growing complication in the common row in transfer from the simple to the more complicated"[39]. According to Lamarck, it was the exchange of fluids between environment and organisms, growth of this exchange in strengthening of the organs' functions that induced changes in the latter. Adaptation of organisms to living conditions strengthens functions of some organs and weakens other ones. The corresponding exchange of fluids with environment causes small changes in all organs. In turn, such changes are inherited, and that, in Lamarck's opinion, could lead to quite considerable reconstruction of organs and appearance of new species in case of long accumulation of changes.

As we can see, the explanation used by Lamarck, to a great extent was initiated by principles, translated from the mechanistic picture of the world.

The function of the mechanist picture of the world as investigation program common to all sciences was displayed not only in studies of various natural processes, but also in knowledge of man and society which attempted to form the science of the 18th century. Certainly, consideration of social objects as simple mechanical systems was the greatest simplification. These objects belong to the class of complicated, developing systems including man and his consciousness. They require special investigation methods. But, to elaborate such methods, science had to go a long road of development. In the 18th century there were no objective premises for that yet. At that epoch scientific approach was identified with those samples which were realized in mechanics, and so it seemed natural to build studies of man and society as some kind of social mechanics on base of application of principles of the mechanistic picture of the world.

Quite a characteristic example of such approach is Lamettrie's and Holbach's thoughts about the nature of man and society.

Basing on ideas developed in the mechanistic picture of the world, Lamettrie and Holbach widely used mechanical analogies in explanation of social phenomena and in discussion of problems of man as a natural and social being.

Considering man as, first of all, a part of nature, a special natural body, Lamettrie presented him as a certain type of mechanistic system. He wrote that man can be presented as a "clockwork", but of enormous size and built so skillfully and ingeniously, that if the second wheel stopped, the minute one would gear and work as if nothing had happened. In the same sense, choking up of several vessels is not enough to destroy or stop the action of the lever of all motions in the heart which is the working part of human machine.[40]

Then Lamettrie indicates that human body is a self-winding machine, the main embodiment of continuous motion.[41] At the same time he denoted singularities of this machine and its complexity in comparison with technical devices studied by mechanics. He wrote that man can be regarded as a very smart machine, so complicated that it is absolutely impossible to form a clear idea of it and, consequently, to give it an exact definition.[42]

Expressing his agreement with Lamettrie in understanding man as a machine,[43] Holbach concentrated his attention at the ideas of universality of mechanistic laws, believing it to be possible to describe human society by means of them.

For him man is a product of nature submitted, on the one hand, to the general laws of nature, on the other hand, to special laws.[44]

According to Holbach, man's specific feature is his desire to self-preservation. Here man resists destruction, feels the force of inertia, is drawn towards himself, is attracted to objects alike and repelled by the ones opposite to him. Everything he does and everything that happens in him is consequence of the inertia force, inclination to himself, attraction and repulsion forces, aspiration for self-preservation, in a word, energy he shares with all beings observed.[45]

When Lamettrie and Holbach use the notions of machine, force, inertia, attraction, repulsion to characterize man, we can clearly trace the language of the mechanistic picture of the world, which, during a long period, determined the strategy of nature, man and society. This strategy can be quite easily detected also in later stages of development of knowledge, for instance, in social conceptions built by H. Saint-Simon and Ch. Fourier. In his "Work on Newtonian Attraction" Saint-Simon said that progress of human mind came to the situation when the most important discourses on politics can and should be deduced from knowledge obtained in higher sciences and the sphere of physics.[46]

In Saint-Simon's opinion, the law of gravity was to become basis of new philosophy, which, in turn, could become foundation of new political science. He wrote that the force of European scientists, joined in a common corporation and linked by philosophy, based on the idea of gravity, would be immeasurable.[47]

Saint-Simon thought that the ideas of gravity could become a base for such a discipline as history. He said that history "still is a collection of facts, more or less exactly known, but in future it should become a science, and, as the only science is classical mechanics, history, in its structure, should approximate to celestial mechanics".[48]

Ideas of the same kind can be found in Ch. Fourier's works; he believed that principles and approaches of mechanics allow us to disclose the laws of social movement. He wrote that there existed two types of laws ruling the world. The first one is the law of material gravity, and the priority of its disclosure belongs to Newton. Regarding himself as successor of Newtonian ideas and disseminating the gravity doctrine to social life, Fourier thought that there was the second type of laws, which regulated social movement. He defined them as laws of gravity by passion, which occupied the central place in his conception as decisive property of human nature.[49]

As a matter of fact, here we face a kind of analogy between existence of gravity of natural bodies and people's bent for each other. To great extent, it is done due to the fact that man is considered as a part of nature, though having some distinctions from other natural objects, but still submitting to general principles of motion formulated in mechanics. The idea of common mechanics of nature and human relations for the most part was initiated by the mechanistic picture of the world, which domineered in the science of the 18th century and partly preserved its positions in the early 19th century.

The role of the ideas of the mechanistic picture of the world was so considerable that they not only determined the strategy of development of scientific knowledge, but also had influence upon political practice. The idea of the world as a regulated mechanical system evidently sufficed over the minds of creators of American constitution, who developed the structure of state machine, whose links were to act as smoothly and exactly as clockwork.[50]

All this presents us evidences of a special place of the mechanistic picture of the world in the culture of technogenic societies of the epoch of early industrialism. Mechanism was one of important origins of formation of corresponding worldview structures, which struck roots in the culture and exerted influence upon various spheres of functioning of social consciousness.

In turn, the spread of mechanist worldview confirmed the belief that the principles of the mechanical picture of the world are universal means of cognition of any objects.

Thus, we may state an important feature in functioning of the mechanistic picture of the world as fundamental investigation program in the science of the 18th century: synthesis of knowledge, realized within it, was connected with reduction of various processes and phenomena to mechanical ones. The correctness of such reduction was justified by all systems of philosophical foundations of science, where mechanistic ideas prevailed.

But, as the mechanistic picture of the world expanded to new and new subject spheres, science more and more often had to take into consideration peculiarities of those spheres, which required new, non-mechanistic ideas. More and more facts hardly could be conformed to the principles of the mechanistic picture of the world.

Up to the late 18th – early 19th centuries a new situation started to arise; it led to appearance of disciplinary natural science, and within it the scientific picture of the world got its special characteristics and functional signs. It was a revolution in science, connected with reconstruction of its foundations, emerging of new forms of its institutional organization and its new functions in the dynamics of social life.

The history of chemistry, biology, technical and social disciplines cannot be understood, if we do not take into account "paradigmatic grafting" which was connected with expansion of the mechanistic picture of the world into new subject spheres.

Let us trace special features of that process. As we have already stated, the first attempts to apply notions and principles of mechanics in chemistry were connected with R. Boyle's program. Analysis of its historical fate shows that Boyle's desire to explain chemical phenomena from positions of notions of motions of "minute particles of matter" (corpuscles) required account of specificity of chemical processes. Under pressure of accumulated facts about chemical interactions, Boyle had to modify the ideas of the mechanistic picture of the world transferred to chemistry, and, as a result, chemistry started to form the picture of specific for chemistry picture of the processes studied.

According to Boyle, the primary corpuscles were to be considered as elements replacing former Aristotle's and alchemic elements. Basing on facts proving that changes of substances allows a scientist both to turn some of the substances into others, and to restore some of them in their initial shape, Boyle concluded that elementary corpuscles, determining properties of the corresponding compound substances, should be preserved in reactions.[51] These corpuscles are presented as qualitatively different elements, which form chemical compounds and mixtures.

Here it is evident enough that Boyle's picture of chemical processes, though conformed to the mechanistic picture of the world, included also specific features. In embryonic state it contained notion of chemical elements as corpuscles, having individuality, which, being physical particles, were as well carriers of properties which let them form various kinds of chemical substances in their compounds.[52]

Mechanics could ignore these properties, considering corpuscles only as masses subject to influence of forces, but in chemistry the properties of corpuscles, which make the chemical elements, are to be the main object of studies.

The mechanist picture of the world (if we take its developed forms), along with elementary objects – corpuscles, picked out the types of bodies built of them: liquid, solid, gaseous. In the picture of chemical reality, offered by Boyle, typology of chemical substances was not entirely reduced to typology of physical objects: together with distinction of liquid, solid and gaseous (volatile) substances Boyle picked out two classes of compound chemical objects – compounds and mixtures, and it was presumed that inside each of them there are special subclasses. Boyle gave these notions in a non-developed and, in many respects, hypothetical form, since concrete empirically fixed features, distinguishing compounds from mixtures, were not yet defined. "A long time yet was taken by the difficult question: what is a chemical mixture and what is compound, what are their nature, properties and differences; it caused contradictory statements of various kinds".[53]

Boyle's program offered an atomistic picture as basis for experimental and theoretical work in chemistry. In its main features it anticipated future Dalton's discoveries, though in the 17th century there were no sufficient conditions for its realization.

At Boyle's time chemistry did not dispose of experimental possibilities to decide which substances are elements, and which are not.[54] Boyle also did not define the idea of atomic weight as a characteristic, which could allow chemists to distinguish substances from each other.[55]

Nevertheless, despite the fact that Boyle's program was not realized, for methodological analysis it can serve as a good example which lets us determine the transfer of principles (in this context, principle of the mechanist picture of the world) from one science to another. The example of this program shows that translation into chemistry normative principle, fixed in the mechanist picture of the world (like the following normatives: all bodies consist of corpuscles, all phenomena can be explained by interaction of indivisible corpuscles which submit to mechanical laws), did not eliminate specificities of chemical investigation. What is more, to apply new principles in a new sphere, they were to be delivered in a special way, with due regard for specificity of objects, studied in chemistry. And that led to construction of a special picture of the reality studied (in this case – the picture of chemical reality), guided by which, the investigator could experimentally find and explain chemical phenomena.

Using the material of history of science, we can state that the formation of most new disciplines was connected with both intradisciplinary development of science and with translation of normative principle from one science to another. In this respect, Boyle's program can be regarded as an attempt of revolutionary transformations in chemistry by transplantation of cognitive directions and principles, taken from the mechanist picture of the world, into it.

Failure of that attempt was connected first of all with the fact that the picture of chemical reality, offered by Boyle, did not include such features of its key object (chemical element), which could be experimentally justified and stimulate new investigation ways in chemistry. That picture also had no experimentally verifiable features, which could allow investigators to clearly distinguish the basic types of chemical objects (element, compound, mixture).

A century and a half later, when chemistry had stored corresponding knowledge, Boyle's attempt was repeated in a more successful variation.

The process of reconstruction of foundations of chemistry in the 18th – 19th centuries was also conditioned not only by inner factors of its development (interaction of theory and experiment). The decisive role here still belonged to the mechanist picture of the world, prevailing at that time. As a universal scheme of explanation of physical phenomena, it introduced the idea of interaction of material corpuscles (bodies) by means of various types of forces. Analogically to this approach, in chemistry there began to establish the notion of "forces of chemical affinity",[56] which determined interaction of chemical elements. This notion was included into the picture of chemical reality, first as a hypothesis, then, in Lavoisier's works, as a thesis justified by experiment.

As Lavoisier noted, probably, one day the exactness of data available will be brought to such degree, that a geometrician will be able to calculate in his study phenomena, accompanying any chemical compound, in the same way, in which he calculates movement of celestial bodies. Laplace's views in this respect and experiments, which we have projected on base of his ideas, to express forces of affinity of various bodies, now do not let us not regard such hope as some chimera.[57]

Lavoisier himself even created a table of oxygen's affinity with various substances and supposed possibility of quantitative measuring affinity.[58]

In his works special attention is paid to defining notions of the main objects – elements. He suggested that the idea of the ultimate limit, reached by analysis, should be connected with the names of elements. In this respect all substances indivisible, in his opinion, at the contemporary state of knowledge, were elements. Before there appear means to divide them, and experiment proves us the contrary, – said he, – we cannot regard them as compound.[59]

Classifying compound substances, Lavoisier, on the one hand, reckoned for these evidently hypothetical substances (such as thermogen), on the other hand, he brilliantly foresaw that a number of substances, which appeared as simple ones, in the nearest future would not be reckoned for simple ones.

Lavoisier's new notions of elements were a decisive "progress of the problem" in establishing the scientific picture of the chemical reality. The results, obtained by Lavoisier, were essential for proof of the law of conservation of substance (1789), which made quantitative study of chemical reactions possible. They exerted influence upon investigations carried out by Dalton, which finished Lavoisier's program of forming a new system of chemistry principles, which would coordinate with domineering physical ideas and base on chemical experiments. The works of Dalton and his followers led to construction of the picture of chemical reality, where chemical elements were presented as atoms different in their form and atomic weight. The latter allowed chemists to explain not only phenomena observed in experiments, but also many laws, discovered at that time and confirmed by experiment (for instance, stehiometry laws discovered by Richter, Proust and Dalton).

Investigators of Dalton's works truly say that Dalton came to construction of stehiometry laws, basing on the atomist hypothesis, and from this position he generalized experimental facts. That hypothesis had its premises in philosophical atomist doctrines, but its direct source was Newton's atomist views, the notions of the mechanist picture of the world of indivisible and indestructible corpuscles.

Dalton's atomist picture, in the process of its development (here the decisive role belonged to A. Avogadro and Ch. Gerhardt), was enriched by the ideas of molecules as integral systems of atoms, and of chemical processes as interaction of molecules when they exchanged atoms. In turn, the notions of atomic-molecular structure of substance started exerting reverse influence upon physical investigations. It is characteristic that the molecular-kinetic theory of heat, which replaced the thermogen theory, was mainly based on the idea that substance consists of moving molecules.

In one of his first works on kinetic theory of gases (1857), R. Clausius created a mathematical model of thermal movement of gas particles and prefaced it with the ideas of molecular structure of substance. In that account, beside translational movement, he singled out also rotating and intramolecular oscillatory movement.[60] Mentioning of the latter is interesting only because it means that a molecule from the very beginning is imagined as a complicated thing, consisting of atoms (this idea entered the scientific picture of the world under the influence of development of chemistry). It is also quite characteristic that in A. Kroenig's work (1856), which preceded Clausius's investigation and initiated the investigation cycle, leading to construction of molecular-kinetic theory of heat, the key moment of justification of hypothesis of heat as kinetic movement of molecules is inference of Avogadro's law. That law, deduced in 1811, was then so entirely forgotten that physical dictionaries did not even include Avogadro's name.[61] But in chemistry Avogadro's law was not only well known, but also it played the decisive role in development of atom-molecular conceptions. Later it was returned from chemistry to physics and there actively used in construction of molecular-kinetic theory of heat.

Thus, we may conclude that in translation of the principles of the mechanist picture of the world into chemistry, they were not merely transplanted into the "body" of chemistry, stipulating purely mechanical view of chemical objects, but were confronted with the features proper to objects studied in chemistry. And that stimulated the development of chemistry as science, with its specific object part, and establishing a new picture of the reality studied, now not reducible to the mechanist one. And though investigators still went on considering transformation of chemistry into a section of applied mechanics or appearance of independent chemical mechanics (D.I. Mendeleev), one could really say that chemistry was becoming constituted in an independent science, under influence of the mechanist picture of the world and regarding specificity of chemical objects. And the most important aspect of that process was the development of a special picture of the reality studied. The physical picture of the world and the picture of the chemical reality acquired subordinational connection, and that connection did not abolish relative independence of each of them.

Similar processes of the development of a special scientific picture of the world also can be traced in the history of biological knowledge.

Above we have mentioned that Lamarck, explaining causes of appearance of life, resorted to the ideas, developed in the mechanist picture of the world of the 18th century, in particular, notions of thermogen and electric fluid as carriers of special forces, which were regarded by the scientist as the main stimuli of life. Though Lamarck did not transfer mechanically the ideas of those hypothetical substances into the field of knowledge developed by him. He emphasized that thermogen and electric fluid, entering a living organism, are transformed into a specific fluid – nerve fluid, proper only to living beings. The nerve fluid, in Lamarck's opinion, was an acting force, as a sort of instrument that produced feelings, ideas, and acts of reason. It is nerve fluid that is able to cause such amazing phenomena, and, to deny its existence and its properties, we would have to give up any investigation of physical reasons of phenomena and again turn to vague, groundless notions to satisfy our curiosity toward this object.[62]

Explaining the nature of living organisms this way, Lamarck, though in a hidden form, accentuated his attention at specificities, proper to living beings, and that circumstance founded for specification of biological science and its special picture the reality studied. Lamarck not only emphasized specificity of biological objects, but also pointed out their interaction with the environment as source of their changes. According to Lamarck, these changes happen due to permanent extraction of fluids from the environment and their transformation inside a living organism. Accumulation of corresponding fluids inside organism causes changes of separate organs and the whole organism, and these changes can be traced, if we consider a row of generations for long enough time. "In the course of time, and under influence of unlimited diversity of permanently changing circumstances, living bodies of all classes and all orders were created".[63]

Thus, the principles of explanation, taken from the mechanist picture of the world, were transformed by Lamarck into the principle of evolutional explanation of features of organisms and species, the principle, fundamental for biology.

The diversity of living organisms, different levels of their organization were the foundation to their arrangement in a certain order, from simple to complicated ones, and the gradation principle, which Lamarck assumed as basis of his evolutional conception. Though, insisting on smooth, imperceptible transitions between species, Lamarck came to the conclusion that there were no real borders between them and, in the final analysis, denied their reality. His idea of changeability and inheritance of accepted changes were the basis of further development of biological knowledge, when it accumulated empirical material which stimulated development of evolutional notions.

Taking into account the fact that ideas of objects and their interactions are aspects of the creation the picture of the world, we may say that Lamarck introduced a new vision of biological reality.

Lamarck's evolutional ideas were heuristically important not only for development of biological knowledge, but for other natural sciences, such as geology, as well.

In his conception, Ch. Lyell strove to solve a difficult and actual (for that time) problem of correlation of modern natural forces and the forces of the past. Solving this problem, Lyell took notice of the ideas, already developed in biological science. He was not satisfied with approaches, applied by "the catastrophists", but in Lamarck's conception he found answers to arising questions. We mean the principles which constitute the foundation of Lamarck's conception: first, the principle of similarity of acting natural forces and those which acted in the past; second, the principle, according to which radical changes are results of gradual small changes, accumulated for a long time.

Lyell employed these principles in his doctrine of geological processes.[64] He transferred normative principles, formed in biology, into geology, and thus constructed a theoretical conception, which later exerted reverse influence upon biology and, along with Lamarck's evolutional ideas, became one of the premises for the shaping of the scientific picture of biological reality connected with the name of Darwin.

When Darwin's conception appeared, biology got the status of independent branch of natural science of full value. At that period the picture of biological reality showed clear features of autonomy and acted as a system of scientific notions disclosing properties of living nature.

Settling of biology as an autonomous branch of knowledge did not mean that its further development took place exclusively by its inner factors. Appearance of new knowledge in disciplinarily organized science always is a complicated and multilateral process, which includes both intradisciplinary and interdisciplinary interactions. Examples of this would be Mendel's discoveries. They resulted not only from development of biology, but were realized through translation of ideas, developed in other sciences, into biology. In his work "Experiments on Plant Hybrids" Mendel formulated his theory of discrete heredity carrier, the "heredity factor", and demonstrated that separate features and properties of organism can be connected with these "heredity factors".[65]

Mendel's experiments became possible due to the development of hybridization in biological practice of the time. At the same time, the empirical material, accumulated in biologists' and practical selectioners' research works, did not by itself lead to the idea of "heredity factors". To formulate this idea, Mendel had preliminarily to dispose some theoretical vision and to accumulate empirical material.

That theoretical vision was being formed not only based on developing biological knowledge, but also under influence of principles of explanation translated from other spheres of knowledge, from mathematics, for instance. Investigators of Mendel's works said that he "joined methods of two branches: mathematics – the probability statistical method (Doppler), and biology – hybridization method (Unger)".[66]

In fact, Mendel ran his experiments as base of the new, only being created at that stage, picture of biological reality, which was constructed thanks to interrelation of intradisciplinary and interdisciplinary knowledge. Gradually did that picture settle the notion of a new biological object – "heredity factors". Exposure of that object and including the idea of it into the picture of the biological reality, on the one hand, let investigators interpret accumulated facts in a new way, on the other hand, contributed to further justification and development of Darwin's theory of evolution and formation of new theories in biology (for instance, the synthetic theory of evolution as joint of the evolution theory and population genetics).

In turn, the new theories and facts exerted reverse influence upon the picture of biological reality, which was corrected and developed under influence of theoretical and empirical material. In the first third of the 20th century Darwin's picture of the biological world was replaced by a new one; there not organism, but population was regarded as the basic unit of evolution, and it introduced the basic organization levels of living nature – molecular heredity carriers, cell, multicellular organisms, populations, biogeocenoses and biosphere (the ideas of the two latter levels were included into the picture of the biologic world mostly due to works of Sukachev and Vernadsky).

Interaction of organisms with each other and with environment was regarded in the contexts of including over-organism structure of the living nature into this interaction. The base of biological processes was reproduction–of–life structures in concordance with their genetic code (heredity) and their changes caused by mutations and natural selection.

Finally, there appeared new ideas of space-time characteristics of biological processes. Even Darwin's picture of the world introduced the notion of evolution time (unlike the mechanist picture of the world, which had an extratemporal character); it consolidated the idea of historicism. Further development of biology corrected these ideas and formed the notion of special space-time structures of the living nature, not reducible to physical space and time. There appeared the idea of biological time of separate living organisms and populations; it became clear that the notions of physical time continuity are not enough to characterize biological systems, and later it contributed to introduction of the idea of "anticipatory reflection".

As a result, the picture of the biological reality became not only autonomous referring to the physical picture of the world, but alternative to it, to some extent. Physics remained non-evolutional science, while biology, starting with consolidation of Darwin's ideas, was based on the evolutional picture of the world of processes studied.

In the historical development of social disciplines we can see similar features of forming of disciplinary knowledge, connected with specificity of the object studied, taken into consideration. The mechanist paradigm, extended to include the sphere of social cognition, was modified, and, in the process of such modification, a break with the mechanist principles became visible. Here a most important part again belonged to new "paradigmatic grafting" in the sphere of social knowledge from biology (as it developed the ideas of evolution), and then, in the 20th century, from the system theory, cybernetics and the information theory.

The first steps to constitution of social studies as a special sphere of disciplinary knowledge entailed modernization of the images taken from the mechanist pictures of the world. Comte, acknowledged as one of the founders of sociology, included the notion of historical development, fundamental, in his opinion, characteristic of society, into his picture of the social reality. Furthermore, his conception first regards society not as a mechanism, but as a specific organism, whose parts form an entity. At this point we can clearly see the influence of biological ideas upon Comte's sociological conception.

Further development of these ideas was connected with H. Spencer's general evolution theory and ideas of social development as a specific phase of evolution of the world. Spencer not only transfers the ideal of biological evolution to the sphere of social knowledge, but also tries to single out some general principles of evolution and their specific concretization as applied to biological and social objects.[67] The idea of society as an integral organism, according to Spencer, should take into account that people as social elements possess consciousness, as if spread over all social aggregate, and not localized in some center.

The further steps, connected with reconstruction of primary paradigmatic images transferred from natural science to social knowledge, were connected with discussions referring to methodology of social cognition. These discussions are still lasting, and their center is the thesis (formulated by Dilthey) of fundamental difference of knowledge of spirit and knowledge of nature. W. Dilthey, W. Windelband and H. Rickert gave this definition to that difference through opposition of understanding and explanation, individualization and generalization, ideographic method, connected with description of unique historical events, and nomothetic method, aimed at finding generalization laws. There emerged two extremes in interpretation of the methods of social and humanitarian cognition: one of them treated them as identical, the other sharply opposed them. But the real scientific practice developed in the space between these two extremes. That development revealed features of the scientific ideal and their specification referring to singularities of the events studied. Reflection over such kind of scientific practice causes methodological approaches, which takes away sharp opposition between explanation and understanding, individualization and generalization. Weber, for example, emphasizing importance of understanding of directions and motives of the active subjects for sociology, also developed the idea of ideal types as generalizing scientific notions which help us to construct explanatory models of social processes.

We should also mention that in the natural scientific cognition it is possible to trace links of understanding and explanation, though in a different accentuation than in social and humanitarian cognition. In particular, understanding is built into the very acts of natural scientific observation and formation of facts. When a modern astronomer observes shining points in the sky, he understands: these are stars, massive plasm bodies analogous to the Sun, while an ancient astrologer could understand the same phenomenon differently: for instance, as celestial light shining through slits in the dome of the sky.

The understanding acts are determined by the cultural tradition, ideological directions, the picture of the world, openly or unwittingly accepted by the investigator. These are common features of understanding in any area of cognition.

In principle, the idea which declares that only in people's activity does the investigator deal with mentalities included into it, and, studying nature, he faces nonliving and spiritless objects, – this is a worldview attitude of the technogenic culture. In other traditions, for example, in traditionalist cultures, which recognize the idea of soul reincarnation, cognition of the nature and man are not opposed as sharply as in the culture of technogenic civilization.

The problem of opposition of individualization and generalization, ideographic method, on the one hand, and nomothetic method, on the other hand, also requires correction. Events, irreproducible individually, occur not only in the history of society, but also in the processes of natural historical development: history of life on the Earth, the history of our Universe.

At the level of separate, empirically fixed events, both social and natural phenomena are irreproducible. But science cannot be reduced to ascertaining irreproducible events empirically. When we speak about historical processes, the aims of science consist in discovering tendencies, logic of their development, connections based on laws, which would allow scientists to reproduce the picture of the historical process on base of the "point-events", uncovered by historical description. In other words, here we deal with historical reconstruction. Each such reconstruction seems purely ideographic knowledge only in outward appearance. In fact, it combines ideographic and nomothetic elements in a specific way, which discloses logic of the historical process, not separated from the gist of its individuality, but woven into it. Historical reconstructions can be regarded as a special type of theoretical knowledge of unique, never repeated historical processes. Weber's studies of Protestant ethics and birth of capitalist spirit are an example of historical reconstruction dealing with theoretical comprehension of history. The same words can be said about K. Marx's works dedicated to revolutionary events of 1848–1852 and 1871 in France. The results of Marx's investigations, presented in his works "Louis Bonaparte's 18th of Brumaire", "The Civil War in France" represent reconstructions, which demonstrate theoretical view through the material of historical description. In principle, one and the same fragment of history can be presented in different reconstructions. In this case each of them presents as a kind of theoretical model aimed at describing, understanding and explaining the historical reality. They compete with each other, and this neither is an extraordinary situation in science. Each new historical reconstruction wants to assimilate larger and larger diversity of accumulated facts and predict the new ones. Prediction as retrodiction (discovery of unknown facts of the past) plays as important role as in historical investigations as in any other types of theoretical cognition.

Certainly, there is specificity of historical reconstructions in sciences and social and humanitarian studies. When an investigator is reconstructing some fragments of spiritual history, he has to understand the corresponding type of cultural tradition, which can radically differ from that of existing in his own culture. In this case the frontier is occupied by the procedures of understanding, movement in hermeneutic circle, when understanding passes from a part to the whole and then from the whole to a part many times, perceiving specificities of other cultural tradition.[68]

At the same time, the very acts of understanding and procedures of historical reconstruction in humanities (though, in natural science as well) are determined by the investigator's disciplinary ontology, the special scientific picture of the world, introducing scheme-image of the object sphere studied. Discussions of ideals and norms of investigation in humanities in many respects refer to the ways of construction of such picture and its philosophic justification. The general principles, commonly accepted, evidently or in hidden form, are three fundamental theses: any notions of man and society should take into consideration historical development, integrity of social life and the fact that social processes include consciousness. These principles mark the scope within which pictures of social reality are constructed.

Their becoming as specific images of the social world, different from paradigmatic models taken from natural science, took place in the second half of the 19th – early 20th centuries. During that time Spencer, Marx, Dilthey, Durkheim, Simmel, Weber offered variants of disciplinary ontologies of social and humanitarian subjects. Though they competed with each other, determining the sphere of acceptable problems and means of their solution, they also interacted. They had common problems, discussed by all investigators, though from different positions. Each of them promoted his ideas of society, correlating with rival investigation programs. All this served as evidence of the final stage of the scientific revolution, which started by transfer of natural scientific paradigms to the sphere of social processes and finished by their reconstruction and forming of social and humanitarian disciplines.

When disciplinary organized science is formed, every discipline acquies its specific foundations and its own impulse of inner development. But sciences do not become absolutely autonomous. They interact, and exchange of paradigmatic principles is an important feature of such interaction. That is why revolutions connected with "paradigmatic grafting", which change the strategy of development of disciplines, at this stage are traced distinctly enough.

In this respect, a characteristic example can be found in transfer from physics into chemistry of a fundamental principle, according to which processes of molecular transformations, studied in chemistry, can be presented as interaction of nuclei and electrons, and therefore chemical systems can be described as quantum system characterized by certain ψ-function.[69] That idea made the foundations of a new trend – quantum chemistry, the appearance of which marked a revolution in modern chemical science and birth of fundamentally new investigation strategies.

We may find examples of translations of paradigmatic attitudes in most different sciences. Thus, notions of self-organization, developed in cybernetics and theory of systems, translated into modern physics, considerably stimulated development of ideas of synergetics and thermodynamics of non-equilibrium systems.

No less productive was the union of biology and cybernetics, based on the ideas of living objects as self-regulating systems with transition of information and reverse connections.

Among numerous examples, which would confirm effectiveness of such interaction, we can mention the theory of biological evolution as a self-regulating process, created by I.I. Shmalgauzen in the 1950s –1960s.

The first step toward the new theory was consideration of biological objects – organisms, populations, and biocenoses – as self-regulating systems. Shmalgauzen wrote: "All biological systems are characterized by greater or smaller ability to self-regulation, i.e. homeostasis. With the help of self-regulation each of these systems maintain its very existence, its composition and structure with its characteristic inner connections, appropriate transformations of the whole system in space and time. Certainly, homeostatic systems are, first of all, a separate individual of each species of organisms, then population as a system of individuals of one species, characterized by its composition and structure with specific intercommunications of its elements, and, finally, biogeocenosis, also having its composition and structure with its intercommunications, often very complicated ones".[70]

Translation of the new paradigm from cybernetics into biology required certain correction of the notions introduced. It was necessary to take into account specificity of biological objects, which belong to a special type of self-regulating systems. It was important to pay attention to their historical evolution. As a result a problem emerged: to what extent can we apply notions of homeostatic systems, which conserve their qualitative stability, to systems which are historically developing, changing qualitatively in the process of evolution.

Shmalgauzen proceeded from the assumption that the basic principles of self-regulation can be used also in description of historically developing systems. He wrote: "Mechanism of control and self-regulation are, naturally, different in different systems. But general principles of regulation can, in all cases, be considered from one point of view, from the standpoint of the doctrine of regulating devices".[71] In principle, it was nontrivial step, since systematic defining notions of mechanisms of self-organization in historically developing objects in natural science started later. Essential aspects here were I. Prigogine's investigations of dynamics of non-equilibrium processes, R. Thom's theory of catastrophes, development of synergetics (H. Hacken, M. Eigen, G. Nicolis et al.). Shmalgauzen's ideas of regulation processes in historical development of biological systems can be regarded as one of preliminary versions of this investigation program, which is now actively being developed.

Using the ideas of self-organization in analysis of interaction of biological systems and considering evolution as a process automatically regulated, Shmalgauzen includes the new notions into the picture of biological reality. Interaction of the main structural units of living beings – organisms, populations and biocenoses – was considered from the point of view of transfer and transformation of information and processes of management.

Applying the ideas of information codes and feedbacks to already established synthetic theory of evolution, Shmalgauzen introduced essential transformations and additions. He uncovered regulating mechanism of evolution with regard to levels of organization of living organisms, considering them as an entity, which includes direct and reverse connections of organisms, populations and biogeocenoses.

Considering each individual as a complicated communication, recoding genetic information of molecular level into a set of phenotypic features, Shmalgauzen presented it as a whole information block, and specific for each individual activity in biogeocenosis regarded as a means of transmission of reverse information.[72]

Translating the theory of evolution into language of cybernetics, he demonstrated that "the very transformation of organic forms is regularly realized within a relatively stable mechanism, lying at biogenetic level of organization of life and acting according to statistic principle".[73] It was "the highest synthesis of the idea of evolution of organic forms with the idea of stability of species and the idea of stability of geochemical function of life in the biosphere".[74] This approach allowed the investigator to formulate the principle of group selection, indicated the role of competition of whole population with each other as condition of creation and maintenance of over-organism systems (species and biogeocenosis).[75] Shmalgauzen's conception also explained many facts of noise-immunity of transmission of hereditary information, opening new ways to apply mathematical methods in the theory of evolution.

Another eloquent example, which demonstrates productivity of translating notions of cybernetics into biology, is the intercellular interaction (A. Turing, 1952; M. Tsetlin, 1964; V. Volterra, 1968; M. Apter, 1970). Comparison of interaction of cells with interaction of a group of automatic devices, where there is no common center, which would deliver commands, allowed the investigators to discover a number of singularities of intercellular regulation. Later it was discovered that this model is applicable to description of processes of regulation not only at the level of cells, but also at organism and population levels.[76]

We may ascertain that notions translated into biology then returned to cybernetics enriched. Elucidation of singularities of regulation of biosystems under decentralized control led to development of the model of intercellular regulation and prepared further use of it in other spheres (its application to systems of developed market economy, to some social systems etc.).

In the 20th century we can see considerable activation of exchange of paradigmatic attitudes not only between various natural sciences, but also between them and social disciplines and humanities.

For instance, we may ascertain that many achievements of modern linguistics were obtained because of application of images of cybernetics, ideas of the theory of information and notions of genetics.

Thus, consideration of natural language in terms of cybernetics and the theory of information, as well as application of notions of genetic code as special language of heredity, turned out quite productive in discussions of the problem of generative grammar. Analogy between sociocode and genetic code (with regard for connections phenotype – genotype) opened new possibilities to generalize the theory of generative grammar developed by N. Chomsky's school. Linguists used to criticize Chomsky's theory from the position that it gave no description of generative models of natural languages, but only a description of general conditions for generative models. Application of analogy phenotype – genotype enabled investigators to put the problem in a new way and to consider under a new angle the results already obtained. They put forward the hypothesis that the real generating process in functioning of languages is analogous to elucidation of the connection phenotype – genotype in development of organisms. In accordance with this new vision, they formulated the problem: to create the theory of generative grammar as a two-level system.[77] The first level is to generate ideal linguistic objects, which form, in their entity, an ideal language (genotypic language). The second one – to provide transformation of objects of genotypic language into objects of a real language (phenotypic language). From that point of view Chomsky's theory was regarded as an attempt to construct conception of genotypic language. Many critical objections to this theory, from the new point of view, were not only disproof of the problem offered, but also a statement of a problem – to find a link between it and the theory of generating models of phenotypic type.[78]

Intercommunication of linguistics, biology and the theory of information, characteristic for development of these disciplines, emerged in the 20th century, to large extent, due to development of semiotics and new interpretation of linguistics as part of semiotics.

Linguistics was sort of proving ground for establishing ideas of semiotics as discipline studying signs and sign communications. Disciplinary ontology of linguistics (picture of language as a special object of investigation) was modernized, when natural languages started to be regarded as a variation of semiotic systems. Then linguistics presented as a special part of semiotics and included investigation of not only natural, but also artificial languages.

Such modernization of object sphere of linguistics, in turn, opened new ways for its interaction with other disciplines, which used ideas and notions of semiotics.

The images of language as a complicated sign system transmitting information are widely used in zoosemiotics, which studies language of animals.

In turn, the results obtained here, make it possible to find new formulations of many linguistic problems. According to prominent linguist Roman Jakobson, "language and other means of people's communication in their various interactions – mutatis mutandis – gave a lot of instructive analogies with transmission of information in other species of living beings. "The adaptive nature of communication", in all its diversity, the essence of which was uncovered by Wallace and Srb, is reduced to two mutually connected classes: adaptation to environment and adaptation of environment to its own needs. Really it became one of "the most disturbing" biological problems; it is hard to overestimate its meaning for modern linguistics. Similar processes in the life of language and in animals' communication are worth thorough investigation and comparison, useful for both ethology and linguistics. In the period between the world wars there appeared the first concord of investigators of the two disciplines, aimed at study of two aspects of evolution: adaptation and convergent evolution. Namely then the linguists' attention was attracted to biological notion of mimicry, and at the same time biologists started examining different types of mimicry as method of communication. Divergent development, as opposed to convergent tendency in spread of communication ... draws more and more attention of both linguists and biologists. The known methods of manifestation of language non-conformism, peculiarity or "narrowness", acquire interesting ethologic analogies, and biologists study and describe what they call "local dialects", according to which animals of the same species, crows or bees, are distinguished".[79]

R. Jakobson emphasizes that parallels between code system, which constitute the array of biological information, and human language open broad possibilities to transfer notions and methods. Referring to the works F. Crick, Janovsky, G. and M. Beadle, F. Jacob, he says that these authors – biologists – consider hierarchical structure of "genetic language", similar to the one discovered by linguists in natural languages, as its most important feature. Jakobson wrote: "Both linguists and biologists attribute hierarchical structure of language and genetic communications to fundamental scientific principles. As Benveniste showed, linguistic unit has only that status which it gets inside a unit of higher level. Transfer from lexical units to syntactic groups of different ranges is parallel to transfer from codons to "cystrons" and "operons"; the two latter levels of genetic sequences are compared by biologists to syntactic groups of different degree of complexity, and limitations for distribution of codons inside such constructions were called "syntax of DNA-chain". In genetic communication "words" are not separated from each other; special signals inside constructions indicate beginning and end of the operon and borders of cystrons inside operon; these signals are metaphorically called "punctuation marks" or "commas". They really correspond to delimitative means used for phonologic distinguishing of phrases inside speech, and simple sentences and word combinations inside phrase".[80]

As one more example of productivity of exchange of paradigmatic models between linguistics and biology, R. Jakobson points at discovering of similarity of synonymy in natural speech and changes "in meanings of codons, caused by their position in genetic communication". He stresses that biologists, investigating singularities of peptide translations, detected some kind of "synonimic codons", and that opened new possibilities to understand flexibility in recording hereditary information.[81]

All these exchange processes of paradigmatic attitudes, notions and methods between various disciplines stipulate some generalized vision of object spheres of each discipline, vision that lets us compare different pictures of reality studied, find there common blocks and identify them, considering as the same reality.

Such vision is determined by a general scientific picture of the world. It integrates notions of objects of different sciences and forms, on base of their achievements, an integral image of the Universe, which includes notions of non-organic, organic and social worlds and their connections. That same picture allows us to determine similarity of object spheres of different disciplines, identify different notions as vision of one and the same object or connections of objects and thus justify translation of knowledge from one discipline into another.

For example, application of notions of atoms in physics, transferred from physics into the general picture of the world, in biology preliminarily stipulated working out a general principle: the principle of atomic structure of matter.

In his lectures on physics, R. Feynman said that, if a world catastrophe destroyed all scientific knowledge, and future generation received only one sentence, carrying most information of disappeared science, that would be the sentence: "all bodies consist of atoms".[82]

However, to use this principle in biology, we are to accept one more notion: to consider biological organisms as a special type of bodies (as living matter). This notion also belongs to the general picture of the world.

But if investigator would put forward the hypothesis that, through notions of atoms and their structure, developed in physics, we could explain, for instance, phenomena of human spiritual life – meaning of works of art, religious and aesthetic principles, – this hypothesis could not find its base in the modern scientific picture of the world, as it does not include spiritual phenomena in the class of bodies and does not regard them as matter.

Thus, the general scientific picture of the world can be considered as such kind of knowledge, that regulates fundamental scientific problems and directs translation of notions and principles from one science into another. In other words, it functions as a global investigation program of science, on base of which its more concrete, disciplinary investigation programs are formed.

By analogy with the already considered process of intradisciplinary integration of knowledge, we may suppose that its interdisciplinary integration is inseparably linked with a heuristic role of the general scientific picture of the world and is provided by processes of translation of ideas, principles and notions from one science into another and further including obtained here new, most fundamental results into the general picture of the world.

The high degree of generalization of such results and aspiration for constructing integral system of notions of the world, including man, his natural and social life, make that special link of developing scientific knowledge, which most closely contacts with meanings of cultural universalities and, consequently, possesses clearly expressed worldview status.

GLOBAL SCIENTIFIC REVOLUTIONS AS CHANGE OF RATIONALITY TYPE

Scientific revolution as choice of new investigation strategies. Potential histories of science

Reconstruction of foundations of investigation means change of the very strategy of scientific search. But any new strategy is established not at once, but in the course of long struggle against previous attitudes and traditional visions of reality.

The process of establishing new foundations in science is determined not only by prediction of new facts and generation of concrete theoretical models, but also by reasons of sociocultural character.

New cognitive attitudes and knowledge generated should be inscribed into the culture of corresponding historical epoch and correlated with values and worldview structures, lying in its foundation.

From this point of view, reconstruction of foundations of science at the time of scientific revolution is the choice of special directions of knowledge growth, which provide both expansion of range of investigation of objects and certain correlation of dynamics of knowledge and values and worldview attitudes of the corresponding historical epoch. At the time of scientific revolution there are several possible ways of growth of knowledge, though not all of them are realized in real history of science. We can distinguish two aspects of non-linearity of growth of knowledge.

The first is connected with competition of research programs within a separate branch of science.[83] Victory of one and degeneration of another program direct development of this branch of science along a certain course, but at the same time close other ways of its possible development.

As an example, let us consider the struggle of two directions in classical electrodynamics: Ampere–Weber's, on the one hand, and Faraday–Maxwell's, on the other hand. Maxwell, while creating the theory of electromagnetic field, did not obtain new results, in comparison with those given by Ampere–Weber's electrodynamics. Outwardly, everything looked like as derivation of already the known laws in a new mathematical form. Only at the final stage of construction of the theory, having discovered fundamental equations of electromagnetism, did Maxwell obtain the famous wave solutions and predicted the existence of electromagnetic waves. When they were experimentally found, it led to triumph of Maxwell's direction and established notions of closeness and force fields as the only true basis of the physical picture of the world.

Though, in principle the effects, which had been interpreted as proof of electromagnetic waves, could be predicted also within Ampere's direction. We know that in 1845 K. Gauss in his letter to W. Weber indicated that for further development of Ampere–Weber's theory it was necessary, in addition to already known forces of action between charges, to assume existence of other forces which spread at finite speed.[84] B. Riemann accomplished that program and derived equation for potential, which was analogous to Lorentz's equations for delayed potentials. In principle, this equation could explain prediction of those effects, which were interpreted in the paradigm of Maxwell's electrodynamics as propagation of electromagnetic waves. But this way of development of electrodynamics stipulated a physical picture of the world, which would postulate propagation of forces in empty space at different speeds. In such picture of the world there is no ether and notion of electromagnetic fields. And then we face a question: how would the theory of electrons look like in that unrealized line of development of physics, what way would lead to the relativity theory?

A physical picture of the world, where interaction of charges would be presented as transition of forces at finite speed, without notion of material fields, is quite possible. It is indicative that just the same image of electromagnetic interactions was used by R. Feynman as base for new formulation of classical electrodynamics, on which he developed the idea of construction of quantum electrodynamics in terms of path integrals.[85] To some extent, Feynman's reformulation of classical electrodynamics can be estimated as reproduction, in modern circumstances, unrealized previously but potentially possible ways of historical development of physics. But, at the same time we are to take into account the fact that modern notions of nature are being developed in different scientific tradition than in classical epoch, in presence of new ideals and norms of explanation of physical processes. The development of quantum-relativistic physics, establishing these norms, made physicists "accustomed" to diversity of various formulations of theory, and each of them is able to express essential characteristics of the object sphere studied. Physicist-theorist of the 20th century treats different mathematical description of the same processes not as anomaly, but as norm, understanding that the same objects can be assimilated in different language means. Different formulations of the same physical theory are condition for progress in investigations. Traditions of modern physics include also estimation of the picture of the world as a relatively true system of notions of the physical world, which can change and improve partly as well as in the whole.

Therefore, when, for instance, Feynman developed ideas of interactions of charges without "field mediators", he was not confused by the fact that he had to introduce, along with delaying, advanced potentials into the new theory, which, in the physical picture of the world, corresponded to appearance of notions of influence of present interactions upon not only the future, but also the past. He wrote: "By that time I was a physicist in degree enough not to say: "No, no, it is impossible". Now, after Einstein and Bohr, all physicists know that sometimes an idea, which seems absolutely paradoxical at first sight, may turn out true after we examine it up to the very minute details, thoroughly and completely and find its connection with the experiment".[86] But "to be a physicist" of the 20th century is not the same as "to be a physicist" of the 19th century. In the classical period a physicist would not have introduced "extravagant" notion of physical world just because he came to a new and perspective mathematical form of theory, details of empirical justification of which can be refined in the future. In the classical epoch the physical picture of the world, before generating new theoretical ideas, had to appear as "visual portrait" of reality, confirmed by experiment. Formation of competing pictures of reality studied stipulated their hard confrontation, in which each of them was regarded by its supporters as the only true ontology.

From these positions it would be correct to estimate Gauss-Riemann's program in physics of the 19th century. To introduce notion of forces, spreading at different speeds, into the physical picture of the world of that time, one needed to justify this notion as a visual image of "real organization of nature". In traditions of physical thoughts of that time force was always connected with material carrier. So its changes in time from point to point (different speeds of spread of force) supposed introduction of material substances, and change of speed of spread of force) was connected with its state. But such notions were proper for Faraday-Maxwell's program and were incompatible with Ampere-Weber's picture (in that picture connection of force and matter was regarded as interconnection between electric and gravity forces, on the one hand, and charges and masses, on the other hand; charges and masses appeared as material carriers of forces, while the principle of instantaneous transmission of forces in space excluded necessity to introduce a special substance, which would provide transmission of forces from point to point). Thus, reasons, due to which Gauss-Riemann's idea has been practically forgotten in the history of classical electrodynamics of the 19th century, lay in the style of physical thinking of that particular historical epoch. That style of thought, with its intention at construction of ultimate true notions of essence of the physical world, was one of the displays of "classical" rationality type, realized in philosophy, science and other phenomena of consciousness of that historical epoch. Such type of rationality supposes that thinking, as if from outside, observes the object, comprehending its true nature this way.

On the contrary, the modern way of physical thinking (within which the unrealized but possible line of development of classical electrodynamics was constructed) appears as display of different, non-classical rationality type, characterized by special attitude to object and itself. Here thinking reproduces object as interweaved into human activity and constructs images of the object, correlating them with notions of historically developed means of its assimilation. Then thinking gropes and more or less clearly understands that it itself is an aspect of social development and, consequently, it is determined by this development. In such type of rationality once obtained images of essence of object are not regarded as the only possible ones (in other language system, in other cognitive situations the image of the object can be different, moreover, in all these varying notions of the object it is possible to express objectively true content).

The very process of formation of modern rationality type is conditioned by the process of historical development of society, transformation of "field of social mechanics", which "moves things up to mind".[87] Investigation of these processes is a special problem. But, in general, we may ascertain that type of scientific thought, developed in the culture of one historical epoch or another, is always correlated with the character of people's communication and activity and determined by the context of its culture. Factors of social determination exert influence upon competition of research programs, activating some ways of their development and impeding others. As the result of "selective work" of these factors within every scientific discipline only part of potentially possible ways of scientific development are realized, while others remain unrealized tendencies.

The second aspect of non-linearity of growth of scientific knowledge is connected with interaction of scientific disciplines, conditioned, in turn, by features of both objects studied and sociocultural environment inside which science is developed.

Appearance of new spheres of knowledge, change of leaders of science, revolutions connected with transformations of pictures of studied reality and normatives of scientific activity in its separate branches, can exert substantial influence upon other branches of knowledge, changing their vision of reality, their ideals and norms of investigation. All these processes of interaction of sciences are mediated by various phenomena of culture and, in their turn, exert active influence upon them.

Taking all these complicated mediations into consideration, in development of every science we may distinguish one more type of potentially possible lines in its history, which is a specific aspect of non-linearity of scientific progress. Singularities of this aspect can by illustrated by means of analysis of history of quantum mechanics.

It is well known that one of the key moments of its construction was N. Bohr's elaboration of a new methodological idea, according to which notions of the physical world should be introduced through explication of an operational scheme, uncovering characteristics of studied objects. In quantum physics this scheme is expressed by means of the principle of complementarity, according to which the nature of micro-object is described by two complementary characteristics correlative to two types of devices. This "operational schemes" was joined with a number of ontological notions, for instance, of corpuscular-wave nature of microobjects, existence of quantum of action, objective interconnection of dynamic and static regularities of physical processes.

But the quantum picture of the world was not an integral ontology in the traditional sense. It did not present natural objects as causatively conditioned interaction of some objects in space and time. Space-time and causative descriptions appeared as complementary (in Bohr's sense) characteristics of behavior of microobjects.

Both types of description were referred to microobject only through explication of operational scheme, which united different, outwardly incompatible fragments of ontological notions. Such ways of construction of the physical picture of the world achieved philosophical justification, on the one hand, through a number of epistemological ideas (of special place of observer as macrooobject in the world, of correlativity between methods of explanation and description of object and cognitive means), on the other hand, due to development of "categorical network", which grasped general features of the object studied (notion of interactions as transformation of possibility into reality, understanding of causality in broad sense, as including probability aspects etc.).

This was how conceptual interpretation of the mathematical apparatus of quantum mechanics was constructed. At the time this theory was being formed the described way was, most likely, the only possible method of theoretical cognition of microworld. But later (in particular, at modern stage) there could be traced a tendency to vision of quantum objects as complicated dynamic self-regulating systems. As we have already mentioned, analysis of the quantum theory shows that its very conceptual structure has two levels of description of reality: on the one hand, notions describing integrity and stability of the system, on the other hand, notions expressing its typically random characteristics. The idea of such splitting of theoretical description corresponds to the idea of complicated systems characterized, on the one hand, by presence of subsystems with stochastic interaction between elements, on the other hand, by some "controlling" level, which provides integrity of the system.[88] Those achievements of the theory of quantized fields, which demonstrate how restricted the traditional notions of localization of particles are, also speak for such vision.

Denoting all these tendencies in development of physical knowledge, we should not forget the very vision of physical objects as complicated dynamic systems connected with the conception, which was formed due to development of cybernetics, the theory of systems and assimilation of large systems in technics. At the time of the development of quantum mechanics this conception had not been established in science yet, and in everyday life of physical thinking notions of objects as large systems were not employed. So it would be proper to ask: could the history of quantum physics follow any other ways in conditions of a different scientific environment? In principle we can assume (as a thought experiment) that cybernetics and corresponding assimilation of self-organizing systems in technics could have appeared before quantum physics and form new type of vision of objects in culture. Under such circumstances, constructing the picture of the world, physicists could imagine quantum objects as complicated dynamic systems and construct their theory in correspondence with this notion. But in this case all further evolution of physics would have been different. On this way of its development, most likely, there would have been not only gains, but also losses, since in such movement it is not necessary that physicists should explicate at once operational scheme of vision of the picture of the world (which means there would be no stimulus to develop the principle of relativity). The circumstance that quantum physics was developed based on the conception of complimentarity, having radically changed classical norms and ideals of physical cognition, directed evolution of science to a special way. There emerged a model of new cognitive movement, and now, even if physics constructs new system ontology (new picture of reality), it will not be a mere return to previously unrealized way of development: ontology should be introduced through construction of operational scheme, while a new theory can be created on base of including operational structures into the picture of the world.

Development of science (as well as any of the process of development) is realized as the transformation of possibility into reality, and not all possibilities are realized in its history. Prognosticating such processes always creates a tree of possibilities, taken into consideration different variants and direction of development. The ideas of the strictly determined development of science emerge only at retrospective look, when we analyze history and already know the result, when we reconstruct logic of movement of ideas, which led to this result. But also possible were such directions, which could be realized in case of other turns of historical development of civilization, but in the realized, real history of science they turned out "closed".

At the time of scientific revolutions, when reconstruction of foundations of science is carried out, culture chooses out of several potentially possible lines of future history of science those ones, which best correspond to fundamental values and worldview structures domineering in this culture.

Global scientific revolutions: from classical to post-non-classical science

In development of science we can distinguish periods, when all components of its foundations were transformed. Change of scientific pictures of the world was accompanied by radical change of structures of investigation and philosophical foundations of science. It would be correct to consider such periods as global revolutions, which can lead to change the type of scientific rationality.

In the history of natural science we can see four such revolutions. The first of them was the one of the 16th century, which marked the establishment of classical natural science.

Its appearance was inseparably linked with the creation of a special system of ideals and norms of investigation, which expressed directions of classical science and realized their concretization with regard to the dominant of mechanics in the system of scientific knowledge of that epoch.

According to the idea, which accompanies all classical natural science since the 17th century, objectiveness and concreteness of scientific knowledge are reached only in case when everything, that refers to the subject and procedures of his cognitive activity is eliminated. These procedures were regarded as given once and for all and unchangeable. The ideal was construction of an absolutely true picture of the nature. The main attention was paid to search for evident, visual, "following from experiment" ontological principle, on base of which investigator could construct theories explaining and predicting experimental facts.

In the $17^{th}-18^{th}$ centuries these ideals and normatives of investigation were fused with a number of concretizing propositions, which expressed attitudes of a mechanist understanding of the nature. Explanation was interpreted as search for mechanist reasons and substances – carriers of forces, which determine the phenomena observed. Understanding of justification included the idea of reduction of knowledge of the nature to fundamental principles and notions of mechanics.

According to these attitudes investigators constructed and developed the mechanist picture of the world, which at the same time appeared as both picture of reality, conformably to the sphere of physical knowledge, and as a general scientific picture of the world.

Finally, ideals, norms and ontological principles of natural science of the $17^{th}-18^{th}$ centuries were based on a specific system of philosophical foundations, where a domineering part belonged to ideas of mechanicism. The role of epistemological component was played by notions of cognition as observation and experimenting with natural objects, which reveal the mystery of their being to comprehending mind. Here mind itself got the status of sovereignty. In the ideal, it was interpreted so as it is at distance from things, as if taking a detached view observing them and investigating them, not determined by any premises besides properties and characteristics of the studied objects.

This system of epistemological ideas was linked with special notions of the objects studied. For the most part, investigators regarded them as small systems (mechanical devices), and, correspondingly, it was employed "categorical network", which determined understanding and cognition of the nature. Let us remember that a small system is characterized by a relatively small number of elements, their force interactions and strictly determined connections. For their assimilation it was enough to suppose that properties of the whole are completely determined by state and properties of its parts, imagine thing as a relatively stable body, and process – as transference of bodies in space in the course of time, interpret causality in Laplace's sense. The corresponding meanings were the ones which were distinguished in categories "thing", "process", "part", "whole", "causality", "space", "time" etc.; they constituted the ontological component of philosophical foundation of natural science of the $17^{th}-18^{th}$ centuries. That categorical matrix provided success of mechanics and predetermined reduction of all other areas of natural scientific investigation to its notions.

Radical changes in this integral and relatively stable system of foundations of natural history took place in the late 18th – the first half of the 19th century. They can be regarded as the second global scientific revolution, which determined transition to a new state of natural science: disciplinarily organized science.

At that time the mechanist picture of the world loses the status of the general scientific one. Specific pictures of reality, irreducible to the mechanic one, are outlined in biology, chemistry and other spheres of knowledge.

At the same time differentiation of disciplinary ideals and norms of investigation occurred. For example, biology and geology give birth to ideals of evolutionary explanation, while physics continues constructing its knowledge, disengaging itself from the idea of development. But here also, along with defining of the field theory, previously domineering norms of mechanical explanations are gradually becoming vague. All these changes referred mainly to the third layer of organization of ideals and norms of investigation, the layer that expressed specificity of studied objects. As to general cognitive attitudes of classical science, they are conserved in this historical period.

Corresponding to features of disciplinary organization of science, its philosophical foundations are transforming. They become heterogeneous, include a quite wide range of meanings of those basic categorical schemes, in accordance with which objects are assimilated (from conservation, within certain limits, mechanist tradition to including "thing", "state", "process", "law" of ideas of development into comprehension). In epistemology the problem of correlation of various methods of science, synthesis of knowledge occupies the central place. Its advance to the forefront is connected with loss of old integrity of the scientific picture of the world, and also with appearance of specificity of normative structures in different spheres of scientific investigation. Search for ways to integrity of science, the problem of differentiation and integration of knowledge turn into one of fundamental philosophical problems, conserving its urgency for all further development of science.

The first and the second global revolutions in natural science went on as development of classical science and its style of thinking.

The third global scientific revolution was connected with transformation of this style and the initiation of a new, non-classical natural science. It envelops the period from the late 19th to the middle of the 20th century. In that epoch we can see sort of chain reaction of revolutionary changes in different spheres of knowledge: in physics (discovery of divisibility of atom, establishing of relativistic and quantum theories), in cosmology (the conception of non-stationary Universe), in chemistry (quantum chemistry), in biology (establishing of genetics). Cybernetics and the theory of system, which played a most important role in development of the modern scientific picture of the world, appear.

In the course of all these revolutionary transformations, ideals and norms of new, non-classical science were formed. They were characterized by rejection of straightforward ontologism and understanding of relative truth of theories and picture of the nature, established at some stage of development of natural science. To counterbalance the ideal of the only true theory, "photographing" studied objects, non-classical science assumes truth of several concrete descriptions of the same reality, different from each other, since each of them may contain a grain of objectively true knowledge. Non-classical science interprets correlations between ontological postulates of science and characteristics of methods used for assimilation of the object; it takes such types of explanation and description, which contain, in evident form, references to means and operations of cognitive activity. The most eloquent samples of such approach were ideals and forms of explanation, description and demonstrative character of knowledge, established in quantum-relativistic physics. While in classical physics the ideal of explanation and description stipulated characteristic of object "in itself", without indicating means of its investigation, in quantum-relativistic physics requirement of clear fixation of features of observation means, which interact with the object, is a necessary condition of objectivity of explanation and description (the classical method of explanation can be presented as idealization, rational aspects of which are generalized within the new approach).

Ideals and norms of demonstrativity and justification of knowledge are changing. Unlike classical models, justification of theories in quantum-relativistic physics stipulated explication of operational foundation of the introduced system of notions (the "observability" principle) and elucidation of connections between new and all previous theories (the correspondence principle).

The new system of cognitive ideals and norms provided considerable expansion of the field of object studied, opening ways to assimilation of complicated self-regulating systems. Unlike small systems, such objects are characterized by level organization, presence of relatively autonomous and variable subsystems, mass stochastic interaction of their elements, existence of controlling level and feedbacks which provide integrity of system.

Just inclusion of such objects into the process of scientific investigation was the cause of sharp reconstructions in the pictures of reality of the leading branches of natural science. The processes of integration of those pictures and development of general scientific picture of the world were now realized on base of notions of the nature as a complicated dynamic system. It was stimulated by discovery of specificity of laws of micro, macro and megaworlds in physics and cosmology, intensive investigation of heredity mechanisms in close connection with studies of over-organism levels of life organization, discovery of general laws of control and feedback in cybernetics. This was the way science created premises for construction of an integral picture of the nature, where one could trace hierarchical organization of the Universe as a complex dynamic whole. Pictures of reality, established in separate disciplines, at that stage still conserved their independence, but each of them took part in forming notions later included into the general scientific picture of the world. The latter, in its turn, was regarded not as an exact and final portrait of the nature, but as a system of relatively true knowledge of the world which is permanently corrected and developed.

All these radical changes of notions of the world and investigation procedures were accompanied by forming new philosophical foundations of science.

The idea of historical changeability of scientific knowledge, relative truth of ontological principles, was developed in science, combined with new ideas of activity of the subject of cognition. It was regarded now not as outward from the world studied, but as the one inside it, determined by it. There emerges understanding of the circumstance that the nature's answers to our questions are determined not only by organization of the nature itself, but also by our method to raise questions, which depends on historical development of means and methods of cognitive activity. That was the base to grow new understanding of categories of truth, objectivity, fact, theory, explanation etc.

Radical transformations also can be seen in "ontological subsystem" of philosophical foundations of science. Development of quantum-relativistic physics, biology and cybernetics was connected with including new contents into categories of part and whole, causality, contingency and necessity, thing, process, state etc. Theoretically it can be showed that this "categorical network" introduced a new image of an object, which appeared as a complex system. Notions of correlation between part and whole, applied to such systems, include ideas saying that states of the whole are not reducible to the sum of states of its parts. An important role in description of dynamics of a system is achieved by categories of contingency, the potentially possible and the actual. Causality cannot be reduced only to Laplace's formula; there emerges the idea of "probability causality", broadening the sense of traditional understanding of this category. New content fills the category of object: it is regarded now not as self-identical thing (body), but as a process reproducing some stable states and changeable in some other characteristics.

All reconstructions of the foundations of science described above, characterizing global revolutions in natural science, were caused not only by their expansion to new object spheres, or the discovering of new types of objects, but also by change of place and time of functions of science in social life.

Foundations of natural science, in the epoch of its establishment (the first revolution), were formed in the context of rationalist views of early bourgeois revolutions, settling of new (with respect to medieval ideology) understanding of people's relations with the nature, new notions of destination of cognition, truth of knowledge etc.

Establishing foundations of disciplinary natural science of the end of the 18th – the early 19th centuries took place in special conditions; the productive role of science grew considerably. Scientific knowledge turned into a special product which had commodity value and gave profits at its production consumption. That was the time of beginning of formation a system of applied and engineer-technical sciences as mediator between fundamental knowledge and production. Different spheres of scientific activity are specializing, and there scientific associations, corresponding to this specialization appear.

Transfer from classical to non-classical natural science was prepared by a change of structures of spiritual production in the European culture of the second half of the 19th – early 20th centuries, crisis of worldview attitudes of classical rationalism, forming of new understanding of rationality in different spheres of spiritual culture, when consciousness, comprehending reality, permanently comes across situations of its being inside that very reality, feeling its dependence on social circumstances, which, to a large extent, determine attitudes of cognition, its value and aimed orientation.[89]

In the modern age, in the last third of the 20th century, we witness new radical changes in foundation of science. These changes can be characterized as the fourth global scientific revolution, which is giving birth to new post-non-classical science.

Intensive application of scientific knowledge in practically all spheres of social life, revolution in means of storage and getting knowledge (computerization of science, appearance of complicated and expensive technical aggregates, which attend to investigation crews and function like means of industrial production etc.) change the character of scientific activity. Along with disciplinary investigations, interdisciplinary and problem oriented forms of investigation activity are more and more advancing to the forefront. While classical science aimed at comprehension of more and more narrowing, isolated fragment of reality, presenting object of such and such scientific discipline, specificity of modern science of the late 20th century is determined by complex investigations, where specialists of various spheres of knowledge take part. Organization of such investigation in many respects depends on determination of priority directions, their financing, manpower training etc. In the very process of determination of science research priorities, along with purely cognitive aims, economical and social problems are playing a more and more important role.

Realization of complex programs engenders a special situation of joining, in a whole activity system, of activity of theoretical and experimental knowledge, intensification of direct and reverse connections between them. The result is: intensification of processes of interaction of the principles and notions of the pictures of reality, established in various sciences. More and more often changes in these pictures take place not so much under influence of intradisciplinary factors, as by way of "paradigmatic grafting" of ideas translated from other sciences. This process is gradually erasing the strict demarcation lines between the pictures of reality determining vision of the object of such and such science. They are becoming mutually dependent and the represent fragments of an integral general scientific picture of the world.

Its development is influenced not only by achievements of fundamental sciences, but also by the results of interdisciplinary applied investigations. In this respect it would be appropriate to remind the reader, for instance, that the ideas of synergetics (which caused a revolt in the system of our notions of the nature) emerged and developed in the course of numerous applied investigations, which brought to light effects of phase transitions and formation of dissipative structures (structures in liquids, chemical waves, laser beams, instabilities of plasma, the phenomena of exhaust and flutter).

In interdisciplinary investigations, science usually faces complicated system objects, which are studied in separate disciplines only in fragments, so effects of their system structure can be not discovered at all in a narrow disciplinary approach, but only in synthesis of fundamental and applied problems in problem oriented search.

Unique systems, characterized by openness and self-development, more and more often become objects of modern interdisciplinary investigations. Objects of such type gradually start determining also character of object spheres of the main fundamental sciences, the aspect of modern post-non-classical science.

Historically developing systems are more complicated, even in comparison with self-regulating systems. The latter are a special state of dynamics of a historical object, a kind of cut, a stable stage of its evolution. Historical evolution itself is characterized by transition from one relatively stable system to another system with a new level organization of elements and self-regulation. The development of each new level of the system is accompanied by its transition through a state of instability (the bifurcation point), and at those moments small random influences can cause appearance of new structures. Dealing with such systems requires fundamentally new strategies. Self-regulating systems are characterized by cooperative effects, fundamental irreversibility of processes. People's interaction with them takes place in such a way that the man's action itself is not something from outside, but is included into the system, every time transforming the field of its possible states. Entering the interaction, man now deals not with hard things and properties, but with certain "possibility constellations". Every time, in the course of his activity, man faces the problem of choice of a certain line of development from a number of possible ways of the system's evolution. Moreover, this choice is irreversible and, in most cases, cannot be simply calculated.

In natural science, the first disciplines, which faced the necessity of taking into account features of historically developing systems, were biology, astronomy and the Earth sciences. They had formed pictures of reality, which included the idea of historicism and notions of unique developing objects (biosphere, the Metagalaxy, the Earth as a system of interaction of geological, biological and technogenic processes). In the last decades physics took this same way. The notion of historical evolution of physical objects is gradually entering the picture of physical reality: on the one hand, through development of modern cosmology (the idea of "Big Bang" and establishing various types of physical objects in the process of historical development of the Metagalaxy), on the other hand – due to refining the ideas of thermodynamics of non-equilibrium processes (I. Prigogine) and synergetics.

It is the ideas of evolution and historicism that became basis for the synthesis of pictures of reality elaborated in fundamental sciences, which fuse them into an integral picture of historical development of the nature and people and make them only relatively independent fragments of the general scientific picture of the world.

Orientation of modern science at investigation of complicated, historically developing systems is considerably reconstructing ideals and norms of investigation activity. Historical character of a system complex object and variability of its behavior stipulate a wide application of special methods of description and prediction of its states – construction of scenarios of possible lines of development of the system in bifurcation points. The ideal of theory structure as an axiomatic-deductive system feels more competition from theoretical descriptions, based on application of the approximation method, theoretical schemes using computer programs etc. Natural science is more and more widely absorbing the ideal of historical reconstruction, which appears as a special type of theoretical knowledge, which previously used to be applied mainly in humanities (history, archeology, historical linguistics etc.).

Samples of historical reconstructions can be found not only in disciplines, which traditionally study evolutionary objects (biology, geology), but also in modern cosmology and astrophysics: modern models, which describe development of the Metagalaxy, can be estimated as historical reconstructions through which one can reproduce the main stages of evolution of this developing object, historically unique.

Notions of strategies of empirical investigation are also changing. The ideal of reproducibility of experiment, conformably to developing systems, should be understood in a special sense. If these systems are typologized, i.e. if it is possible to experiment on many samples, and each of them can be distinguished as the same initial state, the experiment will lead us to one and the same result, with regard for probabilistic lines of evolution of the system.

But, besides developing systems, which outline certain classes of objects, there are also unique, historically developing systems. Experiments based on energetic and force interaction with such system principally will not enable us to reproduce it in the same initial state. The very act of primary "preparation" of this state changes the system, sending it in a new direction, and irreversibility of the processes of development prevents us from reproducing the initial state again. That is why unique developing systems require special strategies of experimental investigation. Their empirical analysis is carried out mainly by method of calculating experiment on a computer, and that allows us to elucidating the diversity of possible structures, which can be born by the system.

Among historically developing systems of modern science a specific place belongs to natural complexes, which include man as a component. Examples of such "man-measured" complexes can be medical-biological objects, including biosphere as a whole (global ecology), objects of biotechnology (first of all, genetic engineering), "man – machine" systems (including complicated information complexes and artificial intellect systems) etc.

In studies of "man-measured" objects, the search for the truth is connected with determination of strategy and possible directions of reorganization of such objects, and that directly affects humanist values. We cannot be free in experimenting with systems of such type. In the process of their investigation and practical assimilation, a special role comes to knowledge of prohibitions on certain strategies of interaction, potentially containing catastrophic consequences.

In this respect, the ideal of neutral value investigation is transformed. Objectively true explanation and description conformably to "man-measured" objects not only assumes, but also prescribes including axiological factors into the explaining statements. It becomes necessary to explicate connections of fundamental intrascience values (search for the truth, growth of knowledge) and extrascience values of social character. In modern program-oriented investigations this explication is realized in the social examination of programs. At the same time, in the course of the investigation activities with "man-measured" objects, the investigator has to solve certain ethical problems, determining limits of possible intrusion into the object. The inner ethics of science, stimulating search for the truth and aiming at augment of new knowledge, constantly correlates, in these circumstances, with general humanist principles and values. Development of all these new methodological directions and notions of the studied objects leads to considerable modernization of philosophical foundations of science.

Scientific cognition is now considered in the context of the social conditions of its existence and its social consequences as a specific part of the life of society, determined at each stage of its development by a general state of culture of the corresponding historical epoch, its value orientation and worldview attitudes. We comprehend historical changeability not only of ontological postulates, but also of the very ideals and norms of cognition. Correspondingly, the content of such categories as "theory", "method", "fact", "justification", "explanation" etc. is developed and enriched.

"Categorical matrix", which provides understanding and cognition of developing objects, is now domineering in the ontological component of philosophical foundations of science. There emerge new understanding of categories of space (consideration of historical time of the system, hierarchy of space-time forms), categories of possibility and reality (the idea of diversity of potentially possible lines of development in bifurcation points), and category of determination (preceding history determines choice of the system reacting to influences from outside) etc.

Three large stages of historical development of science, each of them opened by a global scientific revolution, can be characterized as three types of scientific rationality, changing each other in the history of technogenic civilization.[90] These are: classical rationality (corresponding to classical science in its two states – disciplinary and disciplinarily organized); non-classical rationality (corresponding to non-classical science); post-non-classical rationality. Between them, as stages of development of science, there are "spans". Appearance of every new type of rationality did not reject the previous one, but only limited the sphere of its application by certain types of problems.

Every stage is characterized by a special state of scientific activity, aimed at permanent growth of objectively true knowledge. If we sketchily present this activity as relations "subject – means – object" (value-goal structures of activity, knowledge and practices of application of methods and means included into understanding of subject), the described stages of evolution of science, representing different types of scientific rationality, are characterized by different depths of reflection with respect to scientific activity itself.

The classical type of scientific rationality, concentrating attention at an object, in theoretical description and explanation, tends to eliminate everything, which refers to the subject, means and operations of his activity. Such elimination is considered as a necessary condition for obtaining objectively true knowledge of the world. Goals and values of science, determining strategies of investigation and methods of fragmenting the world, at this stage, as well as at all other stages, are determined by the worldview attitudes and value orientation domineering in the culture. But classical science does not comprehend these determinations.

Sketchily this type of scientific activity can be presented as follows:

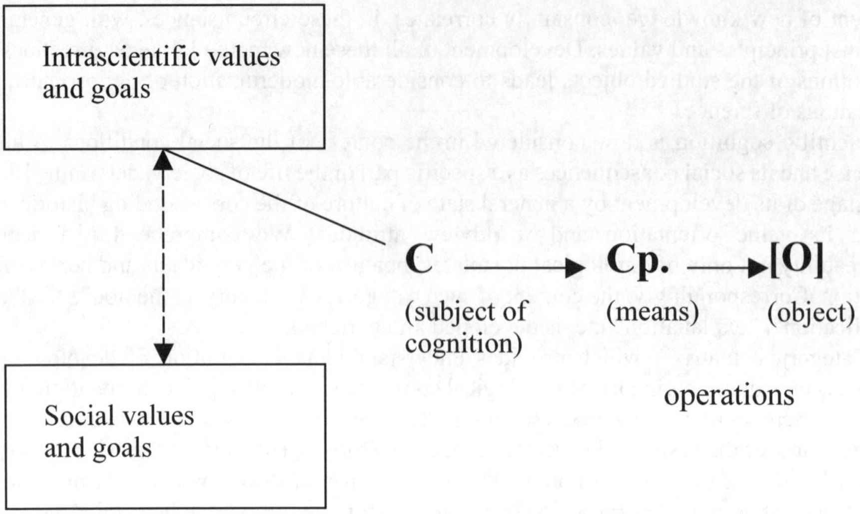

The non-classical type of scientific rationality takes into account connections between knowledge of object and the character of means and operations of activity. Explication of these connections is considered as conditions of objectively true description and explanation of the world. But connections between intrascience and social values and goals still are not subject of scientific reflection, though implicitly they determine the character of knowledge (they determine what and how we distinguish and comprehend the world).

In a scheme this type of scientific activity can be presented this way:

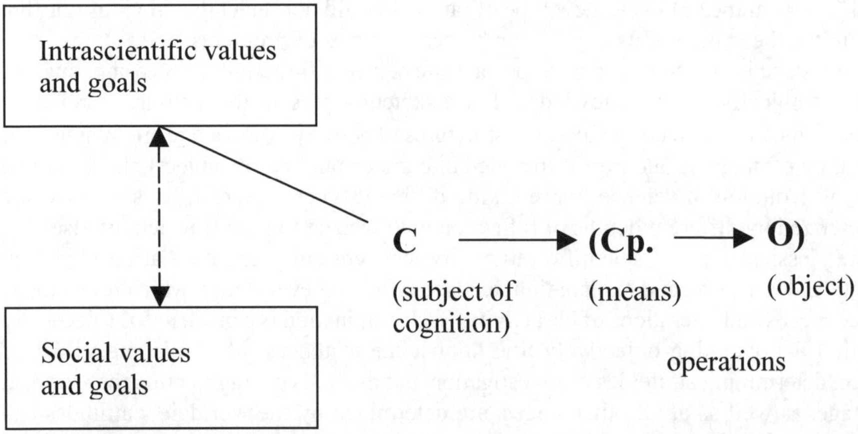

The post-non-classical type of scientific rationality broadens the field of reflection over activity. It takes into account correlation of obtained knowledge of the object not only with specificity of means and operations of activity, but also with value-goal structures. Here we explicate the connection between intrascience goals and extra-scientific, social values and goals.

Here is a scheme depicting this type of scientific cognition:

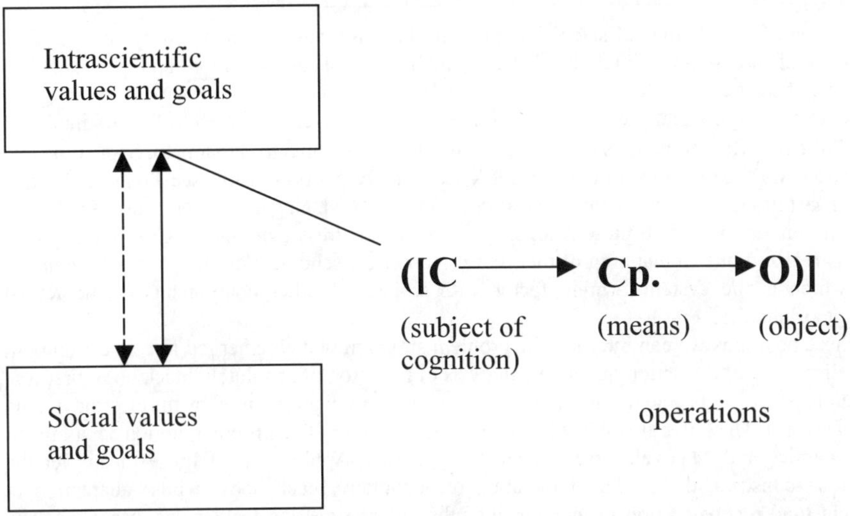

Every new stage of scientific rationality is characterized by specific, proper foundations of the science, which let us distinguish in the world and investigate corresponding types of system objects (simple, complicated, self-developing systems). Appearance of a new type of rationality and a new image of science should not be regarded in a simplified way. On the contrary, there is succession between them. Non-classical science did not at all destroy classical rationality, but only limited the sphere of its applicability. In solving a number of problems, non-classical ideas of the world and cognition turned out redundant; the investigator could be guided by traditional classical models (for instance, solving some problems of celestial mechanics one did not have to use norms of quantum relativistic description; it was quite enough to remain within the scope of classical normatives of investigation). In exactly the same way, establishing post-non-classical science does not cause destruction of all notions and cognitive attitudes of non-classical and classical investigation. They will be used in certain cognitive situations, though they will lose the status of domineering directions, determining the image of science.

Modern science – at the forefront of its search – put into focus research of unique historically developing systems, where man himself is included as a special component, into the focus of investigation. In this situation, the requirement of explication of values not only does not contradict the direction to getting objectively true knowledge of the world, but also is a premise for realization of this attitude. We have all reasons to believe that these processes will strengthen along with the development of modern science. Technogenic civilization is now entering the period of a special type of progress, when humanist reference-points are becoming initial in the determination of strategies of scientific search.

NOTES: CHAPTER 6

[1] We are turning to the analysis of the mentioned fragment of history of physics because

reconstruction of foundations of scientific search in this case was accompanied by change of all components of foundations, including ideals and norms of investigation and philosophical foundations of science.

[2] In Russian methodological literature paradoxes of such type were analyzed as "contradiction of meeting" of two different theories (in this case – mechanics and electrodynamics). Some time ago this approach was realized by M.I. Podgoretsky and Ya.A. Smorodinsky (see Podgoretsky and Smorodinsky (1969)). Later this approach was developed by R.M. Nugaev (see Nugaev (1989)). Not denying the importance of all these results, I would like to stress that "meeting" of the physical theories is realized due to mapping of their core (theoretical schemes) on the physical picture of reality, which is the system-forming factor with respect to other components of theoretical knowledge of physics.

[3] From these positions we can interpret the problem situation, which emerged in connection with Planck's discovery of the action quantum. Analysis of radiation of absolutely black body first was really a quite particular problem in the course of the investigation program set by the electrodynamic picture the world. The latter also shaped the means of solution of this problem: notion apparatus of thermodynamics and Maxwell-Lorentz's electrodynamics. Application of those means let the investigators construct the model of radiation of absolutely black body, whose adaptation to experiment (and reconstruction in the course of that adaptation) led to Planck's discovery. The radiation law, offered by Planck, was coordinated with all experimental data (in this regard the special problem was solved). But in mapping of the model, relatively to which the law was formulated, to electrodynamic picture of the world, there appeared a paradox: the model supposed that oscillators absorb and emit electromagnetic energy in portions multiple of $h\nu$, while in the picture of the world electromagnetic radiation was regarded as continuous medium. Hence there emerged a problem: what was the real nature of electromagnetic field? The solution of that problem was connected with further reconstruction of electromagnetic picture of the world, with introduction of notions of corpuscular-wave character of electromagnetic field (the idea of photons).

[4] See Einstein (1965-1967, vol.4, pp.266-267).

[5] Holton (1979, pp.218-226).

[6] Einstein (1965-1967, vol.4, pp.15).

[7] Ibid, Vol. 1, p.66.

[8] Einstein (1931, p.66).

[9] See, e.g., Mamchur (1975).

[10] Einstein (1965-1967, vol.2, p.120).

[11] Ibid, Vol.4, p.279.

[12] Chudinov (1976, p.40).

[13] See, e.g., Mamchur (1975).

[14] *Classical science* is an accepted term for the period of development of natural science before the revolution of the end of the 19th – the beginning of the 20th centuries, as opposed to science of the 20th century (modern science).

[15] On the difference of classical and modern stages of the development of philosophy and on specificity of cognitive attitudes, characterizing each of those stages, see Mamardashvili, Solovyov and Shvyrev (1972).

[16] For more details see *Ideals and Norms of Scientific Investigation* (1981, pp.37-56).

[17] We find it appropriate to remind the reader that E. Mach sometimes used the term "experiment" in his historical and scientific investigations also in a different sense. He withdrew from interpretation of experiment as complex of perceptions of comprehending subject, and interpreted it as practical action, as experiment providing the receipt of observation data. Most of Mach's constructive ideas were connected with that same, tacitly employed understanding.

[18] Holton (1973).

[19] Ibid, p.225.

[20] Einstein (1965-1967, vol.4, p.277).

[21] See Jakobson (1985, pp.307-308).
[22] Ibid, p.309.
[23] Einstein (1965-1967, vol.1, p.175).
[24] Though all above given justifications of fundamental methodological status of the relativity principle were not explicated by Einstein, understanding of this status can be traced clearly enough in the works of the creator of the relativity theory.
[25] This method of raising problems, as expression of new ideals and normatives of justification of theory, characterized Einstein's work of the period of construction of the relativity. Let us denote that it stimulated not only creation of the special relativity, but transition to the general relativity as well. The process of such transition was connected with generalization of the relativity principle: distinguishing the profound content of this principle as presumption of physical measuring (the laws of nature are displayed in the same way in all frames of reference) and spread the relativity principle over non-inertial frames of reference. The answer, how the nature will look at this new scheme of measuring, led to construction of the general relativity.
[26] Einstein (1965-1967, vol.1, p.7).
[27] Holton (1974).
[28] Einstein (1965-1967, vol.1, pp.7-8).
[29] Ibid, p.179.
[30] Ibid, pp.146-179.
[31] Analyzing the problem of synchronization of clocks, Einstein comes across a seeming contradiction: to measure time, one has to synchronize clocks located in different places of the frame of reference, and that can be done by means of light signals; but in this case it is necessary to know the light speed in its passing from one clocks (point A) to another (point B), while measuring the light speed, in turn, required the notion of time. Here we faced a logical circle (Einstein (1965-1967, vol.1, pp.34, 223)). The way out was found due to application of the postulate of constancy of the light speed.
[32] See Keswani (1973, p.269).
[33] Ibid, p.247.
[34] Tyapkin (1973, p.303).
[35] M.M. Bakhtin called this type of architecture of a work of art a polyphonic novel, emphasizing that Dostoevsky's work serves as consolidation of this fundamentally new form, which destroyed the tradition of monologic (homophonic) novel which used to prevail in European culture (Bakhtin (1979, p.320)).
[36] See more details in Stepin and Kuznetsova (1981. pp.260-279).
[37] Dorfman (1974, p.188).
[38] Lamarck (1807).
[39] Lamarck (1959, p.148).
[40] Lamettrie(1796).
[41] Ibid.
[42] Ibid.
[43] Holbach (1770).
[44] Ibid.
[45] Ibid.
[46] Saint-Simon (1966).
[47] Ibid.
[48] Saint-Simon (1948, p.234).
[49] Fourier (1953).
[50] See Toffler (1986, p.14).
[51] Jua (1975, p.93).
[52] Dorfman (1974, p.188).
[53] Solovyov (1971, p.24).

[54] Ibid, p.24.
[55] Jua (1975, p.93).
[56] I. Newton was one of the first to put this idea forward; it was justified by J. Biot and P.Laplace, then it directed investigations of J. Richter, A. Lavoisier, L. Proust, C. Berthollet et al. See Solovyov (1971, pp.90-99).
[57] Solovyov and Kurashov (1983, p.108).
[58] *A Becoming of Chemistry as Science* (1983, p.108).
[59] Lavoisier (1943, p.362).
[60] Dorfman (1979, p.127).
[61] Ibid.
[62] Lamarck (1809).
[63] Lamarck (1959, p.365).
[64] See Ravikovich (1976, pp.42-43).
[65] Mendel (1959).
[66] See Pastushny (1981, p.17).
[67] See Spencer (1997, pp.282-299).
[68] See Rorty (1985, p.67).
[69] Kuznetsov V.I. (1973, pp.289-293, 295).
[70] Shmalgauzen (1968, p.103).
[71] Ibid.
[72] Ibid, p.147.
[73] Berg and Lyapunov (1968, p.13).
[74] Ibid.
[75] Ibid.
[76] *The History of biology from the beginning of XXth century to nowadays* (1975, pp.591-592).
[77] See Shaumian (1965, pp.97-135, 370-373).
[78] Ibid, pp.370-373.
[79] Jakobson (1985, pp.389-390).
[80] Ibid, pp.393-394.
[81] Ibid, pp.394.
[82] Feynman, Leighton and Sands (1963).
[83] In this case we are talking only about research programs characterized by features of the accepted foundations of the investigation.
[84] See Mandelstam (1948, p.20).
[85] See Feynman (1968, pp.195-196).
[86] Ibid, p.199.
[87] Mamardashvili (1968, p.19).
[88] See Sachkov (1974, pp.71-72), (1994, pp.132-133).
[89] See Mamardashvili, Solovyov and Shvyrev (1972).
[90] Let us mark once again that typology of global scientific revolutions is not identical to distinguishing those stages of development of science, which are characterized as classical, non-classical and post-non-classical science. Within classical science two global scientific revolutions can be fixed: the first one is connected with establishing of classical science itself (appearance of mechanics and further development of science of the 17th century), the second one – with formation of disciplinary organization of science (the late 17th – the first half of the 19th centuries). As to further radical changes of strategies of scientific research, connected with establishing of non-classical, then of post-non-classical science, these changes are also global scientific revolutions. Those revolutions, which determined transition from classical to non-classical, and then to post-non-classical science, at the same time are epochs of establishing the new type of scientific rationality.

CHAPTER SEVEN

STRATEGIES OF THEORETICAL INVESTIGATION IN THE EPOCH OF POST-NON-CLASSICAL SCIENCE

UNIVERSAL EVOLUTIONISM AS FOUNDATION OF THE MODERN SCIENTIFIC PICTURE OF THE WORLD

The transition of science to post-non-classical stage of development has created new premises of establishing a common picture of the world.[1] For a long time this idea of such unity existed as an ideal. But in the last third of the 20th century there appeared real possibilities to unite notions of the three main spheres of being – non-living nature, organic world and social life – into an integral picture on the foundation of basic principles having general scientific status.

These principles do not deny specificity of each concrete branch of knowledge, but, at the same time, are invariant in the diversity of disciplinary ontologies. The construction of such principles was connected with a revision of foundation of many scientific disciplines. Such revision is one of aspects of the great cultural transformation taking place in our age.[2]

If we were to briefly characterize modern tendencies of synthesis of scientific knowledge: they are expressed in desire to build a general scientific picture of the world on base of principles of universal evolutionism, uniting into a whole integrity the ideas of systemic and evolutionary approaches.

The development of evolution ideas has quite a long history. Even in the 19th century they were applied in certain spheres of knowledge, but they were perceived more as exception with respect to the world as a whole.[3]

Most completely was the principle of evolution outlined within the framework of biology and became its fundamental principle since the epoch of Ch. Darwin. But up to nowadays it has not been domineering in natural science. In many respects it happened due to the fact that for a long time the prevailing scientific discipline was physics, that translated its ideals and norms to other branches of knowledge. Traditionally physics investigated fundamental structures of the Universe, so it had always been among those sciences which had a claim on forming basic ideas of a general scientific picture of the world. But in the course of most part of its history, physics did not include – in evident form – the principle of development into the set of its fundamental principles.

As to biology, it has not reached the high status of theoretically developed science, and is now at the stage of theorizing. Its notions dealt with the living nature, which traditionally was not considered as foundation of the Universe. That is why biology, taking part in construction of the general scientific picture of the world, for a long time did not aspire that its fundamental ideas and principles got universal scientific meaning and were applied in all other spheres of investigation.

Paradigmatic incompatibility of classical physics and biology was discovered in the 19th century as contradiction between the theses of Darwin's theory of evolution and the second law of thermodynamics.

According to the theory of evolution, in the world there permanently appear more and more complicated living systems, organized forms and states of living matter. The second law of thermodynamics demonstrated that evolution of physical systems leads to a situation when an isolated system is purposefully and irreversibly shifted to the state of equilibrium.

In other words, biological theory spoke about the evolution process as construction of more and more complicated and organized living systems, while thermodynamics – about destruction, permanent entropy growth. These collisions of physics and biology required settlement, and premises for this could be an evolutionary view of the Universe as a whole, translation of the evolutionary approach into physics, which would lead to reformulation of fundamental physical theories. But this situation is characteristic only for the science of the last third of the 20th century.

The ideas of universal character of the evolution processes in the Universe are realized in modern science in the conception of global (universal) evolutionism. Its principles enable us to describe uniformly the enormous diversity of processes taking place in non-living nature, living matter, society[4].

The conception of universal evolutionism is based on a certain complex of knowledge, grounds in concrete scientific disciplines, and, at the same time, includes a number of philosophical, worldview directions. It refers to that layer of knowledge, which is traditionally designated by the term "scientific picture of the world".

Why is it that the modern stage of functioning of science requires the ideas of universal evolutionism as principal ones, allowing investigators to elucidate the general picture of integral process of development of nature and society? Before answering this question, we have to specify, what universal evolutionism is, and
to understand, what contributed to establishing of its ideas in science, and not at the level of metaphysical speculations, but as generalization of concrete scientific data.

Universal (global) evolutionism[5] is often characterized as a principle, which extrapolates the ideas of evolution, justified in biology, astronomy and geology, over all spheres of reality and consideration of non-living, living and social matters as a united, universal evolutionary process.

It is really a very important aspect in understanding global evolutionism. But it does not exhaust the content of this principle. It is important to take into consideration that in the 20th century the evolutionary approach itself acquired new features, distinguishing it from classical evolutionism of the 19th century, which described phenomenology of development more than system characteristics of developing objects.

When in the 1940s–1950s general system theory appeared and system approach settled, it brought fundamentally new content in the conception of evolutionism. The idea of system consideration of objects turned out quite heuristic first of all within biology, where it led to unfold the problem of structural levels of organization of living matter, analysis of different connections both within a certain system and between systems of different grades of complexity. System consideration of object first of all stipulates uncovering of integrity of the system studied, its intercommunication with environment, analysis of properties of the components and their inter-communication within the system. The system approach, developed in biology, regards objects not as mere systems, but as self-organizing systems of open character. N.N. Moiseev notes that today we see the processes of evolution, self-organization of matter more widely than at Darwin's time, and the notions of heredity, changeability, and selection, for us are filled with another deeper content.[6]

From his point of view, everything happening in the world, the operation of all natural and social laws can be presented as permanent selection, when only several classes and types of states are selected from the diversity of possibilities. In this sense all dynamic systems possess capability "to make a choice" though concrete results of "choice" as a rule might not be predicted in advance.

N.N. Moiseev indicates that we can distinguish two types of mechanisms regulating such "selection". On the one hand, adaptation, under effect of which the system does not obtain fundamentally new properties; on the other hand, so called bifurcation, connected with radical reconstruction of the system. But, besides these mechanisms, to explain self-organization, we are to point out one more characteristic of direction of self-organizing processes, marked by Moiseev as principle of entropy economy, which gives preference to complicated systems compared to simple ones. This principle is formulated as follows: if under conditions given several types of organization of matter are possible, not contradicting to the conservation laws and other principles, the one which has most chances to stability and further development is that one, which allows investigators to utilize outer energy to the largest scale, most effectively.[7]

Forming of self-organizing systems can be regarded as a special period of a developing object, a kind of "synchronous section" at some stage of its evolution. The evolution itself can be presented as transition from one type of self-organizing system to another ("diachronic section"). As a result, analysis of evolutional characteristics is closely connected with systemic consideration of objects.

Universal evolutionism is that very thing which is the combination of the evolution ideas with the ideas of system approach. In this respect, universal evolutionism not only spreads development over all spheres of being (establishing universal connection between non-living, living and social matter), but also overcomes the narrowness of phenomenological description of development, linking this description with the ideas and methods of system analysis.

Many natural scientific disciplines contributed to justification of universal evolutionism.[8]

But the decisive part in its establishing as principle of construction of the modern general scientific picture of the world belonged to three most important conceptual trends in the science of the 20th century: first, the theory of non-stationary Universe; second, synergetics; third, the theory of biological evolution and developed on its base the conception of biosphere and noosphere.

The beginning of the 20th century was marked by a chain of scientific revolutions, among those an essential place belonged to the revolution in cosmology. It played an important role in settling the idea of evolution in non-organic nature and caused radical reconstruction in notions of the Universe.

We are talking about elucidating the theory of expanding Universe. That theory introduced the following ideas of cosmic evolution: about 15–20 billion years ago the Universe started to expand from the point of singularity as a result of "the Big Bang"; the Universe first was hot and dense, but cooled off in the course of expansion, and the matter in the Universe, while cooling off, was condensed in galaxies. The latter ones, in their turn, were broken up into stars, integrated and formed large clusters. In the process of creation and death of the first generations of stars heavy elements were synthesized. After turning of the stars into red giants, they threw out matter condensed in dust structures. Gas-dust clouds formed new stars, and there appear the diversity of cosmic bodies.[9]The 'Big Bang' theory created the picture of evolution of the Universe in general. In its origins there lay discovery of A.A. Fridman, which caused doubts in the postulate of the Universe, stationary in time. Analyzing Einstein's "world equations", describing metrics of four-dimensional curved space, Fridman found their non-stationary solutions and offered three possible models of the Universe. In two of them the radius of curvature was to grow, and the Universe, correspondingly, extends. The third model suggested the picture of pulsating Universe with periodically changing radius of curvature.[10]

The model of the expanding Universe led to three important predictions, which later could be tested by means of empirical observations. First, we mean that, with expansion of the Universe, galaxies are moving away from each other at a speed proportional to distance between them; second, this model predicted existence of microwave background radiation, piercing through the whole Universe and being relic of its hot state of the beginning of its expansion; third, this model predicted formation of light chemical elements out of protons and neutrons at the first minute after expansion had begun.[11]

The model of expanding Universe has essentially transformed our ideas of the world. It required that we should include the idea of cosmic evolution into the scientific picture of the world. This was the way to create a description of the non-organic world in terms of evolution, uncovering common evolutionary characteristics of different levels of its organization and, finally, construct an integral picture of the world on this base.

In the middle of the 20th century the ideas of evolution of the Universe got a new impulse. The theory of expanding Universe, though quite well described events which took place a second after the beginning of the expansion, faced considerable difficulties in its attempts to characterize the most mysterious stages of that evolution from the initial explosion to the world second after it. The answers to these questions, to a large extent, were given within the theory of inflating Universe. This theory emerged at the junction of cosmology and physics of elementary particles. The key element of the inflating Universe was the so called inflation phase – the stage of accelerated expansion. It lasted for 10–32 seconds, and during that time the diameter of the Universe increased 1,050 times. After enormous expansion, the phase with broken symmetry was established once and for all, and that led to change of the state of vacuum and, finally, to creation of various types of elementary particles.[12] In our Universe matter prevails over antimatter, and in this respect we live in an asymmetric Universe. The prediction of asymmetry of matter and antimatter in the Universe was the result of combination of ideas of "Grand Unification" in the elementary particles theory with the model of inflating Universe. Within the program of the "Grand Unification" investigators put forward the idea of initial symmetry, uniting the main types of interaction (strong, electromagnetic, weak and gravitational). It is supposed that in the very beginning of the Universe's evolution (10–46 seconds after "the Big Bang") spontaneous breaking of this symmetry caused a "split" of the initial state and created the four main interactions of the nature. In this approach the types of interaction are presented not as given once and for all, but as emerging in the process of evolution.

Modern science spreads the development ideas to fundamental structures of the Universe, establishing connections between evolution of the Universe and the process of formation of elementary particles. All this enable us to consider the Universe as a unique laboratory for verification of modern theories of elementary particles and their interactions.[13]

The theory of inflating Universe has radically changed our vision of the world: in particular, it changed "the view at the Universe as something homogenous and isotropic, and there emerged a new vision of the Universe as consisting of many locally homogenous and isotropic mini-universes, where properties of elementary particles, amount of vacuum energy and dimensionality of the space-time can vary".[14]

Transforming the established physical picture of the world, the theory of inflating Universe gives a new impulse to defining the general scientific picture of the world on base of ideas of global evolutionism. It requires corrections in philosophical, worldview foundations of science, putting forward a number of important problems of worldview character. The new theory allows us to consider the observable Universe only as a small part of the Universe as a whole, and that means that we have the right to assume existence of quite many evolving universes.[15] In the process of evolution most of them are unable to give create such diversity of organization forms which is proper to our Universe (Metagalaxy). But then there emerges a question: why our Universe is as it is, and how progressive evolution of matter is possible in it? Can we regard appearance of life on the Earth, as well as origin of humanity, random in the existing Universe, or is formation of man a regular process in an evolving Universe? What place belongs to this event in the processes of evolution, how does it influence upon the course of the evolution processes?

One of the variants of answer is based on so called anthropic principle founded on hidden supposition of existence of a multitude of universes, and life appears where there are special conditions. According to one variation of the anthropic principle, what we expect to observe, should be limited by conditions necessary for our existence as observers. Though our position is not necessarily central, it is inevitably privileged, in some sense.[16] This formulation of the anthropic principle let B. Carter concentrate his attention mainly on its two versions: "weak" and "strong", which got quite a broad interpretation. According to the first one, our position in the Universe is inevitably privileged in the sense that it should be compatible with our existence observers. The "strong" anthropic principle states that the Universe is to be shaped so, that at certain stage of evolution it permitted existence of observers.[17] Many times did investigators emphasize wonderful coordination of the main properties of the Universe (A.D. Zelmanov, G.M. Idlis, P. Davies and others). Its physical parameters, such as constants of physical interactions, masses of elementary particles, dimensionality of space, are decisive for the existence of the present structure of the Universe, since any violation of one of them could lead to impossibility of progressive evolution, and our existence as observers would also be impossible. The anthropic principle drives investigators into the sphere of worldview problems, making them think again about the question of people's place in the world, their attitude to this world. New data obtained in cosmology let us suppose that objective properties of the Universe as a whole create possibility of emerging of life, intellect at certain stages of its evolution. What is more, potential possibilities of these processes were present even at the earliest stages of development of the Metagalaxy, when numeric values of the world constants, which determined the character of further evolutionary changes, were formed.[18] All these results can be evaluated as one of the essential factors of settling the idea of global evolutionism in the modern scientific picture of the world.

Not less important role in forming of these ideas belonged to the theory of self-organization (synergetics). It studies any self-organizing systems which consist of numerous subsystems (electrons, atoms, molecules, cells, neurons, organs, complex multicellular organisms, people, people's communities), paying special attention to coherent, coordinated state of self-organization processes in complicated systems of different nature.[19] To be considered as self-organizing, a system should satisfied at least four conditions: 1) the system should be thermodynamically open; 2) dynamic equations of the system are non-linear; 3) deviation from balance exceeds critical threshold; 4) processes in the system occur cooperatively (W. Ebeling). Self-organization here is considered as one of the main qualities of moving matter and includes all processes of self-structuring, self-regulation, self-reproduction. It plays the role of a process leading to formation of new structures.[20]

For a pretty long time self-organization was correlated only with living systems; as to objects of non-living nature, it was believed that if they do evolve, they evolve only toward chaos and disorder; this belief was proved by the second law of thermodynamics. But here we came across a radical problem: how systems of such kind could give creation to objects of living nature, capable of self-organization. There emerged a methodologically important question of interrelation between non-living and living matter. To answer it, we had to change paradigmatic principles of science and, in particular, eliminate gaps between evolutional paradigm of biology and traditional abstraction from evolution ideas in construction of the physical picture of the world.

For a long time functioning of the science of physics excluded the "development factor" from its consideration. Classical science mainly paid attention to stability, balance, uniformity and order. Among its objects there were closed systems. Usually, they were simple objects, and knowledge of laws of their development enabled scientists, on base of information of the state of the system in the present, to undoubtedly predict its future and reconstruct its past. The mechanist picture of the world had timeless character. Time was not an essential element, it was reversible, i.e. states of objects in the past, present and future were practically indistinguishable. In other words, the world is arranged simply and submits to fundamental laws, reversible in time.[21] All these principles and approaches were concrete expressions of non-evolutionary paradigm of classical physics. Processes and phenomena, which did not correspond to this scheme, were regarded as exception; it was believed that they could be neglected.

Gradual eroding of the classical paradigm in physics started as early as in the 19th century. The first important step was the formulation of the second law of thermodynamics, which casts doubt on timeless character of the physical picture of the world. According to the second law, the energy content in the Universe is depleting, and "the world machine in fact has to reduce its activity, approaching the thermal death". Events are not reproducible in principle, and that meant that time had direction. There appears the idea of "arrow of time".[22]

Further development of physics led to understanding of scantiness of idealization of closed systems and description of real physical processes in terms of such systems. The overwhelming majority of natural objects are open systems, which exchange energy, matter and information with the surrounding world, and a decisive role in the radically changed world passes to unstable, non-equilibrium states. Fundamental sciences dealing with non-living nature – physics, chemistry, cosmology – more and more often faced the necessity to take these features into account. But the old theory turned out unfit for their description. The traditional paradigm could not cope with growing multitude of anomalies and contradictions, leaving many discovered phenomena unexplained.[23]

There appeared a need to develop a fundamentally new approach, adequate to objects and processes drawn into the orbit of investigation.

An important contribution to such approach was made by I. Prigogine's school. Researches of that school demonstrated that, moving away from equilibrium, thermodynamic systems get fundamentally new properties and start submitting to special laws. At considerable deviation from equilibrium thermodynamic situation appears a special type of dynamic state of the matter – dissipative structures. According to Prigogine, the type of dissipative structure depends to a large extent on conditions of its formation, and external fields may play a special role in selection of the mechanism of self-organization.[24] This is a conclusion with far consequences, if we take into account its applicability to all open systems which have an irreversible character. Irreversibility is what is characteristic for modern non-equilibrium states. They "carry the arrow of time" and are source of order, engendering high levels of organization.[25]

Prigogine and his colleagues developed ideas that the "arrow of time" is displayed in combination with stochasticity, when random processes are able to cause transition from one level of self-organization to another, radically transforming the system, are getting special heuristic value. Describing this mechanism, Prigogine emphasized a decisive, in the given development process, role of inner state of the system, regrouping its components etc. The situation defined as arising of order through fluctuations – random deviations of magnitudes from their average value – is characteristic for dissipative structures. Sometimes these fluctuations can increase, and in this case the existing organization cannot withstand it and is destroyed. At these breaking points (bifurcation points) it is fundamentally impossible to predict what direction will further development take, whether the system will become chaotic, or will pass to a higher level of ordering.[26]

Stochasticity pushes what remained from the system to a new way of development, and when the way is chosen, determinism again takes effect, and so on, till the next bifurcation.[27]

And here we see that the more complicated the system is, the more sensibility it displays with respect to fluctuations; and this means that even negligible fluctuations, intensifying, can change the structure, and, in this sense, our world is presented as deprived of guarantees of stability.[28]

Prigogine and Glensdorff made an attempt to formulate the universal criterion of evolution (taking the part of a mathematical rule), the core of which was the following: under certain circumstances thermodynamics not only does not contradict to the evolution theory, but can directly predict appearance of new things. Introducing this rule, the authors evidently tried to create a universal law for both living and non-living matter, the law of self-organization and evolution of any open system.[29] Really it was the matter of extending the class of self-organizing systems, when it became possible to apply phenomena of self-organization to non-living nature, biological and social processes.

This aspect of application of the ideas of self-organization was reflected in E. Jantsch's work "The Self-Organizing Universe: Science and Human Implication of the Emerging Paradigm of Evolution".

According to Jantsch, who used Prigogine's results of scientific researches on thermodynamics of non-equilibrium processes, self-organization can be spread over the whole totality of natural and social phenomena. Proceeding from the assumption that self organization is a dynamic principle giving creation to rich diversity of forms, displayed in all structures, Jantsch made an attempt to outline a uniform paradigm able to uncover the all-embracing phenomenon of evolution.[30]

All levels of both living and non-living matter, as well as states of social life – morality, religion – are developing as dissipative structures. From these positions, evolution is an integral process, parts of which are physical, chemical, biological, social, ecological, social cultural processes. The author is not just distinguishing these levels, but tries to find specific features of each of them. Thus, for living systems, the feature of this kind is the function of "autopoesis" as the system's ability to self-reproduction and conservation of its autonomous state with regard to environment.

Uncovering mechanisms of cosmic evolution, Jantsch considers breaking of symmetry as its source. Broken symmetry, prevalence of matter over antimatter in the Universe causes diversity of various kinds of forces: gravitational, electromagnetic weak, strong, and the idea of "Grand Unification" is the program of their investigation in view of their common genesis.

Jantsch presents the next stage in global evolution as appearance of the level of life, which is "fine overstructured physical reality".[31] Jantsch's characteristic of life can be treated differently. At first sight, we can reproach him with reductionism, but at the same time his elucidation of specificity of living matter allows us to come to a different conclusion: here genetic connection between living and non-living matter is meant. If we estimate Jantsch's conception as a whole, it is this aspect which is put forward first of all.

Further complication of the initial living systems, which is now regular, leads to appearance of a new level of global evolution – co-evolution of organisms and ecosystems, and then – to sociocultural evolution. At the level of sociocultural evolution reason is presented as a fundamentally new quality of self-organizing systems. It is capable of reflecting over passed stages of evolution of the Universe and foreseeing its future states. Thus Jantsch defines the place of man in the self-organizing Universe. Inclusion of man in it makes him involved in what is happening there. According to Jantsch, proportionality of people's world to the rest of the world inserts humanist sense into global evolution.[32]

Jantsch's conception can be evaluated as one of quite fruitful attempt to make a sketch of the modern general scientific picture of the world based on ideas of global evolutionism. It offers vision of the world, where all organization levels are genetically interconnected. The bases of such vision are not only philosophical ideas, but also real achievements of concrete sciences synthesized within integral notion of self-organizing Universe.

Modern conceptions of self-organization create real premises for synthesis of this kind. They let us eliminate the traditional paradigmatic gap between evolutionary biology and physics which, in its basic theoretical constructions, abstracts from ideas of evolution and, in particular, solve contradiction between the theory of biological evolution and thermodynamics.

At modern stage these theories do not any longer eliminate, but stipulate each other, in case we consider classical thermodynamics as some particular case of a more general theory – thermodynamics of non-equilibrium processes.

The theory of self-organization, rendered in terms of thermodynamics of non-equilibrium processes, reveals important regularities of development of the world. For the first time we have scientifically grounded possibility to overcome the old gap between notions of living and non-living nature. Life does not any longer look an island of resistance to the second law of thermodynamics. It appears as consequence of general laws of physics with its proper kinetics of chemical reactions which take place at conditions far from equilibrium.[33] It is characteristic that investigators, estimating the role of Prigogine's conception, said that, rediscovering time, opens a new dialogue between people and nature.[34]

The ideas of thermodynamics of non-equilibrium systems and synergetics have fundamental worldview and methodological meaning, since due to them it became possible to justify notions of development of physical systems and include these notions into the physical picture of the world. In its turn, it opened new perspectives for understanding connections between the main floors of the Universe – non-living, living and social matter. If before synergetics there was no conception (which would refer to the class of scientific theories, not philosophical ones), which would allow investigators to collect results, obtained in different spheres of knowledge, into a whole, appearance of synergetics gave us fundamentally new possibilities to form an integral general scientific picture of the world.

Synergetics lets us pass from "linear" thinking, established within the mechanist picture of the world, to non-linear thinking, corresponding to the new stage of functioning of science. Most studied objects (natural, ecological, socionatural complexes, economic structures) are open non-equilibrium systems regulated by non-linear laws. They all display ability to self-organization, and their behavior is determined by preceding history of their evolution.[35]

The notions of open self-organizing systems find confirmation in different spheres of knowledge, stimulating elaboration of evolution ideas there.

Let us, in this respect, mention important results obtained in modern chemistry and, in particular, in the field of evolutionary catalysis. The theory of evolutionary catalysis made considerable contribution to comprehension of what chemical evolution is, what its reasons and regularities are. Within this theory investigators expose special chemical objects with non-equilibrium structural and functional organization, while chemical evolution itself is regarded as process of irreversible continuous changes of elementary catalytic systems. In these chemical objects (chemical systems) with non-equilibrium and functional organization the order of interacting parts and stability are reached due to a permanent interchange of matter and energy.[36]

Synergetics created conditions for intensive exchange of paradigmatic principles between different sciences. In particular, application of the ideas of self-organization in biology enabled scientists to generalize a number of special notions of the theory of evolution and thus extended the sphere of their application, using biological analogies in description of very different processes of self-organization in non-living nature and social life.

As a characteristic example, we can take the application of "Darwin's triad" (heredity, changeability, natural selection) in modern cosmology and cosmogony. We mean such bioanalogies as "natural selection" of universes, galaxies or stars, "cannibalism in the world of galaxies" etc.[37]

It is necessary to denote that conceptual apparatus of biology has traditionally played a special part in the development of the ideas of evolution. As early as in the classical period a tight cooperation of the theory of biological evolution with geology and young social disciplines existed.

Employed in biology of the 20th century, the ideas of cybernetics and the theory of systems stimulated processes of synthesis of evolution notions and system approach, which was a considerable contribution to the outline of methodology of universal evolutionism. Achievements of biology of the 20th century can be regarded as a special block of scientific knowledge, which, together with cosmology and doctrine of self-organization, played a decisive role in defining new approaches to construct an integral general scientific picture of the world.

In the 1920s in biology a new branch of the evolution doctrine started being formed; that branch was connected with the name of V.I. Vernadsky and is called doctrine of evolution of biosphere and noosphere. Certainly, it should be considered as one of essential factors of scientific justification of the idea of universal evolutionism.

According to Vernadsky, biosphere is an integral system, which has the highest degree of self-organization and ability to evolution. It is the result of "long enough evolution in interconnection with non-organic conditions" and can be regarded as a regular stage in the development of matter. Biosphere is presented as a special geological body, whose structure and functions are determined by specific features of the Earth and Cosmos. Regarding biosphere as a self-reproducing system, Vernadsky stated that its functioning is, to considerable extent, conditioned by "existence of living matter in it – totality of all organisms living there".[38]

A specific feature of biosphere, as well as of living matter, is organization. "Organization of biosphere – organization of living matter – should be regarded as equilibria, mobile, permanently oscillating in historical and geological time around an exactly expressible average. Displacements or oscillations of that average continuously became apparent not in historical, but in geological time".[39]

To maintain its existence, the biosphere as a living system needs dynamic balance. But this is a special type of balance. A system which is in absolute equilibrium is unable to develop. Biosphere is a dynamic system, it is always in development. This development, to large extent, is realized under influence of inner interrelations of structural components of biosphere, and influence of anthropogenic factors upon it is constantly growing.

As result of self-development and under influence of anthropogenic factors, in biosphere there can emerge such states, that lead to qualitative change of subsystems, compounding it. In this respect unity of changeability and stability is a result of interaction of its components. Correlation of changeability and stability here plays the part of dialectic unity of constancy and development, and because of this stability itself is stability of process, constancy of development.[40]

Considering the role of anthropogenic factors, Vernadsky noted growing power of people; consequently, their activity causes changes in the structure of biosphere.[41] At the same time, man and humanity are most closely connected with living matter inhabiting our planet, from which they cannot be separated by whatever physical process.[42]

Evolutionary process of living beings, which has embraced biosphere, also exerts influence upon its inert natural bodies and acquires special geological meaning due to the fact that it has created a new geological force – scientific thought of social humanity.[43]

Vernadsky noted that we can more and more clearly see an intensive growth of influence of one species of living matter – civilized humanity – upon transformation of biosphere. Under influence of scientific thought and human activity biosphere is passing to new state – noosphere.[44] "Man is becoming a more and more powerful geological force, and change of his position on our planet coincided with this process. In the 20th century he got to know and embraced all biosphere, by its life humanity has become a whole".[45] In Vernadsky's opinion, "human power is connected with human reason and labour directed by this reason. It should give man base to take measures for preservation the shape of the planet. At the same time the force of reason will let him leave the bounds of his planet, the more so, as biosphere now is getting new understanding, it is regarded as planetary phenomenon of cosmic scale, and, correspondingly, we are to reckon with the idea that life exists not only on our planet".[46] Life "always appeared somewhere in the Universe, where corresponding thermodynamic conditions exist. In this respect we may speak about eternity of life and its manifestations".[47]

In Vernadsky's conception life is presented as integral evolutionary process (physical, geochemical, biological), included into cosmic evolution as a special component. By his doctrine of biosphere and noosphere, V.I. Vernadsky demonstrated indissoluble connection of planetary and cosmic processes.

Understanding of this integrity has imperishable heuristic value, as it in many aspects determines the strategy of further development of humanity. The very existence of man depends on how he will build up his interrelations with environment. It is no mere chance that problems of co-evolution of man and biosphere are gradually becoming domineering problems of not only modern science and philosophy, but also of the very strategy of human practical activity, as "further development of the species homo sapiens, its further well-being require extremely accurate correlation of the character of evolution of human society, its productive forces and development of the nature. But while correlation of processes, taking place in the world of non-living matter, is provided with mechanisms of natural self-organization, correlation of characteristics of natural medium and society can be accomplished only by Reason and will of Man".[48]

We may conclude that the theory of evolution and on its base created the conception of biosphere and noosphere considerably contribute to justification of the idea of universal interconnection of all processes and demonstrate the irreversible character of evolutionary processes, clearly marking a time factor in them.

Thus, we can ascertain that modern science possesses all necessary natural scientific data, which allow us to justify the universal character of evolution. Evolution approach in science of the second half of the 20th century turns out closely connected with system consideration of objects. From these positions global evolutionism, which contains principles of evolution and systemness, presents characterizing interconnection between self-organizing systems of different degree of complexity and uncovering mechanisms of appearance of new structures in the process of development. Such structures emerge in open systems in non-equilibrium state and are formed due to fluctuations and cooperative effects, and thanks to it transition from one type of self-organizing system to another is realized, and evolution finally gets an oriented character.

Universal evolutionism allows us to consider interconnection not only between living and social matter, but also include non-organic matter into integral context of the developing world. It creates base to consider man as an object of cosmic evolution, a regular and natural stage in development of our Universe responsible for the state of the world, in which man himself is immersed.

The principles of universal evolutionism are becoming dominant of synthesis of knowledge in modern science. This is the core idea, which pierces through all existing special pictures of the world and is the foundation for a construction of the integral general scientific picture of the world, where the central place is passing to man.

In view of basic foundations of the modern general scientific picture of the world, the principles of universal evolutionism are demonstrating their heuristic value right now, when science has turned to studies of new types of objects – self-developing systems (unlike simple and self-regulating systems, which were studied at previous stages of functioning of science). Having included a new type of objects into the orbit of investigation, science has to seek also new foundations for their analysis. The general scientific picture of the world, based on the principles of universal evolutionism, is a very important component of such foundations. It plays the part of global investigation program, which determines the strategy of investigation of self-developing system. And this strategy is accomplished at both disciplinary and interdisciplinary levels.

The general scientific picture of the world outlines a preliminary vision of objects studied, taking active part in putting problems, determining initial strategy of investigation. Study of complex, unique developing objects is possible only in the system of interdisciplinary interactions. In this case the general scientific picture of the world as a global investigation program is able "to give a hint", which methods and principles can be translated from one discipline in another, how it is possible to realize joining of knowledge acquired in different spheres of science, how to include this knowledge into culture at corresponding stages of functioning of scientific knowledge.

Setting strategy of investigation of self-developing objects within concrete scientific disciplines and providing strategy of interdisciplinary investigations, whose specific weight is growing in modern science, the general scientific picture of the world takes many functions which used to be performed by special scientific pictures of the world. The latter are losing their autonomy, are transformed under influence of system and evolution ideas and are included as fragments into the general scientific picture of the world and do not have a claim on a separate, independent status any longer.

This aspect of development of modern scientific knowledge should be regarded especially. Here we come across fundamentally new (in comparison with previous states of science) tendencies of historical development of the scientific picture of the world.

What was ideal at the stage of appearance of disciplinarily organized science, is becoming reality in modern conditions. In place of a poorly joined mosaic of pictures of reality studied there appears a common scientific picture of the world, absorbing contents of different disciplinary ontologies.

But this requires that investigators should study the preceding development of pictures of reality studied in different disciplines, include new notions of fundamental objects and structures, of interactions and space-time, which correspond to the ideas of system approach and evolutionism. And when these ideas find support in theories and empirical facts of leading spheres of scientific knowledge – in physics, cosmology, chemistry, geology, biology, technical and social disciplines, – then they start forming vision of objects as complex, historically developing systems. This vision gradually transformed special scientific pictures of the world, intensifying exchange of paradigmatic principles between them. As result, they began to unite naturally into an integral system of notions of the Universe, which, as it developed, gave creation to new levels of organization. Each of the sciences determines the place of its subject in this common picture, connecting it with either certain levels of the world organization, or with common features which determine interrelations and genetic transitions from one level to another.

As a result, relative isolation of special scientific pictures of the world from each other, characteristic for development of disciplinary science of the 19th century, is being replaced by their integration within the general scientific picture of the world. The degree of autonomy of special scientific pictures of the world in the second half of the 20th century has considerably decreased; they are transforming into aspects and fragments of the integral general scientific picture of the world. They join in blocks of this picture, characterizing non-living nature, organic world and social life, and realize, each one in its area, the ideas of universal evolutionism.

At first sight, here we see as if reproduced a situation characteristic for early stages of development of new European science, when a mechanist picture of the world, playing the part of general scientific one, provided synthesis of achievements of science of the 17th – 18th centuries. But behind exterior similarity there is deep interior difference. The modern scientific picture of the world is based not on striving for unification of all spheres of knowledge and their reduction to ontological principles of one discipline, but on unity of different disciplinary ontologies in diversity. Each of them appears as a part of a more complicated whole, and each of them inside itself renders concrete the principles of global evolutionism. But in this case the problem, which was formulated above, in analysis of functions and typology of scientific pictures of the world, achieves a solution. We mean historicity of those typologies. It turns out that, special pictures of the world as relatively independent form of synthesis of knowledge not always existed in this quality. In the age of the development of natural science they did not exist. Appearing at the time of differentiation of science into independent disciplines, they started losing independence and turning into aspects or fragments of a modern general scientific picture of the world. Therefore it is senseless to argue, whether special scientific pictures of the world (pictures of reality studied) exist as independent forms of knowledge, or whether they are only fragments of the whole – the general scientific picture of the world.

Out of historical context any direct answer to these questions may result both right or wrong. Everything depends on to what historical stage of development of science we attribute the corresponding answer.

The destiny of disciplinary ontologies is at the same time destiny of disciplinarily organized science at different stages of its historical evolution. Sometimes the opinion is expressed that one day strengthening of interdisciplinary connections will lead to complete disappearance of independent disciplines. This point of view seems too extreme. It emerges as mere extrapolation of today's situation of considerable growth of specific weight of interdisciplinary investigations to the future. But it does not take into account the fact that different spheres of knowledge have their own specificities which cannot be reduced to each other. Besides, we are to take into consideration that disciplinary organization of science is determined not only by features of different objective spheres of investigation, but also by possibilities of forming subjects of scientific activity, presence of certain limits of "information capacity" of the subject and, consequently, necessary to quantize the body of knowledge, which are to be mastered in order to do scientific search.

Specialization necessary for work in science is still conserved; it is not destroyed even by modern possibilities of computerization of scientific activity, because using base of knowledge requires understanding them, interpretation and mastering methods of work on their content.

It seems that science of the future, at least, nearest future most likely will combine disciplinary and interdisciplinary investigations. Quite another matter is that their direct and reverse links can become far more intensive, and boundary between them less hard. Consequently, the general scientific picture of the world will be comprehended more and more clearly as a global investigation program and necessary horizon of systematization of knowledge.

Intensification of connections between different disciplines and growth of importance of interdisciplinary investigations as factor of development of the general scientific picture of the world affect not only cognitive, but also institutional aspects of modern science.

We may ascertain that modern synthesis of achievements of different sciences is proceeding in conditions, when the role of large complex programs and problem oriented interdisciplinary investigations is increasing.

In his analysis of tendencies of development of science in the first half of the 20th century, V.I. Vernadsky noted that they are classified more in accordance with problems than with subjects.

In science of the late 20th century this tendency attained clearly shaped features, especially in connection with appearance of complicated, often unique complexes as objects of investigation, and their studies stipulate joint work of specialists of different profiles.

Modern practice of social support and financing of "high science" is an evidence of priority of branches which appear on junction of different disciplines. These are, for instance, informatics, ecology and biotechnology, programs of search of energy sources, biomedical investigations etc.

Prestige of branches and programs of such kind is determined first of all by modern search for a way out of global crises caused by the industrial, technogenic development of civilization.

It is just the point of joining two types of factors, which determine development of the modern scientific picture of the world. Social aims and values, changing the shape of science as a social institution, and intrascientific, cognitive factors act in the same direction: they actualize interdisciplinary connections and interactions. Social disciplines actively participate in this process along with sciences, since most of modern trends of investigation study complicated developing complexes which include man and his activity as a component.

All this, on the one hand, reinforces the role of the general scientific picture of the world, which provides an integral vision of complicated developing "anthropomeasured" systems and understanding of the place of each discipline in their possible assimilation, on the other hand, it stimulates "exchange processes" between natural, technical and social disciplines, and that, in its turn, accelerates "building bridges" between corresponding special scientific pictures of the world, their inclusion into the general scientific picture of the world.

At modern stage the general scientific picture of the world, based on the principles of global evolutionism, more and more clearly appears as ontological foundation of future science, uniting sciences on nature and sciences on spirit.

The old opposition of sciences and humanities led investigators to conclude that the gap between them is broadening more and more, and finally it can lead to their isolation and, as a consequence, even to appearance of separate cultures with languages alien for each other.[49]

Actually, for a long time natural science was guided by cognition of "the nature in itself" irrespective of the subject of activity. Its aim was to obtain objectively true knowledge, not burdened with value-meaning structures. The attitude to the natural world was understood as monologue. The main thing which was to be done by scientists – to uncover and explain existence of natural connections in the natural world and, revealing them, reach objectively true knowledge, ascertain the laws of nature.

At the same time humanities were oriented at comprehension of man, human spirit, culture. Priority consisted in uncovering meaning, more in understanding than in explanation. The relation between subject and object (as any cognitive relation) was not monologue but dialogue. To obtain knowledge within humanities, exterior description was insufficient. Method of "objective", or "exterior" investigation of society should be combined with the method of its investigation "from inside", from the point of view of people who have formed social and economic structures and are acting in them.[50]

M.M. Bakhtin quite precisely noted these specific features of methodology of natural scientific and humanitarian knowledge: "Exact sciences are a monologue form of knowledge: intellect contemplates a thing and speaks about it. Here we can see only one subject – comprehending (contemplating) and speaking (uttering). He is opposed only to a voiceless thing. Any object of knowledge (including man) can be perceived and comprehended as a thing. But subject as such cannot be perceived and studied as a thing, since, being subject, he cannot, remaining subject, become voiceless, consequently, cognition of subject can be only dialogue".[51]

It really seemed that an insuperable contradiction arose between sciences and humanities. Moreover, science did not form such general scientific picture of the world, which could integrate them in a single space.

But nowadays there emerge real foundations for solution of this problem. Sciences and humanities can be integrated based on the principles of global evolutionism, which immanently include attitude to objective study of self-developing objects. Correlation of development of such objects with problems of the place of man, man's inclusion and actions in functioning of the overwhelming majority of historically developing systems, assimilated in human activity, introduce new, humanist meaning into scientific knowledge.

The need to join cognitive and value parameters of natural scientific knowledge is more and more clearly understood in natural science itself. An example is the position of representatives of so called biological structuralism, which are attempts to define a new paradigm in biology. Looking for basis, this new paradigm turns not only to "exact" natural science, but also to humanitarian knowledge. Taking into account that biology is closer than any other natural science to study the nature of man, representatives of "biological structuralism" to large extent hope for such changes in the scientific picture of the world, which would attach human dimension to it.[52]

In modern natural scientific cognition new tendencies of people's attitude to the nature arise. The nature, in broad sense of the word, is not any longer presented as "dead mechanism", at which human activity is aimed: man cannot treat it in the way a judge would do, knowing in advance what answers it should give to questions put.

As Prigogine and Stengers note, "it died, that finite, static and harmonious old world, the Copernican revolution destroyed it, having put the Earth into endless space. Our world is not a silent and uniform world of a watch mechanism... The nature was created not for our sake, and it does not submit to our will... It is time to answer for human old ventures, but if we are able to do it, it is only because that such is the way of our participation in cultural and natural settling, such is the nature's lesson, when we take the trouble of listening to it. The time for new alliance came, alliance started long ago, but for a long time unrecognized, between human history, human societies, knowledge and employment of the Nature for our purposes".[53]

To ensure his future, man cannot believe that he has no fundamental restrictions in his attempts to transform the nature in accordance with his own needs; he has adapt his needs according to the requirements put by the nature.[54]

All this means that now it is time to settle new relations of man and nature, not monologue but dialogue. In the past these aspects were characteristic for humanitarian knowledge. Now they penetrate into very different spheres and become priority principle of analysis.

At the same time ideas and principles, developed in natural scientific knowledge, are gradually penetrating to humanities. The ideas of irreversibility, variability in the process of making decisions, diversity of possible lines of development which appear at system's passing bifurcation points, organic connection of self-regulation and cooperative effects – all these and other ideas, justified in synergetics, turn out to be significant for the development of humanities. Constructing various conceptions of development of society, studying man, his consciousness, they cannot any longer ignore these methodological regulations, which are acquiring a general scientific character.

When science assimilates complicated, developing, "anthropomeasured" systems, former insuperable boundaries between methodology of natural scientific and humanitarian cognition are washed away.

We may conclude that, having started investigation of "anthropomeasured objects", sciences are coming closer to "the object field" of investigation in humanities. In this respect we can remind the reader K. Marx's well known statement that "history itself is a real part of the nature, becoming of the nature by man. Afterwards natural science will include science of man to that same extent, to which science of man will include natural science; that will be single whole science".[55]

Thus, in the late 20th century there appeared fundamentally new tendencies of development of scientific knowledge, which led to a reconstruction of the general scientific picture of the world as an integral system of scientific notions of nature, man and society. This system of notions, forming based on the principles of global evolutionism, is becoming a fundamental investigation program of science at the stage of intensive interdisciplinary synthesis of knowledge.

Absorbing the totality of fundamental scientific results and synthesizing them within integral image of development of the Universe, living nature, man and society, the modern scientific picture of the world actively communicates with worldview universalities of culture, in the context of which its development takes place. On the one hand, it adapts to them, but on the other hand, it introduces radical mutations into established cultural mentalities.

Development of the modern scientific picture of the world is one aspect of search for new worldview meanings and responses to historical challenge which modern civilization is facing.

SCIENTIFIC PICTURE OF THE WORLD AND NEW WORLDVIEW REFERENCE POINTS OF CIVILIZATION DEVELOPMENT

Modern science is developing and functioning in a special historical epoch. Its general cultural meaning is determined by inclusion into solving the problem of choice of life strategies of humanity, its search for new ways of civilization development.

Needs of this search are connected with crises of the late 20th century, which led to the appearance of modern global problems. Their comprehension requires that we should re-estimate development of the technogenics civilization, which has existed for four centuries. Many of its values, connected with attitude to nature, man, understanding of activity etc., which used to seem an unshakeable condition of progress, are now cast doubt on.

At our time the technogenics civilization, developing as a kind of antipode of traditional societies, has approached that "bifurcation point", after which transition to a qualitatively new state may follow. What direction the system will choose, what character its development will have – not only status of science in society, but also the very existence of humanity will depend on all that.

The culture of the technogenics civilization has always included scientific rationality as its basic value. Exactly within it the scientific picture of the world as such form of theoretical presentation of knowledge, which embodied the worldview status of science, became, functioned and developed.

In the technogenics civilization employment of science was first of all connected with technologies of transformation of the object world. The scientific picture of the world orientated man not only in understanding the world, but also in transforming activity, aimed at its change.

In fact, from the 17th century to the present, new European culture was regulated by paradigm, according to which man is called for actualization of his creative abilities, when he should direct his activity outwards, at transformation of the world and first of all – nature.

Attitude to nature as opposed to man was a worldview premise of science of the New Age. V.I. Vernadsky wrote: "Copernicus, Kepler, Galileo, Newton in a few decades broke the connection between man and the Universe established in ages. The scientific picture of the Universe, enveloped by Newton's laws, did not leave place for any display of life. Not only man, not only all living matter, but our whole planet were lost in the infinity of Cosmos".[56]

The idea of demarcation between the world of man and the world of nature, which was presented as alien to man, was immanently included in the scientific picture of the world and for a long time served as worldview foundation of its historical development.

This idea found justification in many values of the technogenic civilization, in particular, it correlated with those interpretations of Christianity, which gradually gained dominance in culture beginning with the period of Reformation. This variant of Christianity not only fixed dualism of man and nature, but also insisted on the postulate that it is the God's will that man should exploit the nature to suit his own ends.[57] It gave psychological confidence in man's striving for transforming the nature in the spirit of indifferent attitude to "health" of natural objects. Thus it destroyed bans for exploitation of the nature.[58]

The attitude to transformation, reshaping of the nature and then society gradually turned into the domineering value of the technogenic culture. An investigator, acting within this cultural tradition and guided by some scientific picture of the world, realized himself as an active creator of the new, eliciting the nature's laws from it in order to extend possibilities to bend the nature to people's needs.

The civilization oriented at such type of scientific rationality, achieved indubitable successes: the ideas of progress, democracy, freedom and personal initiative were established in it.[59] It provided constant production growth and improvement of the quality of people's life. But at the same time in the late 20th century, when humanity faced global problems, questions of correctness of the choice of development paths in the Western (technogenics) civilization and, as a consequence, of adequacy of its worldview orientations and ideals arose again.

The search for ways of development of civilization is now attended by the problem of synthesis of cultures and forming of new type of rationality. In this connection arise questions of place and role of the picture of the world in search for new worldview orientations, which will provide the possibility for humanity to survive.

These questions can be formulated in the following way: does the modern scientific picture of the world require any system of values and worldview structures, fundamentally different in comparison with previous stages of development of science, for its justification? Did this picture cause radical transformation of worldview foundations of scientific cognition? What is its concrete contribution to settle worldview reference-points, corresponding to requirements of the new stage of civilization development, called for overcoming global crises and provide survival and further development of humanity?

First of all we are to distinguish those fundamentally new ideas of the modern scientific picture of the world, which concern notions of the nature and man's interaction with it. These ideas do not blend with the traditional technogenics approach understanding of the nature as non-organic world, indifferent to man, and understanding of attitude to the nature as to "dead mechanism", with which one could experiment infinitely and assimilate its parts, transforming it and bending it to man.

In the modern situation we are developing a new vision of natural environment, which we interact with our activity. It is now considered not as a conglomeration of isolated objects, not even as a mechanical system, but as an integral living organism, which can be changed only within certain limits. Violation of these limits leads to change of the system, its transition into a qualitatively new state, able to cause irreversible destruction of the system's integrity.

At previous stages of development of science, from the establishment of natural science to the middle of the 20th century, such "organismic" understanding of the surrounding nature would have been perceived as an atavism, return to half-mythological consciousness, not coordinated with the ideas and principles of the scientific picture of the world. But after notions of living nature as complex interaction of ecosystems had been formed and had entered science, after modern ecology had developed, such understanding of immediate sphere of human vital functions as of organism, not as a mechanical system, became a scientific principle, justified by numerous theories and facts. Ecological knowledge plays a special part in forming scientific system of notions of that sphere of natural processes, with which man interacts in his activity and which is his immediate habitat as a biological species. This system of notions constitutes a most important component, which combines knowledge of biosphere, on the one hand, and knowledge of social processes, on the other hand. It serves as a sort of bridge between notions of development of living nature and development of human society. So it is not surprising that ecological knowledge is getting special importance in solving problems of interactions of man and nature, overcoming the ecological crisis, and so is becoming an important factor in forming new worldview foundations of science.

At the same time the principles, developed in ecology and included into general scientific picture of the world, are also getting a wider worldview character. They exert influence upon worldview of the whole culture, "essentially affect spiritual and intellectual climate of the modern epoch in the whole, determine transformation of value structures of thinking".[60]

In the modern culture more and more clearly shaped contours of the new vision of the world are present, and the scientific picture of the world makes a considerable contribution to its establishment. This vision is based on the idea of interconnection and harmonious relations between people, man and nature, which constitute a single whole.

Within such approach we can trace outlining of new vision of man as an organic part of the nature, not as its lord; science develops the ideas of priority of cooperation over competition.[61]

E. Laszlo speaks of the world, "new vision" of which is, in essence, forming of new worldview system, absorbing achievements of modern science. F. Capra's ideas of "united ecological vision of the world" are keeping such approach. Capra uses this notion in the meaning of "profound ecology", as opposed to "superficial ecology", which is anthropocentric by nature and regards man as towering above nature, sees in him source of values, assigning nature the role of auxiliary means.[62] Unlike "superficial ecology", "profound ecology", in Capra's opinion, does not pick out man from the natural environment, but interprets the world as integral totality of phenomena connected with and dependent on each other. It is oriented at consideration of value of all living beings, and man is regarded as a regular and integral part in the whole diversity of life.[63]

Ecology and, in particular, "united ecology" (A. Ness) demonstrates evidently enough scantiness of anthropocentrism, proving that "man is neither lord, nor centre of the Universe, he is only a being who submits to the laws of reciprocity".[64]

Changes, happening in modern science and fixed in the scientific picture of the world, correlate with intense search for new worldview ideas, which are elaborated and polished in very different spheres of culture. These are searches for new religion, rethinking of the old one, as in works of R. Attfield and L. White,[65] creation of "new ethics", as suggested by E. Laszlo and O. Leopold. Laszlo says that we need new morality, which would base more on necessary requirements of humanity's adaptation as a global system to surrounding natural environment, than on individual values. Such ethics can be created on base of ideal of respect to natural systems.[66]

Similar ideas are developed by Leopold who proposes to distinguish ethics from the philosophical point of view as distinction of social and antisocial behavior and ethics from the ecological point of view as restriction of the freedom of action in the struggle for existence.[67]

Leopold's new ethics is ethics which determines man's relations with the Earth, animals and plants. In his opinion, it should change man's role, converting the conqueror of community into ordinary and equal in rights its member. Ethics of the Earth reflects existence of ecological conscience and, correspondingly, conviction in individual responsibility in health of the Earth. Humanity is facing the goal to form ethical attitude to the Earth, which cannot exist without veneration for its value.[68]

These ideas are in keeping with A. Schweitzer's thoughts expressed in his conception of veneration for life as base of ethical world and life asserting. For him the idea of veneration for life appears as an answer to the question how man and the world are correlated with each other. He notes dual character of relations between man and the world, taking into consideration that man bears both passive and active relations to the world: on the one hand, man has to submit to natural course of events, in accordance of which he builds up his life, on the other hand, he has all the possibilities to exert influence upon life and its change within certain limits. And the only way to attach meaning to human existence is to raise natural connection with the world and make it spiritual.[69]

All these speculations of the prominent philosopher and scientist are developed in the principles of so called biosphere ethics, which includes not only interrelations between people, but also interrelations between man and nature. It contains "veneration for the high (celestial world), compassion to the equal (human world), aid to the low (plant and animal world)".[70]

New worldview ideas appear as sort of resonance of modern science and created pictures of the world with other spheres of cultural creative work. Mutual influence of these spheres accelerates the process of formation of new meanings of cultural universalities and, correspondingly, new system of value priorities, stipulating way to other, non-traditional strategies of human vital functions.

In their turn, new senses and value orientation to larger and larger extent are included into the system of philosophical and worldview foundations of science.

The key moment in their development is notions of the scientific picture of the world of organic involvement of man in an integral cosmos and of proportionality of man, as a result of cosmic evolution, to the world which engendered him.

Ethical ideas of man's responsibility before nature, which appear on this base, make the picture of the world axiologically loaded.

Striving for considering man in his connection with the rest of the world, regarding the world as organic integrity, is an important methodological reference-point, able to lead to change traditional technogenics civilization notions of destination of man and his activity. New worldview ideals of attitude towards nature, based on ethics, rejecting the principle of supremacy over the nature and including the idea of man's responsibility, in their turn, pave the way to new understanding of rationality as dialogue between man and the world.

The principles of openness and self-regulation of complicated systems, developed in synergetics and introduced as the most important principle into the modern scientific picture of the world, lead to the same philosophical and worldview ideas.

As Prigogine and Stengers note, "natural sciences nowadays display need of dialogue with the open world. The time for new concord came, concord started long ago, but for a long time unrecognized, between human history, human societies, knowledge and using the Nature for our purposes".[71]

Comprehending the world, man should not thrust his own language on the nature, but enter dialogue with it. In Prigogine's opinion, modern science has learned how to treat the studied nature with respect, the nature which cannot be described "from outside", from the spectator's position. Description of the nature is a lively dialogue, communication, and it submits to limitations which are evidences, that we are macroscopic beings, immersed in a real physical world.[72]

Dialogue with nature in the new type of rationality is attended by the ideal of openness of consciousness to diversity of approaches, to close interaction (communication) of individual minds and mentalities of different cultures.

This aspect of openness and communication as characteristic of the new type of rationality and corresponding strategies of activity is especially emphasized by J. Habermas. He notes: "instead of relying on reason of productive forces, i.e. finally on reason of natural science and technics, I trust the productive force of communication".[73] Frames and structures of communication, shared activity and openness are continuously changing – both "in themselves and in relation to other spheres of society as such".[74]

Ontology of this new type of rationality is notions of integral cosmos, which organically includes man, notions of the objects of reality as historically developing "anthropomeasured" systems possessing "synergetic" properties.

These ideas, concretized in the modern scientific picture of the world, lead to new consideration of subject and object of cognition, which are now not regarded as alien for each other, but are presented as only relatively autonomous components of a special integral, historically developing system built into the world.

In this approach rationality is already endowed with new distinctive features. It is characterized by openness, reflexive explication of value and meaning structures included in mechanisms and results of objectively true comprehension of the world.

"Open rationality" (V.S. Shvyrev) is now opposed to closed rationality, intraparadigmatic rationality, when an investigator is moving within an adopted rigid conceptual carcass. Open rationality assumes "attentive and respectful attitude to alternative picture of the world, appearing in cultural and worldview conditions different from those of modern science, it assumes dialogue and mutual enrichment of different, but equal in rights cognitive positions".[75]

From this point of view, we are to pay special attention to new and unusual properties of the modern scientific picture of the world. In many aspects it embodies the ideals of open rationality, and its worldview consequences correlate with philosophical and worldview ideas and values, which appear on the soil of different, even in many aspects opposite cultures.

We mean wonderful correspondence of the modern scientific picture of the world not only to those mentalities, which are gradually forming in the Western (technogenic) culture of the late 20th century, but also correspondence to philosophical ideas grown on the soil of distinctive Russian culture and its Silver Age, as well as philosophical and worldview notion of traditional cultures of the East. Up to now, the scientific picture of the world has developed on base of mentalities of technogenic culture, embodied proper only for this culture type of scientific rationality, which occupied one of the first places in the system of its value priorities. When other types of cultures adopted science, it required simultaneous transplantation of certain fragments of Western experience to different ground. Such transplantation have always transformed traditional culture and were realized in the course of catching-up modernizations, aimed at transition of the traditional societies to the way of technogenic development (for example, Peter the Great's reform in Russia). The process of transplantation of science to Russian soil in Peter's epoch is a characteristic example. It became possible only along with adoption of fragments of urban culture, European education, new way of life, which Peter the Great often implanted by force in boyar midst and nobility.[76]

A quite tight connection of new European science with mentalities of the technogenic culture led to fundamental mismatch of the scientific picture of the world, its philosophical and worldview foundations, on the one hand, and prescientific cosmologies of traditional societies, on the other hand.

Scientific knowledge, appearing in traditional cultures, submitted to myth-cosmic and religiously ethical worldview structures, in the forming of which this knowledge did not directly take part. A different situation we can detect in the technogenics civilization. Here scientific rationality claimed for the role of justifying principle of worldview ideas – social, ethical, religious (an example – neo-Thomist philosophy).

No surprise that distinctive opposition of the Western technogenic culture to the culture of traditional societies first of all was displayed in opposition of the scientific picture of the world and its philosophical corollaries to "organismic" notions of the world of traditional oriental cultures.

But such opposition hardly can be applied in respect to modern science. The changes, which took place there in the late 20th century, formed a new picture, which created special philosophical and methodological corollaries. These corollaries are in resonance with fundamental life-sense reference points of cultures of the East and have a lot in common with original philosophical ideas grown on the soil of Russian cultural tradition.

We would like to discuss the latter situation especially, as here we face the fundamentally important, for modern civilizational development, problem of dialogue of cultures, interchange of ideas born by different cultural traditions.

First, let us pay attention to the coincidence of many notions of the modern scientific picture of the world with the ideas of philosophy of Russian "cosmism". These ideas for a long time were taken for some kind of periphery of the world flow of philosophical thought, though they certainly exerted influence upon works of such prominent natural scientists as V.I. Vernadsky.

Traditionally in Russian cosmism at least three trends are distinguished: natural scientific (N.A. Umov, N.G. Kholodny, V.I. Vernadsky, K.E. Tsiolkovsky, A.L. Chizhevsky); religious-philosophical (N.F. Fedorov); poetical (S.P. Dyachkov, V.F. Odoevsky, A.V. Sukhovo-Kobylin).[77]

Russian cosmism appeared as sort of antithesis to classical physicalist paradigm of thinking, based on strict differentiation of man and nature. It made an attempt to revive ontology of integral vision, which organically united man and cosmos. These problems were discussed both in scientist and religious trends of cosmism. In the religious trend N. Fedorov's conception was the most significant. Like other cosmists, he was not satisfied with split of the Universe into man and nature as opposed to each other. Such opposition, in his opinion, condemned nature to thoughtlessness and destructiveness, while man – to submission to existing evil. Fedorov maintained the idea of unity of man and nature, connection between "soul" and cosmos in terms of regulation and resurrection.

He offered a project of resurrection, which was not reduced only to resurrection of ancestors, but contained at least two aspects: raising from the dead – in narrow, direct sense, and in wider, metaphoric sense, which included the nature's ability to self-reconstruction.[78]

Fedorov's resurrection project was connected with the idea of human mind's going out to the outer space. For him, "the Earth is not bound", and "human activity cannot be restricted by the limits of the terrestrial planet", which is only the starting point of this activity.

Critically looking at Utopian and fantastic elements of N. Fedorov's views, which contain a considerable grain of mysticism, nevertheless we distinguish important rational moments of his conception: the quite clearly expressed idea of interconnection, unity of man and cosmos, the idea of correlation of rational and moral elements of man, the ideal of unity of humanity as planetary community of people.

But while religious cosmism was more notable for fantastic and speculative character of its discourses, the natural scientific trend, solving the problem of interconnection between man and cosmos, paid special attention to comprehension of scientific achievements, which confirmed that interconnection.

N.G. Kholodny developed these ideas in terms of anthropocosmism, opposing it to anthropocentrism. He wrote: "Having put himself in place of God, man destroyed his natural connections with the nature and condemned himself to long solitary existence".[79]

In Kholodny's opinion, anthropocentrism passed through several stages in its development: at the first stage man did not extract himself from the nature and did not oppose himself to it, he rather "humanized" the natural forces – this was the attitude of the weak to the strong; at the second stage man, extracting himself from the nature, looks at it as the object for investigation, the base of his well-being; at the next stage man uplifts himself over the nature, basing on spiritual force, he studies the Universe, and, at last, the next stage is characterized by crisis of anthropocentric worldview, which starts to collapse under influence of achievements of science and philosophy.[80]

N.G. Kholodny was right noting that in the past anthropocentrism had played a positive role; it freed man from fright of the nature by means of uplifting him over the latter. But gradually, beside anthropocentrism there appeared sprouts of the new vision – anthropocosmic. Kholodny regarded anthropocosmism as a certain line of development of human intellect, will and feelings, which led people to their aims. An essential element in anthropocosmism was the attempt to reconsider the question of man's place in the nature and of his interrelations with cosmos on base of natural scientific knowledge. Anthropocosmism started to consider man as one of the organic parts of the world and settling the conviction that only on this way we can find the key to understanding the nature of man. Man should strive for unity with the nature, which enriches and broadens his inner life.[81]

N.A. Umov developed similar ideas, emphasizing that "man can understand himself as a part, one of transient links of the Universe". He also believed that anthropocentric worldview was going into ruin, ceding place to anthropocosmism.[82]

The idea of interconnection of man and cosmos was especially emphasized in K.E. Tsiolkovsky's works, one of which was even entitled "Cosmic Philosophy". He wrote: "All cosmos conditions our life ... Everything is continuous and everything is united".[83] "The Universe would be meaningless if it were not filled with organic, intelligent, feeling world".[84]

We can see certain concord of Tsiolkovsky's ideas with the anthrop principle, formulated later.

Tsiolkovsky not merely points at interconnection of man and cosmos, but stresses dependence of the former on the latter. "... It is hard to suppose that any part of it (cosmos) will not exert, sooner or later, influence upon us".[85]

This idea – influence of both nearer and far space upon human life – was at length analyzed by A.L. Chizhevsky, who believed that "our scientific worldview is still far from the historical notion of the role of space radiations for the organic realm".[86] But the number of advances of the 20th century science in Chizhevsky opinion allows to conclude that "in science on nature an idea of the unity and coherence of all phenomena in the world and appreciation of the world as undivided whole become nowadays especially clear and deep ... The structure of the Earth, its physics and chemistry, biosphere are penetration of the structure and mechanics of the Universe".[87]

Chizhevsky opposed his point of view to the existing opinion that "life is a result of random game of only terrestrial forces". For him, to a considerably larger extent life is a cosmic rather than a terrestrial phenomenon. It was created by influence of creative dynamics of space upon inert material of the Earth. He noted that man "is not only a terrestrial being, but cosmic, connected by all his biology, all molecules, particles of the body with the space, with its rays, flows and fields".[88]

In this sense influence of solar energy upon the course of life processes does not seem accidental. Chizhevsky was one of the first investigators who justified this theory by concrete scientific facts. In particular, he analyzed correlations between solar activity and peaks of epidemic deceases and demonstrated that solar activity plays the part of some kind of regulator of the course of epidemic processes. Certainly, it does not mean that "the state of solar activity is the direct reason of epidemic spread of such and such deceases", but activity of the Sun "favors their fast ripening and intensive course".[89]

In the scientist tradition of Russian cosmism the problem of united world and united knowledge of the world was elaborated, in a most significant form, in V.I. Vernadsky's conception. As well as other cosmists, Vernadsky supposed that "anthropocentric notion does not coincide with the real reveal of cosmos, which is enveloped by scientific work and scientific thought of investigator of the Nature".[90] He noted that "science still has no clear understanding that phenomena of life and phenomena of dead nature, taken from geological, i.e., planetary point of view, are displays of the united process".[91] But, as Vernadsky emphasized, biologists should not forget that they study the world of life, which is an integral part of the Earth's crust and exerts active reverse influence upon it, transforming it. They should not consider life out of touch with evolution of the integral cosmos. In his opinion, such direction was caused by the fact that for a long time the Universe had seemed lifeless. The basis for such statements was establishing of Copernicus's principle in science. When in the first half of the 19th century scientists obtained numeric data of size of the Universe, it seemed that life was entirely dissolved in the space, and gradually was settling the opinion that the inconsiderable meaning of life is the proper conclusion from scientific investigations. But, as science developed, there appear reasons to cast doubt on the indisputability of such conclusions.[92]

Vernadsky, like other cosmists, opposes a different point of view to the traditional position. He demonstrated that, in the world evolution, life is not random, but a regular consequence, that character of cosmic development of life processes is conditioned by the cosmic whole. In such consideration life is now presented as a cosmic phenomenon.[93]

V.I. Vernadsky regards humanity as the part of biosphere, which exerts active influence upon this system. Human consciousness, emerging in the course of bioevolution, becomes a special factor of evolution, whose meaning grows in time. The development of biosphere into noosphere is a logical completion of evolution of matter: all parts of the developing world turn out interconnected, and man is naturally fits in with this world.

Russian cosmism quite clearly understood not only man's dependence on cosmos, but also (what is especially important) reverse man's influence upon the surrounding world.

Commensurability of man and the rest of the world served base for the by Russian cosmists developed idea of need to commeasure human activity with the principles of integrity of this world.

Russian cosmism justified the principles of new man's attitude to the nature. In fact, it approached closely enough to understanding the problems, which later received the name of global. At least, the idea of possible ecological crisis, though in hidden form, but quite clearly was expressed in the words of representatives of this trend. It is no mere chance that N.G. Kholodny emphasized that "transformations imposed on the nature by the man, has its limits".[94] As a reasonable being, man should foresee the results of his activity, for which he bears responsibility.

Russian cosmism's intuitive understanding of global contradictions between man's technocratic activity and harmony of cosmos led to searches for a way out of a possible future unfavorable state, in which humanity may find itself.

Practically each of the cosmists offered his own version of humanity's future development. K.E. Tsiolkovsky painted a quite idyllic picture: "... climate will be changed at will of need. All Earth will become inhabited and yield great fruits. There will be total scope for development of human both social and individual qualities. Technics of the future will give people chance to study all planets... imperfect worlds will be destroyed and replaced by own population. The Earth will give heavenly colonies its surplus of people ... Finally, we will see infinite Universe with infinite number of perfect beings".[95]

V.I. Vernadsky in his conception regarded a more realistic scenario. Consideration of man as a special geological force, able to transform the world, where he lives, radically, led to the conclusion of possible negative consequences of human activity, which can be seen as prevision of possible global ecological crises. At the same time, Vernadsky was optimistic when he looked at perspective of humanity, connecting its future with the processes of transition from biosphere to noosphere and growth of the regulating role of human reason.

Original speculations, anticipating the modern situation of global crises, were offered in N. Fedorov's philosophy of "common deed". The thinker brilliantly cautioned against unreasonable treating nature and its possible consequences. "People have, most likely, done all the harm he could do concerning nature (exhaustion, devastation, spoiling), and concerning each other (invention of deadly weapons, just means of mutual annihilation)".[96] All evil of our life, in Fedorov's opinion, proceeds from disharmony of man and nature.

Having drawn a quite bright picture of "all-Earth crisis", he offered his project of solving the problem of "the common deed". This common deed is presented as regulation of spontaneous natural forces. "In regulation, in ruling the forces of blind nature consists that great deed, which can and must become common".[97] In realization of his project Fedorov mostly relied on man's moral force and force of his reason. He wrote: "Cosmos needs reason to be cosmos, not chaos. Cosmos (as it is, but not as it must be) is force without reason, while man is (yet) reason without force. But how can reason become force, and force become reason? Force will become reasonable when reason rules it. So, everything depends on man".[98]

In N. Fedorov's conception "the common deed" was presented as the way leading to unity and renovation on a humanist, moral base.

Thus, the cosmist philosophy quite clearly brought out two aspects of interconnection of man and space: on the one hand, man presented as a fragment of evolving cosmos, its integral part, in all its revelations dependent on the cosmic whole. On the other hand, man himself was regarded as a factor of evolution, developing his abilities in such a way, that, creating new technics and technology, he started exerting active influence upon the surrounding world. Though in the late 19th – early 20th centuries belief in scientific and technical progress was evident enough, and crisis consequences of technocratic attitude to the world were not displayed yet, cosmists warned future generations against possible negative consequences of unrestrained, limitless technological exploitation of the nature.

But still cosmism, though it contained original ideas and possessed considerable prognostic power, did not gain wide spread. In fact, it repeated the destiny of many philosophical conceptions, whose productive ideas greatly outstripped their time.

But in today's situation, when humanity is facing ecological crisis, search for "common deed" as regulation of relations between man and nature is gaining priority meaning.

We may state that cosmism as a special branch of philosophical thought is in tune with modern strivings for new life senses and ideals, harmonization of man and nature.

The coincidence of the main principles of cosmist philosophy and many fundamental ideas of the modern scientific picture of the world and its worldview conclusions is to be especially emphasized. Cosmism returns to an integral vision of the world as unity of man and cosmos. It is able to play a positive role in the synthesis of ideas developed in the Western European cultural tradition and in oriental philosophical systems, where man from the very beginning has been considered as an integral part of cosmos. Correspondingly, the ideas of cosmism are organically included into outlining new metaphysics, which could be philosophical foundation of post-non-classical stage of development of science, providing further development of general scientific picture of the world in the course of global evolutionism ideology, notions of "anthropomeasured", historically developing systems and ideals of "anthropocosmism".

Open character of the modern scientific picture of the world reveals its wonderful commensurability not only with the principles of Russian cosmism, but also to many worldview ideas established in traditional cultures of the East. The clearest display of it we can see in comprehension in terms of synergetic and global evolutionism of numerous fundamental ideas of oriental philosophy, which for a long time had no adequate perception in the European cultural tradition.

First of all it refers to notions of the world as a united organism, different parts of which are in distinctive resonance relation to each other.

This ontology has immanently the ideal of harmony of man and nature and their inner unity. Striving for unity found its expression in the statement "one in all and all in one", which was the domineering principle of Taoism and Confucianism.[99] In Buddhism it is expressed in the doctrine of dharma. "All elements of dharma are something homogenous and equal in force; they all are connected with each other".[100]

For cultures of the East, in particular, the Old Chinese philosophical doctrines, characteristic is the notion of the world as an enormous living organism. It was seen not as dually divided into natural and human worlds, but was perceived as organic whole, all parts of which are correlatively connected and exert influence upon each other. This cosmology excluded opposition of subject and object and was based on adoption of binary nature of things corresponding to the Yin-Yang model.[101] Yin and Yang represented two primary forces, which express bipolarity of existence: Yin acted as the negative pole, which embodied passive (feminine) element, and Yang as positive, active, creative (masculine) element. Being interconnection as light and darkness, Yin and Yang permanently alternate and interact with each other.[102]

The conception of Yin and Yang is set the foundation of understanding of universal interconnection of phenomena and their mutual resonance. "Everything is penetrated by the united way – Tao, everything is interconnected. Life is united, and striving of every part of it should coincide with striving of the whole".[103]

Man, included in the world, should feel the world rhythm, bring his mind into accord with "celestial rhythm", and then he will be able to grasp the nature of things and hear "Music of humanity".[104]

The very idea of rhythms of the world, their influences upon each other, including rhythms of human vital functions in the process of this interaction, was long perceived by the European mind as something without serious base in scientific facts, something mystical and inexpressible rationally. But in the modern scientific picture of the world, assimilating achievements of synergetics, new notions of interaction of parts of the whole and of concordance of their changes are formed. It has been elucidated that non-forced interactions, based on cooperative effects, start playing a special role in complicated systems.

For open, self-organizing systems such interactions are the constituting factor. It is thanks to them the system is able to pass from one state of self-organization to another, creating new structures in the process of their evolution.

Cooperative properties are traced in very different self-regulating systems, which consist of a very large number of elements and subsystems. They can be found, for instance, in behavior of plasma, in laser coherent radiation, in morphogenesis and dynamics of populations, in economic processes of market self-regulation.[105]

For examples, at certain critical levels of energy laser pumping there emerges an effect of emitting of light wave by atoms: they act in a strictly correlative way, each atom emits a purely sinusoidal wave, as if coordinating with behavior of another emitting atom, i.e., here emerges the effect of self-organization.[106]

Similar effects can be observed in processes of embryonic cell division, when each cell in the tissue receives information of its situation from surrounding cells, and so their mutually coordinated differentiation takes place.[107] In the experiments on embryos a cell of the central part of the body, after transplantation into the head sector, developed into eye. These experiments demonstrated that cell do not dispose of information of their further development from the very beginning (for instance, through DNA), but extract it from its position in cellular tissue.[108]

Synergetics generalizes similar situations of cooperative effects of elements and subsystems in complicated self-organizing systems. It regards "resonance" of functioning parts in such systems and presence of cooperative effects as one of the important displays of self-cooperation.

If we turn again, from these positions, to the ideas of oriental philosophies about "resonance" of different parts of united cosmic whole, they will obtain new sounding: in any case, they can be perceived as a worldview guess, which finds response in modern notions of the scientific picture of the worlds, which realizes a "synergetic" approach to description of various processes of the nature, social life and human spirit.

We can give many parallels between cosmological notions of traditional oriental cultures and ideas of synergetics included into modern scientific picture of the world.

In traditional worldview systems of the East a special role belonged to the idea of non-being, which was perceived as all completeness of the world. Non-being was interpreted as reality, wherefrom situations of being (objects, processes, phenomena) emerge, submitting to a strict rhythm of the world development, and then, having exhausted themselves, return to non-being.[109]

It is very interesting to compare these ideas with fundamental synergetics' notions of appearance of structures in non-linear medium. Non-linear medium as potentially possible field of structures, where they appear and disappear, is a special kind of reality, giving creation to the given structures. If we imagine an infinite number of potentially possible structures in an infinitely complicated non-linear medium, it will appear analogue to non-being (with respect to already emerged and disappeared structures), containing all future completeness of the world.

Old oriental notions of the world as an integral organism, in which man is included, of resonance between different parts of this organism, formed an ideal of human activity, different from the one of the Western technogenic culture.

Understanding man as demiurge, who does force transforming objects in order to bend them to his will, was alien to oriental cultures. As H. Hesse stressed, people formed in traditions of those cultures put themselves the same aim – to know how to rule the laws of nature, but they choose entirely different ways. They did not split themselves from the nature and did not try to intrude by force into its mysteries, they never opposed themselves to nature and were not hostile to it, they always remained a part of it and loved it in reverential love.[110]

The Chinese cultural traditional prescribed that human activity toward the nature should not bear character of violence. As J. Needham noted, within this tradition force was always recognized a hardly perspective way of action. In Chinese culture man was associated with the image of a peasant, not that of a sailor or cattle-breeder (who are believed to be inclined to command and submission). "But the peasant-farmer, one has done all that is necessary for the crops, must wait for them to come up. A famous parable in Chinese philosophical literature derides a man of Sung State who was discontented with the growth rate of his plants and started to pull at them to help them to come up".[111]

In Chinese doctrines opposition of force to non-force action was developed in the terms "Wei" and "Wu-Wei"(application of force and non-action). Non-action (Wu-Wei) meant not absence of any action, but such kind of action, which enables nature to develop in its own way. "A perfectly wise, doing deeds, prefers non-action. Realization of non-action always brings calmness".[112]

It is indicative, that the "Wu-Wei" principle, rejecting the way of action based on permanent force intrusion into the course of natural processes, at our age unexpectedly correlates with the ideas of synergetics of possible strategies of regulation of complicated self-organizing systems.

For instance, it becomes clear that such a system, undergoing violent and active force pressure from outside, probably will not give creation to new states and new structures, but will "decline" to old structures. But if it passes a bifurcation point, then a little energetic "influence-prick" in the proper time-space locus is enough to make the system reorganize, and new type of structures will appear.[113]

We have noted above that man's interaction with complex open systems goes on in such a way, that human action itself is not something exterior, it is as if included in the system, transforming every time the field of its possible state.

Hence, it becomes important in the action strategy to determine thresholds of interference in proceeding processes and provide, by means of minimized influence, those directions of development of the system, which would let it avoid catastrophic consequences and achieve people's goals.

The "Wu-Wei" principle is oriented at quite similar behavior and human activity strategies. It required that people should feel natural rhythms of natural world and act in accordance with them, letting the nature itself disclose its interior potentials and choose such ways of development of processes, which would be coordinated with human needs.

The Old Chinese philosophy stressed that only people "ignorant of the true laws of being" understand the "Wu-Wei" principle as absence of action, obedience and submissiveness. But sage, who developed understanding Tao in themselves, by "non-action" meant not absence of action, but natural action, which corresponded to the nature of things.[114]

In discussion of ideals of human activity it is important to distinguish one more extremely significant aspect in oriental doctrines, which has a lot in common with modern searches for new values and strategies of human vital functions.

We mean interconnection between morality and truth, gaining which has always been proclaimed the aim of scientific knowledge.

The question of their correlation has been permanently discussed in Western philosophy, but it was given the following solution: the process of comprehending the truth by itself supposes to be a moral action.

The scientific revolution in Europe, as J. Needham noted, isolated scientific truth from ethics, and the world became more dangerous, while the oriental doctrines never knew such isolation.[115] They developed a more delicate treatment of relation between truth and morality. From the point of view of sages of the East, true knowledge consists not in investigation of objects with the aim to take possession of them, but to reach co-being with the world.[116] One can comprehend this, only following Tao, regarded as natural way of things and at the same time moral way, which one should travel. Tao opens only to moral people, and only this is able to lead people to perfection.[117]

In order to have the truth opened, one needs moral self-training. People's activity, directed at cognition of the exterior world, and their activity, directed at perfection of their interior world, should be coordinated and cannot exist without one another.

One of the oldest and most fundamental ideas of Chinese philosophy was the idea of cosmic importance of man's moral qualities. Thinking of resonance of all parts of cosmos, Chinese sages believed that "it is man's behavior, his morality that the order in cosmos, regular change of seasons, heat and cold depends on".[118] The way the image of Tao, or Heaven, regulates people's actions. But Heaven "can turn to man, and it can turn out from him". It is no mere chance that the Chinese say: "Heaven acts in dependence on people's deeds".[119] The Old China natural calamities were perceived as evidences of untrue ruling, as indicators if immoral behavior of sovereigns.[120]

Certainly, if we understand these ideas literally, they will sound as mystical ones. But they also contain more profound meaning, connected with demand of ethical regulation of people's cognitive and technological activity (including technology of social managing). In this, more profound meaning they are quite consonant to modern searches for new worldview reference-points of civilization development.

Thus, in the late 20th century, when humanity found itself face to face with the problem of choice of new strategies of survival, many ideas, outlined in traditional oriental doctrines, correlate with new values and worldview meanings, which appear within modern technogenic culture, and are formed in different spheres of this culture, including scientific cognition.

Development of the modern scientific picture of the world justifies new methods of cognition of the world, which are consonant to forgotten achievements of traditional cultures, as worldview corollaries.

We may ascertain that development of the modern scientific picture of the world is organically included in the processes of formation of new type of planetary thinking, based on tolerance and dialogue of cultures and connected with the search for a way out of modern global crises.

Getting openness, the scientific picture of the world contributes to the processes of synthesis of different cultures. It unites new approaches, which emerged on base of developing scientific rationality, a constant characteristic of technogenic (Western) civilization, with ideas developed in entirely different cultural traditions, in oriental doctrines and in “cosmic philosophy’.

The modern scientific picture of the world is included in the dialogue of cultures, whose development has up to now passed along parallel lines. It is becoming one of the most important factors of cross-cultural interaction of the West and the East.

NOTES: CHAPTER 7

---

[1] Ideas developed in this chapter are the results of my investigations of last years. Partially those results were published in my works Stepin (1989), (1992), (1992a, pp.177-189), (1998). In a more complete version they are presented also in the book *Scientific Picture of the World in the Culture of Technogenic Civilization* (1994) written together with L.F. Kuznetsova.
[2] Capra (1989, p.113).
[3] Incidentally we mean raising the problem in the framework of science while concerning philosophy there were surmised ideas on global cosmic evolution beyond the science of their own time.
[4] Moiseev (1989, p.53).
[5] In the following discourse we will use these terms as synonyms.
[6] Moiseev (1986, p.25).
[7] Ibid.
[8] See Saushkin (1976), Calvin (1971), Kuznetsov V.I. (1973), Rudenko (1987) and others.
[9] Silk (1982, pp.16-17).
[10] Friedman (1965). On Friedman’s conception see Eremeeva (1985, pp.160-161).
[11] Guth and Steinhardt (1984).
[12] Ibid.
[13] Linde (1984), Gut and Steinhardt(1984).
[14] Linde (1984, p.210).
[15] Gut and Steinhardt (1984).
[16] Carter (1974).
[17] Ibid.
[18] For more details see Kazyutinsky (1989).
[19] Haken (1987).
[20] Klimontovich (1986, pp.56-58).
[21] Prigogine and Stengers (1984).
[22] Ibid.
[23] A characteristic example is the fact that Belousov–Zhabotinsky reaction, which is the striking evidence of synergetic effects, did not get justification at the period of its discovery and was not accepted by the scientific community.
[24] Prigogine and Stengers (1984).
[25] Ibid.
[26] Ibid.
[27] Ibid.
[28] Ibid.
[29] Klimontovich (1986, p.104).
[30] Jantsch (1980, p.19).
[31] Jantsch (1980, p.19).

[32] Ibid, p.19.
[33] Glensdorff and Prigogine (1971).
[34] Toffler (1986, p.17).
[35] See Dobronravova (1991, p.7).
[36] On the role of evolutionary catalysis in settling an idea of chemical evolution see more details in Rudenko (1987, pp.70-78).
[37] Kazyutinsky (1986, p.70).
[38] Vernadsky (1977, p.70).
[39] Ibid, p.15.
[40] Vodopianov (1981, pp.193-194).
[41] Vernadsky (1940, p.47).
[42] Vernadsky (1977, p.13).
[43] Ibid, p.18-19.
[44] Ibid, p.19.
[45] Vernadsky (1944, p.117).
[46] Ibid. P. 114-115.
[47] Vernadsky (1934, p.82).
[48] Moiseev (1990, pp.40-41).
[49] Snow (1971).
[50] Gurevich (1990, pp.30-31).
[51] Bakhtin (1980, p.383).
[52] See Karpinskaya (1992, pp.145-146).
[53] Prigogine and Stengers (1981, p.296).
[54] Such a state of man–nature attitude N.N. Moiseev calls ecological imperative. See Moiseev (1990, p.40).
[55] Marx and Engels (1955-1981, vol. 42, p.124).
[56] Vernadsky (1940, p.176).
[57] White (1967).
[58] Ibid.
[59] See Kara-Murza (1990, pp.3-15).
[60] Zelenkov and Vodopianov (1987, p.81).
[61] Laszlo (1990, pp. 23-31).
[62] Capra (1990, p.33).
[63] Ibid, p.33.
[64] Macey (1990, p.82).
[65] Attfield (1983), White (1967).
[66] Laszlo (1972, p.281).
[67] Leopold (1983)
[68] Ibid.
[69] Schweitzer (1990, p.339), (1992).
[70] See Shipunov (1990, p.450).
[71] Prigogine and Stengers (1981, pp.273, 296).
[72] Prigogine and Stengers (1984).
[73] Habermas (1992, p.85).
[74] Ibid, p.131.
[75] Shvyrev (1992, p.98).
[76] This social experiment was in detail analyzed in N.I. Kuznetsova's works. See Kuznetsova (1982).
[77] Guirenock (1960, p.5).
[78] Fedorov (1982). An analysis of N. Fedorov's conception` see, e.g., Kogan (1990).
[79] Kholodny (1982, p.187).

[80] Ibid, p.175.
[81] Ibid, pp.178-197.
[82] Umov (1916, p.215).
[83] Tsiolkovsky (1986, pp.302, 278).
[84] Ibid, p.378.
[85] Ibid, p.302.
[86] Chizhevsky (1976, p.27).
[87] Ibid, pp.24, 26.
[88] Ibid, pp.33, 331.
[89] Ibid, p.246.
[90] Vernadsky (1978, p.40).
[91] Ibid, p.12.
[92] Ibid, pp.31-33.
[93] Ibid, pp.43, 36.
[94] Kholodny (1982, p.142).
[95] Tsiolkovsky (1986, pp.287-290).
[96] Fedorov (1982, p.55).
[97] Ibid, pp.58-59.
[98] Ibid, p.535.
[99] See Grigorieva (1979, p.119).
[100] Rosenberg (1991, p.128).
[101] Grigorieva (1979, pp.106-112, 148).
[102] East – West (1982, p.244).
[103] Ancient Chinese Philosophy (1972, vol.1, p.26).
[104] Grigorieva (1983, p.127).
[105] For more details, see Haken (1987).
[106] Ibid.
[107] Ibid.
[108] Ibid.
[109] See Grigorieva (1979).
[110] Hesse (1970).
[111] Needham (1964, p.135).
[112] Ancient Chinese Philosophy (1972, vol.1, pp.115-116).
[113] Kurdyumov (1990).
[114] Grigorieva (1983, p.128).
[115] Needham (1988).
[116] Grigorieva (1979, p.75).
[117] Ancient Chinese Philosophy (1972, vol.1, pp.114, 119-121, 128).
[118] Grigorieva (1979, p.112).
[119] Kuo Yü (1987, p.298).
[120] Grigorieva (1979, p.113).

CONCLUSION

Let us summarize the main results of our investigation.

1. Theoretical knowledge emerges as a result of the historical development of culture and civilization. Its primary forms were represented by philosophical knowledge which was the only form of the theoretical at the stage of prescience. The transition from prescience to science resulted in the emergence of scientific theoretical knowledge which in the further evolution of culture represents the theoretical as such.

2. Developed science as opposed to prescience does not confine itself to the modeling of only those thing-oriented (object) relations that are already integrated into the existing practice of production as well as everyday experience. It is capable of going beyond each historically concrete type of practice and of opening new worlds of things to mankind that may become the objects of mass practical assimilation only at future stages of the development of civilization. Leibniz used to characterize mathematics as a science of possible worlds. This characteristic may be in principle applied to any fundamental science.

3. Breakthroughs towards new object worlds become possible in developed science thanks to a new mode of generating knowledge. At the stage of prescience models for the transformation of objects included in practice were created through a schematization of practice. Objects of practical manipulation were replaced in cognition by ideal objects, abstractions belonging to thinking, whereas the relations of ideal objects, the operations involving them were also abstracted from practice, being a kind of scheme of practical actions. Though still applied in developed science, this mode loses its leading positions. A different mode of constructing knowledge comes to the force – one in which models of the object relations of reality are first created as if from above with respect to practice. Ideal objects, acting as elements of such models, are created not from the abstracted properties and relations of objects of actual practice, but on the basis of already existing ideal objects produced at an earlier time. Neither is the structure (network of relations) into which they submerge extracted directly from practice (by abstracting and schematizing the actual relations of objects), but is instead borrowed from earlier forms of knowledge. Models formed in this way serve as hypotheses which later, upon receiving justification, are turned into theoretical schemes of the field of objects under consideration.

It is precisely theoretical research based upon a relatively independent manipulation of idealized objects that can discover new fields of objects before they are assimilated by practice. The theorizing capacity is a kind of indication of developed science.

4. The theoretical mode of investigation and respectively the transition from prescience to science properly speaking was first fulfilled in mathematics, then in natural science and, finally, in technical and social sciences. Each of these stages in the development of science had its own social and cultural preconditions. The birth of mathematics as a theoretical science was linked to the culture of the ancient polis, to the values of public debate and the ideals of validation and proof distinguishing knowledge from opinion, all of them promoted by this culture.

The preconditions of natural science which brought together the mathematical description of nature and experimentation are to be found in the shaping of the fundamental worldview universals pertaining to the technogenic culture, they are: the conception of man as an active practice-oriented being engaged in transforming the world; the conception of activity as a creative process giving man power over objects; the conception of nature as a necessarily ordered field of objects set against man; the interpretation of the ends of knowledge as rational cognition of the laws of nature, etc. All these values and vital meanings that took shape in the age of Renaissance, Reformation and early Enlightenment were radically different from the understanding of man, nature, human activity and cognition which dominated traditional cultures.

The further development of the technogenic civilization, its industrial stage witnesses the emergence of the preconditions for technical and social sciences. The intensive development of industrial production generates the need in inventing and reproducing ever new engineering devices, which stimulates the appearance of technical sciences possessing a theoretical level of research. During the same historical period the relatively rapid transformations of social structures, the destruction of traditional communal bonds replaced by relations of "reified (thing-like) dependence", the emergence of new practices and types of discourse objectifying human qualities, create preconditions for the social sciences and humanities in general. There appear conditions and demands for finding ways of rationally regulating the standardized functions and actions performed by individuals entering one social group or another as well as of managing different social objects and processes. It is in the context of these demands that the first programs of constructing sciences of society and man come into being.

5. Scientific knowledge is a complex historically developing system which over time gives rise to ever new levels of organization. In their turn, they influence the levels of knowledge created earlier and lead to their transformation. This process marked by a changing strategy of scientific exploration witnesses the incessant birth of new devices and modes of theoretical research.

At its developed stages science appears as a form of knowledge organized into disciplines, in which its separate fields – the scientific disciplines (mathematics; natural sciences including physics, chemistry, biology, etc.; technical and social sciences) – function as relatively independent interacting subsystems.

The emergence and development of scientific disciplines is quite uneven. They give birth to various types of knowledge, and while certain sciences have already gone a long way in the direction of theorizing and have demonstrated examples of elaborate mathematized theories, others are only embarking on this path.

The specificity of the subject of each science may result in the fact that certain types of knowledge prevailing in one science may be of a second order in another. They may also become transformed in it.

The primary unit of a methodological analysis of the structure of theoretical knowledge should be associated not with a separate theory in its relation to practice (as was argued by the so-called standard conception), but with a scientific discipline. The cognitive structure of a scientific discipline is determined by the levels of theories of a varying degree of generality, both fundamental and local, by their relations to each other as well as to the complex level of empirical research (of facts and observations), and, last but not least, by their links with the foundations of science. The foundations of science serve as a system-building factor for a scientific discipline. They include: 1) a special scientific picture of the world (a disciplinary ontology) which introduces a generalized image of the subject of a given scientific discipline in its major systemic and structural characteristics; 2) the ideals and norms of research (the ideal and norms of description and explication, of proof and justification as well as the ideals of a structure and organization of knowledge) which determine the generalized scheme of the method of scientific cognition; 3) the philosophical foundations of science that justify the accepted picture of the world as well as the ideals and norms of science, ensuring that the conceptions of reality and of the methods of its cognition devised by science enter the flow of cultural transmission.

The foundations of science, besides a disciplinary component, possess an interdisciplinary one. It is formed by the general scientific picture of the world as a unique form of systematizing scientific knowledge, contributing to a wholesome image of the Universe, of life, society and man (disciplinary ontologies are aspects or fragments with respect to such a general scientific picture of the world), and also by a special layer related to the contents of the ideals and norms of cognition, of the philosophical foundations of science, which places in relief the invariant characteristics of the scientific as such accepted in this or that epoch (these characteristics are specified in accordance with the subject and methods of each scientific discipline). The interdisciplinary component of the foundations of science provides for the interaction between different sciences, the transfer of methods and ideas from one science to another. Theoretical knowledge functions and develops as a complex system of intradisciplinary and interdisciplinary relations.

6. The content structure of scientific theories is defined by the systemic organization of idealized (abstract) objects (i.e., theoretical constructs). The statements of theoretical language are directly formulated regarding theoretical constructs, and only indirectly, due to their relation to an extralinguistic reality, do they describe (this) reality.

In the network of the abstract objects (constructs) of a scientific theory one may single out specific subsystems created from a limited set of basic constructs. By means of their interrelations they form theoretical models of the reality under consideration. These models are integrated into a theory and play the role of its "inner carcass". Theoretical laws are formulated with respect to them. Such models, being the nucleus of a theory, can be named theoretical schemes. They should be distinguished from models-analogues which are employed to build a theory and serve as its "scaffolding" without forming part of its structure.

In a developed theory one may discover a fundamental theoretical scheme in respect to which the basic laws of theory are formulated, as well as local theoretical schemes in respect to which laws of a lesser degree of generality are formulated, being inferred from the first. These schemes and corresponding laws create a hierarchy of levels. Within the framework of the theoretical knowledge of a scientific discipline certain local theoretical schemes and laws may enjoy an independent status. Historically they precede developed theories. Theoretical schemes are projected onto the scientific picture of the world (disciplinary ontology) and the empirical data accounted for by theory. Both of these projections are captured by means of special statements which characterize the abstract objects of theory in terms of the picture of the world as well as of idealized experiments grounded on actual experience. The latter statements are operational definitions. They have a complex structure and are not reduced to the description of actual measuring situations, though such descriptions are incorporated into their very essence.

The relation of a mathematical apparatus to a theoretical scheme projected onto the scientific picture of the world ensures its semantic interpretation, while the relation of the theoretical scheme to experience provides for its empirical interpretation.

7. Theoretical schemes play a most important role in the deployment of a theory which is carried out not only by means of a deductive inference accompanied with formal

operations (the deduction of corollaries from equations), but also in a genetically constructive way by means of intellectual experiments with theoretical schemes. The conception of a theory and its functioning as a hypothetico-deductive system requires serious reconsideration. In theories which do not belong to the class of formalized systems (and theories of this kind constitute an overwhelming majority in natural sciences, in technical and social sciences) the inference, from basic laws of their theoretical consequences presupposes a complicated transformation of the theoretical schemes, a reduction of the fundamental theoretical scheme to a local one. Such reduction combines deductive and inductive modes of investigation and serves as a ground for solving theoretical problems. The unfolding of a theory takes place as the solution of theoretical problems some of which are integrated into theory in the form of "paradigmatic models (patterns)" (T.Kuhn). The conception of the structure of theoretical schemes as well as of the genetically constructive ways of building theories allows to considerably specify Kuhn's problematic of models as an obligatory element in the structure of a theory of experimental sciences.

8. The construction of a theory and its conceptual apparatus regarded as a problem presents itself first of all as the problem of the genesis of theoretical schemes. Such schemes are first created as hypotheses and then are justified experimentally. The construction of theoretical schemes as hypotheses is performed by way of borrowing abstract objects from other fields of theoretical knowledge and of combining these objects in a new "network of relations". This mode of creating hypothetical models can be fulfilled in two ways: by means of a substantive manipulation with concepts and by means of putting forward mathematical hypotheses (in the last instance hypothetical equations implicitly introduce a hypothetical model providing for a preliminary interpretation of those equations).

It is hard to overestimate the role of the foundations of science in articulating the hypothetical variant of a theoretical scheme. They determine the formulation of the problem along with the choice of technical devices indispensable for putting forward a hypothesis. The foundations of science function as a global research program guiding scientific exploration.

9. In the process of constructing hypothetical models abstract objects are endowed with new characteristics since they are introduced within a novel system of relations. An experimental justification of hypothetical models implies that the new characteristics of abstract objects should be acquired as idealizations based on new experiments and measurements – those that were to be accounted for by the hypothetical model. We suggest that this procedure be designated as the method of a constructive justification of a theoretical scheme. Schemes that have undergone this procedure usually acquire a new content in comparison with their initial hypothetical version. Being reflected upon the picture of the world they induce changes in it. Due to all these operations the development of scientific concepts takes place. Not only the formulation, but also the justification of a hypothesis plays a decisive role in shaping a theory's conceptual apparatus. In their turn, the justification of hypotheses and their transformation into theory create means for future theoretical inquiry.

10. The method of constructive justification allows revealing the "weak points" in a theory and thus ensures an effective reconstruction of scientific knowledge. It opens up possibilities for an effective verification of the consistency of theoretical knowledge, allowing to discover a theory's hidden paradoxes before they are unveiled by the chaotic

flow of developing cognition. The method of constructiveness should be regarded as a further elaboration of the rational elements of the principle of observation.

11. The discovery of the procedure of "constructive justification" offers a solution to the problem of the genesis of the "paradigmatic models" of theoretical tasks. The elaboration of a developed theory is carried out through a successive synthesis and generalization of the local theoretical schemes and laws. At each new step of this generalization the intactness of the previous constructive content is verified, which automatically introduces the models of reduction of a generalized theoretical scheme to local ones. At the final stage of the theoretical synthesis when a fundamental theoretical scheme is created and the basic laws of a theory proclaimed, the verification of their constructive meaning is performed by way of an elaboration, on the basis of the newly formulated fundamental theoretical scheme, of all the local theoretical schemes that it encompasses. This results in the emergence of paradigmatic models for the solution of theoretical tasks. The further evolution of a theory and the extension of the field of its application introduce new models into its core. But only those that appeared while a theory was still in the making remain basic with respect to it. A theory preserves the traces of its past history reproducing the main stages of its evolution in the form of paradigmatic tasks and models.

12. The strategies of theoretical investigation are subject to change in the historical development of science. Such changes presuppose a reconstruction of the very foundations of science and are characterized as scientific revolutions. One may sort out two types of these revolutions. The first of them, described by Kuhn, is linked to the appearance of anomalies and crises produced by the expansion of science into new fields. Their mechanisms may be specified on the basis of the foundations of science structure as well as the procedures of an ongoing correlation with the foundations of emerging theories. The second type, rather poorly examined in methodological literature, can emerge without anomalies and crises, springing from interdisciplinary connections. In this case various elements of disciplinary ontologies, ideals and norms and also philosophical foundations are transferred from one science to another. Such "paradigmatic grafting" leads to the reformulation of a scientific discipline's former tasks, to the posing of new problems and the emergence of novel means for their solution. The first type of scientific revolutions is exemplified by the making of the theory of relativity and quantum mechanics. The second – by the coming into being, in the end of the 18th – first half of the 19th century, of a science organized into disciplines as well as by the contemporary "exchange" between cybernetics, biology and linguistics.

13. The reconstruction of the foundations of science at the time of scientific revolutions is performed, on the one hand, under the pressure of new empirical and theoretical data accumulating within scientific disciplines and, on the other, under the influence of socio-cultural factors. Scientific revolutions are specific "points of bifurcation" in the evolution of knowledge when different possible guidelines (scenarios) of scientific development become apparent. However, those guidelines (research programs) are implemented that not only bring about a positive empirical and theoretical shift (I. Lakatos), but that also fit into the culture of the epoch, being in accord with the possible modifications of the meaning of its worldview universals. If the historical development of culture and civilization had taken a different turn, other possible histories of science could actually have happened. In times of scientific revolutions it is as if, from the multiplicity of scenarios of a future history of science, culture sorts out those that best of all correspond to its basic values.

14. In epochs of global scientific revolutions when all the components of the foundations of science are being reconstructed a change in scientific rationality takes place. One can single out three basic historical types of rationality, i.e., classical, non-classical and post-non-classical science. Classical science assumes that true knowledge of an object is conditioned by the elimination, in the process of theoretical explication and description, of all that which has to do with the subject, its goals and values as well as the means and procedures of its activity. Non-classical science (its example being a relativist quantum physics) takes into account the relation between the knowledge of an object and the nature of the means and procedures of the activity in which the object is discovered and cognized. Nevertheless, relations between intrascientific and social goals and values are still outside scientific reflection, though defining implicitly the nature of knowledge (defining what it is that we isolate and conceive in the world as well as the way we do it). The post-non-classical type of scientific rationality extends the field of reflection on activity. It is aware of the relation not only between the knowledge of an object and the specific nature of the means and procedures of activity, but between this very knowledge and the structure of the goals and values of such activity as well. At the same time the relation between intrascientific and extrascientific goals is brought to light. In overall investigations of complex self-developing systems more frequently than ever becoming dominating objects in natural science and technology (including the objects of ecology, genetics and genetic engineering, "man – machine – environment" technical complexes, modern information systems, etc.) the elucidation of the ties between intrascientific and social values is performed through social expertise of respective investigation programs.

The historicism of the objects of contemporary natural science along with reflection on the value-related foundations of research remove the opposition between natural and social sciences. True with respect to 19th century science, at present it considerably loses its significance.

The emergence of a new type of rationality does not eliminate those that preceded it historically, instead it limits their field of application. Every subsequent type of scientific rationality introduces a new system of the ideals and norms of cognition, which provides for the mastering of a respective type of systemic objects, i.e., simple, complex, historically evolving (self-organizing) systems. This brings about a change in the categorical framework of the philosophical foundations of science – in the concepts of thing, process, space, time, causality, etc. (the ontological component), and in the concepts of knowledge, theory, fact, method, etc. (the epistemological component). Finally, a new type of rationality accounts for the alteration in the worldview applications of science. At the classical and non-classical stages of its evolution science was justified only on the values of the technogenic civilization rejecting the values of traditional cultures as contradictory to it. Post-non-classical science markedly extends the field of possible worldview meanings with which its achievements accord. It forms part of the contemporary processes of solving global problems and choosing mankind's vital strategies. Post-non-classical science embodies the ideals of an "open rationality" and actively participates in the search for new worldview guidelines determining the strategies of contemporary civilizational development. It uncovers the proportionality of its own achievements not only to the values and priorities of the technogenic culture, but also to a series of philosophical and worldview ideas elaborated in other cultural traditions (to the worldview ideas of the traditional Oriental cultures and the ideas articulated in the philosophy of Russian cosmism). Post-non-classical science organically enters the contemporary processes of shaping a planetary thinking, of a dialogue of cultures, becoming one of the most important factors of a cross-cultural interaction between the West and the East.

References

Achutin, A.V.,

1976, *History of Principles of a Physical Experiment*, Moscow (in Russian).

1988, *Concept of "Physis" in Antiquity and New Age*, Moscow (in Russian).

Agassi, J.,

1974, 'The logic of scientific inquiry', *Synthese* 26, No 3-4.

Ancient Chinese Philosophy, Vol. 1-2, Transl. from Chinese, Moscow, 1972-73 (in Russian).

A Nature of Scientific Cognition, 1979, Minsk (in Russian).

Arshinov, V.I.,

1973, 'On hierarchy', in: *Some Problems of Dialectics*, Moscow (in Russian).

1974, 'Conception of integrity and hypothesis of hidden variables in quantum mechanics', in: *Physics and Philosophy*, Voronezh (in Russian).

Attfield, R.,

1983, *The Ethics of Environmental Concern*, New York: Columbia University Press.

Bakhtin, M.M.,

1979, *The Problems of Dostoevsky's Poetics*, Moscow (in Russian).

1980, *Esthetics of Verbal Creativity*, Moscow (in Russian).

A Becoming of Chemistry as Science, 1983, Moscow (in Russian).

The becoming of a scientific theory.

Berg, R.L., Lyapunov, A.A.,

1968, *Preface to I.I. Shmalgauzen's Book "Cybernetic Issues of Biology"*, Novosibirsk (in Russian).

Bernal, J.D.,

1954, *Science in History*, London: Watts and Co.

Bibler, V.S.,

1978, *Thinking as a Creative Work*, Moscow (in Russian).

Blauberg, I.V., Sadovsky, V.N., Yudin, E.G.,

1969, *System Approach: Prerequisites, Problems, Difficulties*, Moscow (in Russian).

Bohm, D.,

1952, *Quantum Theory,* New York: Prentice-Hall, Inc.

1971, 'On Bohr's views concerning the quantum theory', in: *Quantum and Beyond*, Cambridge.

Bohm, D., Hiley, B.,

1977, 'On the institutional understanding of nonlocality as implied by quantum theory', in: *Quantum Mechanics: A Half Century Later*, Dordrecht-Boston, pp.207-225.

Bohr, N.,

1963, 'Mathematics and natural science', in: Bohr N., *Essays 1958-1962 on Atomic Physics and Human Knowledge*, Sydney: John Wiley & Sons.

1970-1971, *Selected Papers*, vol. 1-2, Moscow (in Russian).

1971, 'Atoms and human knowledge', in: Bohr N., *Selected Scientific Works*, vol. 2, Moscow (in Russian).

Born, M.,

1956, *Physics in My Generation. A Selection of Papers*, London: New York, Pergamon Press.

Bridgman, P.W.,

1954, *The Logic of Modern Physics*, New York.

1955, *Reflections of a Physicist*, 2 ed., New York (in Russian).

de Broglie, L.,
1960, *Sur le Sentiers de la Science*, Paris: Editions Albin Michel.
Bunge, M.,
1966, 'Are there any operational definitions of physical notions?', *Voprosy Folosofii*, No 11 (in Russian).
Calvin, M.,
1971, *Chemical Evolution*, Moscow (in Russian).
Capra, F.,
1989, *Uncommon Wisdom. Conversations with Remarkable People*, Toronto-N.Y.
1990, 'The shift of paradigm and the shift in value scale', in: *One World for Everybody*, Moscow (in Russian).
1994, *Tao of Physics. An Inquiry of Parallels between Modern Physics and Eastern Mysticism*, St. Petersburg (in Russian).
Carter, B.,
1974, 'Large number of coincidences and the anthropic principle in cosmology', in: *Confrontation of Cosmological Theories with Observational Data*, ed. M. Longair, Reidel, Dordrecht.
Chanyshev, A.N.,
1982, *The Origin of Philosophy*, Moscow (in Russian).
Chernovolenko, V.F.,
1970, *Worldview and Scientific Cognition*, Kiev (in Russian).
Chew, G.F.,
1966, *The Analytic S Matrix. A Basis for Nuclear Democracy*, New-York-Amsterdam: W.A. Benjamin, Inc.
1968, '"Bootstrap": a scientific idea?', *Science* 161, pp. 762-765.
Chew, G.F., Gell-Mann, M., and Rosenfeld, L.,
1965, *Strongly interacted particles*, Moscow (in Russian).
Chizhevsky, A.L.,
1976, *An Earthy Echo of the Solar Storms*, Moscow (in Russian).
Chudinov, E.M.,
1976, 'Einstein's conception of physical reality', in: *Physical Theory and Reality*, Voronezh (in Russian).
Coyré, A.,
1985, *Essays on History of Philosophical Thought*, Moscow (in Russian).
Dirac, P.A.M.,
1963, 'The evolution of the physicist's picture of nature', in: *Scientific Americans*, May.
1963a, 'The evolution of physicist's view on the picture of nature', in: *Voprosy Filosofii*, No 12, pp.83-94.
Dishlevy, P.S.,
1973, 'Natural science picture of the world as a form of knowledge synthesis', in: *Synthesis of Modern Scientific Knowledge*, Moscow.
Dobronravova, I.S.,
1991, *Synergetics: Becoming of Non-linear Thought*, Kiev (in Russian).
Doods, E.K.,
1951, *The Greeks and the Irrational*, Berkeley: University of California Press.

Dorfman, Ya.G.,
1974, *Universal History of Physics from Ancient Times to the End of XVIIIth Century*, Moscow (in Russian).
East – West. Studies. Translations. Publications, 1982, Moscow (in Russian).
Einstein, A.,
1931, 'Maxwell's influence on the development of conception of physical reality', in: *James Clerk Maxwell: A Commemoration Volume*, Cambridge: Cambridge University Press.
1945, 'Autobiographisches (Autobiographical notes)', in: *Albert Einstein – Philosopher-Scientist*, ed. by P.A. Schlipp, Library of Living Philosophers, Evanston (Illinois), pp.1-95.
1965-1967, *Collected Works*, vol. 1-4, Moscow (in Russian).
1967, 'Motives of scientific research', in: Einstein A. *Collected works,* vol. 4 (in Russian).
1967a, 'Physics and Reality', in: Einstein A. *Collected works,* vol. 4, pp.204-213 (in Russian).
Einstein, A., Podolsky, B., and Rosen, P.,
1935, 'Can quantum-mechanical description of physical reality be considered complete?', *Phys. Review* 47, pp.777-780.
Eremeeva, A.I.,
1985, *Astronomic Map of the Universe and Its Creators*, Moscow (in Russian).
Faraday, M.,
1832, 'Experimental researches in electricity', *Philos. Trans.*, Part 1, pp.125-162.
Fedorov, N.F.,
1982, *Works*, Moscow (in Russian).
Feynman, R.,
1968, *The Character of Physical Laws*, Moscow (in Russian).
Feynman, R., Leighton, R., Sands, M.,
1963, *The Feynman Lectures on Physics*, Reading, Massachusetts, Palo Alto, London: Addison-Wesley Publishing Company Inc.
Fok V., and Jordan P.,
1931, *Zeitschrift für Physik*, 66.
Friedman, A.A.,
1965, *World as a Space and Time*, Moscow (in Russian).
Frankfurt, I.U., Frenk, A.M.,
1962, *Christian Huygens*, Moscow (in Russian).
Frolov, I.T.,
1989, *On Human and Humanism*, Moscow (in Russian).
Frolov, I.T. and Yudin, B.G.,
1986, *Ethics of Science Issues and Discussions*, Moscow (in Russian).
Fucault, M.P.,
1966, *Les Mots et les Choses. Une Archeologie des Sciences Humaines*, Paris: Gallimard.
1975, *Surveiller et Punir. Nuissance de la Prison*, Paris.
Fourier, Ch.,
1953, *Textes Choisis*, Paris: Editions Sociales.
Gaidenko, P.P.
1987, *Evolution of the Notion of "Science" (17th-18th centuries)*, Moscow (in Russian).
Gilbert, W.,

1900, *On the Magnet (De Magnete, Magneticisque Corporibus et de Mango Magnete)*, Gilbert club revised English translation, London.
Ginzburg, V.L., Syrovatsky, S.I.,
1967, 'Gamma- and X-ray astronomy', *On What Physicists Think*, vol. 8, Astrophysics, Moscow (in Russian).
Glensdorff, P., Prigogine, I.,
1971, *Thermodynamic Theory of Structure, Stability and Fluctuations*, N.Y.: Wiley-Interscience.
Gorochov, V.G.,
1974, *A Methodological Analysis of Scientific-technical Disciplines*, Moscow (in Russian).
1984, *Methodological Analysis of Scientific-technical Disciplines*, Moscow (in Russian).
Grigorian, A.T.,
1971, *Mechanics from Antiquity to Nowadays*, Moscow (in Russian).
Grigorieva, T.P.,
1979, *Japanese Artistic Tradition*, Moscow (in Russian).
1983, *Japanese XXth Century Literature*, Moscow (in Russian).
Guirenock, F.I,
1960, *Russian Cosmists*, Moscow (in Russian).
Gurevich, A.Ya.,
1972, *Categories of the Culture of the Middle Ages,* Moscow (in Russian).
1983, *Categories of Philosophy and Categories of Culture*, Kiev (in Russian).
1990, 'Social history and historical science', *Voprosy Filosofii* 4 (in Russian).
Guth A., and Steinhardt P.,
1984, 'The inflationary Universe', *Scientific American*, May.
Habermas, J.,
1992, *Democracy – Reason – Morality*, Translations from German and English of lectures delivered in the Institute of Philosophy of the Russian Academy of Sciences in April 1989 in Moscow, Moscow (in Russian)..
Haken, H.,
1987, *Advanced Synergetics*, Springer Series Synergetics, Vol. 20. 2nd corr. Printing, Berlin, Heidelberg: Springer.
Harris, L.,
1990, *Monetary Theory,* Moscow (in Russian).
Hegel, G.W.F.,
1920-1923, 'Wissenschaft der logik', in: *Samtlische Werke*, hrsg. Von G. Lasson, Bd. 3,4, Leipzig: Meiner.
Heisenberg, W.
1969, *Der Teil und das Ganze*, München.
Heraclitus of Ephesus,
1954, *Cosmic Fragments*, Ed. with an introduction and commentary by G.S. Kirk, Cambridge.
Hertz, H.,
1900, *The Principles of Mechanics*, New York.
1948, 'An investigation on distributing electrical power', in: *From the Prehistory of Radio*, Moscow-Leningrad (in Russian).
Hertzen, A.I.,
1946, *Letters on Nature Study*, Moscow (in Russian).

Hesse, H.,
1970, 'Das glasperlenspiel', in: *Gessamelte Werke*, Bd.1-12, Fr./M.
Hilbert, D. and Bernays, P.,
1934, *Grundlagen der Mathematik*, Bd. 1, Berlin.
Holbach, P.,
1770, *Système de la Nature, ou des Loix du Monde Physique et du Monde Moral*, London, Amsterdam: Marc-Michel Rey.
Holton, G.,
1974, 'Einstein, Michelson and "decisive" experiment', in: *Einsteinian Collection*, Moscow (in Russian).
1973, *The Thematic Component in Scientific Thought*, The University of Texas at Austin.
1979, 'Einstein on physical reality', in: *Einsteinian Collection, 1969-1970*, Moscow (in Russian).
1988, *Thematic Origins of Scientific Thought: Kepler to Einstein.* 2nd ed., Harvard: Harvard Univ. Press.
1992, 'What is an "anti-science"?', *Voprosy Filosofii* 2 (in Russian).
Hooker, S.,
1975, 'On global theories', *Philos. Sci.* 42, No 2.
Hufbauer K.,
1982, *The Formation of the German Chemical Community* (*1720-1795*). Berkeley.
Ideals and Norms of Scientific Investigation, 1981, Minsk (in Russian).
Jantsch, E.,
1980, *The Self-organizing Universe: Science and Human Implication of the Emerging Paradigm of Evolution*, Oxford.
Ivanov, B.I. and Cheshev, V.V.,
1977, *Becoming and Development of Technical Sciences*, Leningrad (in Russian).
1981, *Technical Knowledge as an Object of Methodological Analysis*, Tomsk (in Russian).
Jakobson, R.,
1985, *Selected Works*, Moscow (in Russian).
Jordan, P., Pauli, W.,
1928, Zur quantenelektrodynamik landungsfreier felder', in: *Zschr. f. Ph.* 47.
Jua, M.,
1975, *History of Chemistry*, Moscow (in Russian).
Kara-Murza, S.G.,
1990, 'Science and crisis of civilization', *Voprosy Filosofii* 9, pp.3-15 (in Russian).
Karmin, A.S., Khaikin, E.P.,
1971, *Creative Intuition in Science*, Moscow (in Russian).
Karpinskaya, R.S.,
1970, *Biology and Worldview*, Moscow (in Russian).
1992, 'Biology, ideals of scientific and mankind destiny', *Voprosy Filosofii* 11 (in Russian).
Kazyutinsky, V.V.,
1986, 'Conception of global evolutionism in the scientific picture of the world', in: *On Modern Status of an Idea of Global Evolutionism*, Moscow (in Russian).
1989, 'Universe in the scientific picture of the world and social-practical activity of mankind', in: *Philosophy, Natural Science, Social Developments*, Moscow, pp.199-213 (in Russian).
Kessidi, F.Ch.,

1972, *From Myth to Logos*, Moscow (in Russian).
Keswani, G.H.,
1973, 'Beginnings of relativity theory', in: *The Relativity Principle*, Moscow (in Russian).
Kholodny, N.G.,
1982, *Selected Works*, Kiev (in Russian).
Klimontovich, N.Yu.,
1986, *On Synergetics without Formulas*, Minsk (in Russian).
Kogan, L.A.,
1990, 'N.F. Fedorov's philosophy', *Voprosy Filosofii* 11, pp.74-84 (in Russian).
Konopleva, N.P., Sokolik, G.A.,
1972, 'Symmetries and types of physical theories', *Voprosy Filosofii* 1 (in Russian).
Kosareva, L.M.,
1989, *Sociological Genesis of Science of the New Age*, Moscow (in Russian).
Krimsky, S.B., Kuznetsov, V.I.,
1983, *The Worldview Categories in Modern Natural Science*. Kiev (in Russian).
Kuhn, T.,
1962, *The Structure of Scientific Revolutions*, Chicago: Univ. of Chicago Press.
1970, Postscriptum. – 1969, in: *Structure of Scientific Revolutions*, 2 ed. enl., Chicago.
1974, 'Second thoughts on paradigms', in: *The Structure of Scientific Theories*, Urbana, pp.459-482.
1977, 'Objectivity, value judgment, and theory choice', in: Kuhn T., *The Essential Tension: Selection. Selected Studies in Scientific Tradition and Change*, Chicago: The University of Chicago Press, pp.320-339.
Kuo Yü, Transl. from Chinese, 1987, Moscow (in Russian).
Kurdyumov, S.P.,
1982, 'Eigenfunctions of combustion of nonlinear medium and constructive laws of its organization building', in: *Contemporary Problems of Mathematical Physics and Computational Mathematics*, Moscow (in Russian).
1990, *Evolution Laws and Self-Organization of Complex Systems*, Moscow (in Russian).
Kuznetsov, B.G. (ed.),
1967, *Niels Bohr. Life and Works*, Moscow (in Russian).
Kuznetsov, B.G.,
1958, *Principles of Classical Physics*, Moscow (in Russian).
Kuznetsov, I.V.,
1948, *Correspondence Principle in Contemporary Physics and its Philosophical Meaning*, Moscow-Leningrad (in Russian).
1975, *Selected Works on Methodology of Physics*, Moscow (in Russian).
Kuznetsov, V.I.,
1973, *Dialectics of the Development of Chemistry*, Moscow (in Russian).
Kuznetsova, L.F.,
1984, *Picture of the World and its Functions in Scientific Cognition*, Minsk (in Russian).
Kuznetsova, N.I.,
1982, *Science and its History*, Moscow (in Russian).
Lacatos, I.,
1976, 'History of science and its rational reconstruction', in: *Boston Studies*, vol.VIII, pp.91-136.

1970, 'Falsification and the Methodology of Scientific Research Programmes', in: *Criticism and the Growth of Knowledge*, Cambridge.

Lamarck, J.-B.,
1809, *Philosophie Zoologique*, Paris.
1959, *Selected Works*, Vol.2, Moscow (in Russian).

Lamettrie, J.O.,
1796, *Ouevres Philosophiques*, vol 1-3, Paris.

Landau, L.D. and Lifshitz, E.M.,
1960, *Field Theory*, Moscow (in Russian).

Landau, L.D., Peierls, R.,
1965, 'Extension of uncertainty principle on relativistic quantum theory', in: Landau, L.D., *Selected Works*, Vol.1, Moscow, pp.56-70 (in Russian).

Laszlo, E.,
1972, *Introduction to System Philosophy*, New York.
1990, 'A new understanding of the evolution. Introduction to the global epoch', in: *One World for Everybody*, Moscow (in Russian).

Laudan, L.,
1977, *Progress and Its Problems*, Berkeley: University of California Press.
1984, *Science and Values*, Berkeley-Los Angeles-London: Univ. of California Press.

Lavoisier, A.,
1943, 'Preliminary reasoning from "Elementary Course of Chemistry"', *Progress of Chemistry* 5, No 12, Moscow (in Russian).

Lenk, H.,
1993, *Interpretationskonstrukte*, Frankfurt am M.

Leopold, A.A,
1983, *A Sand County Almanac*, New York: Ballantine Books.

Linde, A.D.,
1984, 'The inflationary Universe', *Advances of Physical Sciences* 144, Ser.2, pp.177-214 (in Russian).

Liozzi, M.,
1970, *A History of Physics*, Moscow (in Russian).

Lorentz, G.A.,
1953, *The Theory of the Electron and its Application to Phenomena of Light and Heat Radiation*, Moscow (in Russian).

Losev, A.F.,
1968, *A History of Antique Esthetics, vol. 1* (*Early Classics*), Moscow (in Russian).
1977, *Ancient Philosophy of History*, Moscow (in Russian).

Lotman, Yu.,
1973, 'On two models of communication in the system of culture', *Proceedings on Sign Systems* 6, Tartu (in Russian).

Luria, A.,
1974, *On Historical Development of Cognitive Processes. Experimental-psychological Investigations*, Moscow (in Russian).

Macey, D.,
1990, 'Ecology unified', in: *One World for Everybody*, Moscow (in Russian).

Mamardashvili, M.K.,
1968, 'An analysis of consciousness in works of Marx', *Voprosy Filosofii* 6 (in Russian).
Mamardashvili, M.K., Solovyov, E.Yu., Shvyrev V.S.,
1972, 'Classics and contemporaneity: two epochs in development of bourgeois philosophy', in: *Philosophy in the Modern World. Philosophy and Science*, Moscow (in Russian).
Mamchur, E.A.,
1975, *The Problem of Theory Choice*, Moscow (in Russian).
Mandelshtam, L.I.,
1948, 'Introduction', in: *From Prehistory of Radio*, Moscow (in Russian).
1972, *Lectures on Optics, Relativity Theory and Quantum Mechanics*, Moscow (in Russian).
Margenau, H.,
1950, *The Nature of Physical Reality*. N.Y., L.
Marx, K. and Engels, F.,
1955-1981, *Works*, vol. 1-50, Moscow (in Russian).
Mattis, M.,
1967, *The Theory of Magnetism*, Moscow (in Russian).
Maxwell, J.C.,
'A dynamical theory of the electromagnetic field', *Philos. Trans. Roy. Soc.*, 1865.
Mendel, G.,
1959, *Versuche über Pflanzenhybriden*, heraus von Erich Tschermak, Leipzig, 1901. – Facsimile Ausgabe, Berlin.
Michailovsky, V., Khon, G.,
1989, *Dialectics of Formation of the Modern Scientific Picture of the World*, Leningrad (in Russian).
Mirsky, E.M.,
1980, *Interdisciplinary Investigations and Disciplinary Organization of Science*, Moscow (in Russian).
Mitskevich, N.V.,
1976, 'Cosmology, relativistic astrophysics and physics of elementary particles', in: *Philosophical Problems of Astronomy in the 20th Century*, Moscow (in Russian).
Moiseev, N.N.,
1986, 'Strategy of reason', *Znanie-Sila* 10 (in Russian).
1989, 'Logics of universal evolutionism and cooperativity', *Voprosy Filosofii 8* (in Russian).
1990, 'Man in the Universe and on the Earth', *Voprosy Filosofii* 6 (in Russian).
Moscovici, S.,
1998, *Gods Making Machine*, Moscow (in Russian).
Mostepanenko, M.V.,
1969, *Philosophy and Physical Theory*, Leningrad (in Russian).
Motroshilova, N.V.,
1981, 'Scientific norms and orientations of scientists', in: *Ideals and Norms of Scientific Investigation*, Minsk (in Russian).
Needham, J.,
1964, 'Society and science in East and West', in: *The Science of Science. Society in the Technological Age*, Ed. by. M. Goldsmith and A. Mackay, London, Toronto: Souvenir Press.
1988, 'Precursors of modern science', in: *UNESCO Courier*, November.

Neugebauer, O.,
1968, *Exact Sciences in Ancient Times*, Moscow (in Russian).
New Ideas in Physics, Coll. 2. Ether and Matter, 1913, St. Petersburg (in Russian).
Newton-Smith, W.H.,
1981, *The Rationality of Science*, London-New York.
Niven, W.D. (ed.),
1890, *The Scientific Papers of James Clerk Maxwell*, Cambridge.
Nordin, I.,
1979, 'Determinism and Locality in Quantum Mechanics', *Synthese* 42, No 1.
Nugaev, R.M.,
1989, *Reconstruction of the Process of Change of Fundamental Scientific Theories*, Kiev (in Russian).
Omelyanovsky, M.E.,
1973, *Dialectics in Contemporary Physics*, Moscow (in Russian).
Parsons, T.,
1968, 'Systems analysis: social systems', *International Encyclopedia of Social Science*, N.Y.
Pastushny, S.A.,
1981, *Genetics as a Subject of Philosophical Analysis*, Moscow (in Russian).
Pauli, W.,
1956, 'Exclusion principle and quantum mechanics (Nobel Prize Lecture Delivered December 13, 1946 in Stockholm)', in: *Theoretical Physics of XXth Century*, Moscow (in Russian).
1964, *Collected Scientific Papers by W. Pauli*, in Two Volumes, ed. by R. Kronig and V. Weisskopf, New York-Sydney: Interscience Publishing.
Pauli, W. (ed.),
1955, *Niels Bohr and the Development of Physics*, London: Pergamon Press Ltd.
Petrov, M.K.,
1991, *Language, Sign, Culture*, Moscow (in Russian).
Petushkova, E.V.,
1983, *Reflection on Animate Nature. Dynamics of Theoretical Models*, Minsk (in Russian).
Philippidis, C., Dewdney, C., Hiley, B.,
1979, 'Quantum interference and the quantum potential', *Nuovo Cimento* 52, No 1, pp.15-28.
Pivovarov, D.V.,
1971, 'Practice and forming of cognitive image', in: *Leninist Theory of Reflection*, vol. 5, Sverdlovsk (in Russian).
Planck, M.,
1958, 'Die einheit des physikalischen weltbildes', in: *Max Planck, Physikalische Abhandlungen und Vorträge*, Bd. III, Braunscheweig, Friedr. Viewen & Sohn.
Plutarch,
1961, *Parallel Lives*, Vol. 1, Moscow (in Russian).
Philosophy of the Epoch of Early Bourgeois Revolution, 1983, Moscow (in Russian).
Philosophy of Technics: History and Nowadays, 1997, Moscow (in Russian).
Podgoretsky, M.I., and Smorodinsky, Ya.A.,
1969, 'On axiomatic structure of physical theories', in: *Questions of the Cognition Theory*, Issue 1, Moscow (in Russian).
Popper, K.R.,

1935, *Logic der Forschung*, Wien.

1968, *Conjectures and Refutations. The Growth of Scientific Knowledge*, New York, 1968.

Povarov, G.N.,

1972, 'To Daidalo Ptero (On cognition of scientific-technical progress)', in: *Systemic Investigations. Annual. 1971*, Moscow (in Russian).

Price, D.,

1963, *Little Science, Big Science*, New York, Columbia University Press.

Prigogine, I., Stengers, I.,

1981, *La Nouvelle Alliance: Metamorphose de la Science*, Paris.

1984, *Order out of Chaos. Man's New Dialogue with Nature*, Foreword by A. Toffler, New York-Toronto: Bantam Books.

1994, *Time, Chaos, Quantum. On Solution of the Time Paradox*, Moscow (in Russian).

Ravikovich. A.I.,

1976, *Charles Lyell*, Moscow (in Russian).

Reichenbach, H.,

1961, *Experience and Prediction*, Chicago.

Rokitsky, P.F., Savchenko, V.K., and Dobina, A.I.,

1977, *Genetic Structure of Population and its Change during Selection*, Minsk (in Russian).

Rorty, R.,

1985, 'Historiography of philosophy. Four genres', in: *Philosophy in History. Essays on the Historiography of Philosophy*, Cambridge etc.

Rosenberg, O.O.,

1991, *Works on Buddhism*, Moscow (in Russian).

Rosenberger, F.,

1937, *History of Physics*, Part II, Moscow-Leningrad (in Russian).

Rosenfeld, L.,

1962-1965, 'Newton and the law of gravitation', *Archives for the History of Exact Sciences* 2, pp.365-386.

Rozin, V.M.,

1989, *Specifics and Formation of Sciences, Techniques and Humanities*. Krasnoyarsk (in Russian).

Rozin, V.M. and Moskayeva, A.S.,

1967, 'On analysis of the structure of knowledge system of "elements" by Euclid', in: *New Investigations in Pedagogical Sciences*, vol. IX, Moscow (in Russian).

Rudenko, A.P.,

1987, 'Chemical evolution and biogenesis', in: *Philosophy and Sociology of Science and Technics. Annual*, Moscow (in Russian).

Rutherford, E.,

1972, *Selected Works. Atomic Structure and Artificial Transformation of Elements*, Moscow (in Russian).

Sachkov, Yu.V.,

1974, 'Problems of the style of thinking in natural science', in: *Philosophy and Natural Science. On 70th Anniversary of Academician B.M. Kedrov*, Moscow (in Russian).

1994, 'A forming contingency', in: *Self-organization and Science*, Moscow (in Russian).

Sadovsky, V.N.,

1981, 'Philosophy of science in searching of the new ways', in: *Scientific Investigation's Ideals and Norms*, Minsk (in Russian).
Saint-Simon, C.A.,
1948, *Selected Works*, vol 1, Moscow-Leningrad (in Russian).
1966, *Ouevres Complètes*, vol 1-6, Paris.
Saushkin Yu.G.,
1976, *History and Methodology of Geographical Science*, Moscow (in Russian).
Shachnovich, M.,
1961, *Primeval Mythology and Philosophy*, Leningrad (in Russian).
Shapere, S.,
1974, 'Scientific theories and their domains', in: *Structure of Scientific Theories*, N.Y.
Shaumian, S.,
1965, *Structural Linguistics*, Moscow (in Russian).
Schrödinger, E.,
1971, *New Ways in Physics. Papers and Talks*, Moscow (in Russian).
Schtoff, V.A.,
1966, *Modeling and Philosophy*, Moscow-Leningrad (in Russian).
Schweitzer, A.,
1990, 'Veneration for life as a base for ethical world and life asserting', in: *Global Problems and Common Human Values*, Moscow (in Russian).
1992, *Veneration for Life*, Moscow (in Russian).
Shipunov, F.Ya.,
1990, 'Biosphere ethics', in: *Ecological Alternative*, Moscow (in Russian).
Shmakov. V.S.
1990, *The Structure of Historical Knowledge and Picture of the World*, Novosibirsk (in Russian).
Shmalgauzen, I.I.,
1968, *Cybernetic Issues of Biology*, Novosibirsk (in Russian).
Shubas, M.L.,
1982, *Engineering Thinking and Scientific-Technical Progress: The Style of Thinking, Picture of the World Worldview*, Vilnius, (in Russian)
Shvyrev, V.S.,
1992, 'Rationality as a culture value', *Voprosy Filosofii* 6. (in Rusian).
Silk, J.,
1982, *Big Bang: Birth and Evolution of the Universe*, Moscow (in Russian).
Smirnov, V.A.,
1962, 'The genetic method of scientific theory construction', in: *Philosophical questions of contemporary formal logic*, Moscow (in Russian).
Smirnova, R.S.,
1991, *The Nature of Social Reality*, Minsk (in Russian).
Snow, C.P.,
1971, 'The two cultures', *New Statesman*, 6 October, 1956, Public Affairs, Macmillan, London.
Solovyov, Yu.I.,
1971, *Evolution of the Main Theoretical Problems of Chemistry*, Moscow (in Russian).
Solovyov Yu.I., Kurashov V.I.
1983, *Chemistry on the Science Crossroad*, Moscow (in Russian).

Spassky, B.,
1977, *History of Physics, Part II*, Moscow (in Russian).
Spencer, H.,
1997, *The Synthetic Philosophy by Herbert Spencer in a Short Exposition by Howard Collins*, Kiev (in Russian).
Stapp, H.P.,
1971, 'S-matrix interpretation of quantum theory', *Phys. Rev.*, D., Vol.3, No 4, pp.1314-1319.
Stepin, V.S.,
1970, 'Subject and object problem in experimental science', *Voprosy Filosofii* 1, pp.80-91 (in Russian).
1971, 'Genesis of theoretical models of science', *Filosofskie Nauki* 3, pp.50-59 (in Russian).
1972, 'Genesis of theoretical models of science', in: *Philosophy. Methodology. Science*, Moscow (in Russian).
1976, *Becoming of Scientific Theory*, Minsk (in Russian).
1979, 'Structure and evolution of theoretical knowledge', in: *A Nature of Scientific Cognition*, Minsk (in Russian).
1981, *The Ideals and Norms of Scientific Investigation*, Minsk (in Russian).
1981a, 'Ideals and norms in dynamics of scientific conquest', in: *Ideals and Norms of Scientific Investigation*, Minsk (in Russian).
1982, 'The structure of theoretical knowledge and historical-scientific reconstruction', in: *Methodological Problems of Historical-Scientific Investigations*, Moscow (in Russian).
1986, 'On the prognostic nature of philosophical knowledge', *Voprosy Filosofii* 4, pp.39-53 (in Russian).
1987, 'Scientific revolutions as points of bifurcation in knowledge development', in: *Scientific Revolutions in Culture Dynamics*, Minsk (in Russian).
1989, 'Scientific knowledge and values of technogenic civilization', *Voprosy Filosofii* 10, pp.3-16 (in Russian).
1992, 'Perspectives of civilization: from cult of power to dialogue and consent', in: *Ethic Thought, 1991*, Moscow, pp.182-200 (in Russian).
1992a, *Philosophical Anthropology and Philosophy of Science*, Moscow (in Russian).
1998, 'Philosophy in dialogue of cultures', in: *Man – Philosophy – Science. Main Papers and Reviews of the First Russian Philosophical Congress*, Vol.9. St.Petersburg, pp.23-34 (in Russian).
Stepin, V.S., Gorochov, V.G. and Rozov, M.A.,
1996, *Philosophy of Science and Technique*, Moscow (in Russian).
Stepin, V.S., Kuznetsova, L.F.,
1981, 'Ideals of explanation and the problem of interactions of sciences', in: *Ideals and Norms of Scientific Investigation*, Minsk, pp.260-279 (in Russian).
1994, *Scientific Picture of the World in the Culture of Technogenic Civilization*, Moscow (in Russian).
Stepin, V.S. and Tomilchik, L.M.,
1970, *Practical Nature of Cognition and Methodological Problems of Modern Physics*, Minsk (in Russian).
Suppes, P.,
1977, 'The search for philosophic understanding of scientific theories', in: *The Structure of Scientific Theories*, Urbana.

The History of Ancient Dialectics, 1972, Moscow (in Russian).
The History of Biology from the Beginning of the XXth Century to Nowadays, 1975, Moscow (in Russian).
Timiryazev, K.,
1939, *Works*, vol. VIII, Moscow (in Russian).
Toffler, A.
1986, 'Science and change', Preface to: Prigogine I., Stengers I., *Order out of Chaos*, Moscow (in Russian).
Tomilchick, L.M., Fyodorov, F.I.,
1987, 'Preconditions and mechanisms of scientific revolution', in: *Scientific Revolutions in Dynamics of Culture*, Minsk (in Russian).
Tsiolkovsky, K.E.,
1986, *Dreams on Earth and Heaven*, Tula (in Russian).
Tulviste, P.,
1977, 'On interpretation of parallels between ontogenesis and history-minded thinking', *Proceedings on Sign Systems VIII*, Tartu (in Russian).
Tyapkin, A.A.,
1973, 'On history of formation of ideas of special relativity', in: *The Relativity Principle*, Moscow (in Russian).
Umov, N.A.,
1916, *Collected Works*, Vol. 3, Moscow (in Russian).
Vavilov, S.I.,
1956, 'On mathematical hypothesis', in: Vavilov, S.I., *Selected Works*, vol. 3, Moscow (in Russian).
van der Warden, D.,
1962, 'Exclusion principle and spin', in: *Theoretical Physics of XXth Century*, Moscow (in Russian).
Vernadsky, V.I.,
1934, *Issues of Biogeochemistry*, Moscow (in Russian).
1940, *Biogeochemical Essays*, Moscow-Leningrad (in Russian).
1944, 'A few words on a noosphere', *Advances of Modern Biology*, Vol. XVIII, Ser. 2 (in Russian).
1977, *Reflections of Naturalist. Scientific Thought as the Planetary Phenomenon*, Moscow (in Russian).
1978, *Living Substance*, Moscow (in Russian).
1981, *Selected Works on History of Science*, Moscow (in Russian).
Vodopianov, P.A.,
1981, *Stability and Dynamics of Biosphere*, Minsk (in Russian).
Vygodsky, M.Ya.,
1967, *Arithmetics and Algebra in Ancient World*, Moscow (in Russian).
Wartofsky, M.,
1976, 'Metaphysics as heuristic for science', in: *Boston Studies in the Philosophy of Science*, Vol. III, pp.123-172
1988, *Models: Representation and the Scientific Understanding*, Dordrecht-Boston-London, D. Reidel Publishing Company.

White, L.Jr.,
1967, 'The historical roots of our ecologic crisis', *Science*, 155 (3767), 10 March 1967, 1203-1207.
Wiener, N.,
1948, *Cybernetics or Control and Communication in the Animal and Machine*, Cambridge, Mass.
Yanovskaya, S.A.,
1972, *Methodological Problems of Science*, Moscow (in Russian).
Zaitsev, I.,
1985, *Cultural Revolution in Ancient Greece*, Moscow (in Russian).
Zelenkov, A.I., and Vodopianov, P.A.,
1987, *Biosphere Dynamics and Sociocultural Traditions*, Minsk (in Russian).

SUBJECT INDEX

SYNTHESE LIBRARY

1. J. M. Bochénski, *A Precis of Mathematical Logic*. Translated from French and German by O. Bird. 1959 ISBN 90-277-0073-7
2. P. Guiraud, *Problèmes et méthodes de la statistique linguistique*. 1959 ISBN 90-277-0025-7
3. H. Freudenthal (ed.), *The Concept and the Role of the Model in Mathematics and Natural and Social Sciences*. 1961 ISBN 90-277-0017-6
4. E. W. Beth, *Formal Methods*. An Introduction to Symbolic Logic and to the Study of Effective Operations in Arithmetic and Logic. 1962 ISBN 90-277-0069-9
5. B. H. Kazemier and D. Vuysje (eds.), *Logic and Language*. Studies dedicated to Professor Rudolf Carnap on the Occasion of His 70th Birthday. 1962 ISBN 90-277-0019-2
6. M. W. Wartofsky (ed.), *Proceedings of the Boston Colloquium for the Philosophy of Science, 1961–1962*. [Boston Studies in the Philosophy of Science, Vol. I] 1963 ISBN 90-277-0021-4
7. A. A. Zinov'ev, *Philosophical Problems of Many-valued Logic*. A revised edition, edited and translated (from Russian) by G. Küng and D.D. Comey. 1963 ISBN 90-277-0091-5
8. G. Gurvitch, *The Spectrum of Social Time*. Translated from French and edited by M. Korenbaum and P. Bosserman. 1964 ISBN 90-277-0006-0
9. P. Lorenzen, *Formal Logic*. Translated from German by F.J. Crosson. 1965 ISBN 90-277-0080-X
10. R. S. Cohen and M. W. Wartofsky (eds.), *Proceedings of the Boston Colloquium for the Philosophy of Science, 1962–1964*. In Honor of Philipp Frank. [Boston Studies in the Philosophy of Science, Vol. II] 1965 ISBN 90-277-9004-0
11. E. W. Beth, *Mathematical Thought*. An Introduction to the Philosophy of Mathematics. 1965 ISBN 90-277-0070-2
12. E. W. Beth and J. Piaget, *Mathematical Epistemology and Psychology*. Translated from French by W. Mays. 1966 ISBN 90-277-0071-0
13. G. Küng, *Ontology and the Logistic Analysis of Language*. An Enquiry into the Contemporary Views on Universals. Revised ed., translated from German. 1967 ISBN 90-277-0028-1
14. R. S. Cohen and M. W. Wartofsky (eds.), *Proceedings of the Boston Colloquium for the Philosophy of Sciences, 1964–1966*. In Memory of Norwood Russell Hanson. [Boston Studies in the Philosophy of Science, Vol. III] 1967 ISBN 90-277-0013-3
15. C. D. Broad, *Induction, Probability, and Causation*. Selected Papers. 1968 ISBN 90-277-0012-5
16. G. Patzig, *Aristotle's Theory of the Syllogism*. A Logical-philosophical Study of *Book A* of the *Prior Analytics*. Translated from German by J. Barnes. 1968 ISBN 90-277-0030-3
17. N. Rescher, *Topics in Philosophical Logic*. 1968 ISBN 90-277-0084-2
18. R. S. Cohen and M. W. Wartofsky (eds.), *Proceedings of the Boston Colloquium for the Philosophy of Science, 1966–1968, Part I*. [Boston Studies in the Philosophy of Science, Vol. IV] 1969 ISBN 90-277-0014-1
19. R. S. Cohen and M. W. Wartofsky (eds.), *Proceedings of the Boston Colloquium for the Philosophy of Science, 1966–1968, Part II*. [Boston Studies in the Philosophy of Science, Vol. V] 1969 ISBN 90-277-0015-X
20. J. W. Davis, D. J. Hockney and W. K. Wilson (eds.), *Philosophical Logic*. 1969 ISBN 90-277-0075-3
21. D. Davidson and J. Hintikka (eds.), *Words and Objections*. Essays on the Work of W. V. Quine. 1969, rev. ed. 1975 ISBN 90-277-0074-5; Pb 90-277-0602-6
22. P. Suppes, *Studies in the Methodology and Foundations of Science. Selected Papers from 1951 to 1969*. 1969 ISBN 90-277-0020-6
23. J. Hintikka, *Models for Modalities*. Selected Essays. 1969 ISBN 90-277-0078-8; Pb 90-277-0598-4

SYNTHESE LIBRARY

24. N. Rescher *et al.* (eds.), *Essays in Honor of Carl G. Hempel.* A Tribute on the Occasion of His 65th Birthday. 1969 ISBN 90-277-0085-0
25. P. V. Tavanec (ed.), *Problems of the Logic of Scientific Knowledge.* Translated from Russian. 1970 ISBN 90-277-0087-7
26. M. Swain (ed.), *Induction, Acceptance, and Rational Belief.* 1970 ISBN 90-277-0086-9
27. R. S. Cohen and R. J. Seeger (eds.), *Ernst Mach: Physicist and Philosopher.* [Boston Studies in the Philosophy of Science, Vol. VI]. 1970 ISBN 90-277-0016-8
28. J. Hintikka and P. Suppes, *Information and Inference.* 1970 ISBN 90-277-0155-5
29. K. Lambert, *Philosophical Problems in Logic.* Some Recent Developments. 1970 ISBN 90-277-0079-6
30. R. A. Eberle, *Nominalistic Systems.* 1970 ISBN 90-277-0161-X
31. P. Weingartner and G. Zecha (eds.), *Induction, Physics, and Ethics.* 1970 ISBN 90-277-0158-X
32. E. W. Beth, *Aspects of Modern Logic.* Translated from Dutch. 1970 ISBN 90-277-0173-3
33. R. Hilpinen (ed.), *Deontic Logic.* Introductory and Systematic Readings. 1971 *See also* No. 152. ISBN Pb (1981 rev.) 90-277-1302-2
34. J.-L. Krivine, *Introduction to Axiomatic Set Theory.* Translated from French. 1971 ISBN 90-277-0169-5; Pb 90-277-0411-2
35. J. D. Sneed, *The Logical Structure of Mathematical Physics.* 2nd rev. ed., 1979 ISBN 90-277-1056-2; Pb 90-277-1059-7
36. C. R. Kordig, *The Justification of Scientific Change.* 1971 ISBN 90-277-0181-4; Pb 90-277-0475-9
37. M. Čapek, *Bergson and Modern Physics.* A Reinterpretation and Re-evaluation. [Boston Studies in the Philosophy of Science, Vol. VII] 1971 ISBN 90-277-0186-5
38. N. R. Hanson, *What I Do Not Believe, and Other Essays.* Ed. by S. Toulmin and H. Woolf. 1971 ISBN 90-277-0191-1
39. R. C. Buck and R. S. Cohen (eds.), *PSA 1970.* Proceedings of the Second Biennial Meeting of the Philosophy of Science Association, Boston, Fall 1970. In Memory of Rudolf Carnap. [Boston Studies in the Philosophy of Science, Vol. VIII] 1971 ISBN 90-277-0187-3; Pb 90-277-0309-4
40. D. Davidson and G. Harman (eds.), *Semantics of Natural Language.* 1972 ISBN 90-277-0304-3; Pb 90-277-0310-8
41. Y. Bar-Hillel (ed.), *Pragmatics of Natural Languages.* 1971 ISBN 90-277-0194-6; Pb 90-277-0599-2
42. S. Stenlund, *Combinators, γ Terms and Proof Theory.* 1972 ISBN 90-277-0305-1
43. M. Strauss, *Modern Physics and Its Philosophy.* Selected Paper in the Logic, History, and Philosophy of Science. 1972 ISBN 90-277-0230-6
44. M. Bunge, *Method, Model and Matter.* 1973 ISBN 90-277-0252-7
45. M. Bunge, *Philosophy of Physics.* 1973 ISBN 90-277-0253-5
46. A. A. Zinov'ev, *Foundations of the Logical Theory of Scientific Knowledge (Complex Logic).* Revised and enlarged English edition with an appendix by G. A. Smirnov, E. A. Sidorenka, A. M. Fedina and L. A. Bobrova. [Boston Studies in the Philosophy of Science, Vol. IX] 1973 ISBN 90-277-0193-8; Pb 90-277-0324-8
47. L. Tondl, *Scientific Procedures.* A Contribution concerning the Methodological Problems of Scientific Concepts and Scientific Explanation. Translated from Czech by D. Short. Edited by R.S. Cohen and M.W. Wartofsky. [Boston Studies in the Philosophy of Science, Vol. X] 1973 ISBN 90-277-0147-4; Pb 90-277-0323-X
48. N. R. Hanson, *Constellations and Conjectures.* 1973 ISBN 90-277-0192-X

49. K. J. J. Hintikka, J. M. E. Moravcsik and P. Suppes (eds.), *Approaches to Natural Language.* 1973 ISBN 90-277-0220-9; Pb 90-277-0233-0
50. M. Bunge (ed.), *Exact Philosophy.* Problems, Tools and Goals. 1973 ISBN 90-277-0251-9
51. R. J. Bogdan and I. Niiniluoto (eds.), *Logic, Language and Probability.* 1973 ISBN 90-277-0312-4
52. G. Pearce and P. Maynard (eds.), *Conceptual Change.* 1973 ISBN 90-277-0287-X; Pb 90-277-0339-6
53. I. Niiniluoto and R. Tuomela, *Theoretical Concepts and Hypothetico-inductive Inference.* 1973 ISBN 90-277-0343-4
54. R. Fraissé, *Course of Mathematical Logic* – Volume 1: *Relation and Logical Formula.* Translated from French. 1973 ISBN 90-277-0268-3; Pb 90-277-0403-1 (For *Volume 2* see under No. 69).
55. A. Grünbaum, *Philosophical Problems of Space and Time.* Edited by R.S. Cohen and M.W. Wartofsky. 2nd enlarged ed. [Boston Studies in the Philosophy of Science, Vol. XII] 1973 ISBN 90-277-0357-4; Pb 90-277-0358-2
56. P. Suppes (ed.), *Space, Time and Geometry.* 1973 ISBN 90-277-0386-8; Pb 90-277-0442-2
57. H. Kelsen, *Essays in Legal and Moral Philosophy*. Selected and introduced by O. Weinberger. Translated from German by P. Heath. 1973 ISBN 90-277-0388-4
58. R. J. Seeger and R. S. Cohen (eds.), *Philosophical Foundations of Science.* [Boston Studies in the Philosophy of Science, Vol. XI] 1974 ISBN 90-277-0390-6; Pb 90-277-0376-0
59. R. S. Cohen and M. W. Wartofsky (eds.), *Logical and Epistemological Studies in Contemporary Physics.* [Boston Studies in the Philosophy of Science, Vol. XIII] 1973 ISBN 90-277-0391-4; Pb 90-277-0377-9
60. R. S. Cohen and M. W. Wartofsky (eds.), *Methodological and Historical Essays in the Natural and Social Sciences. Proceedings of the Boston Colloquium for the Philosophy of Science, 1969–1972.* [Boston Studies in the Philosophy of Science, Vol. XIV] 1974 ISBN 90-277-0392-2; Pb 90-277-0378-7
61. R. S. Cohen, J. J. Stachel and M. W. Wartofsky (eds.), *For Dirk Struik. Scientific, Historical and Political Essays.* [Boston Studies in the Philosophy of Science, Vol. XV] 1974 ISBN 90-277-0393-0; Pb 90-277-0379-5
62. K. Ajdukiewicz, *Pragmatic Logic*. Translated from Polish by O. Wojtasiewicz. 1974 ISBN 90-277-0326-4
63. S. Stenlund (ed.), *Logical Theory and Semantic Analysis.* Essays dedicated to Stig Kanger on His 50th Birthday. 1974 ISBN 90-277-0438-4
64. K. F. Schaffner and R. S. Cohen (eds.), *PSA 1972. Proceedings of the Third Biennial Meeting of the Philosophy of Science Association.* [Boston Studies in the Philosophy of Science, Vol. XX] 1974 ISBN 90-277-0408-2; Pb 90-277-0409-0
65. H. E. Kyburg, Jr., *The Logical Foundations of Statistical Inference.* 1974 ISBN 90-277-0330-2; Pb 90-277-0430-9
66. M. Grene, *The Understanding of Nature.* Essays in the Philosophy of Biology. [Boston Studies in the Philosophy of Science, Vol. XXIII] 1974 ISBN 90-277-0462-7; Pb 90-277-0463-5
67. J. M. Broekman, *Structuralism: Moscow, Prague, Paris*. Translated from German. 1974 ISBN 90-277-0478-3
68. N. Geschwind, *Selected Papers on Language and the Brain.* [Boston Studies in the Philosophy of Science, Vol. XVI] 1974 ISBN 90-277-0262-4; Pb 90-277-0263-2
69. R. Fraissé, *Course of Mathematical Logic* – Volume 2: *Model Theory.* Translated from French. 1974 ISBN 90-277-0269-1; Pb 90-277-0510-0 (For *Volume 1* see under No. 54)

70. A. Grzegorczyk, *An Outline of Mathematical Logic.* Fundamental Results and Notions explained with all Details. Translated from Polish. 1974 ISBN 90-277-0359-0; Pb 90-277-0447-3
71. F. von Kutschera, *Philosophy of Language.* 1975 ISBN 90-277-0591-7
72. J. Manninen and R. Tuomela (eds.), *Essays on Explanation and Understanding.* Studies in the Foundations of Humanities and Social Sciences. 1976 ISBN 90-277-0592-5
73. J. Hintikka (ed.), *Rudolf Carnap, Logical Empiricist.* Materials and Perspectives. 1975 ISBN 90-277-0583-6
74. M. Čapek (ed.), *The Concepts of Space and Time.* Their Structure and Their Development. [Boston Studies in the Philosophy of Science, Vol. XXII] 1976 ISBN 90-277-0355-8; Pb 90-277-0375-2
75. J. Hintikka and U. Remes, *The Method of Analysis.* Its Geometrical Origin and Its General Significance. [Boston Studies in the Philosophy of Science, Vol. XXV] 1974 ISBN 90-277-0532-1; Pb 90-277-0543-7
76. J. E. Murdoch and E. D. Sylla (eds.), *The Cultural Context of Medieval Learning.* [Boston Studies in the Philosophy of Science, Vol. XXVI] 1975 ISBN 90-277-0560-7; Pb 90-277-0587-9
77. S. Amsterdamski, *Between Experience and Metaphysics.* Philosophical Problems of the Evolution of Science. [Boston Studies in the Philosophy of Science, Vol. XXXV] 1975 ISBN 90-277-0568-2; Pb 90-277-0580-1
78. P. Suppes (ed.), *Logic and Probability in Quantum Mechanics.* 1976 ISBN 90-277-0570-4; Pb 90-277-1200-X
79. H. von Helmholtz: *Epistemological Writings. The Paul Hertz / Moritz Schlick Centenary Edition of 1921 with Notes and Commentary by the Editors.* Newly translated from German by M. F. Lowe. Edited, with an Introduction and Bibliography, by R. S. Cohen and Y. Elkana. [Boston Studies in the Philosophy of Science, Vol. XXXVII] 1975 ISBN 90-277-0290-X; Pb 90-277-0582-8
80. J. Agassi, *Science in Flux.* [Boston Studies in the Philosophy of Science, Vol. XXVIII] 1975 ISBN 90-277-0584-4; Pb 90-277-0612-2
81. S. G. Harding (ed.), *Can Theories Be Refuted?* Essays on the Duhem-Quine Thesis. 1976 ISBN 90-277-0629-8; Pb 90-277-0630-1
82. S. Nowak, *Methodology of Sociological Research.* General Problems. 1977 ISBN 90-277-0486-4
83. J. Piaget, J.-B. Grize, A. Szeminśska and V. Bang, *Epistemology and Psychology of Functions.* Translated from French. 1977 ISBN 90-277-0804-5
84. M. Grene and E. Mendelsohn (eds.), *Topics in the Philosophy of Biology.* [Boston Studies in the Philosophy of Science, Vol. XXVII] 1976 ISBN 90-277-0595-X; Pb 90-277-0596-8
85. E. Fischbein, *The Intuitive Sources of Probabilistic Thinking in Children.* 1975 ISBN 90-277-0626-3; Pb 90-277-1190-9
86. E. W. Adams, *The Logic of Conditionals.* An Application of Probability to Deductive Logic. 1975 ISBN 90-277-0631-X
87. M. Przełęcki and R. Wójcicki (eds.), *Twenty-Five Years of Logical Methodology in Poland.* Translated from Polish. 1976 ISBN 90-277-0601-8
88. J. Topolski, *The Methodology of History.* Translated from Polish by O. Wojtasiewicz. 1976 ISBN 90-277-0550-X
89. A. Kasher (ed.), *Language in Focus: Foundations, Methods and Systems.* Essays dedicated to Yehoshua Bar-Hillel. [Boston Studies in the Philosophy of Science, Vol. XLIII] 1976 ISBN 90-277-0644-1; Pb 90-277-0645-X

90. J. Hintikka, *The Intentions of Intentionality and Other New Models for Modalities.* 1975 ISBN 90-277-0633-6; Pb 90-277-0634-4
91. W. Stegmüller, *Collected Papers on Epistemology, Philosophy of Science and History of Philosophy.* 2 Volumes. 1977 Set ISBN 90-277-0767-7
92. D. M. Gabbay, *Investigations in Modal and Tense Logics with Applications to Problems in Philosophy and Linguistics.* 1976 ISBN 90-277-0656-5
93. R. J. Bogdan, *Local Induction.* 1976 ISBN 90-277-0649-2
94. S. Nowak, *Understanding and Prediction.* Essays in the Methodology of Social and Behavioral Theories. 1976 ISBN 90-277-0558-5; Pb 90-277-1199-2
95. P. Mittelstaedt, *Philosophical Problems of Modern Physics.* [Boston Studies in the Philosophy of Science, Vol. XVIII] 1976 ISBN 90-277-0285-3; Pb 90-277-0506-2
96. G. Holton and W. A. Blanpied (eds.), *Science and Its Public: The Changing Relationship.* [Boston Studies in the Philosophy of Science, Vol. XXXIII] 1976 ISBN 90-277-0657-3; Pb 90-277-0658-1
97. M. Brand and D. Walton (eds.), *Action Theory.* 1976 ISBN 90-277-0671-9
98. P. Gochet, *Outline of a Nominalist Theory of Propositions.* An Essay in the Theory of Meaning and in the Philosophy of Logic. 1980 ISBN 90-277-1031-7
99. R. S. Cohen, P. K. Feyerabend, and M. W. Wartofsky (eds.), *Essays in Memory of Imre Lakatos.* [Boston Studies in the Philosophy of Science, Vol. XXXIX] 1976 ISBN 90-277-0654-9; Pb 90-277-0655-7
100. R. S. Cohen and J. J. Stachel (eds.), *Selected Papers of Léon Rosenfield.* [Boston Studies in the Philosophy of Science, Vol. XXI] 1979 ISBN 90-277-0651-4; Pb 90-277-0652-2
101. R. S. Cohen, C. A. Hooker, A. C. Michalos and J. W. van Evra (eds.), *PSA 1974. Proceedings of the 1974 Biennial Meeting of the Philosophy of Science Association.* [Boston Studies in the Philosophy of Science, Vol. XXXII] 1976 ISBN 90-277-0647-6; Pb 90-277-0648-4
102. Y. Fried and J. Agassi, *Paranoia.* A Study in Diagnosis. [Boston Studies in the Philosophy of Science, Vol. L] 1976 ISBN 90-277-0704-9; Pb 90-277-0705-7
103. M. Przełęcki, K. Szaniawski and R. Wójcicki (eds.), *Formal Methods in the Methodology of Empirical Sciences.* 1976 ISBN 90-277-0698-0
104. J. M. Vickers, *Belief and Probability.* 1976 ISBN 90-277-0744-8
105. K. H. Wolff, *Surrender and Catch.* Experience and Inquiry Today. [Boston Studies in the Philosophy of Science, Vol. LI] 1976 ISBN 90-277-0758-8; Pb 90-277-0765-0
106. K. Kosík, *Dialectics of the Concrete.* A Study on Problems of Man and World. [Boston Studies in the Philosophy of Science, Vol. LII] 1976 ISBN 90-277-0761-8; Pb 90-277-0764-2
107. N. Goodman, *The Structure of Appearance.* 3rd ed. with an Introduction by G. Hellman. [Boston Studies in the Philosophy of Science, Vol. LIII] 1977 ISBN 90-277-0773-1; Pb 90-277-0774-X
108. K. Ajdukiewicz, *The Scientific World-Perspective and Other Essays, 1931-1963.* Translated from Polish. Edited and with an Introduction by J. Giedymin. 1978 ISBN 90-277-0527-5
109. R. L. Causey, *Unity of Science.* 1977 ISBN 90-277-0779-0
110. R. E. Grandy, *Advanced Logic for Applications.* 1977 ISBN 90-277-0781-2
111. R. P. McArthur, *Tense Logic.* 1976 ISBN 90-277-0697-2
112. L. Lindahl, *Position and Change.* A Study in Law and Logic. Translated from Swedish by P. Needham. 1977 ISBN 90-277-0787-1
113. R. Tuomela, *Dispositions.* 1978 ISBN 90-277-0810-X
114. H. A. Simon, *Models of Discovery and Other Topics in the Methods of Science.* [Boston Studies in the Philosophy of Science, Vol. LIV] 1977 ISBN 90-277-0812-6; Pb 90-277-0858-4

115. R. D. Rosenkrantz, *Inference, Method and Decision.* Towards a Bayesian Philosophy of Science. 1977 ISBN 90-277-0817-7; Pb 90-277-0818-5
116. R. Tuomela, *Human Action and Its Explanation.* A Study on the Philosophical Foundations of Psychology. 1977 ISBN 90-277-0824-X
117. M. Lazerowitz, *The Language of Philosophy.* Freud and Wittgenstein. [Boston Studies in the Philosophy of Science, Vol. LV] 1977 ISBN 90-277-0826-6; Pb 90-277-0862-2
118. Not published 119. J. Pelc (ed.), *Semiotics in Poland, 1894–1969.* Translated from Polish. 1979 ISBN 90-277-0811-8
120. I. Pörn, *Action Theory and Social Science.* Some Formal Models. 1977 ISBN 90-277-0846-0
121. J. Margolis, *Persons and Mind.* The Prospects of Nonreductive Materialism. [Boston Studies in the Philosophy of Science, Vol. LVII] 1977 ISBN 90-277-0854-1; Pb 90-277-0863-0
122. J. Hintikka, I. Niiniluoto, and E. Saarinen (eds.), *Essays on Mathematical and Philosophical Logic.* 1979 ISBN 90-277-0879-7
123. T. A. F. Kuipers, *Studies in Inductive Probability and Rational Expectation.* 1978 ISBN 90-277-0882-7
124. E. Saarinen, R. Hilpinen, I. Niiniluoto and M. P. Hintikka (eds.), *Essays in Honour of Jaakko Hintikka on the Occasion of His 50th Birthday.* 1979 ISBN 90-277-0916-5
125. G. Radnitzky and G. Andersson (eds.), *Progress and Rationality in Science.* [Boston Studies in the Philosophy of Science, Vol. LVIII] 1978 ISBN 90-277-0921-1; Pb 90-277-0922-X
126. P. Mittelstaedt, *Quantum Logic.* 1978 ISBN 90-277-0925-4
127. K. A. Bowen, *Model Theory for Modal Logic.* Kripke Models for Modal Predicate Calculi. 1979 ISBN 90-277-0929-7
128. H. A. Bursen, *Dismantling the Memory Machine.* A Philosophical Investigation of Machine Theories of Memory. 1978 ISBN 90-277-0933-5
129. M. W. Wartofsky, *Models.* Representation and the Scientific Understanding. [Boston Studies in the Philosophy of Science, Vol. XLVIII] 1979 ISBN 90-277-0736-7; Pb 90-277-0947-5
130. D. Ihde, *Technics and Praxis.* A Philosophy of Technology. [Boston Studies in the Philosophy of Science, Vol. XXIV] 1979 ISBN 90-277-0953-X; Pb 90-277-0954-8
131. J. J. Wiatr (ed.), *Polish Essays in the Methodology of the Social Sciences.* [Boston Studies in the Philosophy of Science, Vol. XXIX] 1979 ISBN 90-277-0723-5; Pb 90-277-0956-4
132. W. C. Salmon (ed.), *Hans Reichenbach: Logical Empiricist.* 1979 ISBN 90-277-0958-0
133. P. Bieri, R.-P. Horstmann and L. Krüger (eds.), *Transcendental Arguments in Science.* Essays in Epistemology. 1979 ISBN 90-277-0963-7; Pb 90-277-0964-5
134. M. Marković and G. Petrović (eds.), *Praxis.* Yugoslav Essays in the Philosophy and Methodology of the Social Sciences. [Boston Studies in the Philosophy of Science, Vol. XXXVI] 1979 ISBN 90-277-0727-8; Pb 90-277-0968-8
135. R. Wójcicki, *Topics in the Formal Methodology of Empirical Sciences.* Translated from Polish. 1979 ISBN 90-277-1004-X
136. G. Radnitzky and G. Andersson (eds.), *The Structure and Development of Science.* [Boston Studies in the Philosophy of Science, Vol. LIX] 1979 ISBN 90-277-0994-7; Pb 90-277-0995-5
137. J. C. Webb, *Mechanism, Mentalism and Metamathematics.* An Essay on Finitism. 1980 ISBN 90-277-1046-5
138. D. F. Gustafson and B. L. Tapscott (eds.), *Body, Mind and Method.* Essays in Honor of Virgil C. Aldrich. 1979 ISBN 90-277-1013-9
139. L. Nowak, *The Structure of Idealization.* Towards a Systematic Interpretation of the Marxian Idea of Science. 1980 ISBN 90-277-1014-7

140. C. Perelman, *The New Rhetoric and the Humanities.* Essays on Rhetoric and Its Applications. Translated from French and German. With an Introduction by H. Zyskind. 1979 ISBN 90-277-1018-X; Pb 90-277-1019-8
141. W. Rabinowicz, *Universalizability.* A Study in Morals and Metaphysics. 1979 ISBN 90-277-1020-2
142. C. Perelman, *Justice, Law and Argument.* Essays on Moral and Legal Reasoning. Translated from French and German. With an Introduction by H.J. Berman. 1980 ISBN 90-277-1089-9; Pb 90-277-1090-2
143. S. Kanger and S. Öhman (eds.), *Philosophy and Grammar.* Papers on the Occasion of the Quincentennial of Uppsala University. 1981 ISBN 90-277-1091-0
144. T. Pawlowski, *Concept Formation in the Humanities and the Social Sciences.* 1980 ISBN 90-277-1096-1
145. J. Hintikka, D. Gruender and E. Agazzi (eds.), *Theory Change, Ancient Axiomatics and Galileo's Methodology.* Proceedings of the 1978 Pisa Conference on the History and Philosophy of Science, Volume I. 1981 ISBN 90-277-1126-7
146. J. Hintikka, D. Gruender and E. Agazzi (eds.), *Probabilistic Thinking, Thermodynamics, and the Interaction of the History and Philosophy of Science.* Proceedings of the 1978 Pisa Conference on the History and Philosophy of Science, Volume II. 1981 ISBN 90-277-1127-5
147. U. Mönnich (ed.), *Aspects of Philosophical Logic.* Some Logical Forays into Central Notions of Linguistics and Philosophy. 1981 ISBN 90-277-1201-8
148. D. M. Gabbay, *Semantical Investigations in Heyting's Intuitionistic Logic.* 1981 ISBN 90-277-1202-6
149. E. Agazzi (ed.), *Modern Logic – A Survey.* Historical, Philosophical, and Mathematical Aspects of Modern Logic and Its Applications. 1981 ISBN 90-277-1137-2
150. A. F. Parker-Rhodes, *The Theory of Indistinguishables.* A Search for Explanatory Principles below the Level of Physics. 1981 ISBN 90-277-1214-X
151. J. C. Pitt, *Pictures, Images, and Conceptual Change.* An Analysis of Wilfrid Sellars' Philosophy of Science. 1981 ISBN 90-277-1276-X; Pb 90-277-1277-8
152. R. Hilpinen (ed.), *New Studies in Deontic Logic.* Norms, Actions, and the Foundations of Ethics. 1981 ISBN 90-277-1278-6; Pb 90-277-1346-4
153. C. Dilworth, *Scientific Progress.* A Study Concerning the Nature of the Relation between Successive Scientific Theories. 3rd rev. ed., 1994 ISBN 0-7923-2487-0; Pb 0-7923-2488-9
154. D. Woodruff Smith and R. McIntyre, *Husserl and Intentionality.* A Study of Mind, Meaning, and Language. 1982 ISBN 90-277-1392-8; Pb 90-277-1730-3
155. R. J. Nelson, *The Logic of Mind.* 2nd. ed., 1989 ISBN 90-277-2819-4; Pb 90-277-2822-4
156. J. F. A. K. van Benthem, *The Logic of Time.* A Model-Theoretic Investigation into the Varieties of Temporal Ontology, and Temporal Discourse. 1983; 2nd ed., 1991 ISBN 0-7923-1081-0
157. R. Swinburne (ed.), *Space, Time and Causality.* 1983 ISBN 90-277-1437-1
158. E. T. Jaynes, *Papers on Probability, Statistics and Statistical Physics.* Ed. by R. D. Rozenkrantz. 1983 ISBN 90-277-1448-7; Pb (1989) 0-7923-0213-3
159. T. Chapman, *Time: A Philosophical Analysis.* 1982 ISBN 90-277-1465-7
160. E. N. Zalta, *Abstract Objects.* An Introduction to Axiomatic Metaphysics. 1983 ISBN 90-277-1474-6
161. S. Harding and M. B. Hintikka (eds.), *Discovering Reality.* Feminist Perspectives on Epistemology, Metaphysics, Methodology, and Philosophy of Science. 1983 ISBN 90-277-1496-7; Pb 90-277-1538-6
162. M. A. Stewart (ed.), *Law, Morality and Rights.* 1983 ISBN 90-277-1519-X

163. D. Mayr and G. Süssmann (eds.), *Space, Time, and Mechanics.* Basic Structures of a Physical Theory. 1983 ISBN 90-277-1525-4
164. D. Gabbay and F. Guenthner (eds.), *Handbook of Philosophical Logic.* Vol. I: Elements of Classical Logic. 1983 ISBN 90-277-1542-4
165. D. Gabbay and F. Guenthner (eds.), *Handbook of Philosophical Logic.* Vol. II: Extensions of Classical Logic. 1984 ISBN 90-277-1604-8
166. D. Gabbay and F. Guenthner (eds.), *Handbook of Philosophical Logic.* Vol. III: Alternative to Classical Logic. 1986 ISBN 90-277-1605-6
167. D. Gabbay and F. Guenthner (eds.), *Handbook of Philosophical Logic.* Vol. IV: Topics in the Philosophy of Language. 1989 ISBN 90-277-1606-4
168. A. J. I. Jones, *Communication and Meaning.* An Essay in Applied Modal Logic. 1983 ISBN 90-277-1543-2
169. M. Fitting, *Proof Methods for Modal and Intuitionistic Logics.* 1983 ISBN 90-277-1573-4
170. J. Margolis, *Culture and Cultural Entities.* Toward a New Unity of Science. 1984 ISBN 90-277-1574-2
171. R. Tuomela, *A Theory of Social Action.* 1984 ISBN 90-277-1703-6
172. J. J. E. Gracia, E. Rabossi, E. Villanueva and M. Dascal (eds.), *Philosophical Analysis in Latin America.* 1984 ISBN 90-277-1749-4
173. P. Ziff, *Epistemic Analysis.* A Coherence Theory of Knowledge. 1984 ISBN 90-277-1751-7
174. P. Ziff, *Antiaesthetics.* An Appreciation of the Cow with the Subtile Nose. 1984 ISBN 90-277-1773-7
175. W. Balzer, D. A. Pearce, and H.-J. Schmidt (eds.), *Reduction in Science.* Structure, Examples, Philosophical Problems. 1984 ISBN 90-277-1811-3
176. A. Peczenik, L. Lindahl and B. van Roermund (eds.), *Theory of Legal Science.* Proceedings of the Conference on Legal Theory and Philosophy of Science (Lund, Sweden, December 1983). 1984 ISBN 90-277-1834-2
177. I. Niiniluoto, *Is Science Progressive?* 1984 ISBN 90-277-1835-0
178. B. K. Matilal and J. L. Shaw (eds.), *Analytical Philosophy in Comparative Perspective.* Exploratory Essays in Current Theories and Classical Indian Theories of Meaning and Reference. 1985 ISBN 90-277-1870-9
179. P. Kroes, *Time: Its Structure and Role in Physical Theories.* 1985 ISBN 90-277-1894-6
180. J. H. Fetzer, *Sociobiology and Epistemology.* 1985 ISBN 90-277-2005-3; Pb 90-277-2006-1
181. L. Haaparanta and J. Hintikka (eds.), *Frege Synthesized.* Essays on the Philosophical and Foundational Work of Gottlob Frege. 1986 ISBN 90-277-2126-2
182. M. Detlefsen, *Hilbert's Program.* An Essay on Mathematical Instrumentalism. 1986 ISBN 90-277-2151-3
183. J. L. Golden and J. J. Pilotta (eds.), *Practical Reasoning in Human Affairs.* Studies in Honor of Chaim Perelman. 1986 ISBN 90-277-2255-2
184. H. Zandvoort, *Models of Scientific Development and the Case of Nuclear Magnetic Resonance.* 1986 ISBN 90-277-2351-6
185. I. Niiniluoto, *Truthlikeness.* 1987 ISBN 90-277-2354-0
186. W. Balzer, C. U. Moulines and J. D. Sneed, *An Architectonic for Science.* The Structuralist Program. 1987 ISBN 90-277-2403-2
187. D. Pearce, *Roads to Commensurability.* 1987 ISBN 90-277-2414-8
188. L. M. Vaina (ed.), *Matters of Intelligence.* Conceptual Structures in Cognitive Neuroscience. 1987 ISBN 90-277-2460-1

189. H. Siegel, *Relativism Refuted.* A Critique of Contemporary Epistemological Relativism. 1987 ISBN 90-277-2469-5
190. W. Callebaut and R. Pinxten, *Evolutionary Epistemology.* A Multiparadigm Program, with a Complete Evolutionary Epistemology Bibliograph. 1987 ISBN 90-277-2582-9
191. J. Kmita, *Problems in Historical Epistemology.* 1988 ISBN 90-277-2199-8
192. J. H. Fetzer (ed.), *Probability and Causality.* Essays in Honor of Wesley C. Salmon, with an Annotated Bibliography. 1988 ISBN 90-277-2607-8; Pb 1-5560-8052-2
193. A. Donovan, L. Laudan and R. Laudan (eds.), *Scrutinizing Science.* Empirical Studies of Scientific Change. 1988 ISBN 90-277-2608-6
194. H.R. Otto and J.A. Tuedio (eds.), *Perspectives on Mind.* 1988 ISBN 90-277-2640-X
195. D. Batens and J.P. van Bendegem (eds.), *Theory and Experiment.* Recent Insights and New Perspectives on Their Relation. 1988 ISBN 90-277-2645-0
196. J. Österberg, *Self and Others.* A Study of Ethical Egoism. 1988 ISBN 90-277-2648-5
197. D.H. Helman (ed.), *Analogical Reasoning.* Perspectives of Artificial Intelligence, Cognitive Science, and Philosophy. 1988 ISBN 90-277-2711-2
198. J. Woleński, *Logic and Philosophy in the Lvov-Warsaw School.* 1989 ISBN 90-277-2749-X
199. R. Wójcicki, *Theory of Logical Calculi.* Basic Theory of Consequence Operations. 1988 ISBN 90-277-2785-6
200. J. Hintikka and M.B. Hintikka, *The Logic of Epistemology and the Epistemology of Logic.* Selected Essays. 1989 ISBN 0-7923-0040-8; Pb 0-7923-0041-6
201. E. Agazzi (ed.), *Probability in the Sciences.* 1988 ISBN 90-277-2808-9
202. M. Meyer (ed.), *From Metaphysics to Rhetoric.* 1989 ISBN 90-277-2814-3
203. R.L. Tieszen, *Mathematical Intuition.* Phenomenology and Mathematical Knowledge. 1989 ISBN 0-7923-0131-5
204. A. Melnick, *Space, Time, and Thought in Kant.* 1989 ISBN 0-7923-0135-8
205. D.W. Smith, *The Circle of Acquaintance.* Perception, Consciousness, and Empathy. 1989 ISBN 0-7923-0252-4
206. M.H. Salmon (ed.), *The Philosophy of Logical Mechanism.* Essays in Honor of Arthur W. Burks. With his Responses, and with a Bibliography of Burk's Work. 1990 ISBN 0-7923-0325-3
207. M. Kusch, *Language as Calculus vs. Language as Universal Medium.* A Study in Husserl, Heidegger, and Gadamer. 1989 ISBN 0-7923-0333-4
208. T.C. Meyering, *Historical Roots of Cognitive Science.* The Rise of a Cognitive Theory of Perception from Antiquity to the Nineteenth Century. 1989 ISBN 0-7923-0349-0
209. P. Kosso, *Observability and Observation in Physical Science.* 1989 ISBN 0-7923-0389-X
210. J. Kmita, *Essays on the Theory of Scientific Cognition.* 1990 ISBN 0-7923-0441-1
211. W. Sieg (ed.), *Acting and Reflecting.* The Interdisciplinary Turn in Philosophy. 1990 ISBN 0-7923-0512-4
212. J. Karpiński, *Causality in Sociological Research.* 1990 ISBN 0-7923-0546-9
213. H.A. Lewis (ed.), *Peter Geach: Philosophical Encounters.* 1991 ISBN 0-7923-0823-9
214. M. Ter Hark, *Beyond the Inner and the Outer.* Wittgenstein's Philosophy of Psychology. 1990 ISBN 0-7923-0850-6
215. M. Gosselin, *Nominalism and Contemporary Nominalism.* Ontological and Epistemological Implications of the Work of W.V.O. Quine and of N. Goodman. 1990 ISBN 0-7923-0904-9
216. J.H. Fetzer, D. Shatz and G. Schlesinger (eds.), *Definitions and Definability.* Philosophical Perspectives. 1991 ISBN 0-7923-1046-2
217. E. Agazzi and A. Cordero (eds.), *Philosophy and the Origin and Evolution of the Universe.* 1991 ISBN 0-7923-1322-4

218. M. Kusch, *Foucault's Strata and Fields.* An Investigation into Archaeological and Genealogical Science Studies. 1991 ISBN 0-7923-1462-X
219. C.J. Posy, *Kant's Philosophy of Mathematics.* Modern Essays. 1992 ISBN 0-7923-1495-6
220. G. Van de Vijver, *New Perspectives on Cybernetics.* Self-Organization, Autonomy and Connectionism. 1992 ISBN 0-7923-1519-7
221. J.C. Nyíri, *Tradition and Individuality.* Essays. 1992 ISBN 0-7923-1566-9
222. R. Howell, *Kant's Transcendental Deduction.* An Analysis of Main Themes in His Critical Philosophy. 1992 ISBN 0-7923-1571-5
223. A. García de la Sienra, *The Logical Foundations of the Marxian Theory of Value.* 1992 ISBN 0-7923-1778-5
224. D.S. Shwayder, *Statement and Referent.* An Inquiry into the Foundations of Our Conceptual Order. 1992 ISBN 0-7923-1803-X
225. M. Rosen, *Problems of the Hegelian Dialectic.* Dialectic Reconstructed as a Logic of Human Reality. 1993 ISBN 0-7923-2047-6
226. P. Suppes, *Models and Methods in the Philosophy of Science: Selected Essays.* 1993 ISBN 0-7923-2211-8
227. R. M. Dancy (ed.), *Kant and Critique: New Essays in Honor of W. H. Werkmeister.* 1993 ISBN 0-7923-2244-4
228. J. Woleński (ed.), *Philosophical Logic in Poland.* 1993 ISBN 0-7923-2293-2
229. M. De Rijke (ed.), *Diamonds and Defaults.* Studies in Pure and Applied Intensional Logic. 1993 ISBN 0-7923-2342-4
230. B.K. Matilal and A. Chakrabarti (eds.), *Knowing from Words.* Western and Indian Philosophical Analysis of Understanding and Testimony. 1994 ISBN 0-7923-2345-9
231. S.A. Kleiner, *The Logic of Discovery.* A Theory of the Rationality of Scientific Research. 1993 ISBN 0-7923-2371-8
232. R. Festa, *Optimum Inductive Methods.* A Study in Inductive Probability, Bayesian Statistics, and Verisimilitude. 1993 ISBN 0-7923-2460-9
233. P. Humphreys (ed.), *Patrick Suppes: Scientific Philosopher.* Vol. 1: Probability and Probabilistic Causality. 1994 ISBN 0-7923-2552-4
234. P. Humphreys (ed.), *Patrick Suppes: Scientific Philosopher.* Vol. 2: Philosophy of Physics, Theory Structure, and Measurement Theory. 1994 ISBN 0-7923-2553-2
235. P. Humphreys (ed.), *Patrick Suppes: Scientific Philosopher.* Vol. 3: Language, Logic, and Psychology. 1994 ISBN 0-7923-2862-0
    Set ISBN (Vols 233–235) 0-7923-2554-0
236. D. Prawitz and D. Westerståhl (eds.), *Logic and Philosophy of Science in Uppsala.* Papers from the 9th International Congress of Logic, Methodology, and Philosophy of Science. 1994 ISBN 0-7923-2702-0
237. L. Haaparanta (ed.), *Mind, Meaning and Mathematics.* Essays on the Philosophical Views of Husserl and Frege. 1994 ISBN 0-7923-2703-9
238. J. Hintikka (ed.), *Aspects of Metaphor.* 1994 ISBN 0-7923-2786-1
239. B. McGuinness and G. Oliveri (eds.), *The Philosophy of Michael Dummett.* With Replies from Michael Dummett. 1994 ISBN 0-7923-2804-3
240. D. Jamieson (ed.), *Language, Mind, and Art.* Essays in Appreciation and Analysis, In Honor of Paul Ziff. 1994 ISBN 0-7923-2810-8
241. G. Preyer, F. Siebelt and A. Ulfig (eds.), *Language, Mind and Epistemology.* On Donald Davidson's Philosophy. 1994 ISBN 0-7923-2811-6
242. P. Ehrlich (ed.), *Real Numbers, Generalizations of the Reals, and Theories of Continua.* 1994 ISBN 0-7923-2689-X

243. G. Debrock and M. Hulswit (eds.), *Living Doubt*. Essays concerning the epistemology of Charles Sanders Peirce. 1994 ISBN 0-7923-2898-1
244. J. Srzednicki, *To Know or Not to Know*. Beyond Realism and Anti-Realism. 1994 ISBN 0-7923-2909-0
245. R. Egidi (ed.), *Wittgenstein: Mind and Language*. 1995 ISBN 0-7923-3171-0
246. A. Hyslop, *Other Minds*. 1995 ISBN 0-7923-3245-8
247. L. Pólos and M. Masuch (eds.), *Applied Logic: How, What and Why*. Logical Approaches to Natural Language. 1995 ISBN 0-7923-3432-9
248. M. Krynicki, M. Mostowski and L.M. Szczerba (eds.), *Quantifiers: Logics, Models and Computation*. Volume One: Surveys. 1995 ISBN 0-7923-3448-5
249. M. Krynicki, M. Mostowski and L.M. Szczerba (eds.), *Quantifiers: Logics, Models and Computation*. Volume Two: Contributions. 1995 ISBN 0-7923-3449-3
Set ISBN (Vols 248 + 249) 0-7923-3450-7
250. R.A. Watson, *Representational Ideas from Plato to Patricia Churchland*. 1995 ISBN 0-7923-3453-1
251. J. Hintikka (ed.), *From Dedekind to Gödel*. Essays on the Development of the Foundations of Mathematics. 1995 ISBN 0-7923-3484-1
252. A. Wiśniewski, *The Posing of Questions*. Logical Foundations of Erotetic Inferences. 1995 ISBN 0-7923-3637-2
253. J. Peregrin, *Doing Worlds with Words*. Formal Semantics without Formal Metaphysics. 1995 ISBN 0-7923-3742-5
254. I.A. Kieseppä, *Truthlikeness for Multidimensional, Quantitative Cognitive Problems*. 1996 ISBN 0-7923-4005-1
255. P. Hugly and C. Sayward: *Intensionality and Truth*. An Essay on the Philosophy of A.N. Prior. 1996 ISBN 0-7923-4119-8
256. L. Hankinson Nelson and J. Nelson (eds.): *Feminism, Science, and the Philosophy of Science*. 1997 ISBN 0-7923-4162-7
257. P.I. Bystrov and V.N. Sadovsky (eds.): *Philosophical Logic and Logical Philosophy*. Essays in Honour of Vladimir A. Smirnov. 1996 ISBN 0-7923-4270-4
258. Å.E. Andersson and N-E. Sahlin (eds.): *The Complexity of Creativity*. 1996 ISBN 0-7923-4346-8
259. M.L. Dalla Chiara, K. Doets, D. Mundici and J. van Benthem (eds.): *Logic and Scientific Methods*. Volume One of the Tenth International Congress of Logic, Methodology and Philosophy of Science, Florence, August 1995. 1997 ISBN 0-7923-4383-2
260. M.L. Dalla Chiara, K. Doets, D. Mundici and J. van Benthem (eds.): *Structures and Norms in Science*. Volume Two of the Tenth International Congress of Logic, Methodology and Philosophy of Science, Florence, August 1995. 1997 ISBN 0-7923-4384-0
Set ISBN (Vols 259 + 260) 0-7923-4385-9
261. A. Chakrabarti: *Denying Existence*. The Logic, Epistemology and Pragmatics of Negative Existentials and Fictional Discourse. 1997 ISBN 0-7923-4388-3
262. A. Biletzki: *Talking Wolves*. Thomas Hobbes on the Language of Politics and the Politics of Language. 1997 ISBN 0-7923-4425-1
263. D. Nute (ed.): *Defeasible Deontic Logic*. 1997 ISBN 0-7923-4630-0
264. U. Meixner: *Axiomatic Formal Ontology*. 1997 ISBN 0-7923-4747-X
265. I. Brinck: *The Indexical 'I'*. The First Person in Thought and Language. 1997 ISBN 0-7923-4741-2
266. G. Hölmström-Hintikka and R. Tuomela (eds.): *Contemporary Action Theory*. Volume 1: Individual Action. 1997 ISBN 0-7923-4753-6; Set: 0-7923-4754-4

267. G. Hölmström-Hintikka and R. Tuomela (eds.): *Contemporary Action Theory.* Volume 2: Social Action. 1997 ISBN 0-7923-4752-8; Set: 0-7923-4754-4
268. B.-C. Park: *Phenomenological Aspects of Wittgenstein's Philosophy.* 1998 ISBN 0-7923-4813-3
269. J. Paśniczek: *The Logic of Intentional Objects.* A Meinongian Version of Classical Logic. 1998 Hb ISBN 0-7923-4880-X; Pb ISBN 0-7923-5578-4
270. P.W. Humphreys and J.H. Fetzer (eds.): *The New Theory of Reference.* Kripke, Marcus, and Its Origins. 1998 ISBN 0-7923-4898-2
271. K. Szaniawski, A. Chmielewski and J. Woleński (eds.): *On Science, Inference, Information and Decision Making.* Selected Essays in the Philosophy of Science. 1998 ISBN 0-7923-4922-9
272. G.H. von Wright: *In the Shadow of Descartes.* Essays in the Philosophy of Mind. 1998 ISBN 0-7923-4992-X
273. K. Kijania-Placek and J. Woleński (eds.): *The Lvov–Warsaw School and Contemporary Philosophy.* 1998 ISBN 0-7923-5105-3
274. D. Dedrick: *Naming the Rainbow.* Colour Language, Colour Science, and Culture. 1998 ISBN 0-7923-5239-4
275. L. Albertazzi (ed.): *Shapes of Forms.* From Gestalt Psychology and Phenomenology to Ontology and Mathematics. 1999 ISBN 0-7923-5246-7
276. P. Fletcher: *Truth, Proof and Infinity.* A Theory of Constructions and Constructive Reasoning. 1998 ISBN 0-7923-5262-9
277. M. Fitting and R.L. Mendelsohn (eds.): *First-Order Modal Logic.* 1998 Hb ISBN 0-7923-5334-X; Pb ISBN 0-7923-5335-8
278. J.N. Mohanty: *Logic, Truth and the Modalities from a Phenomenological Perspective.* 1999 ISBN 0-7923-5550-4
279. T. Placek: *Mathematical Intiutionism and Intersubjectivity.* A Critical Exposition of Arguments for Intuitionism. 1999 ISBN 0-7923-5630-6
280. A. Cantini, E. Casari and P. Minari (eds.): *Logic and Foundations of Mathematics.* 1999 ISBN 0-7923-5659-4 set ISBN 0-7923-5867-8
281. M.L. Dalla Chiara, R. Giuntini and F. Laudisa (eds.): *Language, Quantum, Music.* 1999 ISBN 0-7923-5727-2; set ISBN 0-7923-5867-8
282. R. Egidi (ed.): *In Search of a New Humanism.* The Philosophy of Georg Hendrik von Wright. 1999 ISBN 0-7923-5810-4
283. F. Vollmer: *Agent Causality.* 1999 ISBN 0-7923-5848-1
284. J. Peregrin (ed.): *Truth and Its Nature (if Any).* 1999 ISBN 0-7923-5865-1
285. M. De Caro (ed.): *Interpretations and Causes.* New Perspectives on Donald Davidson's Philosophy. 1999 ISBN 0-7923-5869-4
286. R. Murawski: *Recursive Functions and Metamathematics.* Problems of Completeness and Decidability, Gödel's Theorems. 1999 ISBN 0-7923-5904-6
287. T.A.F. Kuipers: *From Instrumentalism to Constructive Realism.* On Some Relations between Confirmation, Empirical Progress, and Truth Approximation. 2000 ISBN 0-7923-6086-9
288. G. Holmström-Hintikka (ed.): *Medieval Philosophy and Modern Times.* 2000 ISBN 0-7923-6102-4
289. E. Grosholz and H. Breger (eds.): *The Growth of Mathematical Knowledge.* 2000 ISBN 0-7923-6151-2

290. G. Sommaruga: *History and Philosophy of Constructive Type Theory*. 2000 ISBN 0-7923-6180-6
291. J. Gasser (ed.): *A Boole Anthology*. Recent and Classical Studies in the Logic of George Boole. 2000 ISBN 0-7923-6380-9
292. V.F. Hendricks, S.A. Pedersen and K.F. Jørgensen (eds.): *Proof Theory*. History and Philosophical Significance. 2000 ISBN 0-7923-6544-5
293. W.L. Craig: *The Tensed Theory of Time*. A Critical Examination. 2000 ISBN 0-7923-6634-4
294. W.L. Craig: *The Tenseless Theory of Time*. A Critical Examination. 2000 ISBN 0-7923-6635-2
295. L. Albertazzi (ed.): *The Dawn of Cognitive Science*. Early European Contributors. 2001 ISBN 0-7923-6799-5
296. G. Forrai: *Reference, Truth and Conceptual Schemes*. A Defense of Internal Realism. 2001 ISBN 0-7923-6885-1
297. V.F. Hendricks, S.A. Pedersen and K.F. Jørgensen (eds.): *Probability Theory*. Philosophy, Recent History and Relations to Science. 2001 ISBN 0-7923-6952-1
298. M. Esfeld: *Holism in Philosophy of Mind and Philosophy of Physics*. 2001 ISBN 0-7923-7003-1
299. E.C. Steinhart: *The Logic of Metaphor*. Analogous Parts of Possible Worlds. 2001 ISBN 0-7923-7004-X
300. P. Gärdenfors: *The Dynamics of Thought*. 2005 ISBN 1-4020-3398-2
301. T.A.F. Kuipers: *Structures in Science Heuristic Patterns Based on Cognitive Structures*. An Advanced Textbook in Neo-Classical Philosophy of Science. 2001 ISBN 0-7923-7117-8
302. G. Hon and S.S. Rakover (eds.): *Explanation*. Theoretical Approaches and Applications. 2001 ISBN 1-4020-0017-0
303. G. Holmström-Hintikka, S. Lindström and R. Sliwinski (eds.): *Collected Papers of Stig Kanger with Essays on his Life and Work*. Vol. I. 2001 ISBN 1-4020-0021-9; Pb ISBN 1-4020-0022-7
304. G. Holmström-Hintikka, S. Lindström and R. Sliwinski (eds.): *Collected Papers of Stig Kanger with Essays on his Life and Work*. Vol. II. 2001 ISBN 1-4020-0111-8; Pb ISBN 1-4020-0112-6
305. C.A. Anderson and M. Zelëny (eds.): *Logic, Meaning and Computation*. Essays in Memory of Alonzo Church. 2001 ISBN 1-4020-0141-X
306. P. Schuster, U. Berger and H. Osswald (eds.): *Reuniting the Antipodes – Constructive and Nonstandard Views of the Continuum*. 2001 ISBN 1-4020-0152-5
307. S.D. Zwart: *Refined Verisimilitude*. 2001 ISBN 1-4020-0268-8
308. A.-S. Maurin: *If Tropes*. 2002 ISBN 1-4020-0656-X
309. H. Eilstein (ed.): *A Collection of Polish Works on Philosophical Problems of Time and Spacetime*. 2002 ISBN 1-4020-0670-5
310. Y. Gauthier: *Internal Logic*. Foundations of Mathematics from Kronecker to Hilbert. 2002 ISBN 1-4020-0689-6
311. E. Ruttkamp: *A Model-Theoretic Realist Interpretation of Science*. 2002 ISBN 1-4020-0729-9
312. V. Rantala: *Explanatory Translation*. Beyond the Kuhnian Model of Conceptual Change. 2002 ISBN 1-4020-0827-9
313. L. Decock: *Trading Ontology for Ideology*. 2002 ISBN 1-4020-0865-1

314. O. Ezra: *The Withdrawal of Rights*. Rights from a Different Perspective. 2002
ISBN 1-4020-0886-4
315. P. Gärdenfors, J. Woleński and K. Kijania-Placek: *In the Scope of Logic, Methodology and Philosophy of Science*. Volume One of the 11th International Congress of Logic, Methodology and Philosophy of Science, Cracow, August 1999. 2002
ISBN 1-4020-0929-1; Pb 1-4020-0931-3
316. P. Gärdenfors, J. Woleński and K. Kijania-Placek: *In the Scope of Logic, Methodology and Philosophy of Science*. Volume Two of the 11th International Congress of Logic, Methodology and Philosophy of Science, Cracow, August 1999. 2002
ISBN 1-4020-0930-5; Pb 1-4020-0931-3
317. M.A. Changizi: *The Brain from 25,000 Feet*. High Level Explorations of Brain Complexity, Perception, Induction and Vagueness. 2003 ISBN 1-4020-1176-8
318. D.O. Dahlstrom (ed.): *Husserl's* Logical Investigations. 2003 ISBN 1-4020-1325-6
319. A. Biletzki: *(Over)Interpreting Wittgenstein*. 2003
ISBN Hb 1-4020-1326-4; Pb 1-4020-1327-2
320. A. Rojszczak, J. Cachro and G. Kurczewski (eds.): *Philosophical Dimensions of Logic and Science*. Selected Contributed Papers from the 11th International Congress of Logic, Methodology, and Philosophy of Science, Kraków, 1999. 2003 ISBN 1-4020-1645-X
321. M. Sintonen, P. Ylikoski and K. Miller (eds.): *Realism in Action*. Essays in the Philosophy of the Social Sciences. 2003 ISBN 1-4020-1667-0
322. V.F. Hendricks, K.F. Jørgensen and S.A. Pedersen (eds.): *Knowledge Contributors*. 2003
ISBN Hb 1-4020-1747-2; Pb 1-4020-1748-0
323. J. Hintikka, T. Czarnecki, K. Kijania-Placek, T. Placek and A. Rojszczak † (eds.): *Philosophy and Logic In Search of the Polish Tradition*. Essays in Honour of Jan Woleński on the Occasion of his 60th Birthday. 2003 ISBN 1-4020-1721-9
324. L.M. Vaina, S.A. Beardsley and S.K. Rushton (eds.): *Optic Flow and Beyond*. 2004
ISBN 1-4020-2091-0
325. D. Kolak (ed.): *I Am You*. The Metaphysical Foundations of Global Ethics. 2004
ISBN 1-4020-2999-3
326. V. Stepin: *Theoretical Knowledge*. 2005 ISBN 1-4020-3045-2
327. P. Mancosu, K.F. Jørgensen and S.A. Pedersen (eds.): *Visualization, Explanation and Reasoning Styles in Mathematics*. 2005 ISBN 1-4020-3334-6
328. A. Rojszczak and J. Wolenski (eds.): *From the Act of Judging to the Sentence*. The Problem of Truth Bearers from Bolzano to Tarski. 2005 ISBN 1-4020-3396-6